The Branches of Ecology

The Branches of Ecology

A 20th Century History

Frank N. Egerton

CRC Press
Taylor & Francis Group
Boca Raton London New York

CRC Press is an imprint of the
Taylor & Francis Group, an **informa** business

Cover: Edward Birge and Chancey Juday, ESA Membership chart, Lake Erie Food Web, NOAA.

First edition published 2023
by CRC Press
6000 Broken Sound Parkway NW, Suite 300, Boca Raton, FL 33487–2742

and by CRC Press
2 Park Square, Milton Park, Abingdon, Oxon, OX14 4RN

CRC Press is an imprint of Taylor & Francis Group, LLC

ISBN: 978-0-367-35702-3 (hbk)
ISBN: 978-0-367-69115-8 (pbk)
ISBN: 978-0-429-34220-2 (ebk)

DOI: 10.1201/9780429342202

Contents

At age 87, I leave historical gaps in my coverage of 1900s ecology, for others to fill: organizing ecologists after 1945, biogeography, 1950–2000, and physiological ecology of plants. There are also new branches of ecology, some arising in the late 1900s, that I have not discussed.

Frank N. Egerton

Dedicated to

Michael Mitchell
Eric L. Mills
Michael J. Lannoo

In memory of

Arthur D. Hasler
Robert P. McIntosh
Howard T. Odum
F. John Vernberg

Preface

The Branches of Ecology: A 20th Century History is a sequel to my *Roots of Ecology: Antiquity to Haeckel* (2012). *Branches* is a concise revision of parts 48 to 64 of my "History of Ecological Sciences" that appeared in the online Bulletin of the Ecological Society of America. In 1900, there were four inconspicuous ecological sciences: plant ecology, animal ecology, limnology, and marine ecology. There were only a few professors of these sciences, few students, and modest funds. If we compare that situation with those sciences in the year 2000, one sees a dramatic increase in professors, students, funding, and organizations. That increase was due to gradual increases in those four aspects of ecology, with a steady increase in knowledge and theoretical sophistical sophistication. My account of that progress includes details on all those aspects of ecological sciences. My goal in Chapters 1 and 2 was to limit Chapter 1 to before 1900 and Chapter 2 to after 1900. In practice, I told stories in both chapters that exceded those boundaries. Because of the details in *Branches,* I think most readers will only read one or two chapters at a time, and so I have introduced some biologists in more than one chapter. This is a grand story, and I hope to convey that to readers.

Cover: Birge & Juday: Fig. 2.7, p.42; ESA Membership: Fig. 3.1, p.59; Lake Erie food web: NOAA

Acknowledgments

I thank Michael Lannoo, Eric Mills, and Greg Mayer for critiques of parts of the text; my editor, Charles R. Crumly, and his assistant, Ana Lucia Eberhart, for their efforts to produce a useful book; Jerry Hershberger for computer assistance, Laura Mason for reproduction of illustrations, and Marybeth Zuhlke for help handling longevity. I thank Professor Emeritus Everett Mendelsohn, Harvard University, for his contribution to the history of biology, most notably his founding and editing of the *Journal of the History of Biology*.

Published previously: *Hewett Cottrell Watson: Ecologist and Evolutionist* (2003), *Roots of Ecology: Antiquity to Haeckel* (2012), *A Centennial History of the Ecological Society of America* (2015), *Guncrazy America: a History and Critique of Our Gun Culture* (2018).

Edited: *History of American Ecology* (1977), Edward L. Greene, *Landmarks of Botanical History* (2 vols., 1983).

I also thank Ganesh Pawan Kumar Agoor for taking care of the book production process.

1 Introduction

Before 1900

From antiquity to 1900, there were many observations published that were relevant to the science of ecology, which I discussed in *Roots of Ecology* (2012). By the 1870s, those observations were being made by specialized biologists in four different subjects: botany, zoology, limnology, and marine biology. Phytogeography received an effective foundation from Humboldt in 1807 and continued throughout the 1800s. It became the dominant source for plant ecology, culminating in August Grisebach's *Die Vegetation der Erden ach klimatischen Andordung* (1872) and in German botanists Adolf and Oscar Drude's edited *Die Vegetation der Erde* (15 vols., 1896–1930). Plant physiology arose during the 1700s and continued throughout the 1800s and also was a source of plant ecology. The organizing text was Danish Eugenius Warming's text on ecological plant geography (1895).

German zoologist Ernst Haeckel named and defined a new science, *oekologie*, in 1866 and made his own contributions in six works (1866–77). Zoogeography also progressed during the 1800s. Belgian zoologist-parasitologist Pierre-Joseph van Beneden's *Les commensoux et le règne animal* (1875) discussed ecological subjects of predation, parasitism, commensalism, and mutualism. Animal ecology was organized by German zoologist Carl Semper in public lectures in Boston, published as *Animal Life as Affected by the Natural Conditions of Existence* (1880).

Limnology was single-handedly organized by Swiss zoologist François Alphonse Forel, who lived on the shore of Lake Geneva (Lac Léman). He was the first to study animal life in that lake and published his first limnological article in 1866; he named and defined limnology in 1892 in his *Lac Léman: monographie limnologique* (3 vols., 1892–1904).

Because of commercial fishing, marine biology, during the 1870s–1920s, seems to have received more funding than plant ecology, animal ecology, or limnology. Van Beneden founded the first marine research station in 1843, on his in-laws' oyster farm at Ostend, Belgium, which he named Laboratorie des Dunes. It attracted biologists from Belgium, France, and Germany. Other stations were soon established in Europe, Russia, and North America. English zoologists Edward Forbes began dredging from ships for bottom animals in the 1840s, and that practice also spread, leading to a worldwide expedition in 1872, lasting 3.5 years, on naval vessel *Challenger*, which essentially founded oceanography. Its findings filled 50 large volumes.

The International Council for Exploration of the Sea was organized in 1902 by European countries. British plant ecologists organized the British Ecological Society in 1912, and it established its *Journal of Ecology*. British animal ecologists soon joined and founded the *Journal of Animal Ecology*. Midwestern ecologists organized the Ecological Society of America (ESA) in 1915, with its *Bulletin* established in 1916 to keep members informed about Society activities, *Ecology* in 1920 to publish scientific contributions, and *Ecological Monographs* in 1931 for longer works. The Limnological Society of America was organized in 1936, and in 1948 it expanded and became the American Society of Limnology and Oceanography. Other organizations and journals followed.

Symbiosis is a term that identifies persistent relationships between species. Examples appeared in natural history since antiquity. In late 300 BC, Theophrastos wrote two botanical treatises that used Greek *sumbiosal* to describe ivory growing on olive trees. No tradition continued such studies until in 1877 and 1878 two German botanists published terms similar to *symbiosis*, concerning lichens. There are textbooks and collections of papers on symbiosis, and a history of the concept, but there is no symbiosis organization known to me.

British observers of bogs, by the mid-1600s, began publishing discoveries about changes in vegetation from one kind to another. French naturalist Georges-Louis de Buffon went further in 1742 by describing environmental conditions leading to the emergence of forests. Swedish botany professor Carl Linnaeus described in 1749 environmental factors that allowed plant succession from lichens on rocks to forests, to decay of dead trees into soil. He also discussed changes in peat bogs in some two dozen publications (1742–73). German botany professor Karl Willdenow discussed succession in the first botany textbook (1792). Genevan naturalist Jean-André Deluc first used *succession* to indicate changes in vegetation. Danish zoology professor Japetus Steenstrup compared cross-sections of Dutch and Danish bogs and indicated a sequence of vegetation types (1842). This European tradition of plant succession studies continued down to Warming (1895), who integrated them into ecological science. That tradition came to America in plant ecology studies during the 1890s by Cowles and Clements.

Animal ecologist Victor Shelford studied under Cowles and published *Animal Communities in Temperate America* (1913), which described plant–animal interactions in nine different environments. He also provided two complex diagrams of both terrestrial and aquatic communities. Clements spoke at a 1916 ESA meeting on biotic communities. He collaborated with Shelford in publishing *Bio-Ecology* (1939), the first textbook covering both plants and animals. Animal ecologist Lee Dice followed with *The*

DOI: 10.1201/9780429342202-1

Biotic Provinces of North America (1943) and *Natural Communities*. Ecologist Robert Whittaker wrote a dissertation on the vegetation of Great Smoky Mountains (1948), but he had also studied insects there, and he published his insect findings (1952) before his vegetation findings (1956).

American zoologist Stephen Forbes' "The Lake as a Microcosm" (1887) discussed a lake as a biological unit, which was close to a community concept. German limnologist August Thienemann also used *microcosm* in a similar way (1925). English animal ecologist Charles Elton went on three Oxford University expeditions to Spitzbergen before publishing *Animal Ecology* (1927). His perspective was close to an ecosystem concept, seen in titles of his chapters: "The Distribution of Animal Communities," "Ecological Succession," "Environmental Factors," and "The Animal Community."

Oxford University plant ecologist Arthur Tansley explained the ecosystem concept in 1935 as interactions of particular plants, animals, and environments, with different ecosystems named. He placed the ecosystem within a series of systems from atoms to the universe, but he did not lead ecologists in this new ecological direction. That step came with American new ecologist Raymond Lindeman, whose dissertation had been on succession at Cedar Bog Lake, Minnesota. His understanding developed further during a "postdoc" at Yale University under Hutchinson. Sadly, he died before he could publish his new ecosystem understanding, but Hutchinson had it published in *Ecology* (1942).

The ecosystem concept was explored in the work of Odum brothers Eugene and Howard. Eugene earned his Ph.D. in 1939 and moved to the University of Georgia, where he had an outstanding career. He ran a very productive school of ecology and published an influential *Fundamentals of Ecology* (1953). He and his younger brother Howard made good use of the ecosystem concept. Howard had studied under Hutchinson. On the title page of the second and third editions of *Fundamentals* (1959, 1971), Eugene acknowledged Howard as a collaborator. Howard was a very prolific author, who established an ecology school at the University of Florida.

Systems ecology was both a perspective and methodology, promoted by Eugene Odum and Stanley Auerbach, at Oak Ridge National Laboratory (ORNL), both having research funding from the Atomic Energy Commission (AEC) for radiation ecology research. That led to research on radiation pollution from military experiments on atomic weapons, which was not likely AEC's goal. Howard Odum introduced engineering symbols into his and his students' research, and into his very influential *Environment, Power, and Society* (1971), a revolutionary book.

New Yorker Bernard Patten earned degrees at Cornell (1952), Rutgers (1954), and the University of Michigan (1959). His highly mathematical "An Introduction to the Cybernetics of the Ecosystem: the Trophic Dynamic Aspect" (1959) undoubtedly came from his doctoral dissertation. He had conducted his research at ORNL. He also worked there from 1963 to 1968 and published "Systems Ecology: a Course Sequence in Mathematical Ecology" (1966). In 1968, he went to the University of Georgia and became a very productive author.

In 1983, Howard Odum published an almost encyclopedic *Systems Ecology: an Introduction* (xvii + 644 pages), and Australian insect ecologist Robert Kitching (B.Sc., University of London, Ph.D., Oxford University) published a textbook, *Systems Ecology: an Introduction to Ecological Modeling* (xx + 280 pages).

International Biological Program (IBP) attracted worldwide attention, and five historical works discuss it from American perspectives. National governments mostly provided new research funds for this occasion, 1968–74. There were ten ecosystem studies in North America, such as on grasslands, desert, tundra, deciduous forests, and coniferous forests. For one biome, grasslands, David Coleman (2010: 33) reproduced a diagram showing functional components interacting—very complex. IBP generated enormous amounts of new data, some new understanding, but no theoretical innovations. There were also some lasting advances, including the Long-Term Ecological Research Program (LTER) and National Ecological Observatory Network (NEON).

The rise of limnology, 1870 to 1920s, is summarized here. Now, I summarize North American limnology. Biological stations were very important research facilities, and there are both directories and histories of them. There were two dominant university schools of limnology, at the University of Wisconsin and Yale University. Limnology, like plant ecology and animal ecology, arose in the mid-West. New York Stater Edward Birge earned his Ph.D. at Harvard in 1878, but he had already begun his career at Wisconsin in 1876. He was a productive researcher-publisher of limnological papers but was less committed to teaching. His colleague Chancy Juday earned an M.A. degree at the University of Indiana in 1897, went to Madison in 1905, and by 1925 he had taken over educational responsibilities for students. Birge and Juday did not achieve American limnological leadership, but one of Juday's doctoral students, Arthur Hasler, who replaced Juday, did. Juday retired from teaching in 1937, and Hasler was hired in that year. Birge and Juday were good observational limnologists, but Hasler's research and teaching involved experimentation. His and his students' most noteworthy discoveries were on homing mechanisms that enable salmon to return to their home stream after one or more years in an ocean. Hasler also obtained funds from the National Science Foundation (NSF) which helped build a spacious laboratory on the university campus at Lake Mendota.

Yale's school was created by English limnologist G. Evelyn Hutchinson, who began teaching at Yale in 1928. He attracted eastern students, who were often more advanced than mid-Western students who went to Wisconsin. Both Hutchinson and Hasler were very productive authors. Hutchinson's *Treatise on Limnology* (4 volumes, 1957–93) is the longest ecology treatise by one person (volume 4 was completed by others after he died). Yet Hasler had the lead in having the most students earn doctorates. He began teaching at Wisconsin nine years after Hutchinson began at Yale; both retired from teaching in 1971. Hasler advised 53 students who earned doctorates; Hutchinson advised 34.

Great Lakes, on the border between Canada and the United States, contain over 95% of North America's fresh water. Half of those lakes' water is from precipitation, and half from rivers and streams. Glaciers gouged five lake basins from river beds; the glaciers melted 8,000 years ago. Before civilization, these lakes had virtually no pollution, and Native American fishing had insignificant impacts on fish populations. A few land animals did decline somewhat from hunting, and humans likely exterminated megafauna. French fur trade began in the late 1600s and continued until the early 1830s, when French traders and their Indian trappers decimated the small mammal population, and the fur trade moved further West.

Forests around Great Lakes were spectacular before Europeans arrived. Logging caused the first environmental changes. The lakes provided cheap transportation of lumber in ships to cities on lake shores. Logged lands were sold to farmers, but only land south of Lake Erie proved to be fertile for more than a few years. The five lake basins contained about 300,000 people by 1800. The coming of the Industrial Revolution to three basins of eastern lakes proved profitable, as the lakes provided cheap transportation for both ores and manufactured goods. By 1900, the lake basins contained 37 million people. Today, there are 105 million people, who have 46 million jobs. Logging, farming, mining, and manufacturing all discharged pollutants into these lakes or rivers flowing into them. Droughts often followed logging, and destructive fires followed droughts. On 8 October 1871, a Chicago fire killed 300 people, and another at Pestigo, Wisconsin, killed 2,200 people.

Wildlife fared poorly as environments were degraded, and after the Civil War, commercial fishing led to overfishing in some parts of these lakes. Spencer F. Baird became concerned about the same problem in North Atlantic waters, and he persuaded Congress in 1871 to create a U.S. Fish Commission under his supervision, with no increase in his salary, to monitor both fish stocks and fishing intensity in America's oceans and Great Lakes. Coastal states followed the Federal example and created state fish commissions. Some state commissions provided fishery biologists with funds to study state fishes. Canada created the Fisheries Research Board of Canada in 1901, which funded a research station on Georgian Bay, available for use by not only the Federal biologist but also the University of Toronto faculty and students. The United States and Canada established an International Joint Commission in 1912 and a Great Lakes Fish Commission in 1945, headquartered in Ann Arbor, Michigan, where there was a U.S. Bureau of Fishery office.

Canals between lakes were built in the 1800s, many with locks. A binational St. Lawrence Seaway was built in the 1950s, costing $1 billion, which allowed commerce between Great Lakes and European ports. It was expected to increase prosperity significantly around Great Lakes, but foreign commerce never achieved anticipated magnitudes, and foreign vessels accidentally introduced aquatic pests that had negative impacts on native species and on commercial fishing.

Invasive species became a nightmare. Sea lampreys found their way into Lake Ontario in the early 1800s but were slow to reach Lake Erie because Niagara Falls was a barrier. After Canada built Welland Canal, 1824–29, sea lampreys reached Lake Erie, then Lakes Huron, Michigan, and Superior. They devastated fisheries in Huron and Michigan and also had a harmful impact on the three other lakes. Federal biologists developed a poison specific to sea lampreys, which restricts their population in breeding streams, but without eliminating them. Other invasives also had a negative impact, especially after the Seaway opened. Most destructive were zebra and quagga mussels from the Black and Caspian Seas. These bottom dwellers consumed food on which some native fish depended and also clogged water-intake pipes. By 1990, fishery biologists realized global warming was having a negative impact on fish stocks. That became another problem for them to investigate.

ASPECTS OF ANIMAL ECOLOGY

A. Physiological Animal Ecology

Plant ecologists began studying physiological ecology long before animal ecologists began studying physiological ecology. Dwight Billings began "The Historical Development of Physiological Plant Ecology" (1985) with the discovery of photosynthesis in the 1770s. Much later, Frederic Clements had a "Vision of a Unified Physiological Ecology." Malcolm Nicolson stated: "Physiological ecology constituted a major programme of research with American plant ecologists in the early decades of the twentieth century." There were several decades between origins of plant and animal physiological ecology.

Danish August Krogh attended the University of Copenhagen and studied animal ecology (Ph.D. 1899). His lifetime researches were on respiratory mechanisms and gas exchanges in animals and humans, first expressed in *The Respiratory Exchange in Animals and Man* (1916). He won a Noble Prize for Physiology and Medicine in 1920. His *Osmotic Regulation in Aquatic Animals* (1937) was surprisingly comprehensive, in 242 pages. His final synthesis was *The Comparative Physiology of Respiratory Mechanisms* (1940).

Swiss Max Kleiber, of Zurich, attended its Federal Institute of Technology and obtained a diploma in 1920 and an Sc.D. degree in 1924, with a dissertation on "the energy concept in nutrition." In 1929, he joined the faculty at the Animal Husbandry Department, University of California, Davis. Continuing his dissertation research, he published "Body Size and Metabolism" (1932), still cited 70 years later. It was the first synthesis of animal physiological ecology. In 1934, he developed a way to predict the basal metabolism rate (BMI) of animals of different sizes and their nutritional needs—Kleiber's Law. His later *The Fire of Life: an Introduction to Animal Energetics* (1961) was a more recent synthesis of physiological ecology.

Bostonian Lawrence Irving earned a B.A. (1916) at Bowdoin College in Maine, an M.A. (1917) at Harvard, and a Ph.D. (1924) at Stanford University. He remained three years at Stanford and then joined the Department of Physiology, University of Toronto, where he studied the physiology of fish, diving birds, and aquatic mammals. In 1937, he became chair of the Zoology Department, Swarthmore College, Swarthmore, Pennsylvania. He spent summers teaching and researching at Woods Hole. To study northern animals, he went on expeditions to Alaska and northern Canada. In 1939, he obtained a Rockefeller grant to enable Krogh's protégé, Per Frederick Scholander, to come to Swarthmore and collaborate in studying the physiology of diving mammals and escape World War II. For Scholander, this was an opportunity of a lifetime. They became a very productive research team, seen in Scholander's *Experimental Investigations on the Respiratory Function in Diving Mammals and Birds* (1940).

Scholander's Swedish father was an engineer, his Norwegian mother, a pianist. His parents divorced, and he went with his mother to Norway. At the University of Oslo medical school, he found a course on physiology very interesting. In Greenland, he became fascinated with diving seals and obtained a grant to study them, whales, and birds.

During World War II, Irving became a major and chief physiologist at Air Force Proving Ground, Elgin, Florida. Scholander became a captain under Irving. After the war, Irving returned to Swarthmore and also directed a new Naval Arctic Research Laboratory, Anchorage, Alaska.

Norwegian Knut Schmidt-Nielsen was the son of Sigval, a professor of chemistry at the Norwegian Institute of Technology, Trondheim. His Swedish mother, Signe, studied physics in Stockholm and was the third Swedish woman to earn a doctorate. As a child, Knut received on his eighth birthday a copy of Samuel Baker's *Wild Animals and Their Ways*, which "opened a new world" of adventure. In 1934, he entered the University of Oslo to major in zoology. A friend suggested he specialize in physiology, which he did. He bought three textbooks from a departing student; one of them "became the most influential book I ever owned." For him, Richard Hesse was a founder of physiological ecology. In the summers of 1936–38 Knut and his father fished in Lake Grönningen. Knut opened the stomachs of 723 brown trout and char caught. Trout ate mostly insect larvae and plankton, and char ate mostly bottom organisms. Yet there was some overlap in food. That led to Knut's first publication, coauthored by his father (1949).

Sigval wrote Krogh and asked if Knut could go to his laboratory; Krogh agreed. He suggested Kunt study osmotic regulation in freshwater crayfish compared to saltwater crabs. He did, and his findings became undergraduate thesis at the University of Oslo. Krogh had four daughters, and Schmidt-Nielsen married the youngest, Bodil. He graduated in 1941, and a few months later, she became Denmark's first woman Doctor of Dental Surgery. He began graduate work and she taught in Denmark's School of Dentistry. In 1946, Irving and Scholander visited Krogh and met Bodil and her husband, Knut Schmidt-Nielsen. Irving invited them to Swarthmore as research associates. They accepted. He earned his doctorate in June 1946, and in November they immigrated to Swarthmore.

Ohio animal ecologist Charles Kendeigh earned his B.A. and M.A. degrees (1926, 1927) at Oberlin College before moving to the University of Illinois, Champaign, to earn a Ph.D. degree (1930) under Shelford. He taught at Western Reserve University, Cleveland, 1930–36, before returning to the University of Illinois, 1936–73. Of 65 masters and 51 doctoral students he advised, the most prominent was Eugene Odum, who summarized his career:

> While still a student, he began his monumental long-time study of physiological ecology of the House Wren at the Baldwin Research Laboratory—He worked there something like 15 summers. This work (reported in numerous papers and three monographs) pioneered ecological energetics.

Kendeigh was ESA president in 1950, and in 1978 he was awarded ESA's Eminent Ecologist Award.

Irving asked Knut to develop a method to determine chlorine levels for field studies. He did, and Irving decided to use it to test animals in Arizona deserts. The Schmidt-Nielsens had never seen a desert, and Kunt wanted to stay and conduct research there. The U.S. Forest Service had a station for experimentation south of Tucson they used. He studied kangaroo rats, trapped and raised on barley without water. He discovered that they used less water to excrete urea than do other mammals. They spent days in burrows and were active at night. They published their findings in *Physiological Reviews* (1952) and in *Scientific American* (1953). Both continued their research based upon findings about kangaroo rats. She did comparative studies on physiological differences between different species of rats, focused on kidney function. She earned her Ph.D. at the University of Oslo in 1955.

In fall 1952, Kunt joined the Zoology Department of Duke University, Durham, North Carolina. In 1962, he became James B. Duke Professor. In his autobiography, he stated his marriage began to unravel in 1950. They were raising their three children, and they remained together until 1964. In 1953, his wife and children went with him to Sahara Desert for a year to study camel physiology. She collaborated with him on his first paper on camel physiology, but not afterward. In 1956, he returned to a question he had wanted to study in 1939, interrupted by World War II: do seabirds drink seawater? Bodil had ceased sharing her kidney research with him, so he invited two postdocs to collaborate. He forced seawater into a cormorant's stomach. "Within a minute or two, I made the fastest discovery I ever made. I noticed that the bird, with a quick movement of the head, shook off droplets of fluid that appeared at the tip of its beak . . . revealing a high concentration of chloride." He discovered glands in its head that secreted salt. They published a paper on it, and he published an article on it in *Scientific American*.

Physiological studies were conducted on humans and white rats until Krogh, Lindeman, Irving, and Hess studied other animals. Their studies attracted little broader notice, but the Schmidt-Nielsens attracted broader attention with their studies. They also integrated field and laboratory studies. Bodil wanted to leave the marriage, partly because she wanted to be known for her own research and not as Knut's assistant. Their divorce was in 1966. Both found new spouses. She became president of the American Physiological Society in 1975–76, and he became president of the International Union of Physiological Scientists in 1980–86. In 1992, the Tokyo Academy of Sciences awarded Knut Schmidt-Nielsen its International Prize for Biology.

Raymond Lindeman, with a University of Minnesota Ph.D. (1941), received a post-doc fellowship to study under Hutchinson at Yale, to explore theoretical aspects of his doctoral research on community succession. The outcome was his posthumous "The Trophic-Dynamic Aspect of Ecology" (1942), which explained "The trophic-dynamic viewpoint . . . emphasizes the relationship or 'energy availing' relationships with the community-unit to the process of succession." Lindeman's paper became the beginning of ecosystem research in animals.

George Bartholomew earned B.A. and M.A. degrees (1940, 1941) at the University of California, Berkeley. His earliest research on cormorant behavior in San Francisco Bay was reported in three papers. He earned a Ph.D. (1947) at Harvard University and then joined the Zoology Department of the University of California, Los Angeles (UCLA). He next studied the social and reproductive behavior of elephant seals, Alaska fur seals, and California sea lions. In Galapagos Islands came his sea lion study, which are challenged by heat. He also studied thermoregulation in colonial nesting birds on tropical and subtropical islands. He and associates studied physiology and behavior of animals to determine temperatures that each species could tolerate. After he and associates published eight specialized papers, he published a more general "Interaction of Physiology and Behavior under Natural Behavior" (1966). Two decades later, he published a retrospective essay on his career (1987).

Bartholomew and K. Schmidt-Nielsen were contemporaries who made important contributions to teaching and research. Joel Hagen compared them in "Camels, Cormorants, and Kangaroo Rats" (2015). His summary: "The differences [are] between Schmidt-Nielsen's engineering approach and Bartholomew's explicitly evolutionary and ecological approach." Another comparison can be drawn between them as professors. Both trained graduate students, who earned masters and doctoral degrees, and both were productive researchers and publishers. Schmidt-Nielsen's stronger publication influence may have been balanced by Bartholomew's having produced more students with graduate degrees.

Early use of the term *physiological ecology* referring to animals was by Kendeigh and his student Eugene Odum, in his animal ecology dissertation (1941) at the University of Illinois. A decade later, John Vernberg wrote a Ph.D. dissertation at Purdue on *Physiological Ecology of Salamanders*.

An academic couple from Midwest, John and Winona Vernberg became associated with both environmental physiology and marine biology. They married in 1945, and they earned Ph.D.s in 1951, she at Purdue and he at DePauw, both universities in Indiana. He rose from instructor to professor, 1951–69, at Duke University. While there, they both used Duke's marine station on North Carolina's coast. In 1966, he participated in a conference on the physiological ecology of marine animals. In 1969–96, John was Baruch Professor of Marine Biology and founding Director of Baruch Coastal Research Institution, University of South Carolina. Winona became Research Professor of Marine Biology. In 1972, they published *Environmental Physiology of Marine Animals*. John Vernberg's other publications included *The Animals and the Environment* (1970), *Physiological Ecology of Estuarine Organisms* (1975), *Functional Adaptations of Marine Organisms* (1981), and autobiographical *It's a Long Story* (2018).

Physiological ecologist Brian McNab, from Chicago, at an early age went with his mother and brother to Portland, Oregon, where he later saw a newspaper notice of a bird walk at Washington Park Zoo. Later, he decided that experience was the most important event of his life, for it took him from "an uneducated, lower middle-class family to install a strong interest in birds and mammals." After joining the Oregon Audubon Society, he appreciated fauna in a unique rainforest at the junction of Willamette and Columbia Rivers and in the Cascade Mountains. In 1951, he entered Oregon State University and majored in wildlife management, until he learned fellow students were all hunters. McNab switched to a zoology major and decided in his senior year to become an animal ecologist. After graduating (1955), he joined an expedition to Point Barrow, Alaska, to study lemming populations. He measured lemming kidneys and adrenal glands. McNab earned graduate degrees (1958–62) at the University of Wisconsin–Madison. His dissertation was on three species of deer mice, living in mesic, intermediate, and xeric environments. After completing his doctorate, he saw an ad for a one-semester position at the University of Florida. He got it, and spent his career there. He alone published two substantial papers, and a third with dissertation advisor Peter Morrison. He found that desert mice had lower basal rates of metabolism than mesic mice, and intermediate mice had intermediate rates. His later work is discussed here.

One of Bartholomew's top students was Bernd Heinrich, whose family fled to western Europe after World War II. They settled into a hut in a forest, which was a paradise of birds and insects for Bernd. His entomologist father,

Gerd Heinrich, gave him a book on beetles, and he collected 447 beetles of 135 species. Gerd was world authority on ichneumon wasps, and he sold bird and mice skins to museums. Bernd earned two degrees at the University of Maine (1964–66) under advisor Dick Cook, who suggested a thesis on *Euglena*, which led to three coauthored papers in *Cell Biology* and *Protozoology*. Cook had ties at UCLA, to whom he recommended Heinrich's acceptance with research fellowship. He then earned a Ph.D. (1970) at UCLA, under Bartholomew, with a dissertation on desert animals. He read Schmidt-Nielsen's works on camels and kangaroo rats. After exploratory research, he wrote an 89-page dissertation on *Flight Energetics and Temperature Regulation by Blood Circulation in the Sphinx Moth* (1970). He then joined UCLA faculty, 1971–80, before settling at the University of Vermont. He continued research on the physiological ecology of insects during the 1970s and 1980s, culminating in *The Hotblooded Insects: Strategies and Mechanisms of Thermoregulation* (1993). He also published two autobiographies—*In a Patch of Fireweed* (1984) and *The Snoring Bird: My Family's Journey through a Century of Biology* (2007)—with his drawings in both.

Long Beach native Glenn Eric Walsberg earned his B.S. degree (1971) at California State University and Ph.D. (1975) at UCLA, maybe under Bartholomew. At Arizona State University, Tempe, he advanced from assistant professor in 1978 to professor in 2004. In 2000, he published "Small Mammals in Hot Deserts: Some Generalizations Revisited." Other zoologists since the Schmidt-Nielsens had studied desert rodents, and Walsberg was synthesizer and critic of what was known in 2000. He mapped the hottest deserts in the Southwest. The Schmidt-Nielsens had conducted their studies slightly east of that region, so their findings were not quite applicable to the entire Southwest.

McNab synthesized *The Physiological Ecology of Vertebrates: a View from Energetics* (2002). Chapters 3, 5, 11, 13, 14, and 19 drew upon his own work. Investigations by others included chapters on ectotherms, osmotic regulation, water–salt exchange, gas exchange, and more. Its 80-page bibliography was an important resource for anyone wanting to explore any topics covered. McNab's culminating work is *Extreme Measures: the Ecological Energetics of Birds and Mammals* (2012), which included subjects studied by most physiological ecologists. He thinks his publications have been under-appreciated.

Springfield, Massachusetts-native Barbara A. Block earned her B.A. degree (1980) at the University of Vermont, then studied marine biology at Duke University (Ph.D. 1986). She was assistant professor at the University of Chicago, 1988–93, and then professor at Stanford University's Hopkins Marine Station, Pacific Grove, California. Hopkins Marine Station and Monterey Bay Aquarium cofounded an endowed chair for her. Her research is on the physiological ecology of tuna and other large fish. She has published over 200 papers and edited *Tuna: Physiology, Ecology, and Evolution* (2001).

B. Ethology

English historians of biology divided the history of ethology into three phases: (1) formative, until about 1930, when pioneering studies were undertaken; (2) classical, 1930–50, when coherent traditions were established by Lorenz and Tinbergen; (3) mature, since 1950, indicated by Tinbergen's *The Study of Instinct* (1951). American historian of biology Eileen Crist acknowledged ethology's dependence upon early natural history but emphasized difference in perspective: naturalists viewed their subjects as living beings, and ethologists viewed their subjects as "natural objects."

French Lieutenant des chasses des Parcs de Versailles Charles-Georges Le Roy rose from being only a gamekeeper to naturalist by writing encyclopedic articles, including on instinct, collected in his *Lettres sur les animaux* (1768). He has been judged the most important early contributor to animal behavior studies. He envisioned a science like ethology.

One can claim Darwin as the effective founder of ethology. His *Journal of Researches into the Geology and Natural History of the Voyage of* H.M.S. Beagle (1839) included numerous observations on the behavior of animals he found worldwide. His *Origin of Species* (1859) had a chapter on instinct. *On the Various Contrivances by which British and Foreign Orchids Are Fertilized by Insects* (1862) is unique in describing mutual interactions of orchid species and particular insect species to insure successful fertilizations, while the orchid morphology excluded other species. His *Expression of Emotions in Man and Animals* (1872) "had a tremendous influence on the development of the study of animal behaviour."

(Jean) Henri Fabre was from a village in southern France and was educated to teach school by 1842. He continued his studies and earned a doctorate in science in 1854. He then studied insects and published books on his findings. His work was cited by Darwin in the *Origin*. However, Fabre never accepted Darwin's theory. His major work, of 50 years, was *Souvenirs entomologiques* (10 vols., 1879–1908). His volumes were accessible, and most were translated into English. Upon his death in 1915, Wheeler, at Harvard, commented: "the world has lost its greatest entomologist."

Swiss psychiatrist-naturalist Auguste-Henri Forel, besides his medical career, had a lifelong fascination with ants. He emphasized ant behavior, but also published *Les Fourmis de la Suisse* (1873). He sent a copy to Darwin, who sent him a thank-you note. Forel retired from Zurich Medical School in 1893 to devote his time to ants. He published *The Senses of Insects* in English (1908, German 1910). His lifetime studies culminated in his monumental *Les monde social des Fourmis* (5 vols., 1921–22, English 1928).

Darwin's influence on amateur behaviorists in the early 1900s is evident in the works of Selous, Kirkman, Howard, and Huxley. Londoner Edmond Selous was educated in law but was more interested in birds, leading to his *Evolution of Habit in Birds* (1933). Frederick Kirkman edited *British*

Bird Book (4 vols., 1910–13), on the behavior of individual species. He conducted a long-time study of black-headed gulls, and he later became active in a new Institute for the Study of Animal Behaviour. Eliot Howard was a successful businessman, whose *Territory in Bird Life* (1920) called attention to territorial defenses of bird species. Julian Huxley, grandson of Thomas Huxley, was also a professional zoologist who began work in a specialization, ethology, but after a few years went on to a broader career in science. He became interested in birds at age 8, and at age 13, he wanted to learn "the language and habits of birds." His main ethological studies were on the reproduction of great crested grebes (1914) and on courtship in the red-throated diver (1923).

Milwaukee entomologist William Morton Wheeler earned his Ph.D. (1892) at Clark University and was at Harvard University after 1907. He began publishing on ant ethology in 1903. His notable discovery was trophallaxis: secretions provided by some individuals and castes that bound the colony together (1918).

Primatologist Robert Yerkes earned his M.A. and Ph.D. (1899, 1902) at Harvard, then joined Harvard's faculty until 1917. In 1911, he founded the *Journal of Animal Behavior*, the first such journal for psychologists and zoologists. It lasted seven years. Yerkes was president of the American Psychological Association in 1916–17. In 1924, he joined Yale University's Institute to Study Primates. In 1929, he founded Yerkes Laboratories of Primate Biology. He sent student Clarence Ray Carpenter to Panama to study monkeys, with important results. Two of Yerkes' students at Yale published monographs in 1937.

Massachusetts native Margaret Morse was the daughter of a history professor at Amherst College. Her mother had graduated from Mount Holyoke College, as did Margaret. She then worked on a graduate degree two years at Clark University before marrying graduate student Blaine Nice. While raising four daughters, she earned an M.A. degree (1915) at Clark, with a thesis on bobwhites. Blaine attended Harvard Medical School, 1911–13, before joining the faculty of the University of Oklahoma. She objected to opening hunting on mourning doves in August, and when she found three nests in use in August, the opening was reset. The Nices collaborated on *The Birds of Oklahoma* (1924). During 1927–36, they were at Ohio State University, Columbus. Her main behavior studies were on song sparrows. Because her article on them was too long for American ornithology journals, it was translated into German and appeared in the *Journal für Ornithologie* (1933–34). There, she read Konrad Lorenz's article on ethology (1931), and he read her article on song sparrows. They met at the 8th International Ornithological Congress, 1934, and became friends. In his "Forward" to her autobiography (1979), he recalled: "I was at once impressed by her deep understanding of ethology, its methods and its approach."

Indianan Warder Clyde Allee was from a Quaker community near Bloomington, and he attended Quakers' Earlham College, Richmond. Then he earned M.S. and Ph.D. degrees at the University of Chicago, under Shelford. He taught at four universities before returning to Chicago for his career. He taught a course on animal behavior and published three relevant books: *Animal Aggregations* (1931), *Animal Life and Social Growth* (1932), and *Cooperation among Animals* (1938, ed. 2, 1951). He also attracted graduate students who researched behavioral subjects. Allee was the senior author of a founding encyclopedia, *Principles of Animal Ecology* (1949). All five authors held prominent positions in the Chicago area.

Englishman William Thorpe entered Cambridge University in 1921 to study agriculture. In 1925, with a B.A. degree, he began work at the Department of Agriculture on insect pests. He also continued graduate studies at Cambridge, earning a Ph.D. in 1929. He then worked at the Imperial Institute of Entomology until 1932, when he became lecturer in entomology at Cambridge. In 1943, he published on insect learning and in 1951 on bird learning. In 1950, he had become the first director of Cambridge Ornithological Field Station, Madingley, which became a preeminent center for animal behavior studies. Thorpe read works by Lorenz and Tinbergen and used new technology to investigate bird song. In 1959, he became Reader in Zoology, and in 1966 he received a chair in zoology.

Viennese Konrad Lorenz was the son of a professor of medicine, who wanted him to also earn an M.D. degree. He did and then earned a Ph.D. in zoology. He was attracted to geese, which he raised from eggs and discov-ered they imprinted on him as their parent. He published five crucial papers, 1931–41, now translated in his *Studies in Animal and Human Behaviour* (2 vols., 1970–71). The Deutsch Gesellschaft für Tierpsychologie was organized in 1936, and in 1937 it began *Zeitschrift für Tierschrift*, with Lorenz as one editor. Lorenz's paper on imprinting (1935) impressed Tinbergen, who contacted him. Tinbergen accepted Lorenz's invitation to spend time during 1938 at Lorenz's home in Altenburg, which was fruitful for both.

Tinbergen's father had a Ph.D., and he taught in a Dutch high school in The Hague. Niko's older brother, Jan, won a Nobel Prize in economics in 1969. These brothers were not close. Niko was unsure of a career choice. After graduating from high school in 1925, his mother sent him to a German field station for two months. There, he decided to study ethology at Leiden University. However, no university taught that in 1925. In 1929, he wandered in a coastal forest in quest of a thesis topic. He watched a bee-wolf wasp dig a hole in sand, fly away, and return with a honeybee, which she dragged to the end of her hole, laid an egg beside it, and covered the hole. He decided this wasp would be a good dissertation topic. In 1932, he earned his Ph.D., received his degree, published the first of three papers from his dissertation, and married a friend's sister.

He and wife Lies joined a Dutch government expedition to Greenland during the International Polar Year, 1932–33. That experience had an enormous influence upon these flatlanders. Niko learned diet and habits of native Inuits and life histories of snow buntings and red-necked phalaropes.

He published a memoir, in which he said that expedition transformed him from "sentimental naturalist to a hunter who saw his animals as objects." In May 1940 Germany invaded the Netherlands, and the Tinbergens turned their house into a hospital. Trains were sabotaged, and German soldiers were shot. Germans arrested 1,900 public figures and confined 1,400 of them in a commandeered Catholic college. Inmates organized activities, and Niko lectured on Greenland and ethology.

Tinbergen's studies were methodologically different from Lorenz's. Lorenz raised free-ranging waterfowl and studied their habits. Tinbergen observed wild birds and experimented on them without interfering with their freedom. Although Tinbergen was Dutch, David Lack brought him to Oxford University in 1949. In 1950, he switched his publishing from German to English. He and Thorpe organized a symposium in 1949, to which they invited two zoologists, each from Germany, the Netherlands, and the United States. Tinbergen's biographer judged this meeting as "a climatic event for Niko," who reunited with Lorenz for first time since World War II. At Oxford, Lack, Elton, and Tinbergen attracted their own students. These students might take courses together, but otherwise there was no communication among the three schools.

Viennese Karl von Frisch was from an academic family. His father, a professor of medicine, wanted Karl to study medicine, which he did for two years. Afterwards, his father let him switch to zoology. In 1911, he read a claim by a professor of ophthalmology that fish and invertebrates were color-blind. Karl assumed brightly colored flowers were attractors of insect collectors of pollen and nectar, while pollinating flowers. He experimented with honeybees in summers and fish during winters and came to understand bee communication. He spent 26 years studying the intricacies of honeybee waggling at hives. Lectures he delivered at Cornell University were published as *Bees: Their Vision, Chemical Senses, and Language* (1950). Later, he and son Otto wrote *Animal Architecture* (1974), which surveyed building abodes in the animal kingdom. Honeybee language was von Frisch's major contribution, and others continued that research.

Before 1973, no ecological science had anyone receive a prize comparable to Nobel prizes given in physiology and molecular biology. Nobel laureate Peter Medawar had befriended Tinbergen, and he was probably the initiator of a Nobel award for three ethologists. The Nobel Prize for physiology and medicine was expanded in 1973 to include ethologists Frisch, 86, Lorenz, 69, and Tinbergen, 66.

Just nine years later, Tinbergen commented: "Ethnology is the name for a phase in the evolution of the behavioral sciences." Other ethologists thought he was too hasty. Yet one of his former students, and biographer, Hans Kruuk, stated: "At the time of Niko's retirement [1974], *behavioural ecology* was born."

Kenyan anthropologist-paleontologist-zoologist Lewis Leakey became facilitator for three young women who studied the behavior of great apes. English Jane Goodall became the most famous living animal behaviorist. Leakey arranged for her to study chimpanzees in Gombe National Park in 1960. In 1962, she participated in a symposium sponsored by the Zoological Society of London, reporting on two years of observing chimpanzees, emphasizing her discoveries of predation and tool use. Solly Zuckerman, who chaired the symposium, realized she contradicted his claim in 1932 that all primates are vegetarians. He rejected her claim. She earned a Ph.D. in ethology (1966) at Cambridge University, with a dissertation on *Behaviour of Free-Living Chimpanzes*. Her research since the 1960s was funded by the National Geographic Society. She celebrated 30 years of her finding with a volume describing them (1980). Other ethologists celebrated her 50 years of research with a celebratory volume (2000).

American Dian Fossey, with Leakey's help, began research on mountain gorillas in Rwanda in 1967. She studied three groups of them and became obsessed with protecting them from poachers. She also enrolled at Cambridge University (1970) and earned a Ph.D. in ethology (1976). She wrote about her experiences in *Gorillas in the Mist* (1983). She was murdered two years later in her camp, apparently by poachers.

Lithuanian parents of Birutė Galdikas moved to Germany after World War II before her birth, and when she was 2, they moved to Canada. Growing up, she read about Goodall and Fossey. In 1962, the Fosseys moved to California, and Biruntė earned a B.A. degree (1966), jointly in zoology and psychology at UCLA and the University of British Columbia. She earned an M.A. degree (1969) in anthropology at UCLA. There she met visiting Leakey and asked him about studying orangutans in nature. He and the National Geographic Society established a research station in Borneo, where she conducted dissertation research. She earned a Ph.D. degree (1978) at UCLA, with a dissertation on *Orangutan Adaptation at Tanjung Putting Reserve, Central Borneo*. She also had to contend with poachers and habitat devastation. She persisted for over 30 years, in one of the longest studies of a single species. She established the Orangutan Foundation International to support orphans and save habitat. She is a professor at Simon Fraser University, Burnaby, B.C. Her publications include *Reflections on Eden: My Years with the Orangutans of Borneo* (1995) and *Great Ape Odyssey* (2005).

C. Animal Population Ecology

Belgian astronomer-mathematician Adolphe Quetelet founded modern statistics in 1835, providing more mathematical methods in 1845. Belgian Pierre-François Verhulst studied under Quetelet and developed the logistical curve in three papers, 1838–47. However, its demographic significance was not appreciated until the curve was rediscovered in 1911 in the context of bacterial growth. It was finally applied to demography by two American mathematicians, Raymond Pearle and Lowell Reed (1920). Since the United States took a census every decade, they had more relevant data than European mathematicians. Pearle subsequently published three books that used the logistic curve.

The Industrial Revolution transformed agricultural machinery, leading to the establishment in the American government of a Department of Agriculture (USDA) in 1862, with similar agencies arising in western Europe. In 1887, Fruit Growers Convention asked USDA entomologist Charles Riley for help against cottony-cushion scale insects, infesting citrus orchards. He replied, if they got Congress to provide funds, he would send an entomologist to Australia, home of this pest, to find a predator. The Convention did, and Riley sent an entomologist to Australia, where he discovered a predator and a fly parasite of these insects. By releasing these enemies in infested orchards, this pest was controlled.

Research on insect pests expanded, and in 1897 it became mathematical and theoretical. In 1897, three studies appeared—two French, one American—on insect populations. Most ambitious mathematically was Ad. Bellevoye and J. Lorent's "Les plantations pins dans la Marne et les parasiteds qui les attaquent." Appearing in a regional science journal, its audience was French-speaking Europeans. Professor Paul-Alfred Marchal, Institut National d'Agronomie, Paris, was France's leading entomologist. His brief note, "L'equilibre numérique des espèce set ses relations avec les parasites chez les insects," argued that oscillations in crop pests—Hessian fly, oat midge, army worm—are due to an increase in parasite population following an increase in host population. A crash in host population is followed by a crash in parasite population, which allows the host population to increase again.

A third 1897 study was by USDA chief entomologist Leland Howard, who had earned his degrees under John Comstock at Cornell University. Washington, D.C., in 1895 "suffered from an extraordinary outbreak" of *Orgyia leucostigma* attacking fruit trees. English sparrows had spread throughout American cities, ridding shade trees of canker worms, and tussock worms apparently filled the void. Howard studied known and potential parasites of tussock worms, reported in "Study in Insect Parasitism: of the White Tussock Moth" (1997). Lotka later provided a graph illustrating Howard's data.

With W.K. Fiske, Howard undertook an investigation of both gypsy and brown-tail moths. Robert Tamarin's observations (1978) on their study (1911) included this comment: "They believed in the balance of nature a view prevalent among those who believe in density-dependent population regulation."

USDA entomologist Dwight Pierce, Dallas, led a team of three investigating boll weevil parasites. Their report resembled Howard and Fiske's but went further with a notable diagram (1912) that indicated parasites and hyperparasites.

In response to overfishing in the North Atlantic, U.S. Congress established a U.S. Fish Commission in 1871 to monitor fish population in U.S. waters. In 1902, European nations established an International Council for the Exploration of the Sea to monitor Europe's waters.

Norway built a research vessel, *Michael Sars*, in 1900, and appointed fisheries biologist Johan Hjort (1869–1948) as founding director of Norway Fisheries, 1900–16, to monitor populations. He also became Norway's representative at the International Council, 1902–48. He was the Council's vice-president, 1920–38, and president, 1938–48. In 1904, German fisheries biologist Friedrich Heincke published a study on age groups of cod and plaice. Hjort adopted Heincke's methods to census cod and monitor its age classes. With John Murray, Hjort coauthored *The Depths of the Sea* (1912), a significant reference book.

In the 1920s, animal ecologists were not trained much in mathematics, when two mathematicians, Alfred Lotka and Vitro Volterra, published guides to calculate animal populations. Their procedures are valid for conditions specified, but not for life in nature. Lotka was a European with American parents, who took graduate courses in mathematics and physical sciences at the universities in Birmingham, England, Leipzig, Germany, and Ithaca, USA, published *Elements of Physical Biology* (1925), and then became a demographer at a New York insurance company. Volterra was an Italian mathematician who taught at the University of Rome until 1931, when he was dismissed for refusing to take a loyalty oath to Mussolini's government. He then lectured in western European countries. In1926, with no knowledge of Lotka's book, he published "Varizioni Fluttazioni del Numero d'Indivividui in Specie animali conviventi" (1926). He also published a three-page summary in *Nature*: "Fluctuations in the Abundance of a Species Considered Mathematically" (1926). Lotka saw the summary in *Nature* and wrote its editor, claiming priority, which he sent to Volterra, and then published Lotka's letter and Volterra's response. Lotka also wrote to Volterra, and they had a cordial correspondence.

Italian mathematician Giorgio Israel contrasted their contributions: "Volterra's approach was strictly adherent to the classical physico-mathematical paradigm, while Lotka's point of view was more eclectic and open-minded as regards the news developments of physics." Kingsland thought Lotka had published too early and lacked a biological audience. True, but his book motivated ecologists to learn enough mathematics to use his book. A major concern of Volterra was reconciling calculations with real fisheries data from D'Acona, his son-in-law.

Russian zoologist Vladimir Alpatov obtained a Rockefeller grant to study at Johns Hopkins University under statistician Raymond Pearl, 1927–29. Back in Moscow, he discussed his new understanding with graduate student Georgii Gause, who had surveyed grasshoppers in the Caucasus Mountains and developed a mathematical formula relating species abundance and environmental factors, which appeared in *Ecology*.

Gause began laboratory experiments on competition and predation with populations of microorganisms. In 1932, he wrote to Volterra, thanking him for a letter and a copy of a recent paper. He mentioned his own research and his limited mathematical ability, and invited collaboration. Volterra collaborated with a few colleagues but not with Gause. In 1933, Gause wrote to Volterra about his pending book, *The Struggle for Existence*, and stated he had confirmed experimentally some of Volterra's mathematical projections, and he requested copies of any papers he published since 1931. Gause confirmed that competition between two similar species for food led to the extinction of one species. This *competitive exclusion principle* was also called "Gause's principle."

Gause did not mention to Volterra his finding on predation:

> We expected . . . to find "classical" oscillations in numbers arising in consequence of the continuous interactions between predators and prey as was assumed by Lotka and Volterra. But it immediately became apparent that such oscillations are impossible in the population studied, and that this holds true for more than our special case.

His book, according to Hutchinson, became a "foundation stone of ecology." Fortunately, he published in America because in 1934, at the Ecological Conference of the Academy of Sciences' Botanical Institute, a fanatical Stalinist, I.I. Prezent, denounced Alpatov and Gause for using mathematics in their "formalist, mechanist school." Gause decided it would be expedient for him to switch to research in microbiology.

Two ecologically oriented entomologists, Canadian William Thompson and American Royal Chapman, wanted mathematical assistance from Lotka and Volterra. Thompson earned a B.A. degree (1909) at the University of Toronto. In summer 1908, he had worked at USDA Gypsy Moth Parasite Laboratory, and in 1909, USDA sent him to Cornell University to earn an M.S. degree (1911) under Professor Comstock. In 1913, he resigned from USDA and went to Paris to study under Professor Maurice Caullery. In 1914, he studied bacteriology and protozoology at Cambridge University. During World War I, he joined the Royal Naval Service. In1918, he returned to Paris and earned a Ph.D. (1921). Leland Howard at USDA rehired him as head of a European Parasite Laboratory. In 1929, he went to England to run Farnham House Laboratory and study insect parasites and predators. During World War II, he returned to Canada to run Imperial Parasite Service. In 1948–58, he established Biological Control Laboratories in California, Trinidad, Uruguay, Switzerland, India, and Pakistan. Culminating Thompson's publications was "Biological Control and the Theories of the Interactions of Populations" (1939). He was wary of zoologists who substituted mathematical calculations for field data, which was commendable, but he also rejected Darwin's theory of evolution.

Minnesotan Royal Chapman earned B.A. and M.A. degrees at his state university and his Ph.D. (1917) at Cornell under Comstock. He returned to the University of Minnesota's faculty and developed the first ecology course taught there. His course emphasized "controlled experiment and quantitative field studies." His "The Quantitative Analysis of Environmental Factors" (1928) cited two of Volterra's papers, showing ecologists methods they needed to master. His textbook *Animal Ecology with Especial Reference to Insects* (1931) had as an appendix an English version of a 40-page Volterra article. Since his textbook never used mathematics, Kingsland suggested he reprint the article to add an aura of exactness to an insecure science. However, providing an example for others need not indicate insecurity.

English A(lexander) J(ohn) Nicholson grew up in Birmingham and earned B.Sc. and M.Sc. degrees at its university. He joined the Royal Field Artillery during World War I. In 1920, he became a lecturer in entomology at the University of Sydney, Australia. In his course, he itemized factors limiting insect populations, including food. In 1929, the University of Sydney awarded him D.Sc. for work on mimicry. In 1930, he joined what would become Entomology Division, Commonwealth Scientific and Industrial Research Organization (CSIRO). Howard and Fisk (1911) had begun an informal "biotic school of population dynamics," and Nicholas became its chief defender.

"Nicholson's 'Balance of Animal Populations' (1933) "is probably the most important single paper on population regulation, having sparked the many controversies in this field." That paper was quantitative, with 11 graphs but no mathematical calculations. In 1935, he collaborated with mathematician V.A. Bailey in publishing a mathematical version of the 1933 paper, using the same title. They acknowledged the contributions of Lotka and Volterra, unknown to them when they began.

Illinoisan brothers Orlando Park and Thomas Park earned B.S. and Ph.D. degrees under Allee at the University of Chicago. Both returned to Illinois after teaching a few years elsewhere. Orlando settled at Northwestern University, and Thomas returned to Chicago. Orlando studied ecology of nocturnal animals, and Thomas studied social relations of insects in laboratories. Thomas Park taught the first course in population ecology and coined the term. Both became coauthors of Allee et al.'s *Principles of Animal Ecology* (1949).

1954

Albany native Edward Deevey graduated from Yale University (1934) and became one of Hutchinson's graduate students. His dissertation was on paleolimnology, yet he shared Hutchinson's interest in quantitative studies. He explained "Life Tables for Natural Populations of Animals" and how to construct them (1947). R.F. Morris and C.A. Miller used his instructions for "The Development of Life Tables for the Spruce Budworm" (1954).

London-born entomologists-ecologist Peter Price singled out three recent studies as important: Andrewartha and Birch (1954), Lack (1954), and Morris and Miller (1954), to which I add three by Nicholson (1954a, b, c) and two by Cole (1954a, b). There was probably little communication among these ecologists, yet all were at the advancing edge of an animal ecology specialization.

Australians A.J. Andrewartha and L.C. Birch's encyclopedic *The Distribution and Abundance of Animals* emphasized weather and shortage of time for reproduction. Cole's "The Population Consequences of Life History Phenomena" related reproductive potential to natural selection in a strategy to maximize fitness. A species had either few offspring, raised by parents, or many offspring who coped with the environment independently. Australian immigrant A.J. Nicholson did experiments and consulted Gause's *Struggle for Existence* (1934) and concluded that "populations do not conform to any pattern, such as the logistic curve" (1950). In 1954, he published three papers: "Experimental Demonstration of Balance in Populations," on sheep blowflies; "Compensatory Reactions of Populations to Stress, and Their Evolutionary Significance"; and "An Outline of the Dynamics of Animal Populations." The third article was a lengthy new synthesis updating his 1930s papers.

English David Lack's *Natural Regulation of Animal Numbers* (1954) was based upon 11 previously published papers. He acknowledged Nicholson's influence on his thinking, and he also cited Margaret Nice's observations from banding nestling song sparrows on rates of mortality. His claim that "in most animals density-dependent variations affect the reproductive rate less than the death rate" was rather controversial. Lack summarized his 1954 book in an appendix to his *Population Studies of Birds* (1966).

Chicago animal ecologist LaMont Cole earned B.S. and Ph.D. (1944) degrees at the University of Chicago under Allee. He settled at Cornell University. In "The Population Consequences of Life History Phenomena" (1954) he viewed the life history of an animal as a strategy to maximize fitness. He noted that population cycles of small mammals correlated with tree rings since both responded to changes in climate. He learned from Lotka and Volterra and devoted 7.5 pages to "Computation Methods."

R(obert) F(ranklin) Morris and C.A. Miller wrote "The Development of Life Tables for the Spruce Budworm" (1954) to show the utility of life tables for insect pests. They adapted what Deevey did for vertebrates to insects. The value of their work for forest entomology was attested by Price.

All these 1954 studies displayed breadth of population ecology, without resolving controversies.

1957–58

Ray(mond) Beverton and Sidney Holt, at Britain's Fishery Laboratory on East Anglia's coast, published the first synthesis of mathematical studies on fish populations: *Dynamics of Exploited Fish Populations* (1957: 533 pages). Their scope was North Sea fisheries, which several European countries exploited. It was important for countries to share data and agree on harvesting limits. They did not.

Since 1933, Biological Laboratory, Cold Spring Harbor, Long Island, has hosted Symposia on Quantitative Biology. Its 1957 symposium, Population Studies: Animal Ecology and Demography, was funded by three federal agencies and three foundations, and hosted about 150 scientists. The goal was to see what animal population ecologists and demographers could learn from each other. (Answer: not much.) Cole's opening talk, "Sketches of General and Comparative Demography" (1958), gave the history of both sciences, down to Darwin's *Origin of Species* plus Lotka and Volterra, and discussed comparative life histories, population regulation, fluctuations, and age structure.

Andrewartha spoke on "The Use of Conceptual Models in Population Ecology," Birch on "The Role of Weather in Determining the Distribution of Animals," and Nicholson on "The Self-Adjustment of Populations to Change." Discussions followed the talks and were published in 1958 following the talks. Neither Andrewartha nor Birch commented on Nicholson's talk, but he commented on both of theirs. A distribute developed between Andrewartha and Nicholson, becoming "highly charged and personal," to the dismay of Americans.

A. Milne, Agriculture Research Council Unit of Insect Physiology, Durham University, England, cautioned that "Theories of natural control of mammal and bird populations are not likely to be useful for insects" for several reasons. Years later, J.P. Dempster commented:

> Milne's ideas form a halfway house between the views of Nicholson and Andrewartha and Birch. He believes that the latter underestimate the part played by density-related processes in the determination of animal numbers, but thinks that Nicholson overestimates the frequency with which intra-specific competition occurs in nature.

Hutchinson had the last word at the symposium, but he devoted little more than a page to the talks, in his "Concluding Remarks," before explaining "The Formalization of the Niche," maybe the most cited paper in the symposium volume (1958). Bodenheimer, Elton, and Lack were not at the symposium, but Bodenheimer and Elton published relevant books in 1958. Bodenheimer had gone to Australia in 1956, and Nicolson had converted him to his own perspective. Bodenheimer's book devoted a brief paragraph to arguments by Andrewartha and Birch and seven pages to Nicholas' arguments.

Elton's *The Ecology of Invasions by Animal and Plants* broke new ground. A local increase in numbers allowed species to invade new territory. His book had no mathematical calculations; illustrations were mostly maps showing ranges of invasive species. That it initiated a new line of research was challenged by David Simberloff, who claimed that distinction for his own 1981 paper.

Robert MacArthur's father taught in Toronto when he was born but soon moved his family to Marlboro, Vermont. Robert studied mathematics at Marlboro College (B.A., 1951) and Brown University (M.S. 1953), and switched to ecology under Hutchinson at Yale University for Ph.D. (1958). His notable dissertation was "Population Ecology of Some Warblers of Northeastern Coniferous Forests" (1958), which won an ESA prize. He was, like Darwin, a theoretician who synthesized his own and others' data to support theoretical arguments. A notable book he initiated was coauthored by E.O. Wilson, *The Theory of Island Biogeography* (1967).

SINCE 1958

Numerous publications from the 1960s to 1970s indicated rising interest in population ecology, including Lloyd Keith, *Wildlife's Ten-Year Cycle* (1963); William Hazen compiled a large volume of *Readings in Population and Community Ecology* (1964); Maurice Solomon compiled a brief *Population Dynamics* (1969, x + 60 pages); Lowell Adams compiled a small *Population Ecology* (1970, x + 160 pages); and more.

Andrewartha and Birch's *Distribution and Abundance* (1954) enabled Andrewartha to join the Department of Zoology, University of Adelaide, South Australia, where he taught population ecology, based upon that book. It was too lengthy for a textbook, so he published *Introduction to the Study of Animal Populations* (1961). It appeared about when Slobotkin's textbook did. Their two textbooks seem to be the first on population ecology. Lack's *Population Studies on Birds* (1966, appendix) argued against Andrewartha and Birch's weather theory (1954). He studied birds, they insects. However, insect ecologist Nicholson also disagreed with their theory.

New Yorker Lawrence Slobotkin earned his Ph.D. (1951) under Hutchinson at Yale. His dissertation was quantitative experiments on the age structure of *Daphina* to test a theory. His first book was *Growth and Regulation of Animal Populations* (1961), which MacArthur's review in *Ecology* stated was best yet by a theoretical ecologist. Since neither of them mentions Andrewartha's *Introduction* (1961), we may assume they had not seen it. Historian Kingsland discussed both authors, but without comparing their 1961 textbooks.

Milne published "On a Theory of Natural Control of Insect Population" (1962), without mentioning either textbook, though he cited an article Andrewartha coauthored in 1961. Milne argued that the old controversy over density-dependent versus density-independent was "inadequate." His alternative was intermediate between the two. Lack agreed with Milne that "even the best field studies of the problem suffer from ecological deficiency," but Lack saw differences in theories as due to "difference in basic assumptions concerning the extent to which competition is held to be important or unimportant in natural populations."

Milwaukeean Gordon Orians earned a B.A. degree at the University of Wisconsin (1954) and Ph.D. (1960) at the University of California, Berkeley. He settled at the University of Washington, Seattle. He wrote a critique of population theory, "Natural Selection and Ecological Theory' (1962), that contrasted Andrewartha and Birch (1954) with Lack (1954). He thought ecology has descriptive generalizations, such as *competitive exclusion*, but only one theory—*natural selection*.

K. Bakker, Zoölogisch Laboratorium, Leiden, wrote "Backgrounds of Controversies about Population Theories and Their Terminologies" (1964), praising Orians' paper but suggesting a better starting point was Nicholson versus Thompson. He explained.

English zoologist Vero Wynne-Edwards had broad ecological interests. At Oxford University he studied under Huxley and Elton. He was active in ornithological and marine work. On a ship to Canada he discovered that marine birds can be divided into inshore, offshore, and pelagic. During World War II, he taught radar mechanics to members of Canada's Air Force. In 1946, he became a professor of zoology at the University of Aberdeen, Scotland. His landmark work was *Animal Dispersion in Relation to Social Behavior* (1962). Territorially was a population control. His book was controversial. Lack devoted part of "Theoretical Controversies concerning Animal Populations" (in Lack 1966) to refuting Wynne-Edwards' theory. The controversy was over the importance of natural selection and group selection. E.O. Wilson would not dismiss group selection as an impossibility. Wynne-Edwards later published *Evolution through Group Selection* (1986), which received more criticism than praise.

MacArthur and Joseph Cornell published *The Biology of Populations* (1966), which they explained had been planned in 1960. That presumably explained why they ignored works by Slobotkin and Milne.

The Royal Entomological Society of London held a symposium (1967) on insect abundance, with 13 papers. Speakers were from London, Europe, North America, Africa, and Australia. Editors of a volume from it, T.R.E. Southwood and O.W. Richards, in an introduction summarized the history of population studies, 1887–1967. Symposium participants discussed both population theories and methods of investigation.

FISHERIES AFTER 1960

English fisheries biologist D.H. Cushing during World War II served in the British Army and afterwards graduated from Oxford University. He joined Lowestoft Fisheries Laboratory and continued Beverton and Holt's work. In 1963, Hasler invited him to come to the University of Wisconsin to deliver lectures, which became Cushing's *Fisheries Biology: a Study of Population Dynamics* (University of Wisconsin Press, 1968). It covered both freshwater and marine fisheries and addressed fish stocks and fishing pressure worldwide.

A large conference volume also appeared in 1968: *The Future of the Fishing Industry of the United States*, D. Gilbert, editor. In 1977, J.A. Gulland edited *Fishery Population Dynamics*, with 15 chapters. W.E. Ricker, "The Historical Development" (1968), emphasized methods of assessing populations of commercial species. Ricker thought the Swedish clergyman H. Hederström first determined the age of wild fish (1759). He counted rings on fish vertebrae. No one followed his example, or that of Leeuwenhoek, who had counted ridges on fish scales in 1684, published in English in Royal Society's *Philosophical Transactions*.

The University of Toronto Professor Henry Regier explained in his chapter 4 types of fishery population problems biologists faced, identifying each and calculating populations for each type. In 1977, Kenneth Johnstone also celebrated 75 years of achievement in *The Aquatic Explorers: a History of the Fisheries Research Board of Canada*.

With such publications available, did coastal nations prudently control their fisheries? Robert May and four associates studied "Management of Multispecies Fisheries" (1979) and found the common goal of maximum sustained yield (MSY) was an academic exercise that only worked when fishermen were the only factor impacting target species. In the real world, that did not exist, and MSY never was achieved.

With funds from Rockefeller Foundation, the University of Washington's Institute for Marine Studies, Seattle, organized a large multi-authored *The Management of Marine Regions: North Pacific*, E.S. Miles et al. (1982). This report attempted to achieve an understanding among nations having North Pacific commercial fisheries.

Two decades after Cushing published *Fisheries Biology* (1968), he reported a method by which an annual catch could be calculated. Did that solve the problem? Many stocks collapsed. Anchoveta stock failed off Peru after three-year classes failed, while fishing was intense. The Atlantic cod fishery also collapsed.

Anne Platt McGinn, in *Rocking the Boat: Conserving Fisheries and Protecting Jobs* (1998), argued that, in practice, for fisheries and communities, short-term advantage defeated long-term sustainability. Carl Safina, in *Song for the Blue Ocean* (1997), saw that dilemma but saw hope in rise of many organizations during the 1990s dedicated to preserving ocean resources. Jennifer Hubbard explained in *A Science of Scales: the Rise of Canadian Atlantic Fisheries Biology, 1898–1939* (2006) difficulties of monitoring fish populations. *Mother Jones* magazine carried three articles (April 2006) on overfishing and pollution.

WILDLIFE AFTER 1960

Numerous publications in the 1960s–70s showed rising interest in population ecology: Lloyd Keith, *Wildlife's Ten-Year Cycle* (1963); William Hazen compiled a large volume of *Readings in Population and Community Ecology* (1964); Maurice Solomon compiled a brief *Population Dynamics* (1969, x + 60 pages); Lowell Adams compiled a small *Population Ecology* (1970, x + 160 pages); Anthony Alison edited 12 original papers, *Population Control* (1970), on animal and human demography; Ian McLaren reprinted 10 articles in *Natural Regulation of Animal Populations* (1971); P.J. den Boer and G.R. Gradwell edited first of two volumes of a symposium, *Dynamics of Populations* (1971, 611 pages); and more.

Hutchinson, in retirement, wrote *An Introduction to Population Ecology* (1978), after two of his former students had published textbooks on this. He recommended Slobodkin's as more concise than his, and he acknowledged the influence of deceased former student MacArthur. Hutchinson's *Introduction* included many historical discussions and footnotes, possibly both entertaining and distracting. Since Hutchinson was a leading ecologist, his book may have attracted some attention.

By outliving Nicolson, Andrewartha and Birch seemingly had the last word in *The Ecological Web* (1984), which updated their arguments after 30 years. Reviewer Charles Krebs praised their first book, but not this one.

Nicholson's birth centennial was celebrated in 1995 by the Division of Entomology, CSIRC, with an impressive conference of over 200 ecologists from a dozen countries. Andrewartha was dead, and Birch did not contribute a paper to the volume edited by R.B. Floyd et al., *Frontiers of Ecology* (1996: xii + 639 pages).

D. Saving Habitats and Managing Wildlife

The earliest human immigrants came from Asia before 1000 BC. Those who came overland were likely big mammal hunters. In Eurasia, large mammals adjusted to increasing hunter effectiveness gradually as techniques slowly improved. However, those hunters burst upon American herds with efficient technology that killed individuals before herds adjusted to a new danger. Species extinction followed.

HISTORICAL RECORD OF EXTINCTIONS

When Dutch explorers visited Mauritius in 1598, they observed flightless dodos. An English ship visited that island in 1681 and was the last to see dodos alive.

German naturalist Georg Steller, on a Russian expedition in 1741, discovered on Bering Island, in North Pacific, Steller's sea cow, which was exterminated a few years later by Russian fur traders for food. Vertebrates on islands were vulnerable to extinctions when humans settled there because they had not had predators before. Fossil trilobites, belemnites, and ammonites were known in Europe, but they might have still lived in ocean depths. Mammoth bones were known from Siberia, but since vast areas of Eurasia remained unexplored, it was unknown whether mammoths still lived.

The great auk in North Atlantic was superficially similar to penguins, though they lived a world apart. Auks also lived along coasts and had lost the ability to fly. They were valued for meat, fat, and feathers. George Cartwright warned in 1785 that the high rate of slaughter would diminish them "to almost nothing" unless restrained. The last of them were killed in 1844 and sold to museums.

UNITED STATES POLICY

Bison/buffalo lived in vast herds on prairies and plains. Natives, before Europeans arrived with horses and guns, killed some, but their livelihoods did not depend on them. Natives' victory over Custer's U.S. troops in 1876 was their last conflict before being confined to reservations.

When Union Pacific's transcontinental railroad was built after the Civil War, Will Cody was hired to supply fresh meat to construction crews. As "Buffalo Bill," he claimed to have slaughtered 4,280 bison, 1867–68. After western railroads were built, upper-class hunters rode railroads to bison and shot them for fun, leaving their bodies for wolves and coyotes. Commercial hunters killed many more to sell hides to eastern processors. The U.S. government did not discourage this since bison scarcity helped keep natives on reservations. When Congress passed a hunting limit, President Grant vetoed it. In 1886, U.S. National Museum, Washington, sent taxidermist William T. Hornaday to collect bison for an exhibit. That experience awakened his interest in saving the species from extinction. A few westerners had taken home calves to raise after their mothers had been slaughtered. Small herds still existed in the United States and Canada. In 1905, Hornaday and others founded the American Bison Society, to breed them and return them to national and state parks and refuges. In 1907, Canada bought a private herd and moved it to Buffalo National Park, Alberta.

Four birds in eastern USA became rare, then extinct in the late 1800s to the early 1900s due to hunting: Labrador duck, passenger pigeon, heath hen, and Carolina parakeet. All could have been raised in captivity. There were also fishes, reptiles, amphibians, insects, and plants that were reduced to rare or extinct by human activities.

SAVING HABITATS AND SOME MAMMALS

Congress passed the Hatch Act in 1862, enabling citizens to buy federal lands at affordable prices after General Land Office surveyed them.

Vermonter George Perkins Marsh was the son of a Woodstock lawyer. As a boy, he appreciated nature and was upset by clear-cutting forests, followed by wildfires, erosion, flooding, and scant restoration. He later became a lawyer, and he served in Congress, 1843–49. Afterwards, he joined the U.S. Diplomatic Service as ambassador to Turkey, then Italy. Observing the impacts of humans on Mediterranean lands for 2,000 years, he remembered the human impact on Vermont lands. He wrote *Man and Nature: or, Physical Geography as Modified by Human Action* (1864), which, after America's Civil War, became the founding document for the environmental movement.

Physician-forester-statistician Franklin B. Hough, from Martinsburg, New York, was head of the 1870 U.S. Census. He found evidence of declining availability of timber. In 1875, the American Forestry Association (AFA) was founded, with Hough an active member. Members collected data on forests, timber production, and reforestation. In 1876, AFA encouraged Congress to establish an office to oversee U.S. forests. Congress created an agent in USDA and appointed Hough as agent. He traveled widely and published *Report on Forestry* (1877, 659 pages). In 1881, a Division of Forestry was established in USDA, with Hough head, 1881–83. He published America's first textbook on *Elements of Forestry* (1882), with chapters on soils, climate, reproduction, fires, and insects.

In 1886, professional forester Bernhard Fernow became USDA head forester. From Germany, he had attended Prussia's Forest Academy and then worked in Prussia's Forest Service for seven years. He came to the United States in 1876 to attend an AFA meeting, an excuse to rejoin his American fiancée, whom he married, and became an American citizen. He headed the U.S. Forest Service, 1886–98. In 1898, he became the first dean of New York State College of Forestry at Cornell University. In 1907, he became the first head of the Faculty of Forestry, University of Toronto.

Three successive wealthy Roosevelts—related, but not fathers and sons—in New York State became concerned for habitat reform. First was Robert B. Roosevelt, who hunted and fished, and saw habitat decline. He expressed these concerns in two books on native fish and one on game birds (1862–66). He was elected to Congress for New York City, 1871–73, and enthusiastically supported the new U.S. Commissioner of Fish and Fisheries. In 1877, he became president of the New York Association for the Protection of Game and lobbied the New York Legislation on behalf of wildlife.

Scottish John Muir, his father, and two siblings sailed to America in 1849. They bought a Wisconsin farm, which was isolated. Muir educated himself, then attended the University of Wisconsin for a year, taking courses in botany, geology, and writing. In 1868, he sailed to California and walked 100 miles to Yosemite State Park, where he felt at home. He began publishing influential articles expressing environmental concerns.

Congress created Yellowstone National Park in 1872. In 1889, magazine editor Robert Underwood Johnson visited author Muir and toured Yosemite Valley, where they decided to campaign for a large Yosemite National Park. Muir's two 1890 articles and Johnson's lobbying Congress succeeded.

New Yorker George Bird Grinnell grew up with Audubon's grandsons while they attended a school run by his widow, Lucy. George enjoyed shooting neighborhood birds and mammals, a common sport. At Yale University, he studied mineralogy and in 1870 assisted paleontologist Marsh in exploring fossils. In 1874–75 he was a naturalist on two expeditions. In 1876 he became editor of *Forest and Stream*. Conserving wildlife became a major concern. He joined Harriman's Alaska Expedition in 1899.

New Yorker Theodore Roosevelt (TR) at age 8 wandered down Broadway and saw a dead seal from NYC harbor on display. Later, he identified that moment as beginning his interest in nature. He began a museum, which justified his hunting—in the name of science. When upper-class men made hunting fashionable, he joined Grinnell in forming Boone and Crockett Club in 1877, promoting manly hunting and conservation.

Roosevelt served three years in New York State Assembly, 1881–83, and as governor, 1898–1900. Fellow Republicans felt he neglected the wealthy and industry, and so stuck him into "dead-end" vice presidency under President McKinley. However, McKinley was assassinated in 1901, and TR became president. His popularity among Republicans declined but increased among American people, ensuring his reelection in 1904.

He was one of America's earliest environmental leaders. Interactions with U.S. head forester Gifford Pinchot were important for developing his environmental outlook. In a controversy in 1903 over killing birds to decorate women's hats, Roosevelt established America's first wildlife refuge, at Pelican Island, Florida. He then spent 66 days touring the West, covering 14,000 miles, with two naturalists, John Burroughs and (for part of the trip) John Muir. When Roosevelt left office in 1909, he had created 50 more refuges and 18 national monuments.

Europeans came to America biased against wolves, which killed unguarded livestock. Governments paid bounties for killing them, and wolf pelts were of value. Canadian naturalist-author-artist Ernest Seton Thompson (later, Ernest Thompson Seton) led a change in attitude through his popular writings and drawings. He saw, in 1892, a coyote die in agony after eating poisoned meat he had left for wolves. That sight helped change his perspective, which he conveyed in writings and drawings, in *Wild Animals I have Known* (1898). He also aspired to be a professional naturalist and wrote *Lives of Game Animals* (4 vols. in 8, 1925–28).

Another Grinnell, on the west side of the continent from George on the east side, Joseph, grew up in Pasadena. He earned a B.A. degree (1897) at Throop (later, California) Institute of Technology, Pasadena. He spent three summers in Alaska before earning his M.A. degree (1901) at Stanford University, and also Ph.D. (1913) there. In 1908–39, he was the first head of the Museum of Vertebrate Zoology, University of California, Berkeley. His publications were mostly on California birds and a few on California mammals.

Prominent men organized the Appalachian Mountain Club in 1876 for hikers in the northeast. In 1894, it began publishing *Appalachia*. It has now expanded from Maine to Washington, D.C., has 12 chapters, headquartered in Boston, with about 275,000 members.

In 1883, three members of the Nuttall Ornithological Club decided to follow Britain's example of having a British Ornithologists' Union and form an American Ornithologists' Union. In September, 21 ornithologists met at the American Museum of Natural History and founded the American Ornithologists' Union. Its scientific journal is *The Auk*.

Muir is correctly remembered as a founder of Sierra Club, 1889, but there were also others: artist William Keith, lawyer Warren Olney, and professors Joseph LeConte, David Starr Jordan, and others. In 1893, Sierra Club began publishing its *Bulletin*. Today, it has 64 chapters nationwide, is headquartered in San Francisco, and has three million members.

In 1906, San Francisco suffered two catastrophes, an earthquake followed by fire. When rebuilding, the city concluded it needed an additional source of water, and wanted to build a dam at Hetch Hetchy Valley, in Yosemite National Park. Sierra Club fought the proposal, but its members living in San Francisco sided with the city—so did Washington and Sacramento, so Congress passed it.

In 1929, Congress appropriated $1 million to add new refuges to the 51 Roosevelt established. The U.S. Biological Survey (USBS) appointed three teams to canvas different regions for favorable purchases. In 1937, land was bought for Aransas National Wildlife Refuge, Corpus Christi, Texas, to protect wintering whooping cranes. By 1955, there were 282 Federal refuges. Presently, there are at least 562 U.S. Fish and Wildlife refuges; two examples are of waterfowl refuges: Lake Mattamuskeet, Atlantic flyway, eastern North Carolina, 50,000 acres, Delta, and Mississippi flyway, near Lacombe, Louisiana, 48, 500 acres. There are books describing national refuges.

Gifford Pinchot, like Theodore Roosevelt, was born into a wealthy family. His father, James Pinchot, owned forests, which led to his interest in forestry. He and other estate owners organized the American Forestry Association (discussed earlier). He interested son Gifford in forestry, who after graduating from Yale (1889), spent a year at France's National Forestry School. Back in the United States, he became forest manager for George Vanderbilt in western North Carolina. Pinchot and Fernow were founders of American forestry. Pinchot and Roosevelt were acquainted but did not discuss forestry until 1899, when Pinchot visited Governor Roosevelt. They had a meeting of minds that lasted for the rest of Roosevelt's life. Pinchot became a father figure to U.S. Forest Rangers.

In 1887, Canada's parliament passed the Rocky Mountains Park Act and created Banff Hot Springs Dominion Park. In 1911, parliament created Yoho and Waterton Lakes National Parks (and the United States added Glacier National Park). Canada's Cabinet created the Dominion Parks Branch in the Department of Interior—the first such agency in world history. James Harkin became Commissioner of Dominion Parks and guided their development.

In America, some national parks were established before national forests were, yet a National Forest System was established before a National Park System was. In 1872, the U.S. Army was put in charge of Yellowstone National Park. Other western national parks were created, but no park service until 1916, when Secretary of Interior Franklin Lane appointed Western businessman Stephen Mather head of U.S. Park Service, assisted by Horace Albright. States soon realized that national parks attracted tourists, so states lobbied Congress to create more. In 1988, national seashores were added, and later National Marine Sanctuaries and National Estuarine Research Reserves.

In 1921, Mather called a meeting of state representatives to encourage the creation of state parks. Places nominated for national parks but not accepted might become state parks. Since 1921, every state has created state parks. Freeman Tilden discussed this process in *The State Parks* (1962).

The third Roosevelt who preserved habitats was Franklin Delano Roosevelt (FDR), who served in New York State Legislature, as governor, and was the longest-serving U.S. President. TR was his inspiration and guide, yet circumstances—four terms—enabled FDR to achieve more than TR had. The Depression required new ideas. For the environment, he had Chicagoan lawyer Harold Ickes as Secretary of Interior. Civilian Conservation Corps (CCC) was FDR's idea to put unemployed young men to work building needed facilities. Most CCC funds Congress sent to the Department of Interior, though the Department of Agriculture also received ample funds.

The Bureau of Land Management (BLM) was created in 1946 by merging General Land Office and U.S. Grazing Service, both of which existed in the West. BLM controlled the most federal land. Americans who visit BLM lands are about the same people who visit National Forests to hunt, fish, and camp.

The American Nature Conservancy arose when ESA members voted in 1945 to keep ESA out of conservation activities. Victor Shelford, former ESA president, led in organizing the Nature Conservancy. It has a chapter in each state. Wisconsin chapter published, Cathy Harrington, ed., *The Places We Save: a Guide to Nature Conservancy's Preserves in Wisconsin* (1997), with accounts of 23 preserves and 11 cooperative projects, and one or two photos for each place, with maps for about half of them.

Canada founded its Nature Conservancy (NCC) in 1962. It is active in southern provinces, where most Canadians live and where nature is most threatened. NCC has collaborated with other nonprofit organizations or with governments to buy and maintain wild lands. Bill Freeman wrote an excellent 50-year history of NCC.

There are also private reserves. Ohio-born craftsman Jack Miner had a waterfowl refuge at Kingsville, Ontario, in 1904, with a hospital for injured birds. He began banding waterfowl in 1909. Professor Don Schueler, University of New Orleans, with partner Willie Brown, bought 80 acres of run-down farmland north of New Orleans and turned most of it into wilderness, and a small part into a home, where they raised chickens and had a garden and orchard. He described it in *A Handmade Wilderness* (1996). Christopher Klyza edited *Wilderness Comes Home* (2001), with 12 chapters by different authors, who explored such themes as "Ecological Reserve Design," "Stewardship and Sustainability," and "Species Reintroduction and Recovery." Wilderness appreciation is important for saving reserves. Dick Frankenberg edited 38 essays describing tours on government lands: *Exploring North Carolina's Natural Areas: Parks, Nature Preserves, and Hiking Trails* (2000).

WILDLIFE MANAGEMENT

Governments conduct most wildlife management, and individuals and private organizations monitor government policies and practices. In 1852, California passed a law limiting deer hunting to six months a year. Since it did not also provide game wardens, the law had little or no effect. Other states did likewise, down to about 1900. Conservation began when game species became scarce. The Federal government developed enforcement agencies in the Departments of Interior and Agriculture. Canada and the United States began selling hunting licenses to fund game wardens by the early 1900s.

New York Stater C. Hart Merriam earned an M.D. degree but was more interested in birds and mammals than in medicine. In 1884, the American Ornithological Union (AOU) asked Congress to establish an office to study birds and mammals. It did, and in 1885 Merriam became head of a division of USDA to study birds and mammals in relation to agriculture. In 1889, he led an expedition to San Francisco Peak and Little Colorado River, Arizona. Merriam is most remembered for his 1890 report on his findings. Most important was his introduction of the concept of *life zones*. As one ascends a western mountain, one enters progressively colder regions, with different faunas and floras in each zone. However, his USDA division became a pest control agency for western farmers and ranchers.

Iowan John Lacey fought in the Union Army during the Civil War; afterwards, he passed law exams and in 1870 was elected to Iowa's Assembly. Much later, 1888–90 and 1892–1906, he served in Congress. He was the first in Congress to advocate wildlife conservation. He is most remembered for Lacey Game and Wild Bird Preservation and Dispositions Act (1900), to end killings to sell meat. Its enforcement provisions were inadequate, and in 1904 he introduced a stronger enforcement, which finally passed in 1913 as Weeks-McLean Bill, protecting migratory birds. By 1918, treaties extended this protection to Canada and Mexico.

Around 1900, Europeans began banding birds. At a meeting of the Michigan Academy of Sciences in 1901, vertebrate zoologist Leon Cole suggested bird banding. He got the idea from U.S. Fish Commission tagging fish to track their movements. In 1909, he and others founded the American Bird Banding Association. In 1920, banding records were given to the U.S. Biological Survey, which hired Frederick Lincoln to manage U.S. banding.

Concern for monitoring wildlife motivated Canadian entomologist Gordon Hewitt to write *The Conservation of Wild Life of Canada* (1921). It became a benchmark against which to judge future progress. He, and later Elton, used Hudson's Bay Company's data on the number of furs collected, 1820–1914, which indicated periodic fluctuations in populations.

Canadian Wildlife Service was founded in 1947, with less than 30 staff, recruited from other wildlife positions. There was no provision for research, but lack of precise mandate allowed it to arise. Canada–U.S. border is long, and there were mutual inclinations to cooperate across borders.

In 1937, the Wildlife Society was organized, which published the *Journal of Wildlife Management*. Midwestern universities began adding wildlife personnel to their faculty, who taught Leopold's *Game Management* (1933). Also in 1937, Senator Key Pittman and Representative Absalom Robertson introduced Federal Aid to Wildlife Act ("Pitman-Robertson Act"), which imposed 11% tax on sale of hunting guns and ammunition. Funds raised were appropriated to states for wildlife restoration. It was the most important wildlife law since the Lacey Act of 1913.

Indianan Waldo Lee McAtee earned two degrees at Indiana University (1904–06). He was a founding member of Wildlife Society and edited its *Journal of Wildlife Management*, 1937–42. He worked for USBS and was founding editor of *Wildlife Review* and wrote 6,500 abstracts for it. His *Wildlife Food Plants* (1939) was a valuable guide.

Illinoisan Herbert Lee Stoddard was self-educated on wildlife and began work for Milwaukee Museum as a taxidermist, 1910–13, 1920–24. USBS hired him in 1924 to study bobwhite quail at Red Hills, Georgia. He was the main author and editor of *The Bobwhite Quail: Its Habits, Preservation, and Increase* (1931) and concluded that hunting and habitat loss were important limits to population increase. It was an amazing book in 1931, likely the most detailed study of a bird species. It was also "the first major landmark in species management." In 1958, he and four colleagues founded Tall Timbers Research Station, specializing in fire ecology. Nearly 40 years later, USF&W biologist Walter Rosene used new advances to write *The Bobwhite Quail: Its Life and Management* (1969).

Iowan R. Aldo Leopold was the son of a hunter who passed on that tradition to him. He studied forestry at Yale University, then joined the U.S. Forest Service and chose Southwest District. There, he managed wildlife for hunting. He improved hunting by killing wolves, then found he had too much of a good thing: arid vegetation could not support a larger deer population. He realized wolves were essential in nature. In 1922, he had 750,000 acres surveyed, and in 1924, he had Forest Service designate a Gila Wilderness Area, the first such designation on Federal land.

Then he became assistant director of Forest Products Laboratory, Madison, Wisconsin, a desk job. Since 1916, he had published papers on wildlife and conservation. In 1928, Sporting Arms and Ammunition Manufacturers' Institute hired him to conduct a national survey of game conditions. After World War I, hunting became popular among veterans. In 1929, he spoke at the University of Wisconsin on game management. His *Report of a Game Survey of the North Central States* (1931) included data he used in writing *Game Management* (1933), the first textbook on this subject. His publications qualified him for appointment as the first professor of wildlife management, at the University of Wisconsin.

Leopold became leader of a new field; his textbook enabled faculty at any university to teach this new course. In 1935, he bought a run-down farm beside Wisconsin River as a weekend retreat for his family. Their only crop was trees, which they planted periodically. In 1948, they saw smoke at a neighboring farm, and he fought it to stop it from spreading to his farm. He died then from a heart attack.

ESA was organized in 1915. By then, three mid-western universities had informal schools of ecology: University of Chicago, terrestrial plants and animals; University of Wisconsin, limnology; and University of Nebraska, prairie plant ecology. Most ESA members have been concerned about managing natural resources. However, in 1945, most members voted not to involve ESA in saving natural areas. Former ESA president Victor Shelford then organized for conservation an Ecologists Union, later renamed The Nature Conservancy.

Minnesotan Murie brothers devoted their careers to wildlife studies, to guide management. Olaus' father died when he was 9; his mother remarried and had another son, Adolph. Olaus guided Adolph in his career. He graduated from Pacific University (1912), then worked two years for the Oregon Fish and Game Commission. Next, he joined staff of Carnegie Museum of Natural History, Pittsburgh, and went on expeditions into northern Canada for bird and mammal specimens, 1914–15, 1917.

In 1916, Olaus Murie was hired to kill predators but felt uncomfortable after killing a cougar. In 1920, he went to Alaska to study caribou for USBS. There he met his future wife, Margaret (Mardy) Thomas, who wrote *Two in the Far North* (1962) alone, followed by *Wapiti Wilderness* (1966), coauthored by Olaus. Later, he wrote *Journeys to the Far North* (1973), beginning in 1914 and posthumously edited by Mardy. In 1923, he wrote his superior that wolves were beneficial for maintaining a healthy herd of caribou. In 1926, USBS published his report, which stated that neither wolves nor sport hunting limited populations. Instead, he blamed competition with reindeer for forage.

In 1927, they moved to Jackson, Wyoming, where he studied elk in Teton National Park and wrote *The Elk of North America* (1951). They raised three children in Moose, near Jackson. Charles Craighead and Bonnie Kreps celebrate Mardy's life in *Arctic Dance: the Mardy Murie Story* (2002).

Adolphe followed Olaus and also condemned USBS for its predator control policy. He studied grizzly bears and wolves in Mount McKinley National Park, for the National Park Service, during 25 summers between 1922 and 1970, and wrote a book on each. He wrote a memoir, *A Naturalist in Alaska* (1961, drawings by Olaus), in which he discussed the different species he studied. He was writing the tenth book in 1974, when he died. His son, Jan, completed it.

Wildlife and habitat were finally a focus of William McShea and William Healy, editors, *Oak Forest Ecosystems: Ecology and Management for Wildlife* (2002), with 34 authors.

BIOGEOGRAPHY, 1700–1980

A. LAND, 1700–1840

Dutch Protestant minister Abraham van der Mijle wrote *De Orogone Animalium et Migration Populorum* (1667) to explain unique species of plants and animals of the Americas and Australia. His answer: different centers of creation. His suggestion was not discredited for almost two centuries.

British Protestant minister John Ray was the main founder of natural history in Britain. He published one of the earliest local floras, *Catalogus plantarum circa Cantabrigiam nascentium* (1660). He stressed the importance of distinguishing between similar species. His *Catalogus* provided data relevant to biogeography but was not organized for it. The same was true for his later catalogs on birds (1676), fish (1686), plants (3 vols., 1686–1704), and insects (1710). He also published two theoretical works: *The Wisdom of God Manifested in the Works of Creation* (1691) and *Miscellaneous Discoursdes concerning the Dissolution and Changes of the World* (1692). He opposed the idea of spontaneous generation but did not explain why different continents have unique species.

French Joseph Pitton de Tournefort became Europe's leading botanist, focusing mainly on taxonomy and systematics. He explored plants in western and southern Europe, and in 1700–02 in the Middle East. Back in Paris, he wrote *Relation d'un Voyage du Levant* but died in a traffic accident before publication (1717, English 1718). In volume 3, letter 19, he described his visit to Mount Ararat (where Noah's Ark supposedly landed after the Flood). He found plants on its sides could be divided into vegetation zones as one ascended, which correlated with vegetation changes as one went from southern to northern Europe. Linnaeus claimed this kind of correlation began with Andrea Cesalpino; he had in mind Cesalpino's *De Plantis* (1583).

Other botanical explorers during the early 1700s also published travel accounts, which became data for phytogeography. One of them, German Johann Grorg Gmelin, was the son of an apothecary who taught at the University of Tübingen. He attended university lectures since age 14 and earned a medical degree (1727). Two of his former professors in 1725 had joined the Russian Academy of Sciences, St Petersburg. In 1727, they invited him to join them. He did, and in 1731 he became an academician. In 1733, the Academy sponsored an expedition to Siberia headed by Danish sea captain Vitus Bering, which Gmelin joined. He returned to St. Petersburg in 1743 and worked on his *Flora Sibirica* (4 vols., 1747–69), which described 1,178 species. He listed species found in six geographical regions.

German zoologist Georg Steller studied medicine and became a physician in Russia's army in 1734 and traveled to St. Petersburg. At the Academy of Sciences, he learned of Bering's Second Kamchatka Expedition, and volunteered to join it, already departed. He was accepted, crossed Siberia to the Pacific Ocean, met Bering, and sailed with him. His travels provided many and varied natural history and geographical recorded observations. When they landed on an island off Alaska, he discovered a jay, now Steller's jay, supplemented by two more birds and two mammals. The return voyage was catastrophic. They landed on an island, now Bering Island, where Bering died. The survivors built a smaller ship, from the remains of their ship, with which they reached Kamchatka. They had survived by eating Steller's sea cow, and his observations were the only ones recorded before sea cows were exterminated by later Russian fur trappers. In 1744, Steller started for St. Petersburgh and died on the way. However, his three books—*De Beastis Marinis, Journal of a Voyage with Bering*, and *History of Kamchatka*—survived and were published.

The foremost leader of natural history in western Europe was Swedish Carl Linnaeus, who, however, was more interested in botany. He founded modern nomenclature and first conceptualized ecology, which he named *oeconomie de naturae* (1749). In 1732, the Uppsala Society of Science provided funds for his natural history exploration of Lapland, which was the "most productive exploratory expedition ever undertaken in Sweden." Later, he undertook similar expeditions to Öland and Gotland (1741), Västergötland (1746), and Skåne (1749). He made phytogeographical observations on these trips, and Swedish ecologist Einar du Ritz listed six discoveries he published. Linnaeus also introduced terms for *station* and *habitation*, indicating environment and geographic range, respectively, in *Statione Plantarum* (1754).

A French contemporary of Linnaeus, Comte de Buffon, devoted his attention to mammals and birds. Unlike Linnaeus, he did not collect his own observations in nature and was more theoretical, which widened his appeal but led him into biased speculation. American animals must be inferior to European animals because America's climate was inferior to Europe's, and species were influenced by their climate. He did not need data to establish this because it was logical. Furthermore, the royal press published his elegant *Histoire naturalle* (15 vols. + 7 supplements, 1749–89), which included his mammal encyclopedia, and *Histoire des oiseaux* (9 vols., 1770–83). These works did include both data and theoretical contributions to zoogeography (unnamed).

One of Linnaeus' former students, Pehr Kalm received funds in 1748 from the Royal Swedish Academy of Sciences to collect useful American plants that could survive in Sweden. His sponsors expected him to search mainly in Canada, but he preferred Philadelphia and vicinity, where most American naturalists lived. Afterwards, he wrote *En Resa til Norra America* (3 vols., 1753–61), added 60 species to Linnaeus' *Species Plantarum* (1753), and also included animals in his publications.

Finnish Petrus Forsskål, another former Linnean student, joined a Danish exploring expedition to Arabia in 1761. They spent a year in Egypt, and he prepared an Egyptian flora. On the way to Arabia, he died of malaria.

He was "one of the most highly talented men ever produced in Finland." His flora was later published in Copenhagen.

Swedish botanist Carl Thunberg, Linnaeus disciple, earned a medical degree in 1770 and went to Holland, where he became a plant collector, in South Africa in 1772. In 1775, pretending to be Dutch, he went to Japan. In 1776, he visited Dutch colonies and, then London. He found and described more new plants than any other botanist during his lifetime. In 1779, he returned to Uppsala and taught botany. There he also published a travel work in four volumes and three important works on plants he had collected.

German professor of mathematics and physics at Brunswick Eberhard Zimmermann, without field work, synthesized animal geography in great detail (Latin, 1777, German 1778–83), a "remarkable zoogeographical publication." His important innovation was to synthesize natural history and travel accounts and to construct a world map. Yet, like Mijle, he believed in multiple centers of creation (excepting humans).

Berliner Karl Willdenow learned plants from his apothecary father and from botanist Johann Gleditsch. He published *Florae Berloinensis* (1787) before earning an M.D. degree. He inherited his father's shop and taught botany informally. His *Grundriss der Kräuterkunde* (1792, ed. 4, 1805, English 1805) became the standard textbook and had a substantial account of plant geography. He distinguished five European floras. His textbook was so successful that he was admitted to the Berlin Academy of Sciences (1794) and became a professor of botany, University of Berlin.

During the 1800s, phytogeography progressed faster than zoogeography, beginning with German Alexander von Humboldt, who studied under Willdenow, and his French colleague, Aimé Bonpland, who spent 1799–1804 exploring Latin America, then over two decades published their findings, beginning with *Essai sur la Géographie des Plantes* (1807). Humboldt became the leading natural scientist of his age, with broad influence. He decided to explore Asia and compare his findings with his from America. He left Berlin on 12 April 1829 and was gone almost six months and traveling 9,700 miles. He was only able to compare the Altai Mountains with the Andes and Siberian steppes with Venezuelan llanos. He wrote two works later, on geology and climate (2 vols., 1831) and on mountains and climate (3 vols., 1843). His final work, *Kosmos* (5 vols., 1845–62, English, 5 vols., 1849–58), was based upon 61 public lectures he had given in Berlin in 1827–28, a distillation of his wisdom.

German naturalist Pyotr Pallas was the son of a professor at Berlin's medical college. He studied the geography of both plants and animals. He earned a medical degree in Leiden, with a dissertation revising Linnaeus' classification of worms. He then studied marine animal collections elsewhere in Holland and England. In 1767, Pallas accepted Catherine the Great's offer of a position at St. Petersburg Academy of Sciences and remained there for over 40 years. He joined an Academy expedition to Lake Baikal, Ural Mountains, and eastern regions in 1768–74. He published an account of his travels, which was important for biogeography (2 vols. in German, 1771–76; trans. into Russian, French, English, Italian). He joined another expedition, 1793–94; he moved to Crimea and wrote another account (92 vols., 1799–1801). He also wrote *Flora Rossica* (2 vols. 1784–1815) and *Zoogeographia Russo-Asiatica* (3 vols., 1811–31).

A medical student strongly interested in botany, Friedrich Stromeyer studied under botanist Gmelin at Göttungen University, writing a dissertation on what he called historical plant geography (1800). He wanted to do for plants what Zimmerman had done for animals. The editor of *Journal für die Botanik* wrote a glowing review of it, but it still faded away. Stromeyer became a chemist, who watched Humboldt become a famous phytogeographer.

Swedish botanist Göran Wahlenberg studied medicine at Uppsala University, where he remained. He began botanical explorations in 1799 before Humboldt influenced him. In 1811, he began publishing floras of central Europe, with plant geographical introductions.

Genevan Augustin-Pyramus de Candolle decided early to become a botanist but followed his father's wishes to study medicine at Genevan Academy. Simultaneously, he became friends with nearby botanists, from whom he absorbed a botanical education. Humboldt was one influence on him, but Linnaeus' ecological writings seem stronger. His most important ecological publication was an encyclopedia article (1820) on phytogeography. He recognized the importance of competition in determining the success of species. He became one of Europe's leading botanists.

Danish lawyer Joachim Schouw was a Humboldt disciple. In 1812, he accompanied Norwegian botanist Christian Smith into Norway's mountains, where Schouw was impressed by vegetation zones up the sides of mountains. In 1816, he earned a doctorate in botany at Copenhagen University, where he remained. He explored Italy on two expeditions (1817–19 and 1829–30). After his first expedition, he published on principles of phytogeography (1822, German 1823, French 1824). He began naming plant associations by adding *etum* to genus names of dominant species. After his second expedition, he published *Tableau du Climat et de la Vegetation de l'Italy* (vol. 1, 1829). His work on a second volume came later as a series of papers.

English William Jackson Hooker, from Norwich, discovered in 1804 a moss not previously known from Britain, which began his lifetime interest in flora. He traveled widely in Britain, and in 1809 he joined a diplomatic mission to Iceland. His collected plants burned in a ship fire, but he published his Iceland in 1811, which was followed by numerous papers on British plants. In 1835, he wrote two encyclopedia articles, on plant geography in general and on botany in England. He acknowledged his general article depended heavily on de Candolle's (1820), whereas his article on England was his own. He became director of the Royal Botanic Garden at Kew in 1841. He was diligent in maintaining a wide variety of species at Kew.

Danish entomologist Johan Fabricius studied under Linnaeus, then traveled widely in Europe. He served as a professor of natural history at the Universities of Copenhagen and Kiel. His best work was *Philosophia Entomologiae* (1778), in which he proposed eight climatic zones for insects. As more data accumulated, his climatic zones were redefined by two French entomologists in 1815 and 1838. Entomology flourished in Europe during these decades.

German zoologist Gottfried Treviranus, from Bremen, studied medicine at the University of Göttingen and returned home to practice medicine and teach at Bremen Lyceum. He was a very productive author, whose greatest work was *Biologie, oder Philosophie der Lebenden Naturfirscher und Aerzte* (6 vols., 1802–22). Volume 2 (1803) discussed the organization of life, distribution of plants and animals, and influence of external conditions. He discussed the geography of both plants and animals in one work before Pallas did so in two works (see previous discussion).

American President Jefferson sent the Lewis and Clark Expedition (1804–06) to explore overland Louisiana Territory, a success. Other land expeditions followed, exploring that vast addition to the United States. Under Jefferson, there were also expeditions by Pike and by Freeman and Custis.

Meanwhile, there were independent naturalist-explorers who published their own observations. Two from Britain were Scotsman Alexander Wilson and Englishman Thomas Nuttall. Wilson became a U.S. citizen and was the founder of American ornithology, with accounts of 264 species he knew, in nine illustrated volumes, 1808–14. Nuttall came to America in two-year-long trips and published accounts of American plants, briefer than Wilson's accounts, in two volumes, 1818. He spent enough time in America to also publish *A Manual of the Ornithology of the United States and Canada*, two volumes, 1832–34. He also kept journals during two explorations out west, which provide some biogeographical information.

There were two French explorers around the same time, whose life patterns were similar to these Britons: John James Audubon, who was a settler, and Charles-Lucien Bonaparte, who spent some years in America, studying birds. Audubon used Wilson's career as a guide for his own. Wilson was a capable illustrator of species he studied, but Audubon aspired to more artistic illustrations, though in a few cases he borrowed Wilson's illustrations (without acknowledgment). Audubon's 435 large illustrations were first published (4 vols., 1827–38) before his *Ornithological Biography* (5 vols., 1831–39). But in a later edition they appeared together (7 vols., 1840–44). He also collaborated with John Bachman on *Viviparous Quadrupeds of North America* (3 vols. plates, 1845–48 and 3 vols. text, 1846–54), both combined in later editions.

Bonaparte decided to study species which Wilson died before studying. He concentrated upon physical features and, unlike Wilson, did not study habits for his *American Ornithology* (4 vols., 1825–33). He published more on European birds than on American ones.

The "father of American entomology" was Philadelphian Thomas Say, author of both *American Entomology* (3 vols., 1824–28) and *American Conchology* (6 vols., 1830–34). He had attended the University of Pennsylvania Medical School. In 1825, six of his science-oriented friends founded the Academy of Natural Science of Philadelphia, and Say was for some years its most active member, especially as author and editor's assistant of the Academy's *Journal*, begun 1827.

Government expeditions continued. Major Stephen H. Long led two expeditions. First, 1819–20, up Platte River to Rocky Mountains, which he named, as he did Long's Peak. Naturalists Say, Titian Peale, and Edwin James were participants. Long called lands west of Missouri River "desert." His second expedition in 1823 was up the Minnesota and Red Rivers and included Say again.

B. 1840–1940s

Dutch ornithologist Coenraad Temmick established the Rijksmuseum in Leiden in 1823, and his own three ornithological works appeared in the 1820s–30s. Dutch naturalist Salmon Müller, who had lived in Dutch East Indies, brought his specimens and notes to Rijksmuseum to write his 82-page "Geographie der Thiere: Ueber die geographiche Verbreitung der Säuethiere im Indischen Archipelagus" (1842), which contributed to both terrestrial and aquatic biogeography. Müller's information was used by geographer Heinrich Berghaus to show what we call "Wallace's Line," which could be called "Müller-Wallace's Line." Temmick wrote the mammalian part of a series of monographs entitled *Fauna Japonica* (1833–50), based upon specimens collected by Philipp Franz von Siebold.

Swiss school teacher Jules Thurmann earned a B.A. degree at the University of Strasbourg and gained scientific recognition from his publications on botany and geology. In 184 he published *Système de Géographie botanique*, and in 1849, he distinguished between flora and vegetation.

Austrian zoologist Ludwig Karl Schmarda graduated from the University of Olomouc in 1841 and from medical school in Vienna in 1843. He took two trips to Adriatic Sea in 1844 and 1846 and published *Zur Natureschichte der Adria* (1852), followed by *Die geographische Verbreitung der Thiere* (1853), in which he defined 21 terrestrial and 10 marine regions of the world. He traveled around the world in 1853–57 and published *Reise um die Erde* (1861). In 1862, he became a professor at the University of Vienna. He published articles on animal geography in *Geographisches Jahrbuch* for 1866.

English lawyer Philip Sclater attended Oxford University, then practiced law in London. His main interests were ornithology and London zoological organizations. His most influential paper was "On the General Geographical Distribution of the Class Aves" (1858). "In this paper Sclater delineated and named the zoological regions that we know today." In 1875, he extended the scope of his survey to all vertebrates. In 1899, he collaborated with his son on *The Geography of Mammals.*

English botanist Joseph Hooker joined a naval expedition to Antarctica, 1839–43. He published extensively on his findings, 1844–60 (see later). One expedition did not cure his itch to travel, and his next expedition was to India, December 1847–February 1851, where he explored vegetation of the Himalayan Mountains. He went after discussions with Darwin on relationships between elevation and vegetation. He published *Rhododendrons of the Sikkim Himalaya* (1849–51) and *Himalayan Journals* (1854). He coauthored a book on India's flora (1855) and later published *Flora of British India* (7 vols., 1872–97). His contributions to phytogeography were so important that they were collected and republished (1953). Hooker's *Journals* is "one of the few and certainly the most accomplished examples for India of a scientific travel narrative comparable to those of Humboldt, Darwin, and Wallace."

Darwin and Hooker's publications after their travels inspired others. English school-teacher-turned-botanist Richard Spruce first traveled to the Pyrenees, 1845–46, and published on Musci and Hepaticae he collected (1850). William Hooker suggested, in 1849, he next travel to Amazon Valley, where he collected 7,000 plant species, many unknown. Britain sent him into the Andes to collect cinchona plants to grow in India. Afterwards, he collected plants along the coasts of Ecuador and Peru and returned to England in 1864. His *Notes of a Botanist in the Amazon* (2 vols., 1908) was posthumously edited by Alfred Wallace.

America's leading botanist during the 1800s was New York State native Asa Gray, who did not develop interest in botany until a medical school teacher introduced him to it around 1830. In 1832, he abandoned medicine and taught school while learning botany under John Torrey in New York City. In 1842, he became a professor of natural history and director of the botanic garden at Harvard University. In 1850, he began taking frequent trips to England and, after completion of transcontinental railroad, frequent trips to California. He emphasized taxonomy because he collaborated with Torrey on *Flora of North America* (2 vols., 1836–43), though later abandoned as too ambitious. In 1840, he reviewed the first part of P.F. von Siebold's *Flora Japanica*, in which he saw 14 species described which resembled American species. In 1848, he published *A Manual of the Botany of the Northern United States.* In the 1850s, he met Hooker and Darwin at the Royal Botanic Gardens, Kew, and became friends with both. In 1855, he wrote to Darwin that he wanted to compare America's flora with those of northern Europe and northern Asia. After Japanese plants arrived, he explained their similarities to American species using Darwin's idea that they had descended from common ancestors but had diverged over time.

Alfred Russel Wallace had little formal education. He was apprenticed to one brother, then to another. He read books by Humboldt, Darwin, and William Edwards, *A Voyage Up the River Amazon* (1847). He and fellow dreamer Henry Walter Bates sailed to Pará, Brazil, to collect animals to sell to English naturalists. They collected many insects and also other animals. Wallace stayed four years and Bates 11 years. Both wrote travel books, Wallace's much briefer than Bates'.

Like Hooker, one expedition was not enough for Wallace. In 1854, he traveled on a British Navy ship to the Malay Archipelago, where he stayed until 1862. He traveled there about 14,000 miles, mostly by ship. He later considered his Malay experience "the central and controlling incident of my life." An important aspect of his experience was studying and collecting orangutans. On 15 June 1856, he sailed from Bali to Lombok, 15–20 miles away. Yet they had strikingly different flora and fauna. On a map in his *Malay Archipelago* (1869), he showed Bali as in an Indo-Malayan Region and Lombok in an Austro-Oriental Region. One Wallace correspondent was Charles Darwin, beginning in the 1850s.

Around 1870, Darwin and Alfred Newman suggested to Wallace that his years spent in the tropics on different sides of the Earth was an important experience for his writing *The Geographical Distribution of Animals* (2 vols., 1876), which was mainly on land vertebrates. Included was a folding map showing zoogeographical regions of the world. In 1880, he remedied a neglect in his two volumes by publishing *Island Life.* By then, Wallace became the most important zoogeographer during the 1800s.

In America, four ornithological societies, with journals, were founded within two decades, and three continue: Nuttall Ornithological Club (1873), AOU (1883), Wilson Ornithological Club (1888), and Cooper Ornithological Club (1893). The first three were organized in the East, with Cooper in the West. AOU replaced Nuttall, and in 2016, Cooper merged with AOU.

Augustin-Pyramus de Candolle's son, Alphonse, followed in his father's footsteps into phytogeography, though he also studied law. His greatest work was his *Géographie botanique Raisonnée* (2 vols., 1855, about 600 pages per volume). His legal training perhaps contributed to his clarity in arguing about causes of distributions. Volume 1 discussed general topics, including temperature, light, humidity, economic botany. Volume 2 discussed the botanical geography of different countries.

The most important phytogeographer during the 1800s was German August Grisebach, who was taught botany as a teen by an uncle and then studied at the University of Göttingen. In 1839–40, he traveled in Balkan Peninsula and northwest Asia Minor and published two books on his findings. During the 1840s, he traveled in western Europe and became a professor of botany at Göttingen in 1847. He studied island phytogeography in *Flora of the British West Indian Islands* (1859–64), which he considered one of his most important works. His most comprehensive work was *Die Vegetation der Erde nach klimatischen Anordnung* (2 vols., 1872, French 1877–78).

Works of Alphonse de Candolle and Grisebach provided a strong foundation for five turn-of-century botanists who advanced phytogeography. Perhaps most influential was

Danish botanist Eugenius Warming, who spent three years in Brazil before earning undergraduate and doctoral degrees at the University of Copenhagen. His *Pflantesamfund: Grundträk af den ekologiske Pflantesgeografi* (1895, German 1896, English 1909) advanced both ecology and phytogeography. French botanist Charles-Henri-Marie Flahault earned a doctorate at Paris in 1878, then traveled with botanist Gaston Bonnier to Sweden to compare alpine and polar climates and plants. Flahault joined the University of Montpellier in 1881, and in 1891 started a school of Mediterranean phytogeography. In 1905, Flahault and Swiss botanist Carl Schröter were asked to establish terminology "that integrated the floristic and physiognomic components of botanical geography." Schröter's great work was *Pflanzenleben der Alpen* (1904–08, 1288 pages).

German botanist C. G. Oscar Drude earned his doctorate under Grisebach at the University of Göttingen (1873) and followed in Grisebach's footsteps with his *Atlas der Pflanzenverbreitung* (1887), *Handbuch der Pflanzengeogrfaphie* (1890), and *Deutschlands Pflanzengeolgraphie* (1896). Silesian H.G. Adolf Engler earned his doctorate at the University of Breslau and became a leading plant taxonomist. However, he also founded (1880) and edited 62 volumes of *Botanische Jahrbücher für Systematik, Pflanzengeschichte und Pflanzengeographie*, which became the leading periodical for phytogeography. In 1884, he returned to Breslau as a professor of botany. He became coeditor with Oscar Drude of *Die Vegetation der Erde* (15 volumes, 1896–1923). Engler wrote the volume on Africa (1910–15).

Andreas Schimper was the son of botanist Wilhelm Schimper at Strasbourg (Strasburgh), where Andreas earned his doctorate. He traveled to Brazil and studied the symbiotic relationship between ants and *Cecropia* trees. His *Pflanzengeographie auf physiologischer Grundlage* (1898; English 1903, 870 pages + 4 folded color maps) was one of the largest ecological treatises written by one author. He thought his approach rendered Grisebach's approach obsolete! Schimper borrowed heavily from Warming with scant acknowledgment.

Plant ecology arose in Britain under the influences of Warming, Flahault, and Schimper, and led by Arthur Tansley, educated at Cambridge University, and then taught at the University of London, 1893–1906, Cambridge, 1906–23, and Oxford, 1927–37. He was founder and editor of *New Phytologist*, 1902–31; chief founder of the British Ecological Society in 1913; editor of its *Journal of Ecology*, 1917–37, *Types of British Vegetation* (1911), and *British Isles and Their Vegetation* (1939). He also initiated and led an International Phytogeographical Excursion (1911) through the British Isles. He was Great Britain's leading phytogeographer of the 1900s.

German zoologist Hans Gadow earned a Ph.D. (1878) under Ernst Haeckel at the University of Jena and went to London in 1880 to catalog British Museum's bird collection. In 1882, he moved to Cambridge University, where he wrote *The Wanderings of Animals* (1913, 150 pages + 16 maps). In chapter 5, "Former Configurations of Land and Water," he commented: "In the majority of cases the present distribution of animals cannot be accounted for without

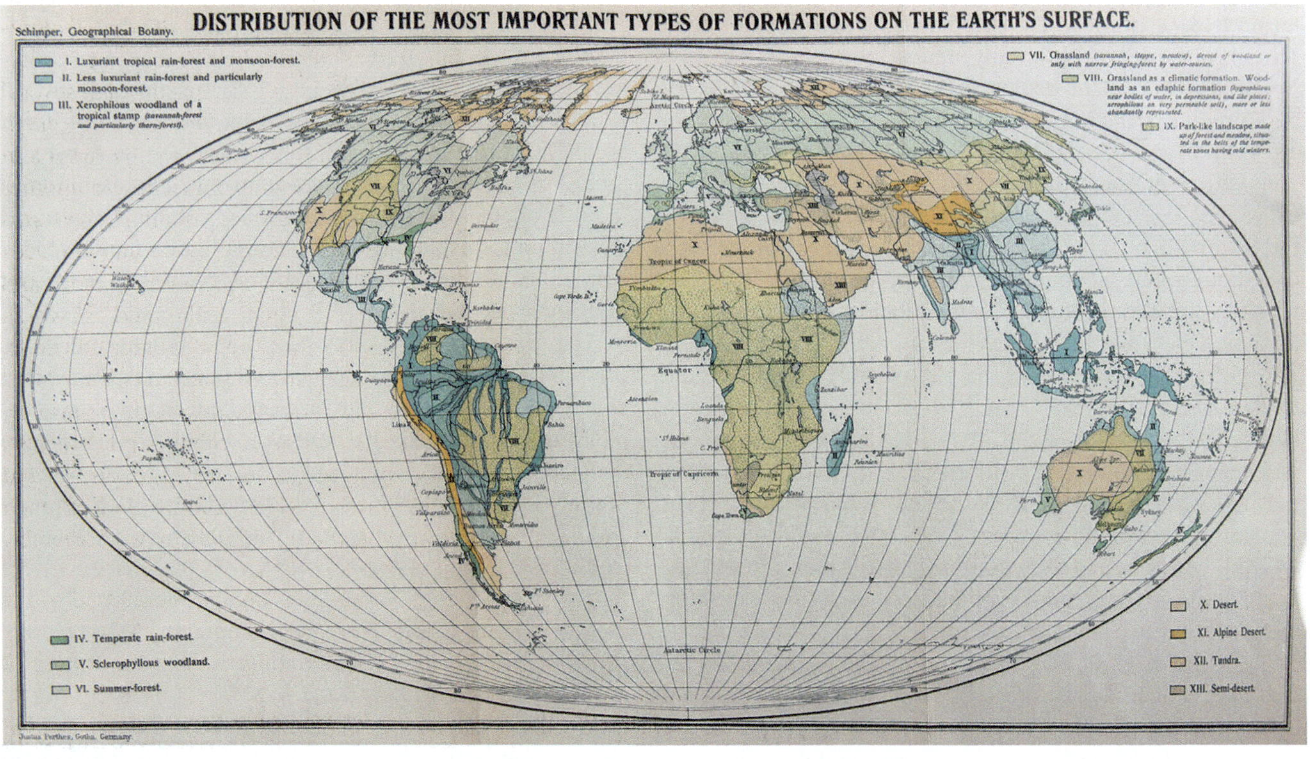

FIGURE 1.1 Distribution of the most important types of vegetation on Earth. Schimper 1903: map 3.

the assumption of enormous changes in the configurations of land and waters." He did not advocate land bridges as much as he advocated the movement of continents.

Berliner Alfred Wegener earned a B.S. degree in astronomy, but then his interests shifted to meteorology and geology. He joined a Danish expedition to Greenland, 1906–08, and later went on three more Greenland expeditions. In 1910, he became impressed by the corresponding shapes of the east coast of South America and the west coast of Africa. In 1911, he learned of similarities between fossils on those coasts, and in 1912, he announced his theory of continental drift. He was wounded in World War I, and during sick leave, he wrote *Die Entstehung der Kontinengte und Ozeane* (1915, new eds., 1920, 1922, 1929). Edition 3 was translated into five languages.

Swedish botanist O. Gunnar Erdtman was an early palynologist who published a text on it in German (1921). His "Pollen Statistics: a New Research Method in Paleoecology" in *Science* (1931) argued that paleoecology would enhance plant ecology. He provided instructions and citations to literature in a Swedish geological journal. Americans accepted his suggestion.

Russian botanist Evgenii Wulff reviewed different suggestions to account for plant distributions and found Wegener's most convincing. He published *An Introduction to Historical Plant Geography* in Russian in 1932 and in English in 1943. He wrote *The Geography of Flowering Plants* before World War II, but publication was delayed by war until 1947.

Indiana plant ecologist Stanley Cain earned a B.S. degree (1924) at Butler University and his M.S. and Ph.D. (1927, 1930) at the University of Chicago. He began his career as a botanist, then served as an administrator in several situations. His first book was *Foundations of Plant Geography* (1944). He published over 100 articles.

C. Marine Biogeography, 1690s–1940s

Marine environments are less accessible than land environments, and studies of marine life are more complex than comparable studies on land. Voyages and research stations are basic assets for such studies.

Best-known early expeditions were three voyages led by British Captain James Cook. His first (1768–71) voyage included wealthy English Joseph Banks, who acquired a large collection of plants and some animals. Instead of publishing lavish volumes describing his collections, he chose to become president of the Royal Society of London and became too busy for scholarly endeavors. On his second voyage (1772–75), Cook took German botanist Johann Forester and, as an assistant, his son Georg. Both he and Georg published accounts of their voyage. On his third voyage (1776–80), Cook did not take a botanist. Instead, he depended upon surgeon William Ambrose, who had been on his second voyage, to collect observations and natural history specimens.

Cook's voyages inspired French, Spanish, Russian, and American governments to send out voyages of exploration during the late 1700s and the early 1800s. British exploration continued. France decided to fill gaps in the geography of the Pacific Basin Cook had not seen. War veteran Captain Françoise de la Pérouse led an expedition of two ships, with naturalists, in 1785. They spent six weeks in Botany Bay, Australia, and left on 10 March 1788. He sent back to France his journal covering observations until then. Both ships struck a coral reef beside an island. No one from the expedition was ever seen again. In 1791, a search expedition of two ships was sent from France, commanded by Captain Joseph-Antoine d'Entrecasteaux, carrying three physician-naturalists and a gardener-botanist. Captains of both ships died abroad, and no evidence of the lost ships was found. When returning ships reached Java in late 1793, they learned of the French Revolution and that France and the Netherlands were at war. Royalist crew identified revolutionary crew, whom Dutch arrested. Both ships were sold to pay for French expenses at Java. The prisoners were later exchanged in a prisoner swap. French collections and recorded observations were retained by the French. Botanist Jacques-Julien de La Billardière published his account of expedition in 1799. His *Hollandiae plantarum specimen* appeared in parts, 1804–07. His "pioneering" *Sertum Astro-Caledonicum* appeared in 1824–25.

In 1800, France and England sponsored expeditions to Australia in hopes of gaining some strategic advantage. Both expeditions encountered difficulties, but also returned with important information on geography and plants, with many specimens. Nicolas Baudin commanded France's expedition; he died on the return voyage. Matthew Flinders commanded Britain's expedition; he was arrested on the return voyage at Mauritius, December 1803, as a possible spy and only released on 14 June 1810. He wrote his memoirs while imprisoned, published on 18 July 1814—a day before he died.

Much more successful was the voyage of British naval vessel *Beagle*, 1831–36, which had one inexperienced naturalist, Charles Darwin. He had just completed his undergraduate education at Cambridge University and was recommended by its botany professor. Darwin rose to the challenge and studied geology and zoology of places visited, both at sea and on land. He collected rock, plant, and animal specimens and wrote very adequate notes on his observations. He wrote afterwards books on geology, zoology, and travel, and he collected plant specimens on Galapagos Islands, which he turned over to his friend Joseph Hooker to describe in print (1851).

Hooker by then had gone on a naval voyage of his own, as botanist, to the Antarctic and other southern lands, 1839–43. Afterwards, he published three botanical works, two volumes apiece: *Flora Antarctica* (1844–47), *Flora Novae-Zelandiae* (1853–55), and *Flora Tasmaniae* (1855–60). These volumes were all on land plants, collected on a naval voyage.

In 1838–42, Lieutenant Charles Wilkes, U.S. Navy, commanded an oceanic expedition around the world. He allowed aboard seven scientists and two artists. It was successful in collecting numerous specimens and observations. Wilkes was a capable naval officer, but he had conflicts with some officers and scientists and was court marshaled back in Washington. Nevertheless, he supervised the publication of its numerous volumes. Expedition specimens were deposited at the U.S. Patent Office—until transferred to Smithsonian Institution, which was established in 1846.

The most important research voyage during the 1800s was that of *Challenger*, a British naval vessel that focused particularly on deep-sea animals and geology, during a 3.5-year voyage around Earth, 1872–76, covering 68,890 miles. Aboard were five naturalists, with different specialties, and a Swiss artist. Numerous notes and specimens were collected, and afterwards the reports ran to 50 large volumes, 1885–95.

German Victor Hensen studied medicine and physiology at Berlin and Kiel, and in 1863, he became interested in marine life. In 1868, he had the Prussian government establish Kommission zur wissenschlaften untersuchung der deutsch Meer. Its goal was to learn why commercial fisheries were declining in the Baltic Sea. Coastal stations were built to monitor fishes and environments between Denmark and Russia. Norway's Voringian Expedition, 1876–78, studied migrations of fishes, especially cod. Hensen studied the metabolic cycle of the sea to determine productivity. He designed a net to collect vertical samples for quantitative data, used in "Über die Bestimmung des Planktons oder des im Meere treibenden Materials an Pflanzen und Thieren" (1887), a quantitative study in which he introduced a term, *plankton*, coined by a colleague to indicate floating organisms. In 1889, he compared the North Sea and Atlantic plankton during an expedition to Greenland and Bermuda, and he edited and contributed to a five-volume report (1892–1912) on findings. Unexpectedly, he discovered plankton are more abundant in northern cold waters than in tropical warm waters. His findings attracted interests of other marine scientists.

By the 1890s, marine biologists realized that European countries doing separate research without considering other countries' research were inefficient. Oceanography Professor Otto Pettersson, University of Stockholm, organized an international survey of the North Sea, 1893–94, and later encouraged multinational surveys to support commercial fishing. Further discussion led to the formation in 1902 of the International Council for the Exploration of the Sea, which soon began publishing its journal.

English Maurice Yonge joined the British Army in 1917–18 and in 1919 enrolled at the University of Edinburgh and earned B.Sc. and D.Sc. (1922, 1927). He helped organize a Great Barrier Reef Expedition with 18 scientists, which he led, in 1928–29, and then, before four volumes of sci. entific reports appeared, he published *A Year on the Great Barrier Reef: the Story of Corals and of the Greatest of Their Creations* (1930), which helped him become a professor of zoology at the Universities of Bristol and, later, Edinburgh.

America's best-known naturalist during the first half of the 1900s was New Yorker William Beebe, who after three years at Columbia University, in 1899, joined the New York Zoological Society (Bronx Zoo) and published *A Monograph of the Pheasants* (4 vols., 1918–22; ed. 2 (condensed), 2 vols., 1926), which indicated ranges of all species. However, in 1923, he led an expedition to Galapagos Islands, which began his shift to marine biology. Although technical reports appeared in *Zoologica*, his popular reputation began with the publication of *Galapagos: World's End* (1924), which was followed by other expeditions and books on his observations. He ventured underwater, with a diving helmet and hose for oxygen, which led engineer Otis Barton Jr. to suggest to him a partnership to build a *Bathysphere* that could be lowered into deep water to study animals living there. They had it built and began using it in 1930 to discover new species, with much publicity.

Prosperity during decades before and after World War I enabled Long Islander Robert Cushman Murphy to begin a career studying marine birds. He graduated from Brown University in 1911 and received there an honorary Ph.D. in 1941. He began his career in 1912 sailing on a whaling vessel to the far south for a year, when he followed Darwin's example of recording observations for a later book, *Logbook for Grace* (1947), but published 67 articles from the voyage before then. His culminating achievement was *Oceanic Birds of South America* (2 vols., 1936).

Norwegian physical oceanographer Harald Sverdrup obtained the collaboration of two marine biologists—American Martin W. Johnson and Canadian-American Richard H. Fleming—to write an encyclopedia, *The Oceans: Their Physics, Chemistry, and General Biology* (1942, xi + 1087 pages). It has a general discussion of zoogeography, with a world map of faunal areas of littoral animals, based on one in Sven Ekman's *Tiergeographie des Meeres* (1935). *The Oceans* was a valuable reference for marine biogeographers, for it surveyed factors influencing distributions.

After World War II, Jacques Cousteau developed his aqualung technology, which led to his extensive exploration of marine life around the world, with accounts in popular books and films.

Biosphere ecology is a geographical, but not a historical, culmination of ecology. Although it would have been difficult to get far into this topic before the data down to 1950 was available, there had been no general aim among ecologists to develop biosphere ecology. However, there were essential roots in other ecological topics.

Experiments to explain plant growth began in 400 AD by pseudo-Clements (only water needed), continued by Nikolaus von Cusa in 1450 (sunlight also needed), doubted by Robert Boyle in 1661 (trace elements in water needed?), and stomates were discovered in vascular plant leaves by Marcello Malpighi (1675) and Nehemiah Grew (1676). Photosynthesis studies began in 1727 with Hales' discovery of transpiration. In 1774, Joseph Priestley heated mercuric oxide, then discovered a particular gas (oxygen), which

FIGURE 1.2 (a) Robert Cushman Murphy. (b) Rachel Carson. Web.

Antoine Lavoisier named *oxygen*. Other chemistry investigators discovered the identity of other gases in the late 1700s and the early 1800s.

A global uniformitarian perspective in geology was defended by James Hutton in 1785, and in more detail in 1795. He had influence in Britain but not in France. Jean Baptiste Lamarck also developed a global uniformitarian geology (1802), which included this comment: "The organic action of living organisms continuously creates combinations of substances which would never have existed otherwise."

English naturalist Richard Bradley published in 1718 a claim that insect predators have parasites, which in turn have their own parasites, "and so to infinity." Carl Linnaeus' *de Oeconomia Naturae* (1749) expanded the ancient concept of balance of nature to include plants: "tree louse lives upon plants. The fly called *Musca aphidivora* lives upon the tree louse." And so on.

Darwin, in his *Journal of Researches* (1839), may have first published a food web, from barren St. Paul's Rocks, where he described multiple species interacting, and not a food chain. German zoologist Karl Semper delivered in 1877 12 lectures in Boston, later published as *Animal Life as Affected by the Conditions of Existence* (1881, German 1880). He explained that when herbivores eat vegetation, there is a loss of organic material, due to oxidation; he estimated that 1,000 units of vegetation could produce 100 units of herbivore, and when predators eat herbivores, it produces 10 units of predators. An Italian zoologist, Lorenzo Camerano, published the earliest two food web diagrams (1880), but no one followed his example.

Illinois state entomologist Stephen Forbes investigated foods of birds and fishes. His "On Some Interactions of Organisms" (1880) explained how changes in species have impacts on other species. His "The Lake as a Microcosm" (1887) explained an ecosystem before that concept was defined.

London-born Austrian geologist Eduard Suess joined the University of Vienna faculty in 1856, where he had an active career of research and publishing. He coined the term *biosphere* in 1875 in a study of the genesis and structure of the Alps. The culmination of his work was *Das Antlitz der Erde* (3 vols., 1883–1909).

America's first food web diagram was by three economic entomologists: W.D. Pierce, R.A. Cushman, and C.E. Hood, "The Boll Weevil Complex" (1912). They traced predators and parasites of an agricultural pest. American animal ecologist Shelford published both land and aquatic food webs in *Animal Ecology in Temperate America, as Illustrated in the Chicago Region* (1913). The earliest food web for a marine community was by Danish fishery biologist Johannes Petersen (1915), co-founder of Dansk Biologisk Station.

Perhaps Berliner Alfred Wegener can be viewed as resuming tradition from Hutton and Lamarck, though he would not have known either's work. In 1911, he originated his controversial theory of continental drift, published in 1915: *Die Entstehung der Kontinrnte und Ozean.*

It seems likely that Russian geochemist Vladímir Vernadsky read Wegener's treatise and used Suess' encyclopedia during his stay in Paris, before publishing his *Biosfera* (1926). This work was theoretical, explaining the roles of atmosphere, ocean, photosynthesis, and living organisms.

In 1841, France's leading chemist Jean-Baptiste Dumas was promoted to professor at the Sorbonne and published "On the Chemical Statics of Organized Beings." Ecologist-historian Eville Gorham considered this paper as beginning biogeochemistry, by discussing chemical interaction between plants, animals, and atmosphere.

German physician Robert Mayer earned his M.D. degree at the University of Tübingen (1838). In 1840–41, he was a physician on a Dutch ship to East Indies. Physiological observations convinced him that motion and heat are manifestations of a force of nature, which is conserved in conversion from one form to the other. In 1845, he published his idea, which received little scientific notice. He published supporting accounts, stating "Plants convert light energy into chemical energy" in photosynthesis. English physicist James Joule attacked his number for the mechanical equivalent of heat, so another English physicist, John Tyndall, examined Mayer's work, found merit in it, and brought it scientific recognition.

Environmental sociologist Eileen Crist saw Darwin's last book, *The Formation of Vegetable Mould, through the Action of Worms* (1881), as a "scientific study about the reciprocal influence of life and environment" (2004). Darwin's uncle, Josiah Wedgewood, had introduced him to actions of earthworms in 1837, and he published three papers on them before this book. Crist thought Darwin's book opened a new field of research, but only later did others begin such investigations.

Ecologist Evelyn Hutchinson introduced his students to Vernadsky's biosphere concept and to biogeochemistry, and he persuaded graduate student Howard T. Odum to write a doctoral dissertation on "The Biogeochemistry of Strontium" (1950). That was before atomic scientists learned radioactive strontium is a waste product of nuclear fission. Howard took his Yale "gospel" to his brother Eugene, an ecologist at the University of Georgia, who was writing a textbook.

Hasler's star limnology student, Gene Likens, joined Dartmouth College's faculty in 1963 and collaborated with forest ecologist Herman Bormann in establishing Hubbard Brook Experimental Forest, New Hampshire. One multidisciplinary treatise they wrote with three other specialists was *Biogeochemistry of a Forested Ecosystem* (1977), followed two years later by Bormann and Likens, *Pattern and Process in a Forested Ecosystem* (1979).

In 1958, the International Society of Soil Science appointed a Soil Zoology Committee to convene a Colloquium on Research Methods, which assembled 47 British and European authors of papers afterwards compiled into a valuable reference work. German zoologist Friedrich Schaller wrote a brief textbook, *Soil Animals* (German 1962, English 1968), with numerous illustrations, but without citations (in English edition). More recently, soil biology became part of biosphere biology. Soil ecologists Daniel Richter Jr. and Daniel Markewitz published *Understanding Soil Change: Soil Sustainability over Millennia, Centuries, and Decades* (2001). Richter's research team later focused upon biogeochemistry (2018).

Independent scientist and thinker James Lovelock had not been a good student since he studied whatever interested him, not assignments. He became a laboratory assistant but later graduated from Manchester University (1941). In the 1950s, he became concerned over pollution from pesticides. In 1965, he learned that Mars and Venus have atmospheres dominated by CO_2. He wondered what kept Earth's complex atmosphere constant. He concluded that respiration of organisms regulated the atmosphere. He published his Gaia hypothesis in 1.5 pages in a scientific journal (1972). His eight references supported his reasoning but were not sources of his ideas. To investigate the role of organisms, he needed a biologist collaborator, who appeared in American cell biologist Lynn Margulis, who in 1970 wanted to consult an expert on atmospheric gases. Her husband, Carl Sagan, suggested Lovelock. In 1971, they met and decided they could work together, and they published together two papers in 1974.

Agreement at the papers level did not last into the book level. Philadelphian physiologist Alfred C. Redfield, with Harvard Ph.D., at Woods Hole, had written *The Biological Control of Chemical Factors in the Environment* (1958), which was similar to what Lovelock explained in his *Gaia: a New Look at Life on Earth* (1979). Margulis disliked the notion of Earth being alive and offered an alternative in *The Symbiotic Planet* (1998): "Gaia is the series of interacting geosystems that compose a single huge ecosystem at the Earth's surface."

Lovelock responded by coining a new term, *geophysiology*. He explained: "biogeochemistry differs from geophysiology as platonic friendship differs from marriage . . . . In geophysiology, the organisms and their environment are so tightly coupled that they constitute a single system, Gaia" (2000). By then, there had been a national conference on Gaia in 1988, with two international ones to come, all with published papers. No consensus emerged.

The dominant aspect of biosphere ecology is global warming, which will retain its dominance because it is a global crisis. Before meteorology became a science, there were several "calendars of flora" published. They started phenology: correlating change in plant growth with the progression of seasons (mostly in spring). An English family, Marsham, kept phenology records (without that name) for over two centuries, 1736–1947, at their estate in the Norfolk region. Naturalist Benjamin Stillingfleet compiled his own "flora" while staying at that estate (1755). He translated one by Linnaeus, *Calendarum florae*, and published his own following that. Naturalist Robert Marsham published his 27 "Indications of Spring" (1789) in Royal Society's *Philosophical Transactions*.

Thomas Jefferson owned a Virginia plantation, Monticello, and sought local climate data for when to plant crops. He recorded weather in his *Garden Book*, 30 March 1766–1816. His *Notes on the State of Virginia* (1787) included state weather.

Harvard medical professor Jacob Bigelow corresponded in 1817 with observers east of the Mississippi and published differences of blooming times in spring between Gulf States and Maine, and Benjamin Silliman summarized those results in the first issue of his *American Journal of Science*. Silliman added data from London, Valencia, and Geneva for comparison to American data.

Besides such data's use in agriculture, disease was then believed to be caused by unfavorable weather/climate and that correlation could be studied. Societas Meteorologica Palatina existed in Germany from 1781 to 1795. In 1823, the Meteorological Society of London arose and began publishing *Transactions* in 1839. In 1840, it published a rainfall map of England, with data from 52 stations. In 1850, the British Meteorological Society arose and established a uniform national system for regular reports. Prussia's Meteorological Institute arose in 1847, and after Prussia unified Germany in 1871, it expanded to all of Germany.

English William Howitt, age 13, wrote "An Address to Spring," which appeared in *Monthly Magazine*. He became a professional author, whose books included *The Book of the Seasons: or, the Calendar of Nature* (1831). In 1836, American Henry David Thoreau read Howitt's *Seasons* and wrote an essay quoting Howitt's phenological goals before giving his own such comments for every month of the year. He began keeping a journal until 1858, with seasonal comments on plants and birds. He published from it in *Walden, or Life in the Woods* (1854). All of his journal entries appeared posthumously, rearranged into four volumes: *Early Spring* (1881), Summer (1884), *Winter* (1888), and *Autumn* (1892). He hoped his data might be of use to others, which occurred in 1947, when Aldo Leopold used it. Franklin Hough, later head of the U.S. Forest Service, also collected phenological data, 1851–59, also used by Leopold.

Iowan Aldo Leopold studied forestry at Yale since wildlife management was not yet an academic discipline. In 1928, he left the U.S. Forest Service to study game conditions for Sporting Arms and Ammunition Manufactures' Institute. His *Report on a Game Survey of the North Central States* (1931) was a basis for his textbook, *Game Management* (1933). That book helped him become the world's first professor of wildlife management at the University of Wisconsin. His mother had given him a copy of Thoreau's *Journal* (14 volumes, 1906), which influenced his interest in phenology. In 1945, he and student Elizabeth Jones organized his phenological data into an article, using data from Thoreau and Hough. Its publication in *Ecological Monographs* in 1947 helped his election as ESA's vice-president (1947) and president (1948).

Global warming is now the most important aspect of biosphere ecology, practically, and will remain so indefinitely. French mathematician Jean-Baptiste Fourier first studied this and concluded there are three sources of Earth heat: sun, stars, and from Earth's interior (1822). He raised the question of global warming by observing that sunlight penetrates the atmosphere better than does heat from Earth reflecting into space. British physicist John Tyndall studied in the 1860s effects of solar and heat radiation on atmospheric gases and found in 1861 it did not block heat escaping from other gases but did inhibit heat escaping from CO_2.

Connecticut native Elias Loomis graduated from Yale (1830) and in 1846 published an original weather map, which was generally adopted for weather prediction. He also became that year professor of mathematics and physics at Western Reserve College, Ohio. He returned to Yale in 1860 and compiled data for New Haven. His *A Treatise on Meteorology* (1868) was the first such textbook written in the United States.

Concern over global warming came first from Swedish physicist-chemist Svante Arrhenius, who studied at Uppsala and Stockholm Universities. He began publishing on greenhouse effect in "On the Influence of Carbonic Acid in the Air upon the Temperature on the Ground" (English version, 1896). He cited the amount of coal mined annually, 500 million metric tons, and estimated how much CO_2 that amount of coal might expel in combustion. He won a Noble Prize in 1903.

Norwegian Vilhelm Bjerknes, as a boy, helped his father Carl Anton Bjerknes with hydrodynamics research. He studied mathematics and physics at the University of Kristiana and earned an M.S. degree in 1888. He went to Paris and attended lectures by Henri Poincaré. Afterwards, he became an assistant to Professor Heinrich Hertz for almost two years, at the University of Bonn, Germany. In 1892, he earned a Ph.D. in Norway with a dissertation on "Movement of Electricity in Hertz's Primary Conductor" (1892). In 1895, he became a professor of applied mechanics and mathematical physics at the University of Stockholm. He and three colleagues published *Dynamic Meteorology and Hydrography* (2 vols., 1910–11).

English engineer Guy Callendar worked for British Electrical and Allied Industries Research Association, collecting temperatures. He concluded world temperature was increasing. CO_2 levels seemed to increase by 10% in 100 years. He concluded the latter caused the former (1938). The atmosphere contained more CO_2 than when Arrhenius published; most scientists saw oceans as safety nets.

Canadian Gilbert Plass earned a B.S. degree at Harvard (1941) and Ph.D. at Princeton (1947). By 1956, few scientists thought CO_2 caused global warming. Gilbert Plass gave a concise history of past studies and called for reexamination of CO_2 theory.

Seattle native Roger Revelle, at Pomona College, California, earned a B.A. degree (1929) and Ph.D. (1936) at the University of California, Berkeley. In 1957, he was at Scripps Institute of Oceanography and collaborated with chemist Hana Suess to study "Carbon Dioxide Exchange between Atmosphere and Ocean and the Question of an Increase of Atmospheric CO_2 during the Past Decades." They found reasons to fear modern fuel consumption might increase global temperatures since ocean capacity to absorb CO_2 seemed limited. They urged: "It therefore becomes of prime importance to determine the way in which carbon dioxide is partitioned between the atmosphere, the oceans, the biosphere and the lithosphere" (1957).

Revelle initiated plans for an International Geophysical Year (IGY) for 1958. He had Pennsylvania chemist Charles David Keeling, B.S. (1948), University of Illinois, Ph.D. (1953), Northwestern University, on Scripps Institute faculty, apply for IGY funds to build a research station on Mauna Loa Volcano, Hawaii, two miles above sea level. In 1958, Keeling invented an instrument to measure CO_2 levels in the atmosphere and received funds for his station. He collected data annually for over five decades, putting CO_2 levels on a diagram, the *Keeling Curve*. "Keeling's data put the capstone on the structure built by Tyndall, Arrhenius, Callendar, Plass, Revelle, and Suess."

Detroit native Reid Bryson, Ph.D. (1948) from the University of Chicago, went to the University of Wisconsin in 1946 and founded its Department of Meteorology and Center for Climatic Research. He collaborated with Thomas Murray on *Climates of Hunger: Mankind and the World's Changing Weather* (1977). They commented that greenhouse effect was a well-known danger, caused by increased CO_2 release from industry and cars. Bryson did not think data were sufficient to declare an immediate danger, since volcanic eruptions and particulate pollution from human activity countered CO_2 effects. English climatologist H(ubert) H. Lamb's *Climate, History, and the Modern World* (1982) did not offer more decisive data.

Iowan James Hansen earned three degrees (B.S. 1963, M.S. 1965, Ph.D. 1967) at the University of Iowa in physics and astronomy. At NASA's Goddard Institute for Space Studies, he studied the atmospheres of Venus and Earth. He headed Goddard Institute in 1981–2013. He and six colleagues published in *Science* (1981) alarming discoveries on global warming. In 1988, he and seven collaborators published global warming could cause extreme events. Congressmen, who never questioned the National Institutes of Health reports, were willing to repeat energy companies' skepticism of conclusions unfavorable to their interests. Most climatologists supported Hansen's alarm.

Also in 1988, the United Nations founded its Intergovernmental Panel on Climate Change (IPCC) for impartial scientific discussions on relevant issues. In 1995, IPCC published "Summary for Policymakers: the Science of Climate Change," which supported Hansen's alarm.

Albert Gore Jr. grew up in Washington, DC, and graduated from Harvard in 1969. He took a course under Revelle which awakened his environmental concerns. In 1976, he won election to Congress, and after three terms he won a Senate seat. In 1992, he published *Earth in the Balance: Ecology and the Human Spirit* and ran successfully as vice-president for Bill Clinton. In 1997 he signed a Kyoto Protocol, which Congress never ratified. After losing a presidential election in 2000, he began speaking on global warming around the world. In 2006, Gore published another book on it, which led to a film that won an Academy Award (2007). He also won a Noble Prize for environmental contributions.

Hansen gave the Keeling Memorial Lecture at the annual meeting of the American Geophysical Union in 2005. His data were more extensive than 17 years earlier when he spoke to Congress. He stated an increase of Earth's temperature by 10C would make Earth hotter than it had been in 1 million years, and if by 20 or 30C, Earth would become a different planet.

One response to people using fuels that cause global warming was environmental education. A leader of this effort was Bill McKibben, who graduated from Harvard in 1982 and was a staff writer at *New Yorker*, 1982–87. In 1988, he published a review of four books on global warming, and in 1989 he published *The End of Nature*, translated into 20 languages.

ESA sponsored almost two decades of planning and constructing 44 terrestrial and 14 aquatic field stations for the National Ecological Observatory Network (NEON). NEON's web site provides a map of locations, names, and other information on stations. Stations monitor global warming and 180 other environmental variables. An inspired National Science Foundation sponsored a Global Lake Ecological Observatory Network.

During the 1900s, scientists and naturalists increased knowledge and understanding of biosphere ecology, while population increased and degraded the environment. Remedial planning is enhanced by biosphere ecological understanding.

2 Formalizing Ecological Sciences, 1870s–1920s

A. PLANT ECOLOGY

European and American plant ecologists were interested in their foreign colleagues' work; discussed here first is Europe and then America.

EUROPE

Plant physiology continued its momentum during the 1800s, but more directly leading to ecology was progress in phytogeography. That subject received an enormous boost from publications by Alexander von Humboldt (1769–1859), based upon his and Bonpland's Latin American explorations, 1799–1804.[1] Those studies were broad in scope, including environmental records, such as the distribution of heat over the globe.[2]

By the end of the 1800s, phytogeographical research in Europe supported a new science, plant ecology, that is seen in a comprehensive treatise by Danish botanist Eugenius Warming (1841–1924).[3] Its first edition was entitled *Pflantesamfund Grunträk of den okologiske Pflantegeograft* (1895), emphasizing plant geography. However, he used ecological literature when revising it for the English edition, which did not mention plant geography in the title: *Oecology of plants: an introduction to the study of plant communities* (1909).

French phytogeographer Charles-Henri-Marie Flahault (1852–1935) earned his doctorate in Paris (1878), then traveled with botanist Gaston Bonnier (1852–1935) to Sweden to study the relationship between plants and climate,[4] published as "Observations sur les modifications des vegetation suivant les conditions physiques du milieu."[5] It was comparable to Kerner's comparison between treeless alpine and prairie regions and plants.[6] Although active when some botanists were making a transition from phytogeography to ecology, Flahault stuck with phytogeography at Montpellier, where he taught for 47 years, leading a school focused upon the Mediterranean coast. He considered Kerner's Germanlanguage phytogeographical terms inadequate, and he published his alternatives in "Projet de nomenclature phytogeographique.[7] However, other botanists did not consider Flahault's terminology definitive either, and the Second International Botany Congress (1905) charged him and German-born Swiss botanist Carl Schröter (1855–1939) to lead a committee to establish terminology "that integrated the floristic and physiognomic components of botanical geography."[8]

The Third International Botany Congress (1910) did not fully accept their terminology, yet their proposal provided in 1913 the basis for the Zurich-Montpellier School of Phytosociology.[9] Schröter had studied at the University of Zurich, and later taught there, 1879–1925. Schroëter, who coined the terms "autecologie" (1896) and "synecologie" (1902), wrote *Planzenleben der Alpen* (1904–08, 1288 pp.), which "contains most of the knowledge of the time on the influence of elevation, soil, climate, and other environmental factors on alpine plants and plant communities."[10] His book was a worthy companion to Kerner's *Pflanzenleben der Danaulaender* (1863). However, as other ecological schools formed, uniform concepts and methods became an elusive dream. Three ecologists responded to this situation: Dutchman Rudy Becking (1957), American Robert Whittaker (1966), and Englishman David Shimwell (1971).

A German phytogeographer, Oscar Drude (1852–1933), who studied under Grisebach, also exerted an influence on the early development of plant ecology through two phytogeographical treatises which contained pertinent information: *Handbuch der Pflanzengeographie* (1890) and *Deutschlands der Pflanzengeographie* (1896).[11] The latter was first known to two ecologically oriented botany students at the University of Nebraska, Roscoe Pound and Frederic E. Clements (see below). Drude adopted Schouw's terminology (1823) in defining groups of plants as "plantae sociales, gregariae, copiosae, sparsae and solitarine."[12] Like Warming, Drude also made the transition from phytogeography to plant ecology—in his case by publishing *Die Ökologie der Pflanzen* (1913). English plant ecologist Arthur Tansley had led an excursion of plant ecologists around the British Isles in 1911, including Drude, and he wrote a 2.5-page review of Drude's textbook in the British *J. of Ecology* (1914).[13] His review was neutral in tone, and he only complained of its price. However, Tansley's review appeared near when World War I began, in which Britain and Germany were enemies, and that war may have caused his loss of memory of Drude's book and his review of it. In 1923, when Tansley published *Practical Plant Ecology*, he stated: "There are, so far as my knowledge goes, only two books in existence which have at all the same objects as this—Dr. Clements's *Research Methods in Ecology* (1905) and Dr. Rübel's *Geobotanische Untersuchungsmethoden* (1922)."[14]

Andreas Franz Wilhelm Schimper (1856–1901) was born in Strasbourg when it was in France but lived there after Germany defeated France in the Franco-Prussian

DOI: 10.1201/9780429342202-2

War (1871).[15] His father was a professor of natural history at the University of Strasbourg, from which he received a doctorate in 1878. In 1880, he became a fellow at Johns Hopkins University, from which he traveled to Florida and the West Indies. In January 1883, he returned to Germany; yet, he took other trips to the tropics. He died from diabetes and malaria. He was a very productive author, whose most important ecological work was his gigantic *Pflanzengeographie auf physiologischer* (1898, English 1903, 870 pp., 502 illustrations, four color maps)—the largest ecological monograph by a single author, surpassed much later by G.E. Hutchinson's *Treatise on Limnology* (4 vols., 1957–93). Schimper's *Plant Geography upon a Physiological Basis* (1903) that illustrates that the synthesis of phytogeography and physiology became a foundation for plant ecology. American plant ecologist Henry Cowles thought Schimper was "a prophet as well as an ecologist of the first magnitude."[16] However, R.J. Goodland discovered that Schimper had borrowed heavily from Warming, with scant acknowledgment.[17]

Nevertheless, Warming's and Schimper's books were complimentary rather than duplicative.[18] Schimper's was encyclopedic and well illustrative, with many photographs and drawings. Warming's was more of a textbook, with no illustrations and half the size of Schimper's. Both books were world-wide in scope. Warming's chapter 3, only 1.5 pages, was a general discussion of plant communities, which oriented readers to the following chapters on different kinds of communities.[19] Equally brief was chapter 24, "Symbiosis of Plants and Animals," which merely identified different kinds of symbiotic relationships, without identifying or explaining them.[20] Chapter 25, on symbiosis among plants and mutualism, is more detailed, discussing parasitism, helotism (fungi), mycorrhiza and endophytes, epiphytes, saprophytes, and lianas.[21] Chapter 26, "Commensalism, plant communities," discussed plants having an aggregate existence, without interactions except a common struggle for existence. Jan Sapp faulted Warming for only devoting nine pages to the subject.[22] However, if one counts all pages in section 2, "Communal Life of Organisms" (chapters 22–26), it is 14 pages, and Warming referred to other literature on that. His book does not meet entirely Thomas Kuhn's concept of a new paradigm, such as Copernicus' *De Revolutionibus* (1543) and Darwin's *Origin* (1859), both of whom took their science in a revolutionary direction[23] but it did perform the other paradigm characteristic of organizing the subject (at least in the English edition (1909) with a useful example treatise.

With the rise of plant ecology, phytogeography did not fade away. German systematic botanist Adolf Engler (1844–1930) collaborated with Drude in editing *Die Vegetation der Erde* (15 vols., 1896–1930). Engler contributed volume 9: *Die Pflanzenwelt Afrikas* (1910–15), and he recruited other botanists to write volumes which matched their expertise.[24] One example of Drude's recruitment was English native botanist and New Zealand settler, Leonard Cockayne (1855–1934). In 1877, he obtained a book on New Zealand ferns by G.M. Thomson, which turned his attention to native flora.[25] In 1890, there was an extensive fire in a subalpine beech forest near Arthur's Pass (930 m). In 1898, Cockayne published "On the burning and reproduction of subalpine scrub and its associated plants, with special reference to Arthur's Pass District." In 1900, he published "A Sketch of the Waimakariri River Basin," followed by a series of ecological surveys and then synthesized them into *New Zealand Plants and their Story* (1910, ed. 3, 1927). Engler asked him to write a New Zealand volume for *Die Vegetation der Erde* series. Cockayne did and

FIGURE 2.1 (a) C.G. Oscar Drude. Web. (b) Karl F. Schimper. Schimper 1903: frontispiece. (c) J. Eugenius B. Warming. Web.

sent his manuscript to Leipzig in 1914, just before World War I began. He worried about its fate during the war, but in 1921 *The Vegetation of New Zealand* appeared (364 pp., ed. 2, 1928, 456 pp.). He also published 184 articles, 40 on ecology and 25 on plant distributions.[26]

Swedish plant ecologists R.E Fries (1950) and Hugo Sjörs (1981, 10-page paper) published histories of Swedish ecology, much of which they had lived through.[27] Swedish historian-sociologist Thomas Söderqvist published *The Ecologists: from Merry Naturalists to Saviors of the Nation* (1985), which had few similarities to Fries' or Sjörs' histories.[28] Fries book has illustrations and two maps, but no bibliography. Söderqvist's subtitle explained that the book was *A Sociologically Informed Narrative Survey of the Ecologization of Sweden, 1895–1975*, which lacks illustrations, excepting the cover, but does have a bibliography. Sjörs' paper is a valuable summary for its time, with bibliography.

Swedish phytogeographer Rutger Sernander (1866–1944), at the University of Uppsala, followed Linnaeus' footsteps when he studied the vegetation of Gotland Island for his doctoral dissertation (1894).[29] In 1895, he became docent of phytogeography at Uppsala and by 1908, he was a professor of plant ecology. In 1908–1910, his seminar on plant ecology conducted a survey of Swedish vegetation similar to the one which Arthur Tansley led in Britain.[30] Thore Fries (1886–1930) was the son and grandson of Uppsala professors of botany.[31] He followed in their profession, but at the University of Lund in 1927. He went on an expedition to Africa, where he died of pneumonia, 31 December 1930. In plant ecology, he studied synecology, but he also researched phytogeography in India and Africa and lichenology. He attempted to identify and name plant communities inductively.

Another Sernander's student was Gunnar Samuelsson (1884–1944), who judged Fries' communities "artificial groupings." From Sernander's seminars and graduate-student researches emerged the Uppsala School.[32] This school gave up on Fries synecology in favor of plant sociology and Fries left ecology for systematics.[33] Still, another Sernander student, Einar Du Ritz (1895–1967) led the Uppsala School from its founding in 1921 and he became a botany professor at Uppsala when Sernander retired.[34] Du Ritz published *Zur methodologischen Grundlage der modern Pflanzensoziologie* (1921).

In contrast to Sernander's phytogeographical emphasis at Uppsala, Henrik Hesselman (1874–1943) at the University of Stockholm followed the Germanic example of laboratory studies relating to ecology.[35] He had his undergraduate degree from Uppsala, but for graduate school, he studied under Stockholm phytogeographer Gunnar Andersson. Hesselman established a plant ecology school at Stockholm that rivaled Sernander's at Uppsala.[36] Hesselman trained two outstanding students: Henrik Lundegårdh (1888–1969) and Lars-Gunnar Romell (1891–1981). Lundegårdh followed Hesselman into ecophysiology, believing he could combine experimental physiology with phytogeography.[37] He wrote *Klima und Boden in irher Wirkung auf das Pflanzenlebens* (1925) that was praised in Germany and attacked in Uppsala.

Ecophysiologist Romell at Stockholm engaged in a "Great Polemic" with Du Ritz and the Uppsala School.[38] Romell (whose portrait is on Söderqvist' cover) thought plant sociology could attain some value if combined with ecophysiology, but du Ritz only did field descriptions using a quadrat. Du Ritz thought Romell's ecophysiology was a vicious circle. The descriptive-empirical ideal prevailed at Uppsala and the hypothetical-experimental ideal at Stockholm.

England's first phytogeographer, Hewett C. Watson (1804–81), was an independent scientist who published on plant geography for four decades, but had no followers.[39] Instead, plant ecology arose in Britain due to influence from publications by Warming, Schimper, and Flahault.[40]

Francis (Frank) Oliver (1864–1951) was son of a botany professor at University College, London (UCL) and became one himself.[41] In summer 1885, he studied in Bonn, Germany, and became friends with Schimper, who interested him in ecology.[42] Oliver's article, "On the Effects of Urban Fog upon Cultivated Plants" (1891) in London may have been the first scientific study of what is now called "smog." He cautiously commented that this study was preliminary, and that more data should be collected.[43] Only after the great London smog of 1952 that killed 12,000 Londoners, did studies on the effects of smog lead to legislation against this pollution.[44]

Another Londoner, Arthur Tansley (1871–1955), at age 12, wrote on two species of *Potenilla* in his school newspaper.[45] He attended UCL for 18 months, 1889–90, then transferred to the Botany School, Cambridge University, where he graduated with honors in 1894.[46] He then returned to UCL to teach until 1907, when he returned to Cambridge University. While at UCL, he helped Oliver translate Kerner's *Pflanzenleben* (1890–91) as *The Natural History of Plants: Their Forms, Growth, Reproduction, and Distribution* (1896). In 1898, Tansley studied the German edition of Warming's book and compared English plant communities with those described in Denmark. In 1899, Tansley taught a plant ecology course based upon Warming's book. In 1903–07, Oliver took UCL students and staff on field trips to England and Northern France, which set an example for Tansley.[47] In 1894, Tansley graduated from Cambridge and founded a new journal, *The New Phytologist*. (*The Phytologist* was a London journal, 1840s-60s, which Watson dominated, but did not edit.) Tansley edited *The New Phytologist* for 30 years. In volume 1, he suggested that full-time botanists coordinate local botanical or natural history clubs in making local surveys. After several vegetation surveys were published, Tansley suggested in a commentary, "The Problems of Ecology" (1904), that authors of surveys form a British Vegetation Committee to achieve uniformity of methodology and compare findings.[48] They did, under his leadership.

At Cambridge, Tansley had met fellow botany students and Londoners, Vernon and Frederick (Fritz) Blackman.

In 1905, American plant ecologist Frederic Clements published *Research Methods in Ecology*. Fritz Blackman and Tansley published a 27-page review in *The New Phytologist*. It merited a long review because "Dr. Clements' book is at once the most ambitious and most important general work on Ecology that has been published during the last seven years."[49] They shared Clements' impatience with botanists who "did ecology" without training, and his belief that ecology was "the central and vital part of botany."[50]

The Vegetation Committee flourished and Tansley had it in mind when he attended the International Congress of Geography at Geneva in 1908. One aspect of that congress was a tour of Swiss types of vegetation which Schröter organized.[51] Tansley was

> much impressed by the benefits of the contacts made and the amount that could be learned from the examination of vegetation under the guidance of native botanists who had studied it, besides the stimulative effects of the comments of foreign visitors: and on my return to England I suggested to the Committee that we should try to organize a similar institutional excursion in the British Isles.[52]

Tansley collected and edited the findings of the British Vegetation Committee as *Types of British Vegetation* (1911), which attracted much interest. Tansley organized and led an International Phytogeographical Excursion in Britain, for which *Types* served as introduction and reference.[53] American participants were Clements, Cowles, and their wives.[54] In 1913, Tansley led the Vegetation Committee in its transformation into the British Ecological Society—first such in the world.[55] Animal ecologists were invited to join and eventually did. Tansley was its first president, and the only one to serve two years. He also started its *Journal of Ecology*, in 1916, after his presidency. It was primarily a plant ecology journal. In 1923, he published a brief *Practical Plant Ecology: a Guide for Beginners in Field Study of Plant Communities for British Students*.

Western Europe is where science arose and where it continued to flourish. From there, it spread to North America in the west and to Russia in the east. Peter the Great traveled to Western Europe and realized that Russia needed to import western scientists into Russia if Russia wanted to compete with the West.[56] Russia began importing German scientists in the mid-1700s, who trained Russian scientists.

Phytosociology arose in the 1890s with the work of Jozef Paczoski (or Pachoskii, 1864–1942), born in Bilohorodka (Byalohorodka), southwestern Ukraine, to Polish parents.[57] He attended the University of Kiev, 1888–94. He then worked in the administration of several organizations before becoming a professor. He introduced the terms and concepts of phytosociology in 1896. In 1921, he synthesized his studies in *The Principles of Phytosociology* (346 pp., in Russian). He applied these principles in "Phytosociology of Forestry" (1925, in Polish). He then became professor of plant systematics, sociology, and geography at the University of Pozan, Poland, 1925–31, and there he founded the first Institute of Plant Sociology which studied the social structure of plant communities. He was dismissed in 1931 for joining a political protest, and his willingness to draw parallels between plant and human societies was controversial. He bought land and established an experimental orchard. His career showed one possible limitation of western science under a totalitarian government.

North America

In contrast with the Russian experience, transfer of science to the New World, with its greater freedom, was a very positive endeavor. Americans responded to the emergence of plant ecology on the Continent at about the same time as Britain, in response to the same literature.[58] Plant ecology spread more rapidly in America than in the Old World because higher education was still expanding in America during the late 1800s and early 1900s, and so it was easier to add plant ecologists to faculties than in stable European universities. Nor did American universities copy the European example of having only one professor in a department. Two professors of botany who first trained plant ecologists were Bessey of Ohio and Coulter of Indiana.[59]

Charles Edwin Bessey (1845–1915) earned B.S. and M.S. degrees from Michigan Agricultural College and taught at Iowa Agriculture College that awarded him a Ph.D., and in 1884, he became a professor of botany at the University of Nebraska.[60] He trained there Conway MacMillan, Roscoe Pound, and Frederic E. Clements. John Merle Coulter (1851–1928) was the botanist on F.V. Haden's geological survey of the Yellowstone region in 1872 and later collaborated on floras of Colorado and Indiana.[61] He earned a B.A. degree from Hanover College (1870), M.A. (1873) and Ph.D (1883) from Indiana University, and taught at four institutions before going to the University of Chicago in 1896. He trained plant ecologist Henry Chandler Cowles, who remained there and trained himself other plant ecologists.

Bessey's student, Conway MacMillan (1864–1929), from Michigan, earned B.A. (1885) and M.A. degrees at Nebraska and studied further at Harvard and Johns Hopkins universities.[62] He went to University of Minnesota as a botany instructor in 1887 and was promoted as professor in 1893. His *Metaspermae of the Minnesota Valley* (1892) was a capable survey, in the German tradition. He was an active botanist until 1906, when he left academia for better pay in business.[63] Another Bessey's student who did not remain in botany was Roscoe Pound (1870–1964), from Lincoln, Nebraska, and he is a judge's son.[64] He spent an academic year, 1889–90, at the Harvard Law School, then returned home to practice law while also enrolling at the University of Nebraska for a doctorate in botany. He published "Symbiosis and Mutualism" (1893), two years before Warming's *Plantesamfund* (1895) appeared with the discussion of the topic. Sapp complained that Pound overemphasized struggle and minimized mutualistic symbiosis.[65] Pound collaborated with another Bessey student, Frederic E. Clements (1874–1945),[66] also from Lincoln, on

The Phytogeography of Nebraska, for which they obtained guidance from Drude's published methods.[67] They submitted it as a joint doctoral dissertation. Pound received his Ph.D. in 1896; Clements was delayed in obtaining his Ph.D. until 1897 by a language requirement.[68] They published it in Lincoln in 1898, and some copies were distributed before the publisher, Jacob North, lost everything in a fire, including most copies. The University funded a second edition in 1900, and the authors stated in the Preface that:

> "We have availed ourselves of the opportunity to make a thorough revision, made necessary in several particulars by the progress of the Geological Survey since 1897 and by further investigation of the vegetation of the state in the Botanical Survey, and to insert some additional matter."

A review of their *Phytogeography* by a fellow plant ecologist, Henry Cowles, asserted that it was an indispensable reference for American ecologists, but also complained about their terminology which "used Greek derivatives in place of the simpler and more expressive English equivalents."[69] That became a persistent complaint about Clements' publications, whose attitude seemed to be: I went to the trouble to learn Greek and Latin, so I am going to use them! Pound then followed his legal inclinations, and eventually became Dean of the Harvard Law School, while Clements became a leader of plant ecology.

Clements earned three degrees from the University of Nebraska (1894–98) and then joined its faculty, 1894–1906. During that time, he married botany student Edith Gertrude Schwartz (1877–1971). She persisted and earned a Ph.D. in 1904.[70] She accepted her position as companion, assistant, and secretary, without expressing resentment in her autobiography (1960) for not being coauthor of any of his publications. Clements was a very productive author, which led to his prominence in American plant ecology. It was very impressive that he published three books early in career: *The Development and Structure of Vegetation* (1904), *Research Methods in Ecology* (1905), and *Plant Physiology and Ecology* (1907). His bibliography shows that he continued to surpass other plant ecologists in publications.[71] In 1907, he replaced MacMillan at University of Minnesota and stayed a decade before accepting his final position as research ecologist at the Carnegie Institution of Washington.

On their honeymoon, 1899, Edith and Frederic visited Pike's Peak, Colorado, and he decided to establish an Alpine Laboratory at Minnehaha, near a cog railroad to the top of Pike's Peak. After spending three summers there, he published in *Science* what today is called an "infomercial," to attract graduate students for summer research.[72] Using the cog railroad, one could traverse eight formational zones, which he described and referred readers to eight of his publications, 1904–13, for more details.

Meanwhile, plant ecologist Henry Cowles (1869–1939), son of a Connecticut market gardener, earned a B.A. degree from Oberlin College (1893) and Ph.D. (1898) under Coulter from the University of Chicago, where he joined as a faculty.[73] Coulter was the founding editor of *Botanical Gazette*. When a review copy of Warming's *Ökologischen Pflangengeographie* (German ed., 1896) arrived, Cowles wrote a synopsis which Coulter used for his review in *Botanical Review*. Warming's book had the same strong impact on Cowles that Drude's books had on Pound and

FIGURE 2.2 (a) Frederic E. Clements. *Ecology* 35: 108. (b) Henry B. Cowles. *Ecology* 16: 281.

Clements, and in the same year. Subsequently, Cowles reviewed books in almost every issue of *Botanical Gazette*. He wrote a notable dissertation, which he published in three articles (1898–1901), which secured his retention at Chicago.[74] Later, he published "The Causes of Vegetative Cycles" (1911), but his main efforts went into teaching and advising plant ecology students.

B. ANIMAL ECOLOGY

Animal ecology progressed simultaneously with plant ecology, though perhaps with fewer ecologists.

Europe

Two steps toward an animal ecology treatise appeared in the 1870s, and a general animal ecology treatise appeared in 1880.[75] Belgian zoologist-parasitologist Pierre-Joseph van Beneden (1809–1894) published *Les commensaux et les parasites dans le règne animal* in 1875 (English 1876), in which he classified the relations between animal species as predation, parasitism, commensalism, and mutualism. He defined commensalism as an animal sharing the food of a neighbor without doing it physical harm and mutualism as a mutually beneficial relationship.[76] Being Catholic, he disliked Darwin's concept of struggle for existence.[77]

French philosopher Alfred Victor Espinas (1844–1922) wrote a doctoral dissertation on *Des Sociétés Animales* (1877, published 1878, English 1935). He wrote to Darwin (1 July 77) and asked if coral should be considered social animals; he thought not.[78] Espinas drew upon van Beneden's book in discussing parasitism, commensalism, and mutualism. His book was controversial for suggesting that human societies evolved from animal societies.[79] Being a philosophical treatise, it had little immediate influence upon zoology, but that situation changed during the 1900s with the rise of ethology and sociobiology.[80]

German zoologist Carl (or Karl) Semper (1832–1893) attended University of Würzburg and wrote a doctoral dissertation (1856) on anatomy and physiology of snails.[81] He had an itch to travel, and did so in Europe, beginning in 1857, but then in the Philippines, 1858–65, where he contracted malaria and dysentery, in addition to collecting stupendous invertebrate zoological specimens. Back at Würzburg, where he became a professor of zoology and Director of the Zoological Institute, he worked over his collections to write four volumes (1. Holothuria, 3. Terrestrial Mollusks, 5–6. Lepidoptera) of *Reisen im Archipel der Philippinen* (10 vols., 1868–1905). However, he did not get lost in the details, and he went to Boston and gave a series of lectures in 1877 on *Animal Life as Affected by the Natural Conditions of Existence* (German 1880, English 1881), which seems to be the first textbook on animal ecology, though he did not claim it to be. He devoted pages 40–329 to "The Influence of Inanimate Surroundings," including light, temperature, water, and food. Before switching to zoology, he had studied engineering and was accustomed to thinking quantitatively when few zoologists did. Under food, he explained that when herbivores convert vegetation into flesh, there is a loss of mass, due to oxidation of organic material, and that the same holds when carnivores convert flesh of prey into their own flesh.[82] To illustrate, he arbitrarily assumed a 10:1 ratio of food to flesh, a ration in accord with Willis Pequegnat's more recent estimate.[83] Semper's discussion of "The Influence of Living Surroundings," is in two chapters. In the first (pp. 330–359), he discussed what we call symbiosis, commensalism, and parasitism. In the second (pp. 360–405), he discussed predation and competition. He was a hostile rival of Haeckel and did not use Haeckel's term, "oekologie;" instead, he used "natural conditions of existence."[84]

More narrowly focused, but also influential, was a study by German zoologist Karl Möbius (1825–1909) on oysters.[85] From a poor family, he was educated enough to teach primary school in a small town and later qualified to teach high school in Hamburg. He also began working at the Hamburg Museum of Natural History, became co-founder of the Hamburg Zoo, and planned Germany's first public aquarium. He studied at the University of Berlin, and in 1853 became a teacher in a school in Kiel. He co-authored a treatise on three classes of Mollusca (sea slugs, marine snails, and clams) in Kiel Bay, with a general introduction, ecological in scope and methodology, with "Biocönose" indicating the forms of life with something in common.[86] He explained in the introduction that five regions (zones) extended from the sandy beach to mud at 10 fathoms and also compared Kiel Bay fauna with fauna in other parts of the North Sea. This treatise helped him become a professor of zoology at the University of Kiel in 1868.

The Prussian government in 1869 commissioned him to study French and English oyster culture and promote oyster culture in German waters. Möbius explained in *Die Auster und die Austerwirthschaft* (1877, English 1880) that "An Oyster-Bank Is a Biocönose, or a Social Community."[87] Oyster beds occur in shallow sea waters on sandy bottoms.[88] His term "biocönose" gained currency in Germany and elsewhere, though "community" prevailed in English-language usage.[89] Clements and Shelford questioned whether Möbius intended biocenose to include plants, since his discussion of plants was limited to *Zostera* and some Florideae which grew upon oyster banks and that oysters consumed desmids and diatoms.[90] However, his limited discussion of plants was because his subject was oyster beds but not all organisms within a specified environment. A modern commentator argued that Möbius' biocenose concept synthesized three lines of thought: ancient balance of nature belief, Humboldt's associations concept, and Darwin's theory of natural selection within a web of life.[91]

Italian entomologist Lorenzo Camerano (1856–1917) studied at the Royal University of Torino and then worked at the University's Museum of Zoology.[92] He also became president of Accademia delle Scienze di Torino. He was a very productive scientist, publishing 341 works, 1877–1917; many papers were descriptive, but in 1880 he published

"Dell' equilibrio dei viventi mercè la reciproca distruzione" (English, Camerano 1994: 375), which included four complex food web diagrams—earliest known web diagrams.[93] However, his paper stimulated neither imitators nor controversy nor papers testing his conclusions.[94]

Better-known works of Beneden, Espinas, Semper, and Möbius were not enough to develop a formal science of animal ecology in Europe. Entomology and parasitology were developing[95] with modest influence on progress in animal ecology. In 1923, English invertebrate zoologist Lancelot Borradaile (1872–1945) published his third college textbook, *The Animal and Its Environment: A Text-book of the Natural History of Animals*. He did not use the term "ecology," three decades after its common use in America.

North America

Like American plant ecology, American animal ecology arose mostly in the Midwest (Merriam and Lotka were exceptions), for the same reason: Midwestern universities were still growing and could add ecologists to their faculty, whereas it was more difficult in Eastern universities that were rather stable. New ecology faculty members then recruited students to ecology. Michael Lannoo's *This Land Is Your Land: The Story of Field Biology in America* has significant discussions concerning animal ecology before 1900 in the Midwest.[96]

Illinois zoologist Stephen A. Forbes (1844–1930) became the curator of the Museum of the State Natural History Society in 1871, which he transformed into the Illinois State Laboratory of Natural History in 1877.[97] He was probably the first to make systematic studies on the foods of birds and fishes and he rose above his massive data to reach ecological generalizations. He cautioned in 1878 those who wanted to control "harmful species" of birds:

> That the balance of nature should be disturbed only after full knowledge and reflection should be evident to anyone who realizes the complicated relations of living things and the consequent numerous and remote results of any changes in these relations.[98]

Forbes is most remembered for an 1887 talk, "The Lake as a Microcosm."[99] He had expressed some of its ideas earlier in "Some Interactions of Organisms" (1880).[100] A lake is a natural "ecosystem" (coined 1935), which he indicated with his "microcosm," and its relationship to a watershed.[101] He mentioned "diatoms and other microscopic algae" and "weeds and lily-pads upon the shallows and around the margin, the Potamogeton, Myriophyllum, Certaophyllum, Anacharis, and Chara, and the common Nelumbian," but only once did he explore the relationships between plants and animals, focusing instead on relationships among different aquatic animals.[102] Only 37 of Illinois' 132 fish species lived in interior lakes, and when common Lake Michigan fish were transferred to inland lakes, they did not flourish.[103] Inland lakes had few bivalve mollusks but numerous univalves. Those lakes contained about 40 species of crustaceans, several worms, and insect larvae. He discussed in detail Entomostraca, minute crustaceans that were fish foods. Black bass ate Entomostraca and insects, but crustaceans themselves consumed Entomostraca, algae, and protozoa. "At only the second step [of a food chain], therefore, we find our bass brought into dependence upon nearly every class of animals in the water."[104] His example of a relationship between plants and animals was unusual: bladderworts (*Utricularia*) are flowering plants with hundreds of bladders that trap minute animals that they digest and therefore bladderworts compete with fish for food. He concluded (rather oddly) that "beneficent order is maintained in the midst of a conflict seemingly so lawless."[105] The "beneficent order" was that if a species consumed too many individuals of its food, its population would then decline due to food scarcity. More on Forbes is given under Limnology.

In 1887, the Federal Hatch Act provided for an agricultural experiment station in each state, and the Illinois station opened in 1888, with Forbes as its entomology consultant.[106] During the 1880s, he began to study the bacterial and fungal diseases of insects. A fungus he found, *Beauveria globulifera*, was fatal on cinch bugs, and in early 1890s, he distributed both the fungus and infected bugs to farmers, with some success, though he eventually concluded the fungus spread on its own well enough. In his *Nineteenth Report of the State Entomologist on the Noxious and Beneficial Insects* (1896), he described ecology as "lately distinguished as a separate subject, and economic entomology as a division of this science."[107] He doubted an argument by Dr. Louis W. Sambon that *Simulium* biting flies transmitted microorganisms that caused pellagra. Forbes found no correlation between *Simulium* abundance in July and new cases of pellagra.[108] That pellagra is a niacin-deficiency disease was shown in 1934–38. Forbes was the president of the Ecological Society of America in 1921.[109]

Maine entomologist Alpheus Packard, Jr. (1839–1902), whose father was a professor at Bowdoin College, Brunswick, Maine, attended it, 1857–1861 (B.A.), then studied under Louis Agassiz at Harvard until 1864.[110] He held a variety of jobs while writing an illustrated *Guide to the Study of Insects* (1869), which propelled him to become America's leading entomologist. It also enabled him to teach at Massachusetts Agricultural College, 1870–78 and at Bowdoin,1871–74; in 1878, he became a professor of zoology at Brown University, Providence, Rhode Island, where he remained. Packard's "The Cave Fauna of North America" (1888) can be viewed as a contribution to ecology, though he wanted it to support his Lamarckian theory of evolution.[111] He continued publishing books on entomology, with his *magnum opus* being *Monograph of the Bombycine Moths* (3 vols., 1895–1914), with the third volume being completed by T.D.A. Cockerell.

New York State native C. Hart Merriam (1855–1942) was from a prosperous farming family.[112] In 1781, his father was in Congress and took him to chat with Spencer Baird at

FIGURE 2.3 (a) Stephen A. Forbes. Mills 1958: 95. (b) C. Hart Merriam. Osgood & Grinnell 1947: 2.

the Smithsonian Institution. Baird examined the birds and mammals he had stuffed and suggested having a taxidermist in New York to train him which was done. In 1872, Baird arranged for him to join Fernand Hayden's U.S. Geological Survey of the Rocky Mountains.[113] Merriman published a "Report on the Mammals and Birds of the Expedition."[114] He attended Yale for three years and then earned a M.D. degree from New York College of Physicians and Surgeons in 1879. He practiced medicine for six years before becoming an ornithologist of the Entomology Division, U.S. Department of Agriculture. He was one of the 24 ornithologists who organized the American Ornithologist Union in 1883.[115] He also helped establish the Washington Academy of Sciences, American Society of Mammalogists, and the National Geographic Society. In 1905, his office became the Bureau of Biological Survey.

Merriam's USDA Division of Ornithology and Mammalogy mainly studied the life histories of species having economic significance and provided a pest control service for farmers.[116] Yet, he could also draw generalizations from his data. In 1889, his division surveyed the San Francisco Peak and the nearby Little Colorado River, Arizona:[117]

> Recent explorations in the West . . . led to the belief that may facts of scientific interest and economic importance would be brought to light by a biological survey of a region comprehending a diversity of physical and climatic conditions, particularly if a high mountain were selected, where, as is well known, different climates and zones of animal and vegetable life succeed one another from base to summit.

He described birds and mammals collected there, and he had a specialist describe amphibians and reptiles collected.[118]

He had his collected data on elevation at different points ascending the Peak, where vegetation changed from hot to cold characteristic species. Oddly, he did not also collect temperatures. Still, he identified seven "life zones from hot Desert Zone at 1200–1800 m (4000–6000 ft.) to Alpine Zone, above 3500 m (11,500 ft.)." Alexander von Humboldt had discovered the existence of such zones on Chimborazo and Cotopaxi mountains in northern South America in 1807.[119] Humboldt was on an expedition and did not expand upon his insight, being busy publishing further discoveries. Merriam realized that the vegetation of these seven zones could also be matched with changes in vegetation as one traveled north from the desert northward, and he compiled a North American map showing such zones. As he obtained more data, he published two revised versions of that continental map.[120] Merriam remained at USDA until 1910 and then he conducted private research until 1939.[121]

In 1912, three entomologists, W.D. Pierce, R.A. Cushman, and C.E. Hood, produced a food web of enemies of cotton boll weevils in a USDA bulletin. Hood had worked on parasites of boll weevil between 1906–11.

Rural New York State native, Victor Shelford (1877–1968) entered University of West Virginia in 1899 for premedical studies.[122] In 1901, he transferred to University of Chicago (B.S. in zoology, 1903). He continued graduate studies there under two zoologists and plant ecologist Henry Cowles, which led him to animal ecology. He studied tiger beetles in the same Lake Michigan dunes where Cowles

FIGURE 2.4 Third version of the map of North America showing life areas. Merriam 1894: facing 228.

had studied vegetation succession. He found that each beetle species lived in particular vegetation and breeding environments (Ph.D. 1907).[123] He did not continue studying insects but instead turned to three studies of succession among fishes in streams and ponds and two on succession in land communities,[124] all of which he folded into *Animal Communities in Temperate America: as Illustrated in the Chicago Region* (1913), which drew upon his observations and experiments. It included terrestrial and aquatic food webs, drawings, photographs, tables, maps, and diagrams. He rejected emphasis on a single factor, such as temperature or food, to explain species distributions and defended a "law of toleration," indicating species are most abundant where factors affecting their lives are optimum and less abundant where those factors are less favorable.[125] One historian evaluated Shelford's *Animal Communities* as:[126]

> . . . the first description of temperate animal communities, and was also the first attempt to classify those communities. His was a broad survey that encompassed a range of habitats—scrub and woodland, grassland, rivers and streams, ponds and lakes, marshes, shores and bogs—stressing in each case the relations between the plant and animal communities. Though it lacked detail, as a general reconnaissance with a clearly defined underlying plan, it was a remarkable achievement.

Shelford was active in founding the Ecological Society of America in 1915 and was its first president in 1916.[127] In 1917, he suggested an ESA committee be formed to collect information on natural areas. A Committee on Preservation of Natural Conditions of about 25 members was established, with him as chairman. Many others were recruited to describe North American natural areas, and by 1926, he could edit many contributions from North and Central Americas for publican as *Naturalist's Guide to the Americas*.[128]

In the same year appeared the *Guide to the Study of Animal Ecology* (1913) by Illinoisan Charles Adams (1873–1955), who had earned a B.S. degree from Illinois Wesleyan (1895), M.S. from Harvard (1899), and Ph.D. from University of Chicago in zoology and animal ecology (1908).[129] He served on the faculty of several places before settling at Albany. His *Guide* has the earliest organization of animal ecology, with helpful bibliographies. The author he most cited was Stephen Forbes and the second-most cited was Shelford. He initiated the establishment of the Roosevelt Wild Life Forest Preserve and its Experiment Station, where a joint Mount Marcy study was conducted (1919–20), in which he participated and coauthored the report.[130] He was the ESA's eighth president (1923).

Warder Clyde Allee (1885–1955) grew up on his family's farm near Bloomingdale, Indiana.[131] From a Quaker family, he earned a B.S. degree from Quaker Earlham College and his M.A. and Ph.D. from University of Chicago under Shelford. Allee taught at four other academic institutions before returning to the University of Chicago, 1921–1950. He developed an informal school of animal behavior.[132] He was a very productive ecologist, using extensive laboratory experiments, focused mostly on behavior among social insects. His publications in the 1920s bore fruit in three books during the 1930s, which brought him prominence.

His doctoral students included the Park brothers, Orlando (Ph.D. 1929) and Thomas (Ph.D. 1932). Orlando Park joined the faculty of Northwestern University and Thomas Park remained at the University of Chicago. Allee was ESA's president in 1929, Orlando Park was in 1943, and Thomas Park was in 1959. Allee and the Park brothers were three of the five authors of the *Principles of Animal Ecology* (1949), the first encyclopedia of animal ecology. The second issue of *Ecology* in 1956 was an Allee Memorial Number.

Synthesis

Italian mathematician Vitto Volterra (1860–1940) and American mathematician Alfred Lotka (1880–1949) realized almost simultaneously that animal ecologists lacked mathematical tools that could facilitate their studies. Lotka, the younger, published first, in 1925: *Elements of Physical Biology*. He had realized the value of using demographer Raymond Pearl's logistic curve.[133] Oddly, he knew no biologists, and it was several years before animal ecologists realized the usefulness of his book. He saw no future in biology

FIGURE 2.5 (a) Vito Volterra. Web. (b) Alfred Lotka. Web.

for himself and turned his attention to human demography. His book was reprinted three decades later, with *Physical* changed to *Mathematical*. Volterra was a versatile mathematical physicist.[134] In 1925, his daughter, Luisa, interested him in commercial fish population dynamics. Unaware of Lotka's book, Volterra published briefly in *Nature* "Fluctuations in the Abundance of a Species Considered Mathematically" (1926). When animal ecologists realized the value of the mathematical tools Lotka and Volterra had explained, they still needed decades of experience to fully master them.[135]

Two animal ecology textbooks appeared during the 1920s one by American Pearse (1926) and another Englishman Elton (1927). Arthur Sperry Pearse (1877–1956) was born in a sod shack at Crete, on the Nebraska frontier, and he earned his B.S. (1900) and M.A. (1904) degrees at the University of Nebraska, and his Ph.D. at Harvard University. He also earned a degree from the London School of Tropical Medicine (1926). He taught at several institutions before a longer stay at University of Michigan, 1908–11, then a year in the Philippines. He stayed at University of Wisconsin, 1911–27, before settling at Duke University in 1927. His most significant publication was *Animal Ecology* (1926). Whatever he published further on ecology is unknown to me, excepting his obituary in *Science* on Chancey Juday.[136] Otherwise, he published historical literary notes on such topics as "The Loves of Plants" (1927) on ancient Greece and "A Sketch of the Development of Ancient Botany" (1952).[137] He also self-published a 68-page booklet, *Adventure .... Trying to be a Zoologist* (1952), for friends and relatives.[138]

Charles S. Elton (1900–1991) was a son of an English professor at Owen's College, Manchester.[139] He attended Oxford University, majored in zoology, studied under Julian Huxley, and was Huxley's assistant during Oxford's Spitzbergen Expedition in summer 1921. He also participated in other Spitzbergen expeditions in 1923 and 1924 and on one to Lapland in 1930.[140] He graduated in 1922 and remained at Oxford during his career. British botanists organized the British Ecological Society (BES) in 1913 and invited animal ecologists to join. BES also sponsored the *Journal of Ecology* (1916) that plant ecologists dominated. British animal ecologists were rather inconspicuous before Elton published a textbook, *Animal Ecology* (1927). It seems likely that he wrote it and perhaps published it before he became aware that an American animal ecologist had published a textbook with the same title in 1926. Elton read Shelford's *Animal Communities* before the 1921 expedition, in which terrestrial and aquatic food webs were essentially independent of each other. At Spitzbergen, however, Elton was impressed by transfer of matter from ocean to land as foraging sea birds brought back fish to their young, and he drew a food web showing that.[141] His reading of Charles Hewitt's *Conservation of Wildlife in Canada* (1921), which discussed lynx and snowshoe hare population cycles, plus his experiences on three Spitzbergen expeditions led to Elton's interest in Arctic ecology.[142] That interest was expressed mainly in the Arctic examples of ecological principles in *Animal Ecology* (1927).[143]

Animal Ecology was more influential than textbooks typically are because Elton used it to synthesize important concepts: animal communities, food chains and webs, niche, and pyramid of numbers.[144] Three chapters (2, 5, and 7) were on *animal numbers*, and a chapter on *ecological succession* also discussed communities. He related *food chains and webs* to both *niche* and *pyramid of numbers. Niche* had been discussed by zoologist Joseph Grinnell (1877–1939), Director of University of California's Museum of Vertebrate Zoology, in reference to place in "The Niche-relationships of the California Thrasher" (1917).[145] Elton used the term to indicate function.[146] Thomas Park's history of animal ecology (1949) compared Pearse and Elton's textbooks. Pearse "laid a general background of physical and biotic factors and then classified animals ecologically according to their major habitats. The treatment was primarily descriptive."[147] Elton organized animal ecology around principles, emphasizing community and population. He was interested in environmental factors limiting species. He also stressed the quantitative aspects in connection with the number of animals that occupy a community and the impact these numbers make on their total environment. He viewed food chains as the most important integrating factor of the community and his treatment of this subject is outstanding.

The contrast between Grinnell's and Elton's niche concepts was rejected by Thomas Schoener after reviewing five of Grinnell's publications on niche, 1913–28. He judged the differences between Grinnell's and Elton's niche concepts were rather slight and a significant contrast existed between their concepts versus G.E. Hutchinson's (1957).[148] The success of *Animal Ecology* boosted Elton's status, and in 1932 he founded the *Journal of Ecology* and established the Bureau of Animal Population at Oxford. Both are discussed in chapter 4, as is Hutchinson's concept.

Life histories of species were part of natural history since antiquity, but natural history was descriptive: this is what lions do and this is what squirrels do. Lynne Houck and Lee Drickamer included only three extracts from the 1800s in *Foundations of Animal Behavior: Classic Papers with Commentaries* (1996). First, a 25-page chapter, "Instinct," from the 1872 edition of Darwin's *Origin of Species*, followed by extracts from George Romanes' *Animal Intelligence* (1882) and from E.L. Thorndike's *Animal Intelligence* (1898).[149] Insect natural history-ethology developed rather independently of vertebrate studies and has received its own historical study.[150] William Thorpe (1902–86), a professor of zoology at Cambridge University, published "Ethology as a New Branch of Biology "(1956), explaining there were "three scientific ways of probing into the behavior of animals," the ways of naturalists, psychologists, and physiologists.[151] Although ethology as a formal branch of biology was formally organized in the early 1950s, it utilized much earlier contributions.[152] Peter Klopfer and Jack Hailman wrote a survey, *An Introduction to Animal Behavior: Ethology's first century* (1967); their chapter 1 covers 1850–1900. Thomas McGill's compilation *Readings in Animal Behavior* (1965) is topically, not historically, organized and it served as supplemental reading for ethology courses.

Three landmark studies on insect parasites of other insects appeared in 1897, two by French entomologists and one by an American entomologist.[153] They showed entomology as a very active science but no synthesis appeared before 1900.

C. LIMNOLOGY

Europe

Limnology was a science founded by François Alphonse Forel (1841–1912), who was born and died in Morges, west of Lausanne, Switzerland.[154] He graduated from Académie de Geneve, studied medicine in Montpellier, and earned an M.D. degree at University of Würzburg. He joined the faculty at Académie de Lausanne, where he taught anatomy and physiology for 25 years. His academic interests were broad, though emphasizing Lac Léman (Lake Geneva).[155] He compiled his own bibliography,[156] listing 126 limnology titles, 66 on glaciology, 12 on seismology, 19 on meteorology, 25 on natural history, 12 on archeology, 10 on history, and 15 on biography. He published his first limnology article in1866.[157] He named and defined limnology in volume 1 of *Le Léman* (3 vols, 1892–1904).[158] Limnology actually began in 1869, when he published "Introduction à l'étude de la faune profonde du Lac Léman."[159] Forel had looked under a microscope at some mud which stuck to an instrument sent to bottom of the lake and had discovered a nematode, *Mermis aquatilis*. He wondered if other animals also lived in that mud and soon discovered they did.[160] He developed new research goals: determine bottom topography, profundal fauna and flora, lake temperatures, and study deltas of Rhone and other rivers. All the studies were later synthesized in *Le Léman*, volume 1 of which included geography, hydrography, geology, climate, and hydrology; volume 2 included hydraulics (seiches), temperature, and optics, and volume 3 had over 400 pages on fauna, and included history and fishing. Along the way, he also published the first limnology textbook: *Handbuch der Seenkunde: Allgemeine Limnologie* (1901) in which he cited Forbes' "The Lake as a Microcosm."[161]

Illinoian zoologist Stephen A. Forbes (1844–1930) held positions of university professor and state entomologist and his publications were influential on American animal ecology.[162] He is most remembered for his "The Lake as a Microcosm" (1887).[163] German limnologist Hans-Joachim Elster, in his "History of Limnology" (1974) noted that "the first concept of the limnological ecosystem was put forward by Forbes (1887) in his essay 'The Lake as a microcosm.'" Elster thought "Perhaps biology would have found a quicker way to ecology, if those who contended over the doctrine of evolution had been familiar with the work of Forbes."[164]

German zoologist (Emil) Otto Zacharias (1846–1916) was from Leipzig and received his education there. He became a tutor and independent scholar until 1891, when he published, and on April 1 1892, he became the founding director of Germany's first freshwater biological station at Plön.[165] One of those books is relevant here: *Die Tierund Pflanzenwelt des Süsswassers* (1891), to which other limnologists contributed including Forel. Later, Zacharias published two more books on plankton: *Das Süesswwasserplankton* (1907) and *Das Plankton* (1909). In 1893, he began publishing an annual, *Forschungsberichte aus der Biologischen Station zu Plön*, which evolved into *Archiv für Hydrobiologie*.[166] In 1917, after his death, his

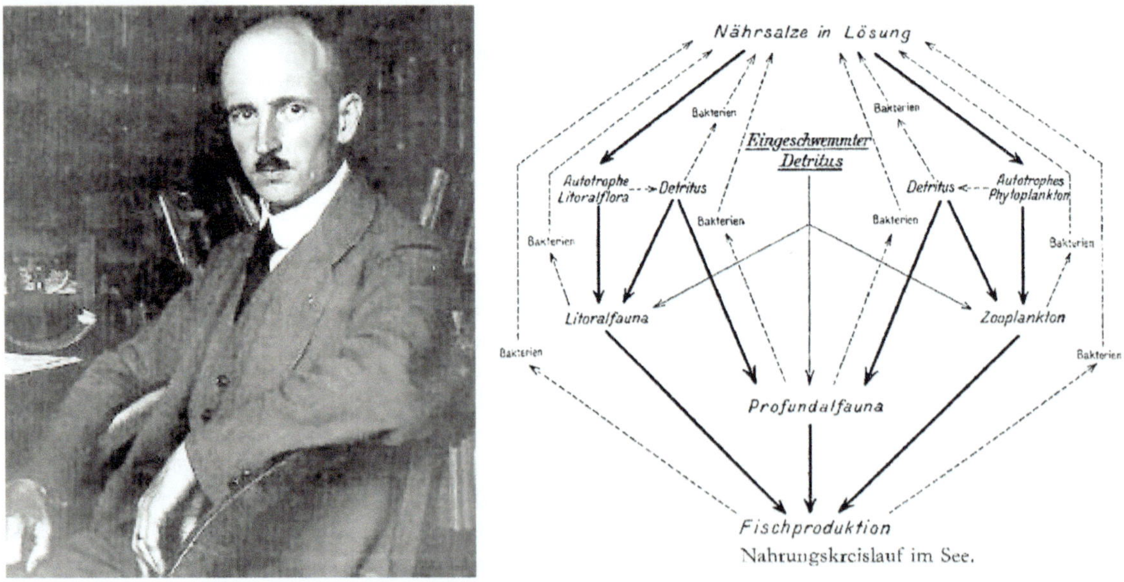

FIGURE 2.6 (a) August F. Thienemann. Web. (b) Food web in a lake. Thienemann 1926: 57.

Biologische Station became affiliated with Kaiser Wilhelm Gesellschaft, under the excellent direction of August Thienemann (1882–1960), who had studied at Heidelberg, under Robert Lauterborn, "Germany's leading river biologist."[167] Thienemann joined University of Kiel faculty and eventually became Europe's leading limnologist.[168] He remained as director at Plön for 40 years and edited *Archiv für Hydrobiologie* for over 40 years.

In Sweden, Finnish immigrant fisheries manager Oscar Nordqvist (1858–1925) organized Sodra Sveriges Fiskeriforening (South Sweden Fishery Association) in 1907, with laboratory at Aneboda.[169] In summer 1910, he hired two assistants, student Gunnar Alm (1891–1962), Uppsala University (Ph.D., 1915), who studied Ostracod Crustacea and Einar Naumann (1891–1934), Lund University (Ph.D., 1917), who studied lake plankton and bottom fauna.[170] Alm's career was in fisheries management, researching biota and lake environments in relation to fish populations. Naumann became a leading limnologist. In 1921, Thienemann went to Sweden to meet Naumann. They developed a close working relationship, and Naumann suggested they organize an international association. Thienemann agreed, and in January 1922, they sent a prospectus to 100 colleagues and received positive responses.[171] A meeting in August organized *Internationale Vereinigung für Limnologie* (later, *Societas Inernationalis Limnologiae*) with 188 members from 28 countries. At Thienemann's suggestion, it published an annual *Verhandungen*. Naumann was active in this society during the remaining 12 years of his life.

In 1907, C.A. Weber, who studied German bogs, introduced terms to indicate what is now known as eutrophic, mesotrophic, and oligotrophic.[172] Naumann appreciated Weber's concepts and introduced them with modern terms into limnology.[173] Thienemann accepted Naumann's terms and added "distrophic" for lakes with brown, humic waters and little lime.

French native of Nantes, Emile Gadeceau (1845–1928), at age 16 participated in a field trip of Société botanique de France, at Nantes, which led to his becoming a disciple of English botanist James Lloyd (1810–86), who had settled in Nantes.[174] Lloyd had published *Flore Loire-infèieure* (1844) and *Flore de l'ouest de la France* (1854). In 1893, Gadeceau met Université de Montpellier plant geographer Charles Flahault (discussed above) at a meeting of Société botanique de France.[175] They began to correspond and Flahault became a major influence on Gadeceau. Flahault introduced the term "ecologiques" into French botany[176] and he wrote preface for Gadeceau's *Le Lac de Grand-Lieu* (1909).[177] That lake is 13 km south of Nantes in Western France and his map indicates it is about 6.5 km wide and 8 km long with an irregular shape. His introduction indicated he had access to Lloyd's unpublished journal, 1837–89, and his own observations, 1879–1908, from all seasons. His subtitle was "monographie phytogeographie," and he devoted 11 of its 120 pages to the lake's birds, fish, and insects, 10 pages to mammals. Fifty-five pages are on aquatic plants—48 pages of annotated lists and seven pages of ecological influences; three chapters discussed physiographic classification, ecological groups, and phytogeographic history. Gadeceau synthesized his and Lloyd's observations with the limnology literature of the time.

Cambridge University zoologist J.T. Saunders taught Britain's first hydrobiology course, 1924–25, and with W. Pearson and F.E. Finch organized in 1926 Freshwater Biological Association, with a biological station at Lake Windermere.[178] In 1901, Eustace Gurney (1876–1927) had built a station on Sutton Broad.[179] Homer Jack included neither station in his world survey of stations around 1940.[180] Robert Hiatt cited the date for the Lake Windermere Laboratory establishment as 1931—perhaps when its building was built.[181]

In July 1928, Thienemann, Austrian limnologist Franz Ruttner (1882–1961), director of a biological station at Lunz, and German zoologist H.J. Feuerborn embarked on an expedition to Sunda Islands—Java, Sumatra, Bali—for 10 months to collect plants, animals, water samples, and environmental data, believing "only knowledge of the pri-mordial, tropical, conditions can give the scientist a deeper understanding of how life proceeds at our latitudes."[182] They studied tropical lakes and running waters. Later, collaborating with over 100 European specialists, they described about 1100 new species, in 11 supplementary volumes to *Archiv für Hydrobiologie*, 1931–58, entitled *Tropische Binengewässer*, in 7920 pages and 348 plates.[183] Ruttner provided hydrographic, physical, and chemical data of 15 lakes that could be correlated with their flora and fauna. Ruttner wrote an introductory textbook (1940), the third edition of which was translated (1963) and used by Art Hasler to teach me limnology in 1963.

Whether or not Thieneman ever saw Stephen Forbes' "The Lake as a Microcosm," he knew of its message and used Forbes' term, "microcosm" when discussing life in a lake. He also used the superorganism metaphor to describe the interacting species. His researches led to three generalizations, two from the 1920s:

1. The more variable the biotope (environmental) conditions, the greater the number of species in the biocenosis (biotic community);
2. The more biotope conditions deviate from optimal for most species, the smaller the number of species, but the greater the number of individuals of species represented.
3. The longer a locality retains the same conditions, the richer and more stable is its biocenosis.

Thienemann, in 1926, first used the concepts of producers, consumers, and reducers to organize data and included a diagram that later influenced Raymond Lindeman in his "Trophic-Dynamic Aspect of Ecology" (1942).[184]

In 1910, two Midwestern American aquatic biologists, Chancey Juday and Charles Kofoid, independently published illustrated surveys of European freshwater and saltwater stations. Juday's was an article and Kofoid's a book.

The earliest founding date of 11 freshwater stations Juday listed was 1888 for the Bohemian Portable Laboratory under Karl Fritish, first located near several lakes and later moved to the Elbe River.[185] Zacharias' station at Plön was possibly the earliest permanent freshwater station. Attempts to have Germany's government fund a freshwater station before World War I failed. In 1918, long before Ruttner's textbook, Dutch professor H.J. Jordan published (1918) one on freshwater animal life.[186]

North America

Limnology in America, like plant ecology and animal ecology, began in the Midwest, and for the same reason, midwestern universities were younger and more flexible than eastern ones, and so could add professors of new sciences more easily than could eastern universities. Stephen Forbes' "The Lake as a Microcosm" (1887) provided "the first concept of the limnological ecosystem."[187] Limnology was ahead of plant ecology and animal ecology in establishing field stations.[188] University of Michigan Professor Jacob E. Reighard (1861–1942), supported by the Michigan Fish Commission, led a team in a biological survey of Lake St. Clair, located between Lakes Huron and Erie.[189] In 1894, the team moved to Traverse Bay, Lake Michigan. The reports from these collaborations were substantial contributions,[190] which surveyed fauna and fisheries. There were no funds for a third year. In 1898–1901, the U.S. Fish Commission supported Reinhard's team in a biological survey of Lake Erie, operating from the U.S. hatchery, Put-in-Bay, Ohio, using hatchery boats.[191] The Erie survey also yielded valuable reports. Ohio also had a hatchery at Put-in-Bay, where Ohio State University provided research support for its own biologist. Eventually, it built its own facility there named Franz Theodore Stone Laboratory.[192]

FIGURE 2.7 Edward Birge and Chancy Juday. Beckel 1987: cover.

Prof. Stephen Forbes (1844–1930)[193] established University of Illinois Biological Research Station at Havana, on the Illinois River, in 1894, America's first river research station.[194] Illinoisan Charles Kofoid (1865–1947), with a Harvard Ph.D., became head of that station in 1894–1900; his specialty was plankton, and he was one of the very few who studied river plankton. His studies on Illinois River plankton extended from 1897–1908, with his most substantial studies appearing in 1903.[195] In 1900, he moved to the Department of Zoology, University of California, Berkeley.[196] What progress occurred in river limnology after 1903? Evidence is in index of Frey, ed., *Limnology in North America* (1963: 729), under heading of "Streams," with numerous references. River limnology did not fade away but it did not become a conspicuous part of limnology. Troy, New York native Edward A. Birge (1851–1950) earned A.B. (1873) and M.A. (1875) degrees at Williams College in Massachusetts and then entered Harvard to earn a doctorate under Louis Agassiz who died three months later. A professor at Williams College, John Bascom, left to become president of the University of Wisconsin. He invited Birge to come be an instructor in natural history and curator of the university cabinet (museum). Birge began teaching there in 1876.[197] Later, he received leave to complete a doctorate at Harvard (1878). In 1879, he was promoted to professor of zoology. Like Forel, he first studied aquatic invertebrates, but later devoted his studies mostly to lake environments.[198] His "Notes on Cladocera" (1879), from his dissertation, represented three years of research in both Madison and Cambridge, Massachusetts. Further research at Lake Mendota, on whose shores the university is located, indicated that both Crustacea and lake waters undergo seasonal stratification changes. Despite Wisconsin's cold winters, he conducted continuous plankton and water temperature studies, July 1894 to December 1896. The published report was the longest paper he ever published alone.[199]

Thermal stratification which he found in July lasted until surface waters cooled to nearly the deep-water temperature in October, when winds caused surface and bottom waters to mix. Crustacea then became more uniformly distributed. As lake waters cooled during winter, the top to bottom temperature remained about the same. In April, with lake ice melting, another mixing of waters occurred, followed by the July—August stratification. While discussing these top-to-bottom temperature stratifications, Birge introduced the term "thermocline" to graph the twice-yearly stratification changes that occur, though he thought he only provided an operation definition it, as was also the case for two other terms he coined, "epilimnion" and "hypolimnion."[200]

Like Stephen Forbes, Birge held both academic and government positions, but with little, if any, diminishment of research and publishing (Birge minimized teaching). However, Birge needed an assistant, and in 1900 hired Indianan Chancey Juday (1871–1944) as biologist for the Wisconsin Natural History Survey.[201] Juday had B.A. (1896) and M.A. degrees from Indiana University. In 1908, he also became instructor in zoology, and in 1909 he began teaching North America's first course in limnology; in 1931, he became professor of limnology and resigned from the survey. Birge and Juday's first important collaboration was on measuring dissolved substances in Wisconsin lakes.[202] A major question was why lakes differed significantly in their ability to support plankton. Birge and Juday devoted the remainder of their careers to finding answers. The quantity and chemical composition of plankton was important, as was penetration of sunlight into lake waters.[203] Lake Mendota is largest of the five interconnected lakes; Birge and collaborators studied Mendota limnology without a biological station. Eventually, Mendota became the most studied lake in the world (9781 acres, maximum depth, 83 ft, mean depth, 42 ft, 5.62 miles [9.04 km] long, 4.11 miles [6.61 km] wide). In 1925, Wisconsin limnologists established a station at Trout Lake, in far north, to facilitate comparative studies.[204]

Evelyn Hutchinson commented on Birge and Juday's achievement:[205]

> Birge's mature point of view is expressed in the concept of the heat budget, which, though derived from ideas of Forel and others, represented a highly original and important contribution because it first called attention to the lake as a natural system with an input and output. This point of view has tended to underlie most of what has been done in lake chemistry and in the study of primary productivity during the past three or four decades.

Illinoian James G. Needham (1868–1957) earned his B.S. (1891) and M.S. (1893) from Knox College, Illinois. He then earned his Ph.D. degree (1898) under entomologist John Comstock at Cornell University.[206] His specialty was aquatic insects. He taught at Knox College and Lake Forest College before returning to Cornell in 1907 as professor of entomology and limnology. He was the senior author of a popular textbook, *The Life of Inland Waters* (Needham & Lloyd 1916). Needham was the president of the Entomological Society of America in 1919 and vice-president of the Ecological Society of America in 1936. In 1918, both Needham and Birge contributed chapters to Henry B. Ward and George C. Whipple, *Fresh-Water Biology* (1121 pages). It covered bacteria, algae, vascular plants, and aquatic invertebrates but not aquatic vertebrates.

Five Great Lakes between Canada and United States contain almost a fifth of the world's fresh water. They became a valuable asset for the development of Euro-American economies. Significant adverse human impacts began with the fur trade in later 1700s, then, after the U.S. Civil War, harvesting timber, and extraction of minerals of the region, both of which were shipped on the lakes, leaving pollution behind. Two devastating fires began on October 8 1871 along Lake Michigan in Chicago and in Pestigo, Wisconin, with the Pestigo Fire causing the most deaths. Commercial fishing arose amid the pollution, and before 1900, several commercially fished species were in decline or gone. The complex environmental history of the lakes and the surrounding region is the subject of numerous studies, some of which I cite here.[207] My 2018 account did not single out any individuals for extensive discussion.

An outstanding limnological collaboration was *Limnology in North America* (1963), which encompassed Canada, USA, Mexico, and Caribbean Islands in 26 chapters by 32 authors.[208] A chapter is on the history of the American Society of Limnology and Oceanography,[209] an organization discussed in chapters 3 and 4. Here, I discuss three lengthy chapters as a sample: New England, Southwest, and Western Canada, 1870–1920. I have already drawn upon David G. Frey's chapter on Wisconsin in the Birge—Juday Era (longest chapter).

John Brooks and Edward Deevey's chapter 4, on New England includes a 4.3-page bibliography. Yet, their focus was mostly on limnological features, with much less discussion on limnologists who documented those features. There were limnological studies conducted by state fish and game agencies, but Brooks and Deevey cited no authors before 1936.[210] They stated that "About 1350 of New England's lakes have been the subject of at least primary biological surveys."[211] It was impracticable to cite authors of such studies. They cited tree species extinct since about 1920, but listed no authors of this information.[212] They reprinted numerous maps, the earliest two being dated 1905 and 1924.

Zoologist Gerald A. Cole, Arizona State University, Tempe, wrote chapter 14 "The American Southwest and Middle America." The first part (map, p. 394), extending from Southern Utah east to Oklahoma and south to the Mexican border and the second part (map, p. 399) extending from Mexico to Panama. That is, five pages each on Southern USA and five on Spanish-American states. The third part, of 31 pages, surveyed the bodies of standing water throughout this 3.8 million km2, with very little on running waters. USA has had a field station on Barro Colorado Island, Panama since 1903, and there is considerable biological literature which Cole did not use in his brief account on Panama.

Both T.G. Northcote and New Zealand-born Peter A. Larkin (1924–96) were at the Institute of Fisheries, University of British Columbia when they wrote "Western Canada,"(1963), which they subdivided into "Western Plains" and "Subarctic," but in practice they named the second division "British Columbia."[213] They also provided a section under each subdivision, "History of Limnological Endeavor," which are helpful. Publications for both regions began in 1922, when marine ichthyologist Archibald Huntsman (1883–1973) published an evaluation of the commercial fishery possibilities of Quill Lakes, Saskatchewan. Subsequent publications continued throughout the 1920s and beyond. In 1925, the Biological Board of Canada began supporting research by Russian immigrant, Alexander D. Bajkov (1894–1955), who joined fishery expeditions in 1914 and 1915 before fleeing from Bolsheviks during the Russian Revolution, earning a bachelor's degree from the Czechoslovakia College of Agriculture and Forestry in 1924 and then immigrating to Canada and earning his M.S. (1927) and Ph.D. (1931) from University of Manitoba. All four of his papers Northcote and Larkin cited (1963: 478) seem related to Manitoban commercial fishing. In 1929, he established a biological station at Gimli, Lake Winnipeg, which has sometimes been claimed to be Canada's earliest freshwater station. However, that honor is now known to have been at the Canadian Ministry of Fisheries' Georgian Bay Station at Go Home Bay, Lake Huron, established 1901.[214] In 1934, Bajkov moved to the Dominion Research Station, New Brunswick.

Donald S. Rawson (1905–1956), from Claremont, Ontario, was Canada's most productive limnologist-fisheries manager, despite living only 51 years.[215] He earned three degrees at the University of Toronto (1926, 1927, and 1929). His doctoral dissertation was on the bottom fauna of the Simcoe Lake, Ontario, on which he published two papers (1928 and 1930). Since his contributions came mainly after 1930, he is discussed in chapter 4.

D. MARINE ECOLOGY

Because of the commercial importance of oceanic fisheries, marine ecology during 1870–1929 seems to have received more financial support in this period than did plant, animal, and limnological ecologies. Much of those funds were invested in building and funding coastal research stations.

EUROPE

In 1910, two American zoologists, Charles Kofoid (1865–1947) and Chancey Juday (1871–1944), independent of each other, published accounts of European aquatic biological stations, fresh and salt-water. Kofoid's was a book—motivated by his planning to build such a station for University of California—published by the U.S. Bureau of Education.[216] Juday's report was a 20-page article + four plates. Both reports were illustrated with photographs of the stations. Later compilations followed: T. Wayland Vaughn, *International Aspects of Oceanography* (1937), on institutions; Homer Jack compressed his Ph.D. dissertation at Cornell University (1940) into *Biological Field Stations of the World* (1945), with a few illustrations and references on almost all stations; Robert Hiatt compiled the *Directory of Hydrobiological Stations and Personnel in North America* (1954) and the *World Directory of Hydrobiological and Fisheries Institutions* (1963), both containing photographs of buildings, station grounds, and ships; C.M. Yonge published a brief survey of marine stations (1956); Margaret Deacon discussed founding marine stations in later 1800s Britain (1993); and Patrick Scaps' *Histoire de la Biologie marine* has a chapter on early stations (2005: 73–88). A few small stations have arisen before 1870s but most arose after 1870.[217] Proceedings of international congresses on oceanographic history provide the histories of other stations.[218]

Professor of Zoology Pierre-Joseph van Beneden (1809–94), University of Louvain, Belgium, established the first marine research station in 1843, at Ostend, on his in-laws' oyster farm, which he named Laboratorie des Dunes.[219] It attracted biologists from Belgium, France, and Germany. In 1859, it was overshadowed when Victor Coste (1807–73), College de France, founded first the coastal station in France at Concarneau.[220] Other coastal stations gradually arose in Belgium and in France.[221]

In Ukraine, Russian Empire, developed one or two stations on Black Sea around 1871. A controversy arose over whether the first station appeared at Novorossiysk University, Odessa,[222] or at Sevastopol University's station.[223] The Sevastopol Station (SBS) had a female director, Sofia M. Pereiaslavtsva, 1878–88, who afterwards remained there as chief zoologist.[224] The early goal of that station was to inventory regional plants and animals. In 1890–91, Nikolai Andrusov organized expeditions to extend inventory into deep waters. In 1892, the Sevastopol Station came under the Russian Academy of Sciences. Sergi Zemov (1871–1945), who headed the SBS in 1902–14, "created the Russian school of hydrobiology and established a chair in hydrobiology at Moscow Agricultural University."[225] Zemov "extended the scope of the research from zoology to ecology and ethology."[226]

Kofoid began his book with a detailed account of Stazione Zoologica, Naples,[227] founded by German zoologist Anton Dohrn (1840–1909). The Dohrn's discussion was justified by the important research done there and its influence upon other marine stations. Dohrn had studied under Carl Gegenbaur and Ernst Haeckel.[228] In 1867, he visited a Scottish biologist living at the seashore, where he realized the advantages of residing at seashore to study marine life.[229] In 1871, he obtained permission from Naples to build a zoological station in a city park at Naples Bay.[230] With financial assistance from public and private interests—Darwin sent money and books-—and charging researchers a rental fee for research space, he built the foremost marine zoology station in the world (begun 1872, opened 1873). It attracted researchers from Europe and America, usually with government or institutional funds.[231] As its popularity grew, he had funds to expand his facilities. Stazione Zoologica founded three periodicals to publish studies done there.

France was the country with the most marine stations in Kofoid's book.[232] The largest of 19 French stations described was founded by Sorbonne zoology professor Félix-Henri de Lacaze-Dthiers (1821–1901) in 1872 at Roscoff, a fishing village in Western Brittany.[233] Later, it received a research vessel from l'Association française pour l'Advancement des Sciences. By 1880, it accommodated 17 biologists, and in 1881, it became affiliated with laboratories of the Faculté des Sciences in Paris, and later it was named Laboratoire

Lacaze-Duthiers.[234] Lacaze-Dethiers died in 1901 and was succeeded by long-time associate, Professor Yves Delagne (1854–1920), an experimental physiologist.[235] The station grew under able leadership until it accommodated over 100 students and investigators by 1908. Lacaze-Duthiers was the cofounder of *Archives de Zoologie experimental et générale* (1872), which published many monographs at Roscoff. Lacaze-Duthiers also founded Laboratorie Arago Banyuls-sur-Mer on the Mediterranean coast (1881).[236]

Kovoid devoted 10 pages to Monaco's Institut Oceanographic on the Mediterranean coast.[237] Prince Albert I (Honoré Charles Grimaldi, 1848–1922) was for a time in Spain's navy, and in 1873, he bought a schooner, mastered navigation, and in 1885 began oceanographic research that lasted the rest of his life.[238] He enlisted the services of Scottish chemist John Buchanan (1844–1925), who had served on the *Challenger* Expedition (see below). He became Albert's close collaborator until World War I.[239] In 1887, Albert sailed to the Azores, where he met Afonso Chaves (1857–1926), a military officer who was interested in sciences relating to the Azores; Albert and Afonso became close friends.[240] In 1910, Albert opened Musée Oceanographique de Monaco, which displayed biological specimens. Its director, Professor Paul Poriter (1886–1962), University of Paris, supervised over 100 Ph.D. dissertations.[241] Inspired by Dohrn's Stazione, Albert also responded to slow development of oceanography and marine biology in France.[242]

In 1886, Spain's navy launched Spain's first oceanographic expedition.[243] A doctoral student at the University of Madrid, Odón de Buen (1863–1945), was an expedition naturalist. That experience changed his interests. He earned his Ph.D. in 1887, and in 1889 he became a professor of zoology at the University of Barcelona until 1911, when he returned to University of Madrid as Professor of Natural History. While at Barcelona, he frequented Lacaze-Duthiers's marine Laboratorie Arago at Banyuls-sur-Mer. Lacaze-Duthiers and his laboratory were important influences on Buen when he established Laboratorio Biológico-Marino (1906) at Porto Pi, Majorca. In 1908, he began corresponding with Jules Richard, another of Albert's collaborators, who became the head of Albert's Insitut Océanographique (1911), and later Buen corresponded with Albert.[244]

Albert inspired King Carlos I de Bragança (r. 1889–1908) of Portugal to study oceanography.[245] He previously knew about oceanographic explorations in or near Portuguese waters.[246] In 1896, he used his vessel, *Amélia*, to begin research in coastal waters. His scientific colleague was Albert Girard, of French ancestry, who had grown up in Portugal and worked at the Lisbon Zoology Museum.[247] Their biological collection was accompanied by measurements of ocean waters. Carlos was interested in fisheries; a transition was underway from sail to steam fishing vessels, since the later had a greater harvesting potential. To study tuna fishery, he distributed questionnaires on fixed nets in 1899 and published his findings. An outstanding zoologist at the time was Augusto Nobre (1865–1946), from a wealthy family in Oporto, who was educated in Portugal, at Sorbonne in Paris, and at Station de Biologie Marine de Sète, University of Montpellier.[248] In 1890, Nobre returned to a place where he had studied in Portugal, Oporto Polytechnic Academy; there, in 1891, he organized a Department of Zoology and a university museum. In 1914, he established a Marine Zoology Station at Foz do Douro, where he published accounts of Portugal's invertebrates and fishes.

Northern Germany had a different situation than elsewhere. Kiel is on Kiel Bay, beside a canal between North and Baltic Seas. Kiel had Christian-Albrechts Universität, and in 1870 German government founded there *Kommission zur wissenschaftlichen Untersuchung der deutschen Meere*.[249] This commission established observation stations used by university zoologists. Stations were established elsewhere.[250] Helgoland Station, founded 1892, was important for zoological, ecological, and fisheries research.[251] The station began publishing *Wissenschaftliche Meereuntersuchungen* in 1894. Other important research came from zoologists at Naaturhistorisches Museum, University of Hamburg.[252] Berlin Aquarium opened a collecting station at Trieste on Adriatic Sea in 1870; in 1891, because of pollution, it moved to Rovinj and expanded its scope to include research.[253] Like Albert 1 and Carlo 1, Friedrich Alfred Krupp (1854–1902), monarch of a German industrial empire, used his own ships for biological exploration in Mediterranean Sea, from his villa on Isle of Capri.[254] Salvatore Lo Blanca, Stazione Zoologica, published Krupp's findings.

England had led the way in dredging from ships, but was slow to build marine stations.[255] A successful International Fisheries Exhibition in London, 1883, motivated zoology professor Ray Lankester (1847–1929), University College, London to encourage colleagues and Royal Society to promote a station. In 1884, a Marine Biological Association of the United Kingdom (MBA) was organized to make it happen.[256] Government land was allocated beside Navy's Citadel at Plymouth. Building began in 1886, and by 1887, the public raised 10,000 pounds, and government contributed 5,000 pounds and promised 500 pounds a year for 5 years. It opened in a fine building in 1888, while commercial fish stocks declined.[257] Annual operating funds were modest, but "early work at the laboratory was prolific and significant."[258] In 1895, Walter Garstang (1868–1949), who taught zoology at Universities of Oxford and Leeds, began bringing classes there to study marine biology. In 1897, he became Plymouth Laboratory Naturalist (with meager salary).[259] He showed that total fisheries catch in 1889–98 had only increased 30%, while trawler capacity had increased 300%.[260] Plymouth Laboratory was fortunate in having E[dgar] J. Allen (1866–1942) as director, 1895–1936.[261] He had studied at Leeds, London, and Berlin. He was a productive and capable researcher. Kiel's school had begun to lose its productive members before World War I and it never regained leadership after World War I.[262]

Plymouth Laboratory assumed world leadership under Allen, doing much with modest funds.[263] It not only

FIGURE 2.8 (a) Johan Hjort, about 1915. Web. (b) Alister Hardy. Web.

collected much diverse data but also achieved meaningful synthesis of it "during the late 1930s and 1940s into a virtually paradigmatic qualitative model involving biological, chemical, and physical processes that took place mainly at temperate latitudes . . . "[264] MBA retained the world leadership in plankton dynamics studies until 1958.[265] In 1892, two other laboratories had opened at Liverpool and Port Erin and later another at Lowestoft.[266]

Norwegian Johan Hjort (1869–1948) was son of a professor of ophthalmology at University of Oslo. He studied zoology at Munich and at the Zoological Station, Naples. He became head of Norway's fisheries research and not only professor of marine biology at Oslo University (1921–39) but also director of the marine station at Drøbak.[267] He is most remembered as coauthor with John Murray of *The Depths of the Ocean* (London, 1912).

Russia had at least six biological stations before 1910, some being freshwater.[268] Konstantin Saint-Hilaire (1866–1941), as student and later, worked at various European biological stations, and in 1903 he became Professor of Zoology at Juriew University (now University of Tartu, Estonia). In 1911, he established a mobile station at Kovda, Kandalaksha Bay, White Sea.[269] It lasted 30 years, without a permanent home, though both teaching and research occurred. German troops occupied Juriew University in 1918 and Russia's Revolution also disrupted the station's activities. All of Saint-Hilaire's scientific papers were destroyed in German attacks during World War II.

Romania lacked a marine station until Professor Ioan Borcea (1879–1936), who had a doctorate from the Sorbonne (1905) and who taught at Alexandru Ioan Cuza University, founded a marine station on the Black Sea coast at Agigea.[270] Borcea died there from septicemia, but in 1938 Homer Jack visited the station and judged "this is an extremely successful zoological station . . . "[271] Cushing summarized European end-of-the-century station foundings.[272]

Marine Stations: United States

In United States, controversy raged over the causes of decline in fisheries. In 1871, Spencer Baird (1823–87) got Congress to create U.S. Commission of Fish and Fisheries with him as Commissioner (without increase in salary at Smithsonian Institution).[273] That was America's first conservation agency.[274] Baird sponsored research at Woods Hole, Massachusetts, where he took his family every summer to escape Washington's heat and humidity.[275] Woods Hole had been a commercial fisheries port, about 1815 to about 1860, and later one home of Pacific Guano Works.[276] In 1871, Baird brought to Woods Hole Yale Professor of Zoology Addison Verrill (1839–1926). Baird asked him to survey the animals and their distributions in Martha's Vineyard Sound.[277] In five or six years, Verrill collected thousands of specimens of over 2000 species.[278] His survey enabled him to publish a substantial report.[279]

In 1882, the Commission built a fish hatchery at Woods Hole, including a hatchery ship, *Fish Hawk*, and in 1883 built a research vessel, *Albatross*, which Paul Galtsoff thought by 1921 had "probably did more significant work in oceanic research than any other vessel."[280] To this claim, German marine biologist Vera Schwach commented: maybe true for America, but untrue if it included Europe.[281] In 1885, the Commission built a Biological Laboratory at Woods Hole—influenced by Dohrn's Stazione.[282] Although Baird had responded to a fishing crisis in the North Atlantic, he knew the Commission had to be a national Agency to persuade Congress to provide funds. Thus, he also sponsored a survey of the Great Lakes fisheries.[283]

Marine Biological Laboratory (MBL) was built at Woods Hole because U.S. Biological Laboratory was there. It also was inspired by Dohrn's Stazione, though Keith Benson thought some historians over-emphasized this influence.[284] A philanthropist made available to Louis Agassiz Penikese Island with buildings, near Woods Hole, for a summer school for teachers of natural history, which opened in 1873.[285] It was successful; among the 28 men and 16 women attending were: Alpheus Hyatt, a founder of MBL; David Starr Jordan, later, leading ichthyologist and president of two universities; Charles Otis Whitman, a founder, and first director of MBL (all three discussed below).

Louis Agassiz died in December 1873, and his son, Alexander Agassiz (1835–1910), who became America's foremost oceanographer,[286] organized the school for summer 1874. Alexander, like his parents, was born in Switzerland. His mother died in 1848, and he went to Boston in 1849, joining his father.[287] Knowing no English, he was still enrolled in Cambridge High School to prepare for college.[288] His father taught natural history there, and physics and chemistry were also well taught. He entered Harvard in 1851, age 15. He graduated in 1855 and entered Lawrence Scientific School; he earned one master's degree in engineering (1857) and another in zoology (1862). He worked at Harvard's Museum of Comparative Zoology (MCZ) under his father,[289] but decided he could not raise a family on his salary. In 1866, he traveled with several Bostonians to Michigan's Upper Peninsula to inspect two copper mines and then bought shares in them.[290] The mine supervisors asked him to run them, which he did, 1866–68; he eventually became a millionaire. In 1874, he and a brother-in-law bought land at Narragansett Bay, Rhode Island, and Agassiz included a laboratory in his home where he could study marine animals.[291]

Alpheus Hyatt (1838–1902), from Washington D.C., studied under Louis Agassiz at Harvard.[292] He graduated in 1862 and joined Union Army during the Civil War. Between 1870–88, he was a part-time professor of zoology and paleontology at Massachusetts Institute of Technology, while also being the curator of Boston Society of Natural History. In 1877, he additionally became professor of biology at Boston University. He did research on living and fossil invertebrates. In 1879, he added a marine biological laboratory to his summer home at Cape Ann, Massachusetts,[293] and in 1880 he had a yacht built for him and his students for collecting and studying marine animals.[294] In 1882, he opened a small laboratory at Lobster Bay but after pollution made removal necessary, Baird persuaded him to move it to Woods Hole in 1887, where it was reorganized into Marine Biological Laboratory.[295]

William K. Brooks (1848–1908) graduated from Williams College in 1870, attended Agassiz's Summer School of Natural History on Penikese Island in 1873, studied under Alexander Agassiz, and earned a Harvard Ph.D. in 1875.[296] He joined Johns Hopkins University faculty, where he headed the Biology Department, 1894–1908. He studied morphology of marine invertebrates. With university funds, he ran the summer Chesapeake Zoology Laboratory, 1878–1906, the first such station in America.[297] Over 30 of his doctoral students worked there and 15 major articles from the station were published.

North Woodstock, Maine native Charles Otis Whitman (1842–1910) earned his B.A. degree at Bowdoin College (1868) and his Ph.D. at Harvard University.[298] He was the first American to work at Dohrn's Stazione, staying for six months in 1881–82. He was a pioneer in ethology.[299] Although he was a capable director, MBL struggled with limited funds during his tenure.[300] Three years of funds did come from the Carnegie Institution of Washington.[301] MBL began the *Biological Bulletin* in 1900 as an outlet for research done there. Its issues were traded for other biological journals published around the world.[302]

Toronto-native Frank Lillie (1870–1947) graduated from University of Toronto in 1891 and went to Woods Hole to work under Whitman.[303] In the fall, he followed Whitman to Clark University and in 1892 he followed him to University of Chicago, where Lillie earned a Ph.D. (1894). After teaching elsewhere, in 1902 Lillie returned to University of Chicago in 1902 as a faculty member. He succeeded Whitman as director of MBL in 1908 and as chairman of the Department of Zoology at Chicago in 1910.

> Lillie is most widely known for his vital roles in development of the Marine Biological Laboratory, one of the strongest influences on the rapid development of field of biology in America.[304]

Dr. Alfred Mayer (changed spelling to Mayor in 1918, 1868–1922), from Frederick, Maryland, earned an M.E. degree at Stevens Institute of Technology (1889) in physics and Sc.D. from Harvard (1897) in zoology.[305] Because he was a skilled animal illustrator, Alexander Agassiz hired him at the Museum of Comparative Zoology (MCZ), 1892–1900. He then became a curator of natural sciences at the Brooklyn Institute of Arts and Sciences, 1900–04. He had collected marine invertebrates for MCZ and he urged the Carnegie Institute of Washington to establish a tropical marine station.[306] He had chosen a site for one in the Dry Tortugas Islands (now a national park). He published two notes in *Science* supporting his idea.[307] The Carnegie Institute agreed and in January 1904 offered him the directorship.[308] He had buildings erected in summer 1904 and a boat built in Maine. In fall, he sailed it to his station, collecting specimens along the way. (It only remained seaworthy through summer 1910, then replaced by a larger, more expensive *Anton Dohrn*.) Tortugas Marine Laboratory was the first tropical marine station in the Americas. Mayer/Mayor invited promising biologists or geologists to study there. He was a productive invertebrate zoologist who set an example for others, and he also initiated *Papers of the Dry Tortugas Laboratory* (35 vols., 1908–42). His most

FIGURE 2.9 (a) William E. Ritter. Web. (b) Charles A. Kofoid. Web.

prominent work was *Medusae of the World* (3 vols, 1910). He broke new ground with quantitative ecological surveys of coral reefs.[309] Tortugas Laboratory was first to systematically study living coral.[310] In June 1922, Mayor was ill with tuberculosis, yet went to the station, where he fainted and drowned.[311]

Mayor's laboratory survived his tragedy. That history I have told elsewhere and do not summarize here.[312] I also omit here my previous brief discussion of Latin American stations.[313] For the U.S. Pacific Coast, only one example is included.

Wisconsinite William Ritter (1856–1944) grew up on a farm and was in college when he read Joseph Le Conte's geology textbook and transferred to University of California and completed his B.A. under Le Conte (1888). He received a scholarship to earn his M.A. and Ph.D. at Harvard (1891, 1893). He had returned to University of California, Berkeley and joined its biology faculty in 1891, until retirement in 1923. In 1903, he met newspaper magnate E.W. Scripps and sister Ellen Scripps, who agreed to fund a Scripps Institute of Oceanography, north of San Diego, which became part of U.C., San Diego, with Ritter as director. Ritter recruited Charles Kofoid (1865–47) as his assistant in running what became the Pacific Coast's most significant marine station.

Research stations: Canada

Canada's concerns for its fisheries arose parallel with U.S. concerns, which led to some parallel developments, one being construction of biological stations for fisheries research. Rev. Dr. Moses Harvey in 1892 was not the first to call for Canadian research stations but his urgings were addressed to the Royal Society of Canada and remained alive afterwards.[314] In 1893, Canada's government appointed English zoologist E(dward) E. Prince (1858–1936) as Commissioner and General Inspector of Fisheries.[315] He had been educated at St. Andrews, Edinburgh, and Cambridge universities. He soon recommended that Canada's government establish a marine station. The British Association for Advancement of Science (BAAS) met in Toronto in 1897, and in 1896, it appointed a committee on that matter to provide a report for the 1897 meeting. In 1898, the BAAS recommended Canada's government that a floating station be established on St. Lawrence River for five years (as in Petersen's portable Danish station). In summer 1899, the station opened under Prince's supervision, but at St. Andrew's on New Brunswick's coast, across St. Croix River from Maine. A group of researchers investigated different topics.[316]

After two summers, the station was towed to Canso, Nova Scotia, the largest fishing center on Canada's coast, where zoologists could buy specimens from commercial fishermen, and Whitman Brothers, fish merchants, lent zoologists their steam tug.[317] A third site to which the station moved was Malpeque, Prince Edward Island, center of oyster fisheries. In 1907, the station was being towed to Gulf of St. Lawrence, when the scow on which it rested leaked and it was abandoned. A fish house attic was rented instead. The second station the Canadian Minister of Fisheries established was the freshwater Georgian Bay Station, Lake Huron, previously discussed.[318]

The first Pacific marine station was on Vancouver Island, established by Josephine Tilden (1869–1957), Botany Department, University of Minnesota, Minneapolis, 1901–07. The only Canadian known to have worked there was McLean Fraser, in summer 1903.[319] The University of

Washington opened a marine station on San Juan Island, near Victoria. Those two stations motivated Canadians to build their own. Englishman George W. Taylor did museum work before immigrating to Canada in 1882.[320] In 1905, he became a member of the British Columbia Fisheries Commission, which recommended establishing a biological station to Canadian Marine Biological Board. In April 1907, Canada's Parliament voted funds for Atlantic and Pacific stations, both of which were begun in 1907 and opened in 1908. The Atlantic station was at St. Andrews.[321]

Zoologist Playfair McMurrich (1859–1939) earned B.A. and M.A. degrees at University of Toronto (1879, 1881) and Ph.D. at Johns Hopkins University (1885).[322] At Johns Hopkins, he conducted research at its coastal Zoology Laboratory, and later he suggested that Canada's Department of Fisheries establish marine stations in Nova Scotia, Gulf of St. Lawrence, and Vancouver Island. McMurrich spent 1884–1907 teaching in America and returned to University of Toronto in 1908. He did research at the new St. Andrews station during summer 1908 on sea anemones and cephalopods. Others appeared to study various taxa, mainly fishes.[323] The Pacific station was built at Departure Bay, near Nanaimo, Vancouver Island, headed by Taylor, who brought his own scientific library. A.W.H. Needler published a history of research at Nanaimo, 1908–58.[324]

Ontario native A(rchibald) G. Huntsman (1883–1972) earned two bachelor's degrees at University of Toronto (1905 and 1907) and then spent his career there.[325] In 1908, he spent two months at the new Nanaimo station for collecting and studying marine animals and he returned in summer 1909 to continue.[326] In 1911, he began a long association, as curator, 1911–19, and as director, 1919–34. During 1914–15, at the Prince's invitation, Norwegian fisheries biologist Hjort went to Canada and organized Canada's Fisheries Expedition of 1915. "Hjort's influence on Huntsman was immense."[327] ESA member Huntsman led other expeditions along the Atlantic coast which were more comprehensive than previous Canadian studies, lasting until 1923.[328] Huntsman obtained funds in late 1920s to hire a physical oceanographer for St. Andrews, Harry Hachey, who also helped Huntsman to publish his findings in 1954.[329] By late 1920s, physical oceanography in Canada was being funded comparably to marine ecology.

Research Voyages

British marine biology, initiated by Edward Forbes (figure 5.4) and Philip Gosse,[330] continued the dredging work of John Gwyn Jeffreys (1809–85), William Carpenter (1813–85), Wyville Thomson (1830–82), and Alfred Norman (1831–1918). Their work became the context for the *Challenger* Expedition.[331] That expedition was an early example of "big science," when the government made a major commitment to scientific endeavor.[332]

The greatest contribution to marine biology was the British global *Challenger* Expedition, 1872–76.[333] It was on a scale far greater than previous exploring expeditions and was an early example of "big science," with a 50-volume report. Although no other country attempted an expedition on that scale, its example encouraged other expeditions, such as that initiated by Professor Victor Hensen (1825–1924), Kiel University, for 15 weeks in 1889, to study plankton in both the north and south Atlantic.[334] Marine biological stations began appearing along the coasts of Europe and North America by 1870–1900.[335] An enormous volume, *Foundations of Biogeography: Classic Papers with Commentaries* (2004, xx + 1291 pp.) was edited by Mark Lomolino, Dov Sax, and James Brown. It includes only one marine ecology paper (Dana 1853) and also one part of *Foundations* is on the biogeography of islands.

It was a giant step from summer dredging around Britain to a 3.5-year voyage around the world, covering 68,890 miles. Strong backing from the Royal Society of London, when transoceanic cables were being laid, provided incentive for government support to gain information on ocean floors.[336] Thomson, Professor of Natural History at University of Edinburgh since 1870, was appointed Director of Civilian Staff. Captain George S. Nares, his 23 officers, and 240 men were highly experienced from previous surveying voyages.[337]

Canadian John Murray (1841–1914), at age 14 immigrated to his parents' homeland, Scotland, where he attended high school and University of Edinburgh.[338] He studied medicine but never earned a degree. He was a last-minute recruit for *Challenger*, yet in the long run, he was as important a naturalist, if not more so, as Thomson, since Murray saw the 50-volume report through the press after terminally ill Thomson resigned in late 1881.

Naturalists besides Thomson and Murray included biologistanthropologist Henry N. Moseley (1844–91), who afterwards became Professor of Anatomy at Oxford;[339] chemist John Buchanan (1844–1925), who afterwards continued oceanographic research (see above, at Albert);[340] German invertebrate zoologist Rudolph von Willemoes-Suhm (or Willimoes-Suhm, 1847–75) was highly regarded but died of erysipelas during the voyage; and Swiss artist John James (Jean Jacques) Wild (b. 1824), who was also Thomson's secretary. Later, Wild published *Thalassa: an Essay on the Depth, Temperature, and Currents of the Ocean* (1877) and *At Anchor, a Narrative* (1878). Wild's drawings illustrated several volumes of the *Challenger Report*.

Willemoes-Suhm sent reports to a former professor, Carl von Siebold, who published them. He wrote letters to his mother, which were also published. Extracts from his letters which Daniel Merriman translated show he was a happy colleague and enthusiastic about their discoveries.[341] He described invertebrate specimens dredged up from a thousand fathoms (6000 feet) and deeper.[342]

Challenger memoirs by Thomson (1878) and Moseley (1880) can sometimes be compared with Darwin's *Journal* (1839) from his *Beagle* voyage (1831–36). An example is that both ships stopped briefly at St. Paul's Rocks in the South Atlantic, lacking land plants, which all three authors discussed. Darwin had only used common names when reporting on the two species of nesting species: booby (a gannet) and noddy tern.[343] Thomson added scientific names: booby *Sula fusca* and noddy tern *Sterna stolida*.[344] Moseley, however, reported that Howard Sa[u]nders (1835–1907), specialist on gulls and terns, had discovered in *Challenger* specimens there were two species of noddy terns: *Anous stolida* and *A. melanogens*. In retrospect, Moseley remembered noddys nesting in two kinds of locations and he speculated that each noddy species lived in one kind of location.[345] Nesting noddys were seen there by Darwin in February, in May by Captain James C. Ross, and in August by *Challenger* naturalists. Moseley concluded noddys breed there all year.[346] He did not consider that each species might have a different breeding season. Thomson saw noddys use seaweed for their nests, which Moseley identified as *Caulerpa clavifera*. Noddys combined seaweed with feathers in nests, bound with cement, possibly from their stomachs.[347] Moseley thought the cement was Noddy dung.

Beagle spent a few hours at Fernando Noronha—some 300 miles to the southwest—where Darwin only had time to study the rocks. *Challenger* was there for over two days, giving Moseley time to collect land plants. Thomson saw boobys and noddys nesting but also frigate birds *Tachepetes aquila*, nesting high up on cliffs.[348] There were also three species of land birds: thrush *Nesocichla eremita*, bunting *Emberiza Braziliensis*, and nearly extinct waterhen *Gallinula nesiotis*. On two smaller islands, Inaccessible and Nightingale, rockhopper penguins *Eudyptes chrysocome* nested.

The *Challenger* Expedition focused primarily on deep-sea exploration and Thomson's memoir contained fold-out maps indicating dredging locations and temperatures at different levels of water columns at those stations. He itemized kinds of data obtained at each station, including 1441 water samples.[349] One could seek correlations between water temperatures and animals dredged up at different locations. Thomson pointed out some such correlations: *Fungia symmetrica* was the only coral found deeper than 1600 fathoms; it lived at temperatures of 1°—20°C; species from great depths are larger and more fragile than related species living in shallow water.[350]

After over six months study of deep-sea life, Murray concluded, "*Challenger* observations seem to prove that life is distributed all over the ocean floor."[351] His collection of plankton at surface, mid-level, and bottom provided clues into diurnal plankton migration.[352] *Challenger Report* announced an inverse relationship between distance from shore and numbers of benthic species. Thomson summarized 8 details of those findings which I quoted elsewhere.[353] Later, Baruch Kimor credited *Challenger Report* with first description of "free-floating animals that inhabit the open sea, later known as plankton."[354] However, in contrast to zooplankton studies, F.J. Taylor claimed *Challenger* Expedition:[355]

> . . . had not generated much useful ecological information on phytoplankton, the sole products in this area being a taxonomic monograph on the diatoms by Castracane (1886) and some passing observations by John Murray on the genus *Pyrocystis* (the first formal descriptions) and the photosynthetic nature of coccolithophorids.

Physical and chemical properties of seawater and bottom sediments were relevant to marine ecology.[356] *Challenger* returned home with 13,000 kinds of plants and animals, 4417 of which were new species. Thomson and Murray had to unpack, sort, and distribute classes or phyla of animals to experts in Europe and America to describe and publish their findings. Thomson invited Alexander Agassiz to come help, which he gladly did, to distract him from family deaths.[357] Murray and Agassiz formed a strong friendship at that time. The 50 large volumes Murray edited, 1885–95, were a monument to the expedition and to Murray.[358] "Murray seems to have outstripped his times, and certainly the statistical resources of his time."[359]

In "On the Structure and Distribution of Coral Reefs and Islands" (1880),[360] Murray challenged Darwin's theory (1842) that coral reefs and islands formed on sinking volcanoes.[361] He had not studied coral reefs and islands any more than Darwin[362] but he was quite interested in sea floor geology. He judged the accumulation of plankton skeletons on shallow elevations and that volcanoes were sufficient for coral substrates.[363] Agassiz found Murray's hypothesis more plausible than Darwin's.[364] Darwin read Murray's article and sent a rebuttal letter to Agassiz in response to observations he had sent Darwin,[365] but Darwin did not publish his rebuttal. However, James Dana (1813–95), at Yale, had accepted Darwin's theory,[366] and he defended Darwin's and his own ideas in "Origin of Coral Reefs and Islands" (1885). Agassiz, at Harvard, was dissatisfied with Dana's rebuttal, though he kept his doubts to himself.[367] It was impossible to determine if Darwin or his opponents were right without drilling to the foundation of a coral reef, which only happened decisively in 1950, which showed Darwin was mostly right.[368]

Thomson's decision to let specialists anywhere describe *Challenger* specimens within their expertise not only ensured high-quality work but also helped build an international community of oceanographers, including marine ecologists.[369] British investigators founded a Challenger Society in 1903, because British government felt it had done its part for oceanography by funding *Challenger* Expedition; later, marine biologists had trouble funding their work.[370] Murray was the only *Challenger* scientist who survived and joined (Buchanan also survived, but did not join).

German oceanography was stimulated by both German unification in 1871[371] and by *Challenger* Expedition conducted by a rival nation. The focus of *Gazelle* Expedition around the world, 1874–76, was on physical sciences. Science includes both rivalry and cooperation. When *Gazelle* met *Challenger* in Montevideo, February 1876, "their captains arranged their homeward courses in such a way that the best possible coverage of the Atlantic Ocean could be achieved."[372] Martin Ehrlich and Ramiro Sánchez provide a modern map of these two cruises in Argentine waters.[373]

Kiel is in Schleswig-Holstein; Prussia annexed Schleswig after wars with Denmark (1848 and 1864) and annexed Holstein after war with Austria (1866). Karl Möbius (1825–1908) developed a concept of biocönose at Kiel from studies on oyster beds (1877, English 1880).[374] Oyster fisheries in Germany were declining, and Möbius went to study British oyster harvesting, but found they also were in decline.[375] His research did not save Germany's oysters from over-harvesting.[376] Möbius left Kiel in 1887 to become director of a natural history museum in Berlin.

Schleswig-native Victor Hensen (1825–1924), who had studied medicine and physiology in Berlin and Kiel, became senior zoologist at Kiel University in 1887.[377] In 1863, he had become interested in marine life. In 1864, he became associate professor of physiology. In 1867–68 he served in Prussia's Landtag for four months, long enough to establish at Kiel a government *Kommission zur wissen-schlaften untersuchung der deutsch Meer.*[378] *Kommission's* goal was to learn why commercial fisheries in Baltic Sea were declining.

A series of coastal stations were built to monitor fish and environment between Denmark and Russia, and a naval vessel enabled Hensen to do research voyages, beginning 1871. Two Prussian North Polar Expeditions by ships during late 1860s and a deep-sea expedition in 1876–78 were discussed. None focused on biology.[379] Norway's Voringian Expedition, 1876–78, studied the migration patterns of fishes, especially cod.[380] Hensen decided to study the metabolic cycle of the sea to determine its productivity. His expedition collected physical, chemical, and biological samples and data. He designed a net to collect vertical samples for quantitative data. His "*Über die Bestimmung des Planktons oder des im Meere treibenden Materials an Pflanzen und Thieren*" (1887) was a quantitative study in which he introduced "plankton," coined by a philologist colleague to indicate floating organisms, plant and animal, living and dead.[381] In 1889, Hensen decided to compare North Sea plankton with Atlantic plankton and led a 15-week expedition from Baltic Sea to Greenland, to Bermuda, to Equator and back to Kiel.[382] He contributed to and edited a five-volume report (1892–1912) on expedition findings. An unexpected discovery was that plankton were much more abundant in cold northern waters than in warm tropical waters—a reverse of the land situation.

Hensen's plankton studies were beginnings of the quantification studies of food chains. However, he also had to respond to Ernst Haeckel's skepticism over the accuracy of his sampling techniques and his assumption that plankton were distributed uniformly within a given area.[383] Haeckel also though it unlikely that the northern waters were more productive than tropical waters.[384] Haeckel thought that food from rivers was adequate to support fish in North Sea, whereas Hensen suggested they depended on a plankton food chain.[385] Nothing wrong with Haeckel's skepticism, except he conducted no studies to test his speculations. Hensen henceforth spent much time and effort demonstrating the reliability of his methods and conclusions.[386] He established a close connection between plankton and fish abundance.[387]

Hensen's important findings attracted other capable marine ecologists—notably Karl Brandt, Friedrich Dahl, and Hans Lohmann—whose contributions I have discussed elsewhere.[388] In addition, for practical reasons, I offer the above discussions of British and German achievements as illustrative of marine ecology being conducted in other countries during that period, which I also have discussed elsewhere.[389]

By the 1890s, marine biologists realized that each European country doing its own research without considering what other countries were doing was inefficient.

FIGURE 2.10 (a) Victor Hensen in 1903. Murray & Hjort 1912. (b) Karl Brandt in 1931. Photo Rebisch. (c) Hans Lohmann in 1935. Photo Klatt. Mills 1989: 11, 45, 133.

Oceanography Prof. Otto Pettersson (1848–1941), University of Stockholm,[390] organized the first multinational survey of the North Sea in 1893–94 and at the Sixth International Geographical Congress, in London in 1895, he offered a resolution to continue and extend multinational surveys to support commercial fishing.[391] After much further discussion, a plan was developed, and was implemented in May 1902 to form an International Council for the Exploration of the Sea.[392]

English war veteran Alister Hardy (1896–1985) studied zoology at Oxford University (B.A. 1921, M.A. 1921, and D.Sc. 1938) and became a zoologist on R.R.S. *Discovery*'s Antarctic Expedition, 1925–27. Besides his observations, skilled illustrations, and specimen collections, Hardy invented his continuous plankton recorder on this voyage.[393] He then became the first professor of zoology of University of Hull, 1928–42. He finally published his notable memoir from the 1925–27 voyage in 1967.

In Brisbane, Queensland, a Great Barrier Reef Committee was formed in 1922 to study the origin, growth, and natural resources of the nearby coastal landmark.[394]

The Committee began by studying geology and geography, but by 1926 it turned to reef biology. It raised funds in Australia and in Britain for a year-long expedition of biologists. Yorkshire native Maurice Yonge (1899–1986) joined the British army for a year in 1917–18 and enrolled in 1919 at University of Edinburgh, where he earned two degrees (B.Sc. 1922 and D.Sc. 1927). He received a Carnegie scholarship to study at Cambridge, where he helped organize a Great Barrier Reef Expedition of 18 scientists, which he led in spring 1928. They reached Brisbane on July 9 and stayed on the reef until August 1929. Scientific findings were detailed later in four volumes but in 1930, Yonge published *A Year on the Great Barrier Reef: the Story of Corals and of the Greatest of Their Creations*. Well written, with numerous photographs, diagrams, and maps, it displayed his talents and helped him become a professor of zoology at Bristol, age 34, and in 1944, Regius Professor of Zoology at the University of Glasgow, and in 1946, Fellow of the Royal Society.

CONCLUSIONS

During the period from 1870s to 1920s, four sciences emerged that went beyond the natural history of previous periods: plant ecology, animal ecology, limnology, and marine ecology. These specializations arose primarily at institutions of higher learning and mostly in Europe and North America. Outgrowths of these developments were organizations that supported these sciences, beginning with the British Ecological Society (1913) and Ecological Society of America (1915). National governments also supported the growth of these sciences in various ways, especially by directly providing funds for research and expeditions, and indirectly by organizing supportive agencies, such as departments of agriculture, fish and wildlife, and more.

Journals were also crucial for advancing these sciences. At first, investigators depended upon existing general science journals for publishing their studies. The earliest still-existing science society, the Royal Society of London (1660), began publishing its *Philosophical Transactions* in 1665.[395] The second-oldest existing society was Académie royale des Sciences of France, organized in 1666.[396] It published two sets of journals: *Histoire de l'Académie royale des Sciences de Paris* and *Mémoires de l'Académie royale des Sciences de Paris*. The success of these societies and their journals inspired other countries to establish similar organizations and journals. In America, Benjamin Franklin suggested in 1740s organizing a comparable learned society, and the American Philosophical Society was organized in Philadelphia, which published both *Transactions* and *Proceedings*.[397] Later, specialized societies, such as the Linnean Society of London, arose and began publishing their own journals. Journals of British and American ecological societies initiated journals that built upon traditions established by earlier scientific societies.

All four of these sciences continued to flourish from 1890s to the present.

NOTES

1. Egerton 2009, 2012: 121–125, Nicolson 1996, 2013.
2. Humboldt 1820–21. See above, chap. 1 on Humboldt.
3. Christensen 1924–26: I, 617–665, 776–806, Rosenvinge et al. 1927, Müller 1976, Matagne 1998: 429–431.
4. Emberger 1936, Jovet 1954: 256–257, Stafleu & Cowan 1976–2009: 1, 843, Matagne 1998: 434–437, 1999: 49–50, Egerton 2013: 344–345, 348.
5. Bonnier & Flahault 1879.
6. Kerner 1863.
7. Flahault 1900.
8. Matagne 1998: 436.
9. Becking 1957: 426–463, Shimwell 1971: 45–47, Lévêque 2003: 18–19.
10. Gigon et al. 1981: 314.
11. Matagne 2009: index, Egerton 2013b: 345–346.
12. Becking 1957: 415, citing German ed. of Schouw 1822.
13. Tansley 1914, Ayres 2012: 83–86.
14. Tansley 1923: 7.
15. Sanders 1975, Cittadino 1990: 97–115, Egerton 2013: 346–347.
16. Cowles 1909: 356.
17. Goodland 1975: 243.
18. Dajoz 1984: 108–111.
19. Warming 1909: 12–13.
20. Warming 1909: 83–84.
21. Warming 1909: 84–91.
22. Sapp 1994: 30.
23. Kuhn 1970.
24. Stafleu & Cowan 1976–2009: I, 757–797, Stafleu 1978.
25. Hill 1935, Moore 1967, Stafleu & Cowan 1976–2009: 1, 515–516, Thomson 1982, Egerton 2013b: 349–350.
26. Thomson 1982, 1983: 15–17.
27. Sjörs 1981.
28. Söderqvist 1986. For a review of it see Egerton 1986.

29. Fries 1950: 70–71, Söderqvist 1986: 46–47.
30. Söderqvist 1986: 54, on Tansley, see below.
31. Fries 1950: 45, Söderqvist 1986: 88–89.
32. Söderqvist 1986: 89.
33. Söderqvist 1986: 90.
34. Söderqvist 1986: index.
35. Söderqvist 1986: 56–58.
36. Söderqvist 1986: 58.
37. Söderqvist 1986: index.
38. Söderqvist 1986: 108–115.
39. Egerton 1976, 1979, 2003, 2010, 2012: 138–143.
40. Tansley 1947, Godwin 1977: 4, 9.
41. Salisbury 1952, Desmond 1977: 472, Fairley 2004.
42. Ayres 2012: 65–66.
43. Oliver 1891: 149.
44. Thorsheim 2006.
45. Ayres 2012: 3.
46. Godwin 1977: 2–19, Ayres 2012: 48.
47. Tansley 1947: 133, Ayres 2012: 65.
48. Sheail 1987: 22–29, Aures 2012: 70.
49. Blackman & Tansley 1905: 199.
50. Blackman &Tansley 1905: 201.
51. Fischedick & Shinn 1993: 110.
52. Tansley 1947: 134.
53. Sheail 1987: 30–35, Fischedick & Shinn 1993: 110–111, Cameron & Matless 2011, Ayres 2012: 83–87 includes map of route.
54. Golley 1993: 14, Cassidy 2007: 51.
55. Pearsall 1964: 1–2, Duff & Lowe 1981: 141–144, Sheail 1987: 36–44, Fischedick 2000, Ayres 2012: 80–100.
56. Weiner 1988: 7.
57. Maycock 1967: 1031, Kajak & Pieczyska 1981: 287, Weiner 1988: index, Kozlowski & Sevedrzuski 2009.
58. Gleason 1975, Kohler 2002: 74–77.
59. Egerton 1976a: 340–341, McIntosh 1976: 353–354.
60. Ewan 1970, Tobey 1976, 1977, 1981: 9–45, Cittadino 1980: 175–178, Ewan & Ewan 1981: 18, Overfield 1993, Burgess 1996: 14–15.
61. Rodgers 1944, Stafleu & Cowan 1976–2009: 1, 558–559, Ewan & Ewan 1981: 49–50, Tobey 1981: 124–126, Cittadino 1993: 527–531, Burgess 1996: 31.
62. Egerton 1976a: 340, 350, Stafleu & Cowan 1976–2009: 3, 230–231, Cittadino 1980: 184–186, Burgess 1996: 70–71.
63. Kohler 2006: 100.
64. Wigdor 1974: chapter 3, Stafleu & Cowan 1976–2009: 4, 387–388, White 1981, Burgess 1996: 89.
65. Sapp 1994: 30–31.
66. Clements 1960, Sears 1973, Burgess 1996: 26–27.
67. Tobey 1981: 51–58, 70–75, Nicolson 1989, Overfield 1993: 148–149.
68. Pound 1954, Clements 1960, Ewan 1971a, Stafleu & Cowan 1976–2009: 1, 510–511, Worster 1977: 208–220, Hagen 1988, 1992: 20–28, 1999, Burgess 1996: 26–27, Cittadino 1997.
69. Cowles 1898.
70. Ewan & Ewan 1981: 43, Tobey 1981: 76–77, Burgess 1996: 26–27.
71. Pool 1954: 110–112.
72. Clements 1913: 327.
73. Sears 1958, Worster 1977: 206–208, Ewan & Ewan 1981: 50–51, Tobey 1981: 104–108, Engel 1983: 137–150, Mitman 1992: 16–19, Cittadino 1993, Burgess 1996: 31–32, Cassidy 2007: 5–29.
74. Cowles 1898, 1901a, b.
75. Egerton 2014.
76. Beneden 1876: 1.
77. Sapp 1994: 20.
78. Brooks 1998: 97–133.
79. Sapp 1994: 20.
80. Allee 1949: 32, Wilson 1975: 16.
81. Mayr 1975, Egerton 2013: 229, 2014: 60.
82. Semper 1881: 51–52.
83. Pequegnat 1958, Egerton 2007: 53.
84. Semper 1881: v—vi, 461–463.
85. Querner 1974, Kölmel 1981, König et al. 1981, Nyhart 2009: 125–160.
86. Meyer & Möbius 1865–72, Querner 1974: 432,
87. Möbius 1880: 721.
88. Möbius 1880: 727.
89. Reise 1980, 1990.
90. Shelford & Clements 1939: 5–6.
91. Reise 1990: 150.
92. Cohen 1994: 351–352.
93. Egerton 2007: 53, 2014: 63.
94. Egerton 2014a: 62–63.
95. Egerton 2013a, b.
96. Lannoo 2018: 2–90.
97. Howard 1932, Mills 1958: 94–97, Winsor 1972, Egerton 1976: 339–342, 2013a: 72–73, Kingsland 1985: 12–17, Hagen 1992: index, Golley 1993: 336–37, Lovely 1995, Burgess 1996: 42, Schneider 2000, Croker 2001.
98. Quoted in Croker 2001: 76–77.
99. Reprinted in 1925, 1977, 1991 in Real & Brown 1991: 14–27.
100. Forbes 1880: 3.
101. Forbes 1925: 539.
102. Forbes 1925: 542.
103. Forbes 1925: 543.
104. Forbes 1925: 548.
105. Forbes 1925: 549.
106. Croker 2001: 102.
107. Quoted in McIntosh 1985: 32.
108. Forbes 1913, Croker 2001: 122–123.
109. Egerton 2015: 205.
110. Howard 1934, Norland 1974, Bocking 1988.
111. Packard 1888, Bocking 1988.
112. Osgood & Grinnell 1947, Shor 1974, Sterling 1977, 1997b, 1999, Burgess 1996: 76–77.
113. Sterling 1977: 8–13.
114. Merriam 1873.
115. Barrow 1998: 50–54.
116. Hamilton 1955: 664–668.
117. Merriam 1890: 1, Sterling 1977: 204–240.
118. Stejneger 1890.
119. Egerton 2009b: 268–269, 2012: 124.
120. Merriam 1894: facing 228.
121. Sterling 1977: 56–90, 1989, Kohler 2006: 94–99.
122. Kimler 1990, Croker 1991: 7, Mitman 1992: 37–45, 1997, Burgess 1996: 99, Egerton 1999.
123. Shelford 1908, Croker 1991: 15–17.
124. Shelford 1911–12.
125. Shelford 1937: 302–304.
126. Cox 1979: 15.
127. Burgess 1977: 2, Croker 1991: 120–121.

128. Shelford 1926: v, Croker 1991: 121–124.
129. Palmer 1956, Mitman 1992: 35–37, Collins 1994a, Burgess 1996: 10, Harmond 1997a, Ilerbaig 1999, Kohler 2002: index, 2006: index, Egerton 2015: 14–15.
130. Adams et al. 1920.
131. Emerson & Park 1955, Schmidt 1957, Kimler 1988, Croker 1991: 31–34, Mitman 1991, 1999, Burgess 1996: 11, Sterling 1997a.
132. Banks 1985.
133. Gridgeman 1973, Kingsland 1982: 41–43, 1985: index.
134. Volterra 1976, Kingsland 1985: index.
135. Kingsland 1985: 116–206.
136. Pearse 1944.
137. Pearse 1927, 1952a.
138. Pearse 1952b.
139. Cox 1979, Southwood & Clarke 1999.
140. Cox 1979: 8–32, Southwood & Clarke 1999: 135–136.
141. Shelford 1913 [but here from edition 2] 1937: diagrams 3 and 6, Summerhayes & Elton 1923, Elton 1927: 58, Egerton 2007: 54–58.
142. Cox 1979: 8–38, Southwood & Clarke 1999: 135–136.
143. Cox 1979: 73–139, Chew 2006: 106–111.
144. Cox 1979: 77–105.
145. Shor 1972, Burgess 1996: 49, Grinnell 1917.
146. Cox 1979: 106–146.
147. Park 1949: 57.
148. Schoener 1989: 79–80, Hutchinson 1957.
149. Darwin 1872: 184–208, Romanes 1882: 360–366, Thorndike 1898: 1–10, Houck & Drickamer 1996: 13–54.
150. Richard 1973.
151. Thorpe 1956, reprinted 1965: 35.
152. Klopfer & Hailman 1967, Egerton 2016.
153. Bellevoye & Laurent 1897, Marchal 1897, Howard 1897.
154. Blanc 1912, Egerton 1962, 1978, 2014: 131–132, Bartola 1999, Vincent & Bertola 2012.
155. Berg 1951, Steleanu 1989: 100–110, 191–196.
156. In Blanc 1912: 133–148.
157. Forfel in Blanc 1912: 133–148.
158. Forel 1892–04, 1: vi.
159. Forel 1869, Lampert 1925: 13, Allee 1949: 41, Steleanu 1989: 191–196.
160. Forel 1892–1904: 3: 232–233.
161. Forel 1901: 239.
162. Croker 2001, Egerton 2014a: 64–65.
163. Forbes 1887, reprinted in 1925, 1977, 1991.
164. Elster 1974: 10–11.
165. Thienemann 1917, Wetzel 2004.
166. Chumley 1910: 750–751, Elster 1974: 9.
167. Nyhart 2009: 321.
168. Thienemann 1959, Ohle 1961, Steleanu 1989: 300–308, 385–404, Egerton 2008, 2014: 134–136.
169. Söderqvist 1986: 80.
170. Thienemann 1937, Söderqvist 1986: 80–81, Steleanu 1989: 382–385.
171. Rodhe 1974, 1975, Steleanu 1989: 405–411.
172. Hutchinson 1969: 17, Rodhe 1969: 50.
173. Naumann 1919, Hutchinson 1969: 17–18, Rodhe
174. Desmond 1977: 390, Perrein 1994: 181.
175. Perrein 1994: 187–188.
176. Flahault 1900: 444–445.
177. Gadeceau 1909.
178. Jack 1945: 43, Hiatt 1963: 194, Slack 2010: index.
179. Juday 1910: 1260, Kofoid 1910: 166–168.
180. Jack 1945: 43–44.
181. Hiatt 1963: 194.
182. Thienemann 1959: 129, translated in Rodhe 1974b: 540.
183. Thienemann 1959: 263.
184. Egerton 2007a: 60–61, 2014a: 136.
185. Juday 1910: 1274.
186. Cramer 1987: 68–70.
187. Elster 1974: 10, Hagen 1992: 7–10, Golley 1993: 36–37.
188. Kofoid 1898.
189. Chandler 1963: 98–99, Bocking 1987: 81–91, 1990: 478–495, Burgess 1996: 91.
190. Reighard 1894, Ward 1895.
191. Stuckey 1988, Bocking 1990a: 485–497.
192. Kofoid 1898: 394–395, Langlois 1949: 1–6, Gerking 1963: 239.
193. Howard 1932, Burgess 1996: 42–43, Croker 2001.
194. Kofoid 1898: 398–406, Bennett 1958: 165–166, Gunning 1963: 167–168, Bocking 1987: 101–117, 1990a: 472–478.
195. Kofoid 1903.
196. Mullen 1973, Burgess 1996: 62.
197. Sellery 1956: 4–14, Noland 1970, Beckel 1987: 1–2, Egerton 1987a: 86–94, 1999a, Burgess 1996: 16.
198. Juday & Hasler 1946: 470–472, Mortimer 1956, Frey 1963b: 15–34, Steleanu 1989: 279–299.
199. Birge 1898.
200. Birge 1910: 1005b, note.
201. Frey 1963: 4–6, Noland 1973, Beckel 1987: 4–5, Burgess 1996: 59–6, Egerton 1999b.
202. Birge & Juday 1911.
203. Birge & Juday 1922, 1934, Juday & Hasler 1946.
204. Frey 1963b: 5.
205. Hutchinson 1969: 109–110.
206. Cattell & Cattell 1938: 133–34, Chandler 1963: 99, Mallis 1971: 174–178, Burgess 1996: 80.
207. Beeton 1965, Beeton & Chandler 1963, Scott & Grossman 1979, Flader 1983, Egerton 1985, 2018, Ashworth 1986, Edwards & Regier 1990, Cronon 1991, Bogue 2000, Grady 2007, Riley 2013, Egan 2017.
208. Frey 1963a.
209. Lauff 1963.
210. Brooks & Deevey 1963: 119.
211. Brooks & Deevey 1963: 120.
212. Brooks & Deevey 1963: 122.
213. Northcote & Larkin 1963: 451, 464.
214. Johnstone 1977: 47–51.
215. Bocking 1990b: 40.
216. Goldschmidt 1951: 1127, Mullen 1973, Shor 1974, Burgess 1996: 62, Day & Mills 200013: 237–238.
217. Yonge 1956, Benson 1988a, b, c, Dexter 1988, Maienschein 1988, Kohler 2002: 42–44, Deacon 2003.
218. Sears & Merriman 1980, Lenz & Deacon 1990, Morcos & Zhu 2004.
219. Florkin 1970, Houvenaghel 1980: 668–669, Ellis 2005: 22, Scaps 2005: 79, Charlier 2013: 107.
220. Hiatt 1963: 73. 50–51, Elster 1974: 10, Steleanu 1989: 382–385.
221. Vaughn 1937: 104–105, Hiatt 1963: 15, Houvenaghel 1980: 669.
222. Bologa 2004.
223. Tokarev 2004: 223.
224. Bolog 2004: 211.

225. Tokarev 2004: 225.
226. Bologa 2004: 211.
227. Kofoid 1910: 7–32 + 8 plates.
228. Vaughn 1937, Ward 1974: 137–141, Oppenheimer 1978, 1980, Groeben 1985, Heuss 1991: 35–36.
229. Allen 1976: 209.
230. Dohrn 1872, Juday 1910, 1275–1277, Heuss 1991: 114–120.
231. Herdman 1923: 135–144, Ebert 1985: 173–174, Ghiselin 2008, Slack 2010: 63–69.
232. Kofoid 1910: 48–143. Also, Joquin Debaz's Ph.D. dissertation, 2005.
233. Kofoid 1910: 95–104 + 3 plates, Juday 1910: 1268–70, Appel 1973, Théodoridès 1990, Matagne 1999: 203–204.
234. Juday 1910: 1268–69, Jack 1945: 46, Hiatt 1963: 74–75.
235. Tétry 1971.
236. Hiatt 1963: 72–73.
237. Kofoid 1910: 37–47, Juday 1910: 1271–72, Herdman 1923: 119–133, Jack 1945: 53–54, Hiatt 1963: 135Deacon 1971: 382, 391–92, Schlee 1973, Mills 2009: 162–191.
238. Petit 1970, Mills 1983: 48–58, 1989: 116–117.
239. Kutzbach 1970, Carpine-Lancre & McConnell 2013.
240. Martins 1997.
241. Monnier 1975: 101, Day & Mills 2013: 250.
242. Mills 1983: 48–58, 2009: chapter 6.
243. Morcos et al. 2013: 269.
244. Morcos et al. 2013: 270–271.
245. Saldanha 1980, 1997, Carpine-Lancre & Saldanha 1992, Amorim 2009: 53–55.
246. Deacon 1997b.
247. Ramos 2006: 180.
248. Almaça 1997.
249. Kofoid 1910: 219–221, Mills 1989: 14, 1990: 20–21.
250. Kofoid 1910: 217–246, Roll 1990.
251. Bulnheim 1990.
252. Scheele & Hünemörder 1990.
253. Zavodnik 1990.
254. Müller 1990.
255. Deacon 1993.
256. Allen 1976: 211–212, Mills 1982: 6–8, 1983: 62–63, 1989: 190–199, Cushing 1988: 191.
257. Juday 1910: 1259, Hiatt 1963: 192.
258. Mills 1989: 195.
259. Hardy 1951, Holland 2011.
260. Cushing 1988: 193.
261. Kemp & Hill 1943, Mills 1989: 203–206, Deacon 2004a.
262. Mills 1989: 258.
263. Mills 1989: 189–257.
264. Mills 1989: 258.
265. Ellis 2005: 32.
266. Mills 1989: 197–201.
267. Sømme 2001.
268. Jack 1945: 59–60.
269. Goryashko & Fokin 2013.
270. Serpoinau & Malciu 2002: 273–274, Bologa et al. 2013.
271. Quoted in Bologa 2013: 175, Jack 1945: 56.
272. Cushing 1988: 191.
273. Allard 1967: 78–86, Smith 1994: 44–47, Egerton 2011: 156.
274. Hobart 1995: 4.
275. Allard 1967: 64–65, Pauly 1988: 128, Rivinus & Youssef 1992: 141–151.
276. Galtsoff 1962: 1–6.
277. Coe 1932, Hedgpeth 1957: 4–5, Verrill 1958, Shor 1976, Dexter 1997.
278. Verrill 1958: 65–69, Shor 1976, Rivinus & Youssef 1992: 89.
279. Verrill 1973.
280. Galtsoff 1962: 38–50, Allard 1967: 340, Spencer 2002: 288–290.
281. Personal communication, 2018.
282. Hiatt 1954: 141–142, 1963: 246, Schlee 1973: 67–73.
283. Allard 1967: 107–109.
284. Benson 1988a, b.
285. Lillie 1944: 15–22, Lurie 1960: 379–381, Maienschein 1989: 8–12.
286. Murray 1911, Goodale 1912, Agassiz 1913: 129–130, Dohrn 1913, Herdman 1923: 99–118, Lane 1969: 47–49, Dupree 1970, Ward 1974: 152–167, Lurie 1999, Dobbs 2005: 114–119.
287. Lurie 1960: 152, 170.
288. Agassiz 1913: 14–15.
289. Winsor 1991: 134–146.
290. Agassiz 1913: 47–85.
291. Agassiz 1913: 151–156.
292. Brooks 1909, Mayer/Mayor 1911, Gould 1972.
293. Dexter 1952, 1956.
294. Hiatt 1954.
295. Lillie 1944: 26–33, Hiatt 1954: 95–96, Dexter 1980, Maienschein 1985, 1989.
296. Conklin 1913, Edds 1970.
297. Benson 1985: 193–197.
298. Morse 1912, Conklin 1944, Ward 1974: 142–146, Mayr 1976, Maienshein 1985b: 187, Pauly 1988: 129–132.
299. Kohler 2002: 31, 45–47, Burkhardt 2005: 19–33.
300. Maienschein 1985c.
301. Lillie 1944: 53–59, Ebert 1985: 174–180.
302. Lillie 1944: 106–107, Russell-Hunter 1985.
303. Watterson 1973.
304. Watterson 1973: 356.
305. Stephens & Calder 2006: 2–26.
306. Davenport 1926, Colin 1980, Stephens & Calder 2006: 34–38.
307. Mayer 1903a, b.
308. Stephens & Calder 2006: 48.
309. Mayer 1918, 1924, Colin 1980: 142.
310. Fautin 2002: 446–447.
311. Stephens & Calder 2006: 147.
312. Egerton 2014c: 367–370.
313. Egerton 2014c: 374.
314. Johnstone 1977: 25.
315. Johnstone 1977: index.
316. Johnstone 1977: 33–34.
317. Johnstone 1977: 36.
318. See above and Egerton 2014b: 146.
319. Johnstone 1977: 63.
320. Johnstone 1977: 61–69.
321. Hachey 1965: 141–143, Johnstone 1977: 52–53, Mills 2014: 13–14.
322. Johnstone 1977: 24.
323. Johnstone 1977: 55–60.
324. Needler 1958, Hachey 1965: 143–145.
325. Hubbard 1997, Mills 2014: 14.
326. Johnstone 1977: 67–68.
327. Mills 2014: 14.
328. Johnstone 1977: 87–90.
329. Mills 2014: 16.

330. Egerton 2010.
331. Herdman 1923: 8–10, 37–68, Lane 1969: 45–46, Deacon 1971: 306–332, 2004a, b, Thomas 1971, Heppell 1973, Schlee 1973, Ward 1974: 95–107, Thomas 1976, Mills 1980b, 1983: 7–10, Rice & Wilson 1980: 378–384, Vanney 1993: 208–2113: 94–103, Rice 2004.
332. Burstyn 1968: 665.
333. Deacon 1971: 333–365, Linklater 1972, Schlee 1973: 99–103, Rice 1999: 290–319, Egerton 2014c: 375–384, 2019: 28–32.
334. Porep 1970, Rothschuh 1972, Schlee 1973: 233–236, Mills 1989: 22–28.
335. Kofoid 1910, Hiatt 1954, 1963, Benson 1988a, b, Egerton 2014: 348–374.
336. Burstyn 1968, 1972, Mills 1975: 6–12, Rozwadowski 2005: 6–13.
337. Rozwadowski 2005: 161–170.
338. Herdman 1923: 69–9, Merriman 1972a, Burstyn 1974, Ward 1974: 113–136, Ashworth & Mills 2004, Price 2004.
339. Bourne 1892, Desmond 1977: 452–453, Corfield 2003: 113–114.
340. Kutzbach 1970, Corfield 2003: 15–16, 251.
341. Merriman 1971.
342. Thomson 1878: II, 20–21.
343. Darwin 1839: 9.
344. Thomson 1878: II, 95.
345. Moseley 1892: 58.
346. Moseley 1892: 63.
347. Thomson 1878: II, 95.
348. Thomson 1878: II, 107–109.
349. Thomson 1878: I, xi—xii.
350. Thomson 1987: II, 133–134.
351. Murray 1895, quoted from Saldanha 20022: 237.
352. Deacon 1997: 342.
353. Thomson 1878: II, 133–134, Egerton 2014c: 382–383.
354. Kimor 2002: 211.
355. Taylor 1980: 516.
356. Mills 1975: 18–20.
357. Dobbs 2005: 160–166.
358. Scaps 2005: 44–53.
359. Mills 1983: 30.
360. Murray 1880a, b, 1889.
361. Egerton 2010b: 419–423.
362. Sponsel 2009: 303.
363. Dobbs 2005: 164–165.
364. Dobbs 2005: 165–166.
365. 5 May 1881, quoted in Dobbs 2005: 181–183.
366. Stanton 1971.
367. Dobbs 2005: 187–188.
368. Yonge 1980: 443–445, Dobbs 2005: 254–256, Sponsel 2009: 318–452.
369. Deacon 1997: 367–369.
370. Deacon 1990.
371. Roll 1990: 4.
372. Roll 1990: 7.
373. Ehrlich & Sánchez 1990: 488.
374. Querner 1974, Reise 1990, Egerton 2014a: 60–62, 2014c: 390.
375. Parsons 2002.
376. Neudecker 1990.
377. Porep 1970, Rothschuh 1972, Schlee 1973: 229–236 et passim, Mills 1989: 10–13 et passim.
378. Schlee 1973: 230–233, Ward 1974: 191–194, Mills 1989: 10–18, Kölmel 1990, Weiner 1990.
379. Roll 1990: 4–9.
380. Schwach 2007: 51–53.
381. Hedgpeth 1957: 4, Taylor 1980: 515, Mills 1989: 19.
382. Schlee 1973: 233–236, Mills 1989: 22–28.
383. Porep 1972, Schlee 1973: 229–237, Lussenhop 1974, McIntosh 1985: 53–54, Mills 1989: 29–31, Kimor 2002: 210–211.
384. Haeckel 1890.
385. Mills 1989: 41.
386. Mills 1989: 29–35.
387. Lenz 2004.
388. Egerton 2014c: 392–394.
389. Egerton 2014c: 394–407.
390. Schlee 1973: index, Mills 1989: index.
391. Went 1972: 3–4.
392. Went 1972: 13–16, Smed & Ramster 2002.
393. Hardy 1967.
394. Yonge 1930: 12–13.
395. Stimson 1948, Purver 1967, Sorrenson 2000.
396. Hahn 1971, McClellan 2000.
397. Bates 1965, McClellan 1985.

3 Some Specializations

A. ORGANIZING ECOLOGISTS BEFORE 1946

The first institution to sponsor natural history research was Aristotle's Lyceum before 300 BC.[1] That example was not duplicated for about 2000 years. What did happen in between was occasional studies on natural history, within different social contexts. An early example was Plinius the Elder (c. 223–79 AD), member of Rome's ruling class, who wanted to make Greek natural history available to a Latin audience.[2] He lacked the knowledge and sophistication to excel, but his *Natural History* fulfilled his goal and remained a popular reference in Antiquity and Middle Ages.

During the Middle Ages, natural history was taken up by some university scholars, including Bavarian nobleman Albertus Magnus (c. 1200–80), who joined the Dominican Order and studied theology in Padua and Paris and later taught in Cologne.[3] He played an important role in the transmission of Geek and Arabic science into European universities. He then built upon that knowledge in his own encyclopedic works, *De vegetabilibus libri VII* and *De animalibus llibri XXVII.*

Early Organizing

The earliest Renaissance scientific organizations arose in prosperous Italy in several larger city-states.[4] However, Italy's states fought each other, which discouraged science. The first permanent ones arose in the nation-states of Western Europe.[5] The very first, still active, was the Royal Society of London, which began in 1660 and received a royal charter in 1662.[6] Its *Philosophical Transactions* began appearing in 1665. France responded to its near rival by founding Académie Royal des Sciences on December 22, 1666.[7] Other European nations slowly followed their examples.

The United States also followed European examples, in the context of a growing nation. First was Benjamin Franklin's American Philosophical Society (APS), founded in 1743 or '44 in Philadelphia, but it did not flourish until it was revived in 1766.[8] Such national scientific societies arose in capital cities, which our British colonies lacked, but Philadelphia was the cultural capital. New York and Boston were rivals of Philadelphia and soon had their own comparable societies. In 1780, John Adams organized the Academy of Arts and Sciences in Boston. Ralph Bates (1963) devoted a chapter to American scientific societies in the later 1700s and James McClellan (1985) did so for Europe and America.[9]

National scientific societies were useful, but specialized sciences emerged and so did their own societies. McClellan mentioned three early ones without giving their founding dates before he discussed England's first, Linnean Society of London, founded in 1788.[10] It was prompted by botanist James Edward Smith's purchase of Linnaeus' manuscripts, herbarium, library, and specimens, from Linnaeus' executors in 1783, arriving in London in 1784.[11] Smith founded the Linnean Society and was its president until he died in 1828.[12] Linnaeus studied animals as well as plants, and the Linnean Society included zoology as well as botany.

More precisely, the Linnean Society was the first permanent scientific society, having a wealthy founder and patron. Insects are common and conspicuous, and entomology became an early focus for hobbyists and naturalists. London's first Aurelian Society was formed in 1745 and continued until March 25, 1748, when a fire destroyed its collections and library.[13] The second Aurelian society lasted till 1762–66, or longer. A Society of Entomologists of London existed till 1780–82. The third Aurelian Society lasted till 1801–06. However, its former members still published *Entomological Society of London Transactions*, 1807–12, first publication by an entomological society.[14] Society revivals occurred in 1822, 1824, and 1826. On May 3, 1833, eight entomologists met at the British Museum to form the Entomological Society of London, which lasted and sponsored its centennial history in 1933, by S.A. Neave and F.J. Griffin, with some illustrations.

In addition to societies specialized by discipline, there were also societies of scientists at a particular location. The Philadelphia Academy of Natural Sciences (PANS) was founded in 1812. APS was not a local society, so local naturalists founded the Academy. Its early history is told in two chapters of Patricia Stroud's biography of entomologist Thomas Say, since he was the sparkplug who ran it during its early years.[15] PANS' journal mainly published papers in biological sciences.

Explorations

A better-known example from the 1700s was the three voyages around the world led by British Navy's Captain James Cook (1728–79), who took along on his first voyage wealthy English botanist, Joseph Banks (1743–1820), with two assistants, who acquired a large collection of plants and some animals.[16] A general expectation was that Banks would publish lavish volumes describing his collections and his experiences. However, Banks' charismatic personality and wealth enabled him to dominate the natural history community in London. He was knighted and became the president of the Royal Society of London and was too busy for scholarly endeavors.

DOI: 10.1201/9780429342202-3

Banks' expectations for his retinue on Cook's second voyage were so grandiose that Cook could not accept it and Banks withdrew in a huff. Instead, a German botanist in London, Johann Reinhold Forester (1726–98), was recruited, and he brought along as his assistant his son Georg (1754–94).[17] Johann Forester was a capable and industrious naturalist with a difficult personality. Both he and Georg published accounts of their global voyage. Cook did not take a naturalist on his third voyage as he wanted to focus on geography; so, he depended upon a capable surgeon William Anderson (1750–78), who had been on his second voyage, to make natural history observations and collections, which he did.[18] However, both Cook and Anderson died during that voyage. The French,[19] Spanish,[20] and Russian[21] governments were inspired by Cook's voyages to send out their own voyages of exploration during late 1700s and early 1800s. British exploration continued.

France decided to fill in some gaps in the geography of the Pacific Basin which Cook had left. An expedition of two ships, was organized with naturalists, under the command of war veteran Captain Jean-Françoise de la Pérouse (1741–1788) in 1785. They stayed six weeks at Botany Bay, Australia, and left from there on March 10, 1788. He sent back to France his journal covering his observations before then. Both ships struck a coral reef beside a Pacific island and sank.[22] No one on this expedition was ever seen again. In 1791, a search expedition of two ships left France commanded by Joseph-Antoine d'Entrecasteaux and carrying three physician-naturalists and a gardener-botanist.[23] The captains of both ships died before the return, and when the returning ships reached Java in late-October 1793, the crews learned of the French Revolution and that France and the Netherlands were at war. The Royalist crew members identified revolutionaries among the crew, whom the Dutch arrested. The two ships were sold to pay for their expenses at Java and the prisoners were later released in a prisoner swap. The French collections and observations were retained by the French. Botanist Jacques-Julien Houton de La Billardière (1755–1834) published his account of the expedition in 1799, which was followed by the English and German editions. His "great botanical work, *Novae Hollandiae plantarum specimen*" appeared in parts, 1804–07. His "pioneering" *Sertum Astro-Caledonicum* appeared in 1824–25.[24]

In 1800, both France and Britain sponsored expeditions to Australia in hopes of gaining some strategic advantages over the other country.[25] Both expeditions encountered difficulties but also returned with important observations on geography and plants and with numerous plant specimens. The French expedition was commanded by Nicolas Baudin (1754–1803), who died at Mauritius on the return journey, and the British expedition was commanded by Matthew Flinders (1774–1814), who received a leaky ship for his expedition. On his return voyage, he stopped at Mauritius on December 17, 1803, a day after the French expedition left there for France. Britain and France were again at war and Flinders was arrested as a possible spy and not released until June 14, 1810. He wrote his memoirs during imprisonment, which was published on July 18, 1814—a day before he died.

There is a vast literature on the history of natural history explorations in America.[26] Lewis and Clark Expedition (1804–06) was an early American example of (Jefferson's) government sponsorship of overland exploration to the Pacific.[27] Its success—except no plan for handling collections afterwards—led to a series of other similar land explorations.[28] In the 1830s, the Americans decided it was time to emulate European voyages of discovery. Congress had not exerted much influence on preparations for land expeditions but an ocean voyage of discovery, 1838–42, was on a larger scale, which required Congressional appropriations.[29] In this case, it was not a matter of squeezing a few naturalists on a ship with other objectives—the naturalists' goals were the ship's goals. The U.S. Navy's commanding officer, Lieutenant Charles Wilkes (1798–1877), took his responsibilities very seriously, but he had a rigid personality and had difficult relationships with some scientists and officers.[30] He limited the number of scientists to seven, plus two artists.

Later Organizations

After the expedition, the specimens were deposited at the U.S. Patent Office. The Smithsonian Institution was established in 1846 and its first secretary (director), Joseph Henry, a physical scientist, resisted accepting expedition specimens because he did not want the Smithsonian to become a warehouse of specimens and he thought its maintenance would consume too much of his annual Congressional appropriation.[31] However, in 1857 Congress provided funds to construct display cases, moving expenses, and annual maintenance and the transfer from the Patent Office to the Smithsonian occurred.

Pennsylvanian Spencer Fullerton Baird (1823–87) earned two degrees from Dickinson College (B.A. 1840 and M.A. 1843), then went to New York to study medicine, but soon abandoned it in favor of self-study to become a zoologist.[32] He succeeded and became a professor of natural history at Dickinson. He was productive in publishing, which helped his successful application to become assistant secretary at the Smithsonian in 1850. He was a strong advocate of collections of animals and was undoubtedly behind Congress' allocation of funds for the U.S. Expedition Collection's transfer to the Smithsonian. Upon Henry's death in 1878, Baird became his successor and remained secretary for the rest of his life.

Aquatic research institutions appeared in Europe before 1870 but spread mostly after 1870.[33] Land-oriented stations developed more slowly as university-based ecologists often investigated lands nearby. Robert Kohler's *Landscapes and Labscapes: Exploring the Lab-Field Border in Biology* (2002) discussed examples.

In 1871, commercial fisheries in the North Atlantic were declining, and in response, Baird proposed to Congress to create the U.S. Fish Commission to sponsor fisheries research, with himself as head, with no increase to his Smithsonian salary. To gain votes from senators from Great Lakes, Gulf of Mexico, and Pacific states, they all became included in his proposal, which Congress passed.[34] That commission published annual reports until the Franklin Roosevelt Administration in 1940 merged it with the Bureau of Biological Survey to establish a new U.S. Bureau of Sport Fisheries and Wildlife. The name was changed to U.S. Fish and Wildlife Service in 1956. Topical reports continued being published and a U.S. fisheries laboratory was established at Woods Hole, on Cape Cod, Massachusetts.

The first organization relating to specific ecological sciences seems to have been the International Council for the Exploration of the Sea (ICES), organized in 1902 by west European nations, motivated by stresses on commercial fisheries in the Atlantic and adjacent waters. The ICES encompassed several oceanographic sciences and was important in advancing marine ecology.[35] Beginning with a few countries, it later expanded on both sides of the Atlantic Ocean and continues to flourish.

In 1908, an International Congress of Geography met in Geneva, and plant ecologist Carl Schröter (1855–1939) organized a tour of Swiss types of vegetation.[36] That tour inspired a British ecologist, Londoner Arthur Tansley (1871–1955).[37] London was the home of the Royal Society of London (founded 1660), the Linnean Society of London (founded 1788), and the British Association for the Advancement of Science (founded 1831).[38]

Tansley had studied botany at both University College London (UCL) and Cambridge University. In 1898, he had studied the German edition of Eugene Warming's *Plantesamfund* (Danish 1895, German 1896), and he compared English plant communities with those Warming described in Denmark. In 1899, he taught a course based on Warming's book. In 1902, Tansley used his own money to find a new journal, *The New Phytologist*, which he edited for 30 years, and it still continues. In its first volume, he proposed that full-time botanists coordinate local botanical surveys through a British Vegetation Committee (BVC). In 1903–07, he accompanied his former teacher at UCL and now colleague, Frank Oliver, on summer field trips in England and France. In 1908, he suggested that the BVC organize a botanical tour of Britain similar to Schröter's in Switzerland.

In 1909, Tansley edited BVC's findings, as *Types of British Vegetation*. *Types* became a convenient introduction for a very successful International Phytogeographical Excursion in August 1911. In 1913, he led the Vegetation Committee members to form the British Ecological Society (BES), the world's first such society. He also founded its *Journal of Ecology* in 1913. Since he was the first BES president, he only became its editor after his term ended in 1916.[39] In 1927, he became a professor of botany at Oxford University.

The United States also had relevant general scientific organizations before ecological ones. The Smithsonian Institution was founded in 1846, and became important for natural history.[40] The American Association for the Advancement of Science (AAAS) was founded in 1847 and first met in 1848. The Botanical Society of America (BSA) was founded in 1893, and in 1906, it merged with the Society for Plant Morphology and Physiology (founded 1896). The BSA celebrated its Golden Jubilee in 1956, and plant ecology was well represented in its jubilee volume.[41] On March 27, 1914, entomologist Robert Henry Wolcott (1868–1934), University of Nebraska,[42] wrote to animal ecologist Victor Shelford, then at University of Chicago, suggesting formation of a Midwestern society of plant and animal ecologists.[43] Shelford discussed it with Chicago plant ecologist Henry Cowles, who was enthusiastic, but suggested a national society that could be organized at the next AAAS meeting.[44] Word was passed around, and there was an organizational meeting on December 30, 1914, with 23 potential members present. All known American and Canadian ecologists were notified of an organizational meeting, December 28, 1915, when about 50 ecologists were present and was supported by about 50 letters from other ecologists.[45] There were 286 charter members, but membership grew slowly until about 1950.[46] Shelford was the first president, William Morton Wheeler, the first vice-president, and Forrest Shreve, the first secretary-treasurer, with 125 members present. Despite USA entering World War I in 1917, ESA then began publishing its *Bulletin*.[47] Daniel MacDougal, at the Carnegie Institute's Desert Laboratory, offered to transfer *Plant World* to ESA.[48] ESA accepted and renamed it *Ecology*. Its first issue was in January 1920 and *Ecological Monographs* began in 1931.[49] In Limnology, two informal schools developed at the universities of Wisconsin and Yale, and two informal schools in Canada at the universities at Toronto and Saskatoon.[50] The Limnological Society of America was organized in 1936 with 221 charter members.[51] In 1948, it expanded and renamed itself the American Society of Limnology and Oceanography.

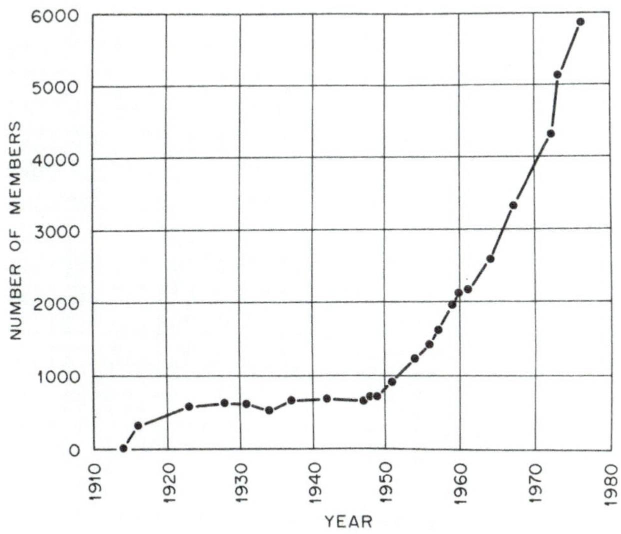

FIGURE 3.1 ESA membership, 1915–76. Burgess 1977: 3.

Besides general ecological organizations, ecology was advanced by educational institutions, biological stations, and voyages of marine exploration. Universities arose in Europe during the Middle Ages, and in modern times, employed ecological faculty and students. In two previous papers, I have discussed the prominent marine stations in Europe and North and Latin Americas.[52]

Botanical gardens, herbariums, zoos, and museums have been less numerous but still important both for the employment of ecologists and as reference institutions. Charles Gager compiled the *Botanic Gardens of the World: Materials for a History* (1937).

The Woods Hole Oceanographic Institution (WHOI) was founded in 1930 at Woods Hole—near the Federal Fisheries Laboratory and Marine Biological Laboratory. It was not primarily oriented toward marine biology, though its first director was Harvard marine biologist, Henry Bigelow (discussed in chapter 4). The Third International Congress on the History of Oceanography was held in 1980 at WHOI, which was WHOI's 50th anniversary. Six participants in the Congress spoke on WHOI history: three on its origin[53] and three on its growth and influence.[54] Later, Vicky Cullen wrote its 75-year history.[55]

The history of the Georgia Institute of Ecology indicates that it was founded in 1940, when animal ecologist Eugene Odum (1913–2001) joined the University of Georgia faculty.[56] The institute developed at Athens with the postwar funding from the Atomic Energy Commission.

Reference works have also been important. James McClellan wrote a historical survey, *Science Reorganized: Scientific Societies in the Eighteenth Century* (1985). Edward Kormondy and Frank McCormick edited the *Handbook of Contemporary Developments in World Ecology* (1981), with 34 chapters on nations written by native ecologists. They failed to find ecologists in Canada, Finland, France, India, Ireland, and Spain who would contribute a chapter. Now, Canada has a historian of ecology, Professor Stephen Bocking, who discussed the history of Canadian ecology, thus emphasizing resource management.[57]

General science reference works also contain information relevant to organizing ecologists: Arne Hessenbruch edited *Readers Guide to the History of Science* (2000);[58] Charles Gillispie edited *Dictionary of Scientific Biography* (16 vols., 1970–78), with later volumes edited by others; *American Men and Women of Science* includes Canada (ed. 1, 1906, many later eds.).

Science arose in some countries while they were colonies. George Basalla wrote a brief, but important article, "The Spread of Western Science," (1967) in which he explained that the level of science in European colonies depended upon the level of science in the mother country. From the 1600s on, Britain and France were leaders in European science and they also had colonies where science was best developed. Spain lagged behind but tried to follow its example in its own colonies. Centre National de la Recherche Scientifique sponsored an International Colloquium in 1956, papers from which appeared as *Les Botanistes François en Amerique du Nord avant 1850*.[59] James McClellan compiled a concise bibliography on "Colonialism and Science" (2000) and Roy MacLeod edited substantial articles on *Colonialism and science* (2001).

The above discussion is mostly about UK and USA, which led the world in developing organizations and institutions that assisted naturalists and ecologists develop ecological sciences before 1946. In a longer discussion on this, it was practicable for me to briefly survey comparable developments in other countries, which readers can find online.[60]

B. SYMBIOSIS STUDIES

Symbiosis is a term that identifies persistent relationships between species.[61] "Symbiosis is the most relevant and enduring biological theme in the history of our planet."[62] A valuable early collection of articles was collected into two volumes by Mark Henry (1966–67): volume 1 on microorganisms, plants, and marine organisms and volume two on invertebrates, birds, ruminants, and more. Its authors lived in Europe, United States, and one in Australia. Jan Sapp's history of symbiosis is useful, though of incomplete coverage.[63] He did not tell us who first commented that birds and mammals that eat fruit are important disseminators of seeds. He did not discuss insects and hummingbirds that pollinate flowers. Two small, not-quite textbooks, entitled *Symbiosis* appeared in 1970, the smaller by William Trager, Rockefeller University, and the larger by Thomas Cheng, Lehigh University. In addition, in 1970, Clark Reed, Rice University, published *Parasitism and Symbiology: an Introductory Textbook*. In 1971, Cheng edited a collection of a dozen articles on diverse aspects of symbiosis.[64] About three decades later, two professors at Worcester State College, Massachusetts, Surindar Paracer and Vernon Ahmadjian, collaborated in writing a comprehensive textbook, *Symbiosis: an Introduction to Biological Associations* (ed. 2, 2000). Joseph Seckbach, Hebrew University of Jerusalem, edited another collection of articles, *Symbiosis: Mechanisms and Model Systems* (2002), which is a virtual encyclopedia of 774 pages of text, containing a multitude of articles by scientists from Europe, America. Israel, and Japan.

In antiquity, there were isolated examples of symbiosis reported, and also a concept to explain it. The earliest written example was in the Egyptian Ebers medical papyrus (ca. 1500 BC) that contained a prescription for killing intestinal round worms.[65] The earliest concept to account for symbiosis was the balance of nature. Examples were cited but the concept remained unnamed. The earliest example known was from traveler-historian Herodotos of Halicarnassus (died ca. 425 BC), whose *History* stated that Divine Providence had contrived to have predatory animals produce fewer offspring than their prey to prevent predators from eating all their prey.[66] He also reported that Egyptian plovers [*Plovianus aegypticus*? *Hoplopterus armatus*?] picked leeches from inside mouths of crocodiles, which in appreciation never harmed the plovers.[67] This illustrated both the symbiosis and balance of nature (neither concept named).[68]

More sophisticated than Herodotos' *History* were zoological works from Aristotle's Lyceum. *Historia Animalium* attributed to Aristotle (384–322 BC),[69] reported *Pinna* clams and *Pinnotheres* crabs together but did not explain why.[70] (Current understanding: clam receives no benefit and crab eats some of clam's food.) Apiculture arose in ancient Egypt, Mesopotamia, and maybe elsewhere.[71] *Historia Animalium* contained a guide to apiculture and stated that when a bee visited one kind of flower, it only visited that kind until it returned to hive.[72] That observation was correct and important but no interpretation was given. Insect parasites were described, and although insect copulations were observed, they nevertheless were said to arise spontaneously. In one example, a parasite supposedly had a compensatory effect: "people with lousy heads get fewer headaches."[73]

Aristotle presided over the Lyseum for about a dozen years and he was succeeded by Theophrastos (ca. 371-ca. 287 BC), who headed the Lyceum much longer.[74] The Lyceum's botanical works were attributed to him. Theophrastos knew mistletoe is a distinct kind of plant, though he did not call it a parasite:[75]

> . . . some plants are unable to sprout—either the seeds or the plants—in the ground, as the ixia [mistletoe, *Loranthus europaeus*], *the stelis* [*Viscum album*], and the hyphéar . . .
>
> Some assert that all of them are a single natural entity, but because they grow on different plants they are also considered to be different: so the hyphéar occurs on the silver-fir and pine, and so too the stelis, whereas the ixia occurs on oak, terebinth, and a number of other trees.

Renowned medieval scholar Albertus Magnus (ca. 1200–80) wrote *De vegetabilis et plantis* after 1250, without access to botanical works by Theophrastus.[76] His observations on oak galls are available in English[77] and he reportedly described mistletoe, (but knowing it was parasitic?), without citation.[78] What is clear is that although parasites were identified as such during medieval times, no helpful insights emerged.

Parasitology is relevant for symbiosis, and elsewhere I discussed its history, 1700–1900, for both plants and animals.[79] Carl Linnaeus (1707–78) provided his students at Uppsala University with information for writing doctoral dissertations, which he published under their names but are now attributed to him. "The Oeconomy of Nature" (translated title) was listed by Isaac Biberg (1749, English, 1775) and was the first paradigm of ecology.[80] Its account discussed seed dispersal, some by wind and others by animals:[81]

> Berries and other pericarps, are by nature allotted for aliment to animals, but with this condition, that while they eat the pulp they shall sow their seeds; for when they feed upon it they either disperse them at the same time, or, if they swallow them, they are returned with interest; for they always come out unhurt.

Linnaeus cited a variety of examples.

A new aspect of symbiosis and balance of nature during mid and late-1700s was studying pollination. Bees and butterflies obviously visited flowers to obtain food but the benefits of flowers was not obvious. German botanist Rudolf Camerarius (1665–1721) published hybridization experiments demonstrating sexuality in plants (1694), which won slow acceptance during 1700s. Arthur Dobbs was an Irish Protestant ruling class, who was a governor of North Carolina, 1754–65, who afterwards returned to his estate, where he read some of René Réamur's *Mémoirs pour server à l'histoire des insects* (6 vols., 1734–42), while also raising honey bees.[82] Réaumur had disagreed with Aristotle's claim that once a bee visits a flower, it further visits only that kind until it returns to the nest. Dobbs found that Aristotle was correct and provided convincing proof in a paper published in the *Transactions of the Royal Society*.[83] Despite being the best journal, it attracted no attention.[84] Dobbs did not claim to have discovered that insects pollinate flowers while gathering food. He suggested that Providence must have intended that.

German botanist Hermann Müller (1829–83) wrote an encyclopedia *Befruchtung der Blumen* (1873, English 1883), with a historical introduction that only began with a school rector and amateur botanist, Christian Sprengel's *Das entdeckte Geheimniss der Naturu im Bau und in der Befruchtung der Blumen* (1793).[85] Sprengel (1750–1816) concluded that some flowers cannot be fertilized without insect assistance and so he studied flower morphology to learn how flowers guide insects to ensure fertilization. Sprengel did not ask why cross-fertilization seemed necessary. Three other botanists subsequently published that it was needed to produce healthy plants.[86] Their findings only attracted interest after Darwin published the *Origin of Species* (1859), with a new context for pollination studies.

Darwinian Revolution

Darwin's theory of evolution by natural selection is logically incompatible with the Linnean economy of nature.[87] However, that incompatibility was overlooked by Darwin and his associates. Darwin noted what we call symbiotic relationships, such as that between flowers and pollinating insects.[88] His hypothetical food chain—clover, bees, mice, cats—illustrates two symbiotic relationships (clover—bees, bees—mice), though aspects of the chain are of limited validity.[89] He made detailed studies of insects fertilizing orchids, which put symbiotic relations within an evolutionary context.[90] The impact of Darwin's revolution upon symbiotic studies was a challenge to see all such relationships as having evolved over time and how might that have happened.

Two contemporary naturalists, who had read Darwin's *Journal of Researches* (1839) Alfred Russel Wallace (1823–1913) and Henry Walter Bates (1825–92), left Britain in 1848 to explore Amazonia; Wallace stayed until 1852 (and traveled to Malaya) and Bates until 1859.[91] Both were quick

FIGURE 3.2 (a) Pyotr A. Kropotkin in 1864. Web. (b) Pierre-Joseph van Beneden. Web. (c) William Morton Wheeler in 1910. Web.

converts after reading Darwin's *Origin*. In 1861, Bates read to the Linnean Society of London a paper on *Leptalis* butterflies mimicking the color patterns of Heliconiidae butterflies, providing new evidence for Darwin's theory.[92] Batesian mimicry occurs when a palatable species evolves to resemble an unpalatable species. Remarkably, Wallace found a parallel case among Papilionidae of Malaya, which he reported to the Linnean Society in 1864.[93] Wallace announced the laws of mimicry based upon his and Bates' discoveries.[94] German zoologist Fritz Müller (1822–97) challenged Wallace's laws based upon his own discoveries in Southern Brazil (where he had settled) and explained in *Für Darwin* (1864, English 1869). Müllerian mimicry differed from Batesian mimicry in that both species were about equally distasteful to, and by coming to resemble each other, predators learned more quickly their distastefulness. Several species of Euglossini orchid bees exhibit Müllerian mimicry.[95]

Recent Times

There is no general consensus on the scope of symbiosis. That social species—some mammals, some birds, some fish, ants, honey bees, termites—have intraspecific cooperation was known since antiquity.

Intraspecific Cooperation

In the second edition of their textbook (2000), Surindar Paracer and Vernon Ahmadjian defined symbiosis as "an association between two or more different species of organisms."[96] Others simply disagreed about excluding intraspecific cooperation.

French philosopher Alfred Espinas (1844–1922) wrote a doctoral dissertation in Paris on *Des Sociétés Animales* (1878, English 1935).[97] Published, it became the first comprehensive survey of mutualism within species: coral, social insects, fish, reptiles, birds, mammals, and humans. Russian Prince Pyotr (later, Peter) Kropotkin (1842–1921), a naturalist and exiled anarchist, wrote "Mutual Aid among Animals" (1890), which he later expanded into a book, *Mutual Aid* (1902), which drew upon his Siberian observations and was widely read.[98] He had been inspired by a lecture on mutual aid by Ichthyology Professor Karl Kessler, St. Petersburg University. Kropotkin's hero was Alexander von Humboldt, whose explorations inspired Kropotkin to explore Siberia for five years. He returned to St. Petersburg University in 1867, where he spent another five years studying, until he was arrested and imprisoned for anarchist activities. After a year, he escaped to England in1876. He saw the same impulse for cooperation among humans he had witnessed in mammals and birds and decided if animals could live without government, so could humans. Kropotkin's autobiography (1899) did not mention Espinas.

Milwaukeean William Morton Wheeler (1865–1937) had an early interest in nature and became involved in several museums and universities before settling at Harvard University in 1908.[99] In 1911, he argued that an ant colony can be considered as an organism, or, in 1923, superorganism.[100] In 1921, he delivered six lectures at the Lowell Institute in Boston. Those lectures became chapters in *Social Insects among the Insects* (1923) on beetles, wasps, bees, ants, parasitic ants, ant guests, and termites. His fifth chapter, on ants and ant guests, showed how intraspecific cooperation can lead to interspecies relationships.

Indianian W. Clyde Allee (1885–1955), Quaker, attended a Quaker college, Earlham (B.S. 1908) and was a graduate student in zoology at the University of Chicago under Victor Shelford (M.S. 1910 and Ph.D. 1912). After holding several positions elsewhere, he returned to Chicago in 1921.[101] He acknowledged that his pacificism influenced his choice

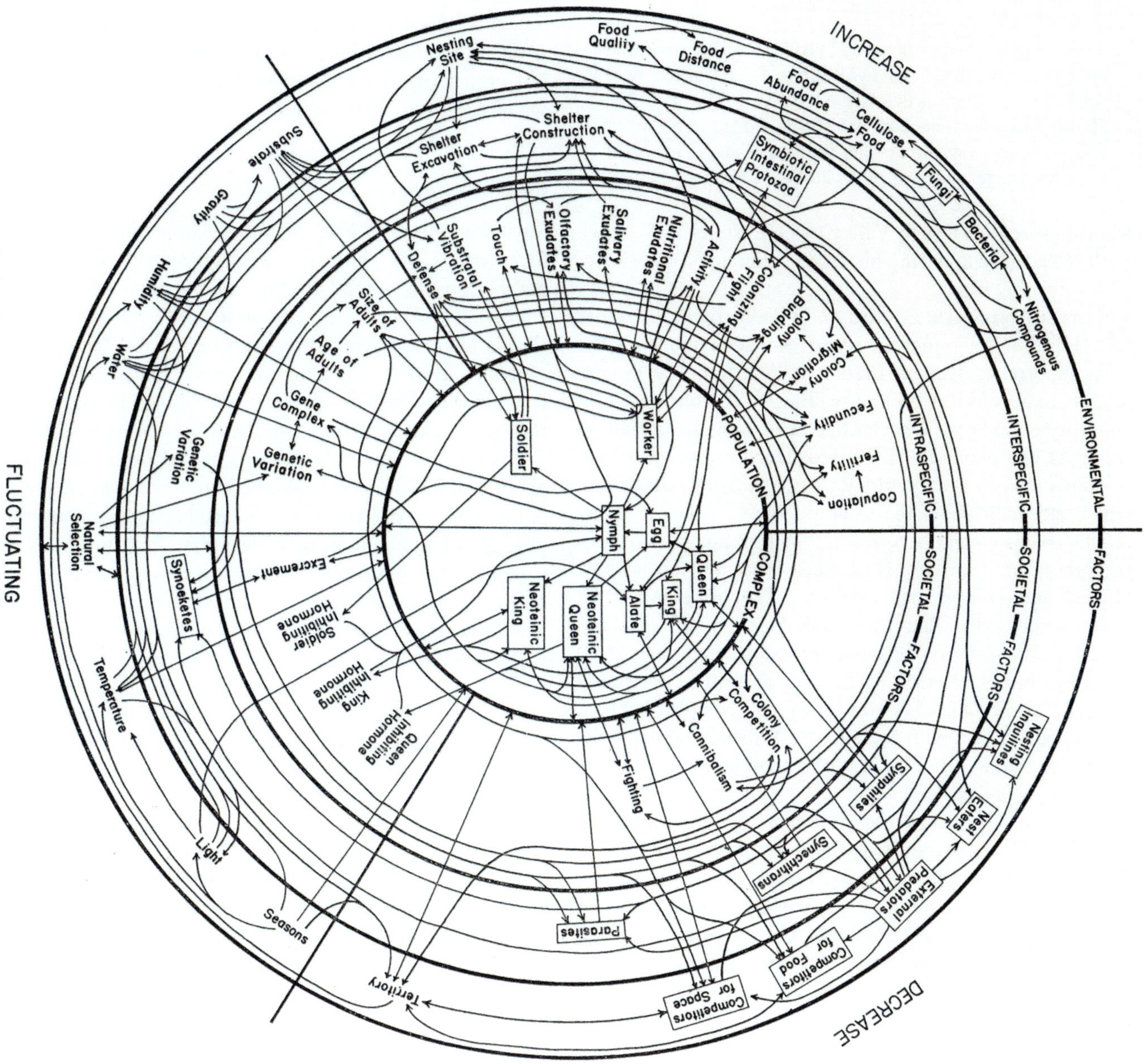

FIGURE 3.3 Factorial complex influencing the population of termites, family Rhinotermite. Allee et al. 1949: 722.

of research topics but claimed that his research was nevertheless objective. He studied social aggregations, which were intermediate between individuals and communities. He published books on it—*Animal Aggregations* (1931), *Animal Life and Social Growth* (1932), and *The Social Life of Animals* (1938)—and he interested students in conducting research on this. His ally at Chicago was Alfred Emerson (1896–1976), who had earned his degrees at the Cornell University (B.S., 1918, M.S., 1920, and Ph.D. 1925).[102] He became the world authority on termites and wrote chapters on insect societies in a five-author encyclopedia, *Principles of Animal Ecology* (1949).[103]

Englishman William Donald Hamilton (1936–2000) studied genetics at Cambridge University, where Sir Ronald Fisher was, but they had few contacts.[104] Still, Fisher's *Genetical Theory of Natural Selection* (1930) became his bible. He had an early interest in the biology of altruism. Not finding a congenial intellectual group at Cambridge, he transferred to London School of Economics but still worked mostly alone. Yet, his graduate student life is the period of his greatest scientific work.[105] He was less a mathematician than Fisher but he built upon Fisher's foundation to develop his own mathematical models: "inclusive fitness was a major conceptual advance in biology wholly original with Hamilton."[106] Edward O. Wilson summarizes his achievements:[107]

> The modern genetic theory of altruism, selfishness, and spite was launched . . . by William D. Hamilton in a series of important articles (1964, 1970, 1971a, b, 1972). Hamilton's pivotal concept is inclusive fitness: the sum of an individual's own fitness plus the sum of all the effects it causes to the related parts of the fitnesses of all its relatives.

Hamilton went to Congo Valley twice to study HIV virus, contracted malaria during his second trip, and died shortly after returning home.

Harvard professor Edward O. Wilson (b. 1929) wrote that Hamilton's seminal theory of "kin selection" was the most important "element" in sociobiological theory, though he did not accept it upon first reading.[108] Wilson is a world authority on ants, as Wheeler had been, and also a leading ecologist and evolutionist. He broadened the scope of his investigations in *Sociobiology: the New Synthesis* (1975), which attracted about the same degree of controversy as had Darwin's *Origin of Species*.[109] Within a narrower context, there are similarities between what Wilson accomplished and Darwin had. Wilson's *The Insect Societies* (1971) discussed in detail intraspecific cooperation; chapter 19 is on symbiosis between ant species and chapter 20 is on symbiosis between ant species and other arthropod species. Sociobiology is a theoretical and complex subject, and David Wilson, with coauthor Ed Wilson, updated their understanding in "Rethinking the Theoretical Foundations of Sociobiology" (2007).

University of Louisville professor Lee Dugatkin (b. 1962) provided one synthesis limited to intraspecific relationships, *Cooperation among Animals: an Evolutionary Perspective* (1997), but his *Principles of Animal Behavior* (2009) also discussed mutualism between species.[110] Stanford University professor Joan Roughgarden (b. 1946) challenged both Richard Dawkins, *The Selfish Gene* (1976) and the aspects of sociobiology in *The Genial Gene: Deconstructing Darwinian Selfishness* (2009). The selfish vs genial genes controversy may not be over.

One aspect of intraspecific cooperation that has received recent attention is assistance in raising young ("cooperative breeding") seen in some bird species[111] and in some mammal species, including wolves.[112] In 1914, Frans de Waal and Gary Stix argued (separately), human ability to cooperate on large scales was a huge asset in human rise to world dominance.[113]

Predation?

Surindar Paracer and Vernon Ahmadjian's textbook *Symbiosis* (2000) includes a discussion of carnivorous plants (pp. 190–192) but not of carnivorous animals. Charles Darwin provided the first synthesis, *Insectivorus Plants* (1875). Aaron Ellison and Nicholas Gotelli provided an admirable recent synthesis, "Energetics and the Evolution of Carnivorous Plants" (2009). Darwin knew four kinds, while Ellison and Gotelli identified 18 kinds.

Carnivorous animals have been observed since prehistoric times, and although one could explore discussions of them from Aristotle to Darwin and beyond, I am unaware of such a history. Insects or deer that feed on leaves are called "herbivores," which perhaps is an intermediate term between "carnivores" and "parasites."

While they were both on the faculty at Stanford University, zoologist Paul Ehrlich (b. 1941) and botanist Peter Raven (b. 1936) collaborated on a notable study, "Butterflies and Plants: a Study in Coevolution" (1964). Later, Joel Kingsolver and Robert Paine put that study in historical perspective:[114]

> The idea that plant secondary compounds evolved as defenses against insect herbivores had been suggested by Fraenkel (1956, 1959); and much of Ehrlich and Raven's paper is a comprehensive survey of patterns of host plant use by different taxa of butterflies. To explain these patterns, they define the process of coevolution.

Ehrlich and Raven used the term "coevolution" without defining it. Michael Allaby's *The Concise Oxford Dictionary of Ecology* did.[115]

FIGURE 3.4 (a) Alex G. Blytt. Web. (b) Ragnar Hult. Collander 1965: facing 80. (c) Leo Ramensky. Web.

Parasitism

Great advances occurred in the understanding of parasitism and diseases during 1800s, of both plants and animals, leading to a germ theory of disease.[116] Histories of animal parasitism,[117] phytopathology,[118] biological control,[119] and microbiology[120] provide details of that progress to germ theory by 1870s.

Entomologist Leland Howard (1857–1950) grew up in Ithaca NY and studied under John Comstock at Cornell University (B.S. 1877).[121] In 1878, he joined the Bureau of Entomology, U.S. Department of Agriculture, where he spent his career. In 1894, he became the Chief Entomologist. He studied parasitism in white-marked tussock moths, a native species. Its increase to plague numbers was soon followed by increase of its native parasite species, which quickly reduced its numbers. A similar cycle was less likely when foreign pests—such as gypsy moths and brown-tailed moths—were introduced and increased to plague numbers, unless their foreign parasites were also introduced. Howard and entomologist Frank W. Fisk, Head of the Gypsy Moth Parasite Laboratory, Melrose Heights, Massachusetts, reported on the success of introducing their European parasites.[122] Their landmark monograph included 69 pages on the history of previous introductions of foreign parasites to target their victim species. It proved impossible to eliminate foreign moths but their native parasites reduced their damage to a level existing in Europe.[123]

Russian mathematician-scientist V. Alexandrovitch ("Anatol") Kostitzin (1882/83-ca. 1963) fought in World War I and in Russian Revolution.[124] His early interests were in physical sciences, but he married a parasitologist, Julie, in mid-1920s, and he followed by 1927. In 1930, he attended Italian mathematician Vito Volterra's lecture in Paris on "mathematical theory of the struggle for life," and in 1931, he was a junior coauthor with Julie Kostitzin of a mathematical analysis of relationship between hermit crab *Eupagurus curanensis* and its barnacle parasite *Chlorogaster sulcatus*. Their study became one example in his *Symbiose, parasitisme et evolution (étude mathématique* (1934). That book, in turn, was absorbed into his more general *Biologie mathématique* (1937, English 1939).

Looking back in 1978, Italian geneticist Francesco Scudo and Californian ecologist James Ziegler assembled a source book, *The Golden Age of Theoretical Ecology: 1923–1940, a Collection of Works by V. Volterra, V.A. Kostitzin, A.J. Lotka and A.N. Kolmogoroff.* In fact, they reprinted only three papers by Lotka and one from Kolmogoroff. In addition, their bibliography cited a page of titles from Lotka and two titles from Kolmogoroff. The papers reprinted from Kostitzin outnumber those by Volterra. Not all papers included relate to parasitism, but a substantial number do.

In the United States, there are state veterinary medical associations. Oklahoma's was established in 1907 and its centennial was commemorated with a history.[125] The Helminthological Society of Washington was organized in 1910. The *Journal of Parasitology* was founded in USA in 1914. The American Society of Parasitologists was founded on December 30, 1924,[126] and by 1927 it had 281 members and by 1931, 550 members. Many states supported a veterinary medical school at state land-grant colleges. In 1948, Oklahoma A & M College (now, Oklahoma State University), in Stillwater, founded its veterinary medical school with Wendell Krull (1897–1971) as head of veterinary parasitology. His biography was written by his successor, Sidney Ewing (2001). Krull was the first secretary-treasurer of the American Association of Veterinary Parasitologists and later its president. Although Krull and then Ewing emphasized livestock and companion parasitology, Krull's "The Life History of two North American Frog Lung Flukes" (1930) was on wildlife.

Elsewhere, I have summarized the careers of American parasitologists William W. Cort, Edward Steinhaus, John Janovy, Jr., and Gerald Esch.[127]

Pollination

Pollination studies began with observations by Dobbs, Sprengel, and Darwin and historical introduction by Hermann Müller in *Die Befruchtung der Blumen durch Insekten* (1873, English 1883), discussed already.[128] Paul Knuth (1854–1900) earned a doctorate at Greifswald University (1876) and became a high school teacher.[129] He aspired to surpass Müller, took a leave of absence, and studied from November 1898 to March 1899 at the tropical garden at Buitenzorg, Dutch East Indies, which had facilities for visiting scholars.[130] There, Knuth studied pollination of over 200 species of tropical plants (and other subjects) in preparation for his *Handbuch der Blütenbiologie* (3 vols., 1898–905, English 1906–09). F. Faegri and L. van der Pijl wrote *The Principles of Pollination* (1966, ed. 2, 1971),

Philip Regal studied "Pollination by Wind and Animals: Ecology of Geographic Patterns," with different modes of pollination for the "most exposed layers of vegetation: trees in the case of forests, and the dominant shrubs for shrub communities" in North and Central Americas and Australasia.[131] He described the ecogeographic trends of different regions and the possible reasons for the trends. Much uncertainty remained, but he provided guidance to existing literature and to research opportunities. Olle Pellmyr's "Pollination by Animals" (2002) is a synthesis focused on evolutionary dynamics.

Lichens

During 1800s, botanists increased their knowledge of both algal and fungal species.[132] Frankfurt-am-Main native Anton de Bary (1831–88) was interested in both algae and fungi.[133] In a textbook on fungi (1866), he speculated that certain algae may have different forms if penetrated by parasitic Ascomycetes fungi.[134] In 1867, Swiss botanist Simon Schwendener (1829–1919) announced that lichens have a dual origin, combining algae and fungi.[135] In 1868, he elaborated his conclusions.[136] In 1869, he claimed lichens are "ascomycetes parasitic on algae.[137]

In addition, in 1867, Russian botanist Andrey Famintzin (1835–1918) and Ukrainian botanist Josif O.V. Baranetsky (1843–1905), both having studied under de Bary,[138] studied a cross-section of a lichen thallus and observed green cells, gonidia (now, photobionts), which they cultured and thought they resembled *Cystococcus* of Nageli.[139] Still, they imagined the alga *Cystococcus* was an immature stage of a lichen. They published their findings in French and German.[140] Baranetsky then found lichen gonidia are "physiologically independent organisms."[141] De Bary thought it might be possible to make an artificial lichen synthesis by "inoculating free-living gonidia with spores of the species concerned."[142] His student Max Reess (1845–1901) attempted this with some success (1872).[143] Another student, Ernst Stahl (1848–1919) attempted to clarify reproduction in lichens. He also met with some success, in his doctoral thesis.[144] "His thesis is a major contribution to lichenology; it deals with pioneering experiments involving developmental history of lichens. He synthesized lichen *Endocarpon pusillum* from single spores and algal material and saw formation of apothecia and thalli."[145]

The theory of lichens being composed of algae and fungi met resistance from lichenologists who had invested thought, time, and ego in lichens as a fundamental group comparable to algae, fungi, and mosses. Finnish lichenologist William Nylander (1822–99), living in Paris for almost 20 years, was the foremost lichen taxonomist of the day.[146] Not surprisingly, he attacked the dual theory beginning in 1870 and ended contact with Schwendener.[147] He urged other lichenologists to do likewise. If the dual theory prevailed, Nylander's classification system would collapse.

The origin of the term "symbiosis" has been the subject of some discussion, details of which lichenology historian Michael Mitchell has provided. In Greek antiquity, Aristotle's successor at the Lyceum, Theophrastos (discussed above) wrote the first two botanical treatises: *Historia Plantarum* and *De Causis Plantarum*.[148] In both, he used the term "sumbiosai" to describe ivy growing on olive trees.[149] Mitchell found that de Bary mentioned in 1853 that he had read Theophrastos, which may have led him to coin the term "symbiose." Botanist Albert Frank (1839–1900) published a very similar term in 1877 before de Bary published his term in 1878. This led some mycologists to wonder if de Bary used Frank's term without acknowledgement.[150] Mitchell provided arguments why that was highly unlikely.

Englishman Hugh Weddell (1819–77) earned his MD degree in Paris in 1841, studied under Adrien Jussieu, and at some point, he became a French citizen.[151] He explored South America in 1843–48 and again in 1851. He began to study lichens in late 1860s, and his 1875 paper "provided the first clear description of the zonation exhibited by lichens on rocky shores."[152]

Besides symbiosis, "the study of lichen ecology was pioneered by Bruce Fink (1861–1927)."[153] He was from Illinois and earned his B.S. and M.S. from the University of Illinois (1887 and 1894), an M.A. from Harvard (1896), and a Ph.D. from University of Minnesota (1899). A professor at Harvard had interested him in lichens and when Prof. Conway Macmillan, University of Minnesota, hired Fink as part of his survey team, Fink focused upon lichens of Minnesota. Mitchell cited eight of Fink's publications, beginning with his report on "Lichens of the Lake of the Woods" (1896) and including *The Lichens of Minnesota* (1910), and finally, *The Lichen Flora of the United States* (1935), which his former student, Joyce Hedrick, completed after he died.[154] He taught at Upper Iowa University and Grinnell College before settling at Miami University where he taught from 1906–1927.

Illinois native Albert Schneider (1863–1928) obtained M.D. degree (1887) in Chicago before earning B.A. (1894) at University of Illinois, M.S. (also 1894!) at University of Minnesota, and Ph.D. (1897) at Columbia University.[155] He began his career as an instructor of cryptogamic botany at Minnesota, 1893–94; lecturer in botany, Brooklyn Institute, 1896; professor of botany, pharmacology, and bacteriology, School of Pharmacy, Northwestern University, 1897–1903; and professor of botany and bacteriology, University of Nebraska, 1919 to at least 1921. Schneider turned his doctoral dissertation into *A Text-book of General Lichenology* (1897), "which condensed practically all that was known about lichens."[156] He also wrote a general article, "The Phenomena of Symbiosis" (1897), covering plants and animals, though published in a botanical journal read by few, if any, zoologists.[157] It included an outline of his conception of symbiosis and definitions of terms used in the outline and a chart indicating how he thought these stages of symbiosis evolved.[158] He made one more contribution to lichenology: *A Guide to the Study of Lichens* (1898).

Fink's lichen papers sparked an interest in a school teacher in Copenhagen, Olaf Galløe (1881–1903), "who published the first notable European study devoted to lichen ecology (1908)."[159] The second such paper Mitchell found was by Uppsala botany professor, Rutger Sernander (1866–1944), who had also studied at Uppsala.[160] His 1912 paper correlated substrates with the kinds of lichens that grew on it.[161]

Eugenius Warming, discussed above as a founder of plant ecology, described, at least in the English edition of *Oecology of Plants* (1909), the ecological zones of lichens in very specific detail, moving from the coast to inland, the width of each zone depending upon the amount of saltwater spray carried there by wind.[162] In doing so, he followed in the footsteps of Weddell, though Weddell's name is not in Warming's bibliography.[163] Following Warming, Irish naturalist Matilda Cullen Knowles (1864–1933), who was an assistant at the Dublin Science and Art Museum, in a substantial paper on "The Maritime and Marine Lichens of Howth"(1913), described five lichen belts.[164]

Across the Irish Sea, at the British Museum (Natural History) in London, Annie Lorrain Smith (1854–1937) worked on fungi and lichens, 1892–1933.[165] She completed a textbook, *Lichens*, in 1917, but there was a war on, that delayed its appearance until 1921. It was worth the wait and included historical discussions. However, just four years later, another comprehensive book appeared, in German, by Swiss botanist, Friedrich Tobler (1879–1957), at the University of Dresden.[166] His own studies emphasized symbiosis. The living Friedrich Tobler, who also studies lichens, is presumably a descendant of the first one.

Commensalism and Mutualism

Invertebrate zoologist Pierre-Joseph van Beneden (1809–94, figure 2.2) was Belgium's foremost zoologist, at Catholic University, Louvain.[167] In "*La Vie sociale des Animaux inférieurs*" (1873), he introduced the terms "commensalism" and "mutualism."[168] He expanded and popularized his concepts in *Les commensaux et les parasites dans le règne animal* (1875, English 1876). He defined a commensal individual as requiring "a simple place on his [host's] vessel, and does not partake of his provisions."[169] Mutualists are:[170]

> . . . animals which live on each other, without being either parasites or messmates; many of them are towed along by others; some render each other mutual services, others again take advantage of some assistance which their companions can give them; some afford each other an asylum . . .

He concluded there was an "almost insensible gradation of differences between parasite, messmate, and free animals," and he provided numerous examples of such relationships.[171] De Bary thought that different kinds of lichens have this gradation of relations.[172] Later, "commensalism has been expanded to include all those ecological unions in which, although both parties do not benefit, as in mutualism, neither one is harmed . . . "[173]

In late 1860s and/or early '70s, English mine supervisor and amateur naturalist Thomas Belt (1832–78), working in Nicaragua, discovered that small stinging red ants *Pseudomyrmex spinicola* and *Acacia collinsii* trees formed alliances with ants found only in Nicaragua and Costa Rica, living in hollow thorns and protecting bull's horn acacias from other vegetarians, such that they alone can benefit from its productions.[174] Discussing army ants *Eciton predator*, Belt saw birds follow them and catch insects taking flight.[175] He did not mention which bird species; perhaps, he saw different bird species involved. Edwin Willis and Yoshika Oniki counted 50 species that do so.[176] Another Belt discovery was that leaf-cutting ants *Oecodoma* use leaf snippets to grow fungi underground for their food. Wheeler's "The Fungus-growing ants of North America" acknowledged Belt's discovery (1874) of ant agriculture.[177]

Nebraskan lawyer-botanist Roscoe Pound introduced many American naturalists to the two concepts of "Symbiosis and Mutualism" (1893).[178] However, Professor Paul Portier (1866–1962), University of Paris, wrote *Les Symbiotes* (1918), the first comprehensive synthesis since Beneden's (1875). Portier was an animal physiologist with broad interests.[179] His concept of symbiosis involved bacteria as symbionts in all multicellular organisms,[180] which Maurice Caullery challenged in his *Parasitism and Symbiosis* (1922, English 1952). Portier's hypothesis never won out; yet, it resembled later ideas on symbiogenesis/ endosymbiosis. Invertebrate zoologist Caullery had earned his doctorate in Paris (1895), taught in Lyons and Marseille, and returned to Paris (1909) as head of the Department of Evolution, Faculty of Sciences.[181] He also headed Station Zoologique de Wimeraux at Straits of Dover. Sapp summarized his perspective:[182]

> Caullery's aims were to rid accounts of teleology and anthropomorphism; to show that all cases of mutualistic symbiosis could be accounted for in terms of individual life struggle, domination, and control; and to show that no natural distinction could be made between parasitism, mutualism, and commensalism.

German Professor Paul E.C. Buchner (1886–1978) studied zoology at University of Munich, where he remained for his career. Since 1911, he began to publish ten articles, which he synthesized into *Tier und Pflanzen in intrazellularer Symbiose* (1921, 464 pages). A 1953 revision was entitled *Endosymbiose der Tier mit pflanzlichen Mikroorganismen* and the English version was *Endosymbiosis of Animals with Plant Microorganisms* (1965, xvii + 909 pages). His book excluded symbiosis between two plant species and between two animal species.

The first synthesis on symbiosis in English likely was in Allee, Emerson. Park, Park, and Schmidt, *Principles of Animal Ecology* (Allee et al. 1949: 243–262, 710–721), was preceded by a French edition and followed by an English edition of Caullery, *Parasitism and Symbiosis* (French 1922 and 1950 and English 1952). Caullery's list of journals omitted Elton's *J. of Animal Ecology*. Clark Read's *Parasitism and Symbiology: an Introductory Text* (1970) omitted Caullery but cited Allee et al. (1949). Douglas Boucher et al. compiled a bibliographical guide: "The Ecology of Mutualism" (1982). There are at least five edited, multi-authored collections of symbiosis studies: Henry (1966–67), Cheng (1971), Boucher (1985), Herrera and Pellmyr (2002), and Seckbach (2002).

Sometimes a species of plant and an animal have developed a tight bond of mutual dependence. Such a bond Stanley Temple (1977) hypothesized between Mauritius' extinct dodo bird *Raphus cucllatus* and the tambalacoque tree *Sideroxylon grandiflorum*, whose seeds are inside hard-shelled fruit, which supposedly only dodos could open.[183] Andrew Mitchell accepted Temple's hypothesis (without citation and naming tree *Calvaia*) in *The Fragile South Pacific: an Ecological Odyssey*.[184] However, the basis of Temple's hypothesis was subsequently discredited.[185]

A better case of mutual benefit was explained by Professor Emeritus Ronald Lanner (b. 1930), Utah State University, in *Made for Each Other: A Symbiosis of Birds and Pines* (1996). Seeds of pine species having small seeds with wings are obviously disseminated by wind. What about species with large seeds with little or no wings? These species participate in a pine-corvid mutualism.[186] Lanner studied a relationship between white-bark pine *Pinus albicaulis* and Clark's Nutcracker *Nucifraga Columbiana*. Only this bird has a beak large enough to open this pine's cones, which it stores underground for later use. A single nutcracker was known to harvest 129,000 seeds in a year, of which 75% were stored, but not all were later recovered.[187] Other investigators documented seed dispersal by other animals.[188]

Mimicry is a form of mutualism discussed above concerning discoveries by Bates and Wallace. Georges Pasteur's "Classificatory Review of Mimicry Systems" (1982) includes a historical chart, indicating landmark studies, 1817–1978, though his article is not historically organized.[189]

Algal Mutualism with Animals

Genevan Abraham Trembley (1710–84) became a tutor in Holland in 1733 and in 1740 discovered green hydra *Chorohydra viridissima*.[190] Not knowing if plant or animal, he cut it in half to see if it regenerated like a plant or died like an animal. It regenerated but its behavior left him uncertain. He initiated a correspondence with French entomologist, René Réaumur, who judged it to be animal. French chemists Pierre-Joseph Pelletier and Joseph-Bienaimé Caventou in 1817 studied green pigments in vascular plant leaves and named it chlorophyll.[191] Biologists later found that it is pervasive in all green plants, from algae to orchids. Since chlorophyll was found to absorb CO_2 and expire O_2 that relationship of green plants and animals was mutualistic.

Coral reefs are impressive, with multicolored variety and abundance of life. Yes, mutualistic reefs are adapted to nutrient-poor environments and algae are more important reef builders than scleractinian corals.[192] When their physical environment is favorable, coral reefs flourish because they function at ever trophic level: algae are primary producers and corals are primary consumers, when utilizing phytosynthetic products. Secondary and tertiary feeders when ingesting herbivorous carnivorous zooplanktons, direct-deposit feeders when absorbing dissolved organic substances from sea water.[193] With such symbiotic versatility, are corals not environmentally flexible? Sadly, no. Global warming leads to ocean acidification, which is fatal to corals.[194]

Elsewhere, I have discussed bacterial and fungal mutualisms with plants, bacterial, and protozoan symbioses in animals and symbiogenesis.[195] Two final discussions here are on cleaning symbiosis and Gaia.

Cleaning Symbiosis among Animals

Herodotos' ancient report of plovers plucking leeches from Nile crocodiles (discussed above) was not a unique cleaning relationship. Birds were the easiest cleaners to observe. Caullery provided examples:[196]

> . . . ungulates and certain birds which follow herds, sometimes resting on the animals and picking off the ticks and the oestrid larvae that infest the hide. Starlings do this, also some wagtails (*Motacilla flava*) and magpies in France. *Crotophagus* in America, and *Buphagus*, the ox-pecker, in Africa.

Randall Breitwisch observed both red-billed and yellow-billed oxpeckers in Kenya's savannas, whose symbionts include zebras, impalas, giraffes, wart hogs, Cape buffaloes, rhinoceros, and Masai cattle.[197] He saw waterbucks and hartebeests intolerant of oxpecker probings. Oxpeckers sound an alarm if predators of their "customers" approach, which Breitwisch suggests could be motivated by oxpeckers being unable to eat when a symbioint moves quickly, and if large mammals trust oxpeckers as lookouts, they can relax while being searched.

Cleaning symbiosis seems best known among marine animals.[198] William Beebe (1877–1962), New York Zoological Society,[199] in Galapagos Islands, saw a red crab remove three red ticks *Almblyomma darwinii* from a sunbathing marine iguana *Amblyrhyncus cristatus*.[200] He also observed several small wrasse-cleaning parrotfish:[201]

> . . . a little school of wrasse darted out and thoroughly cleaned cheeks, lips, teeth and scales of all particles of organic coral debris, the parrotfish remaining quite motionless all the while. It was an aquatic parallel of crocodile and plover, cattle and egret, rhino and tick bird.

and a loyal supporter was American biologist Lynn Margulis (1938–2011), who received the National Medal of Science in 1999 from President Clinton. Her main research was on how simple forms of life combined symbiotically and became more complex forms.[202] She found her conclusions compatible with Lovelock's Gaia theory. Scientists held two symposia on Gaia, and the MIT Press published their reports and findings.[203] At least four books assessed the merits of Lovelock's Gaia: Lawrence Joseph, *Gaia: the Growth of an Idea* (1990); Tyler Volk, *Gaia's Body: Toward a Physiology of Earth* (1998); Jon Turney, *Lovelock and Gaia: Signs of Life* (2003); and John Gribbin and Mary Gribben, *James Lovelock: in Search of Gaia* (2003).

Conclusions

Symbioses are relationships important for survival in most species. Yet, their manifestations are so common, so subtle, and so easily overlooked. Three examples were known in antiquity but were viewed as merely natural history novelties. Linnaeus's economy of nature concept was a version of balance of nature he used to organize examples of mutualism in support of his economy of nature. Darwin's theory of evolution by natural selection placed symbiosis within a context of evolving over time, from casual to significant contacts. Since 1870, many lines of research developed that contributed to the current understanding of symbiosis. Exiled Russian anarchist Pyotr (Peter) Kropotkin published a much-read *Mutual Aid: a Factor in Evolution* (1902), illustrating a claim by Russian zoologist, Karl Kessler that mutual aid was more important for evolution than struggle. Maurice Caullery organized symbiosis into a branch of ecology in *La Parasitisme et Symbiose* (1925). A broader synthesis appeared in Allee et al., *Principles of Animal Ecology* (1949). German biologist Paul Buchner published a very detailed *Endosymbiosis of Animals with Plant Microorganisms* (English ed., 1965). Multi-authored collections of papers on symbiosis began with Mark Henry's two-volume *Symbiosis* (1966–67). English scientist James Lovelock published a novel and a controversial hypothesis, Gaia, which attempted to explain world-wide symbiosis.

C. SUCCESSION, COMMUNITY, AND CONTINUUM

Plant ecologists developed concepts of *succession, community*, and *continuum*, which were later expanded to include animals. Yet, zoologist Karl Möbius developed a concept of *biocönose* to describe animal communities on mudflats which oysters dominated—plants on mudflats were inconspicuous phytoplanktons.[204] The *balance of nature* concept had developed since antiquity with only animal examples, until Carl Linnaeus (1749) named the concept *oeconomia naturae* and included plants in his expanded version.[205]

Ordinary language divides familiar environments into forest, prairie, savanna, marsh, swamp, lake, seashore, and more. These environments can be subdivided, such as into pine forest, oak forest, and rain forest. Science goes beyond ordinary language when its entities are unnamed in ordinary language or it needs definitions more precise than ordinary language provides. The three ecological terms in the above title are borrowed from ordinary language and given scientific meanings. However, different ecologists can use different names for the same thing and so Robert McIntosh compiled "The Succession of Succession: a Lexical Chronology" (1999).

Historical Survey Until 1880

A concept of plant succession began with early observations on bogs; some bogs were known to have come into existence and others to have dried up; some had been drained to create dry land. Englishman at court of Henry VIII, John Leland (*c.* 1503–52) traveled for six years around England, visiting different environments.[206] His posthumous *Itinerary of John Leland the Antiquary* (9 vols., 1710) reported that farmers who dug peat commonly found tree trunks or roots, indicating that vegetation changed over time. He reported two cases in which bogs had burst and flowed away, causing much damage.[207]

Gerrit/Gerard Boate (1604–50), Dutch physician living in London, divided some lands, in *Irelands Naturall History* (1652), into three groups: moory or boggy heaths, dry heaths, and wet bogs.[208] His discussion of different kinds of vegetation did not specify which plant species grew in each kind. Anglican clergyman in Dublin, William King (1650–1729), wrote "Of the Bogs, and Loughs of Ireland" (1685) to assist bog drainage.[209] He divided bogs into quaking bogs and drier turf bogs. He thought when seeds of bog-moss fell on dry land it grew into health. Yet, he also gave "a clear definite statement of succession, beginning with mineral soil, passing through grassy quaking bogs, and growing into turf bog."[210] King saw that the bogs are "generally higher than the land about them, and highest in the middle: the chief springs that cause them being commonly about the middle." Irishman John Honohane (1697) reported to the Royal Society that heavy rains caused a bog near Charleville to slide down a slope and cover a meadow, which is in some areas 16 feet deep. A diagram—perhaps first with ecological relevance—showed the direction of the slide, south to north.[211] Boate, King and Honohane, 1652–97, discussed bogs without identifying bog plants.

France's leading naturalist of the 1700s, Georges-Louis Leclerc de Buffon (1707–88) was initially interested in forestry.[212] His "Mémoire sur la culture des forêts" (1742) contained "the first clear record of the succession of forest dominants, and of the effect of light and shelter on the process."[213] He saw that seedlings did not do well in bare fields. Instead,[214]

> . . . plant shrubs and bushes which can break the force of the wind, diminish that of frost, and moderate the inclemency if the seasons. These bushes are the shelter which guards the young trees, and protects them against heat and cold. An area more or less covered with broom or heath is a forest half made; it may be ten years in advance if a prepared area.

He added that after young trees "passed the first few years in the shade and shelter of the others, they quickly stretch up, and suppress all the surrounding plants."[215]

Sweden's rival to Buffon as a leading naturalist of the 1700s, Carl Linnaeus (1707–1778), was primarily a botanist. Yet, also had broad interests.[216] He wrote essays that his students defended for their doctorate. His *Oeconomia Naturae* (1749) was the first explicit account of the balance of nature.[217] It was also the first summary of what ecological science might encompass. He described plant succession from practically bare rock with lichens through several distinct vegetation stages to a forest, followed by death and decay of trees and use of decaying wood by fungi and beetles to create a revolving cycle of nutrients. In another essay, *Stationes Plantarum* (1754), he distinguished *station* (habitat) from *habitation* (range). " . . . native places or stations of plants respect the country, climate, soil, and situation, nature of the ground, earth, and mould,"[218] which seems clear with habitation including "latitude, altitude, and topography."[219] His list of stations of plants was: maritime or saline, marine, freshwater, damp regions, prairie, cultivated, rocky, sand, sterile soil, rubbish piles, forest, bushes and hedges, subterranean, mountain, parasitic, and saprophytic. In *Politia Naturae* (1760, English 1781, French 1972), competition and avoiding extinction were themes.

Linnaeus realized that species produce more offspring than could survive, which caused "a war of all against all!"[220] Despite that predicament, he believed that balance of nature was preserved with the God providing means of survival by what we call *ecological diversity* as a way of reducing competition between species. Sweden, for example, had about 1300 plant species but only about 50–100 lived in one place.[221] He assumed that all these conditions were the aspects of a static system among static species, until his later years, when he began to doubt species stability.[222]

Linnaeus discussed peat bogs in some two dozen pub-lications, 1742–73. Swedish plant ecologist Einar du Rietz wrote a "Summary of Linnaean Contributions to Paludology" (1957), identifying eight contributions of Linnaeus that to this not-yet-existing science, such as: "Linnaeus seems to have been the first author to describe mire vegetation with lists of species."[223]

Scotsman James Anderson (1739–1788) at age 15 took over a family farm.[224] He took an academic course in chemistry to increase his understanding of farming and he also became interested in the Scottish economy. His writings on peat bogs were paradoxical: he carefully identified plant species and indicated their sequence in succession. Yet, he also believed that "peat moss, as we find it in the natural state is, of itself, a vegetable production, not a congeries of dead plants preserved by some mystical influence, as has been generally supposed . . . "[225]

That Berlin physician-botanist Karl Ludwig Willdenow (1765–1812) knew Linnaeus' works are evident in his *Grundriss der Kräuterkunde* (1792)—the first comprehensive botany textbook,[226] evident in this quotation from an English translation:[227]

> Naked rocky places, on which nothing can grow, are, by the winds, covered with seeds of Lichens, that by means of the accustomed showers in Harvest and Spring are induced to germinate. Here they grow, and the rock is spotted with their coloured frond. In time the winds and weather deposit small dust in the rough interstices of the rock, and even the decaying Lichens leave a thin scurf.

Alan Morton judged Willdenow's textbook as "the first attempt to explain the world distribution of species in a scientific manner."[228]

Genevan natural philosopher Jean-André Deluc (1728–1817) had studied geology, meteorology, and physics.[229] He discussed peat formation in a letter (Jan. 1796) to Rev. Robert Rennie for his *Essays on the Natural History and Origin of Peat Moss* (1807–10). Deluc was "the keenest and most indefatigable of early students of peat-bogs . . . . " He was probably the first to make use of the term *succession* and certainly the first to see it with full recognition of its developmental significance.[230]

French scholar Adolphe-Jules-César-Auguste Dureau de la Malle (1777–1857) managed crops, meadows, and woodlands on his estate.[231] He got the idea of plant succession from English author Arthur Young's writing on crop rotation. He based his own report on plant succession (1825) on his experience during 30 years of planting crops and harvesting timber. Frederic Clements summarized Dureau de la Malle's article and translated his conclusion:[232]

> The alternance or alternative succession in the reproduction of plants, especially when one forces them to live in societies, is a general law of nature, a condition essential to their conservation and development. This law applies equally to trees, shrubs, and under-shrubs, controls the vegetation of social plants, of artificial and natural prairies, of annual, biennial, or perennial species living socially or even isolated.

Later authors cited this article, showing Dureau de la Malle's influence.[233]

German botanist-physician and Humboldt protégé Franz Meyen (1804–40) traveled as physician on the royal cargo ship *Princess Louise* during a world cruise lasting almost two years.[234] In 1834–35, he published an account of his voyage, followed by *Grundriss der Pflangengeographie* (1836, English 1846). He drew a distinction between habitat and range:

> The station (*statio*) of plants denotes the relation in which plants stand to the situation in which they always grow; or we understand by it, the locality . . . The distribution of plants (*extension plantarum*) . . . signifies the whole range of their stations[235]

He described plant stations under two headings, aquatic plants and land plants, each being subdivided.[236] His aquatic stations were: marine, freshwater, submerged, floating, lake, ditch, river, fountain, saline, amphibious and inundated, maritime and shore, and mangrove forests. His subdivision of land plants was even more complex. He called those plant species that grew together as *social* and other species as *unsocial* or *solitary*. Meyen's synthesis was the most practicable to date and it was usable as a textbook.

Swiss biologist Otto Heer (1809–83) grew up collecting plants and insects and never lost interest in either. Instead, he later added an interest in fossils of both kind of organisms.[237] He studied theology at University of Halle, where he found he was more interested in science. He was early associated with the new University of Zurich and became an associate professor in 1835 and a professor of botany and entomology in 1852. His doctoral dissertation was on *Die Vegetationsverhältnisse des südostlichen Teiles des Kantons Glarus; ein Versuch die pflanzengeographischen* (1835), which R. Becking identified as the first application of "numerical values to species abundance and sociability," and E. van der Maarel saw that Heer's monograph "had a clear notion of the plant community as a type, its interrelations with the environment"[238]

Danish naturalist Japetus Steenstrup (1813–97) began his career as a school teacher, without a university degree.[239] In 1842, he published two noteworthy studies on very different topics. He was not the first to suggest alternation between sexual and asexual reproduction in many invertebrates, but he provided enough evidence to establish its importance and earned him a professorship of zoology at University of Copenhagen.[240] By 1842, Europeans had published a good number of studies on peat bogs and their changes, but few of them except Linnaeus studied the plant species involved. Steenstrup dug into Danish forest bogs and discovered distinct layers, each having different vegetation from the layers above and below it.[241] He compared Danish bogs with two in Zeeland, Netherlands. He drew diagrams of the layers in different bogs and identified the dominant species in the bogs and the trees around the bogs. He thus founded paleoecology (unnamed in 1842). No diagrams of ecological significance are known between Honohan's in 1697 and Steenstrup's two in 1842 which indicate layers in bog cross-sections. Most of his 103-page article identified bog layers and plants in each.[242]

French botanist Henri Lecoq (1802–71), native of Clermont-Ferrand, became director of that city's Jardin des Plantes.[243] His study on French grasslands (1844) "developed a quantitative measure which was essentially a combined cover-abundance estimation."[244] Even more important was his *Étude sur la Géographie botanique de l'Europe* (nine volumes, 1854–58). In volume 1, he defined "association végétale" along the lines Braun[-Blanquet] elaborated later, including "use of faithful species."[245] Malcolm Nicolson found that Lecoq "clarified the distinction between *sociability*—many plants of the same species living together—and *association*—many plants of different species living together."[246]

Austrian botanist Siegfried Reissek (1819–71), Royal Botanical Collection, Vienna, studied vegetation on unstable islands in the Danube River (*Flora* 1856, summarized in Clements).[247] Islands grew from sand bars and vegetation arose in a regular pattern. Willows, usually *Salix purpurea*, were first established, and their hummocks caught sand from river and helped islands grow. *S. riparia* and *Myrica germanica* came next, followed by *Alnus incana, Populus alba*, and *Cornus sanguinea*. Still other low plants came until trees overshadowed them, eventually including the willows.[248]

A paleo-ecologist who Steenstrup trained was Danish botanist-forester Christian Vaupell (1821–62), who examined fossil records of Danish forests and found no evidence of beech before recent times. He therefore concluded it was a recent arrival from France and Germany, as Denmark's forests became drier than before.[249] Vaupell published a treatise on the succession of trees in Danish forests (1857) and concluded that beech could displace birch and pine except in marshes or sterile soils.[250]

Massachusetts social critic-naturalist Henry David Thoreau (1817–62) was connected well enough to society but less connected to scientists.[251] In 1860, he spoke before the Middlesex Agricultural Society on "The Succession of Trees," which was published in that society's *Transactions* and in a New York newspaper, neither of which sources naturalists or biologist likely read.[252] He saw that when pines were logged, oaks replaced them, and when the oaks were logs, pines replaced them. He explained that when pines were logged, pines could not replace them, because oak seedlings were already growing there, and pine seedlings could not grow in shade. Squirrels buried more acorns in pine forests than they ever recovered in winter. Pine seeds blew annually into oak forests, and when oaks were cut, pine seedlings flourished in sunlight. He also saw birds eat cherries from trees and later expel seeds elsewhere in their droppings. Stephen Spurr has traced the concept of forest succession back to Jeremy Belknap's *History of New Hampshire*, volume 3 (1792) and to John Dawson in *American Journal of Science* (1847).[253] If Thoreau had known either of them, he very likely would have said so.

Austrian botanist Anton Kerner (1831–98) studied medicine at the University of Vienna and after teaching botany elsewhere, became professor of botany at the University of Innsbruck, 1860–78, before returning to Vienna in 1878 as a professor of botany. Most significant ecologically of his writings was *Das Pflanzenleben der Danauländer* (1863, English 1951).[254] He used Grisebach's *formation* concept to describe the vegetation of Danube Valley: Juniper formation, oakwoods, poplarwoods, riverbank forests, swamps, formations on soils with efflorescent salts, how a swamp becomes a meadow, draining swamps, plant formations of dry lands.[255] For the adjacent Hercynian Mountains, he had another set of formation names. His research and publishing continued into his last year and his later writings also contained ecologically relevant observations.

Norwegian botanist Axel Blytt (1843–98) was a son of botany professor Mathias Blytt (1789–1862) at Christiania University (now University of Oslo).[256] In 1863, Norway bought Matthias Blytt's herbarium and in 1865 Alex Blytt became the conservator of Christiana University Herbarium and in 1880 also a professor. Axel continued work on his father's *Norges Flora* (1861,386 pp.) and published continuations in volumes 2–3 (1874–77, pp. 387–1348). Alex Blytt's most important work, *Haandbog i Norges Flora* (1902–06, 780 pp.), continued a discussion begun in his 1874–77 volumes, on his thesis that alternating vegetation in the Norway paleo-ecological record represented alternating rainy and dry periods in past climates.[257]

After 1880

Finish botanist Ragnar Hult (1857–99) studied at Uppsala University and conducted research for several years at Blekinge, Southern Sweden.[258] In his important *Försök till Anallytisk Behandling af Växtformationerna* (*Attempt at an Analytical Treatment of Plant Formations*, 1881), he introduced a new analytical method of studying plant communities.[259] Clements praised Hult for "being the first to fully recognize the fundamental importance of development in vegetation, and to make a systematic study of a region upon this basis."[260] Clements based his claim upon Hult's "classic investigation of the vegetation of Blekinge" (1885) and upon his tracing "development of the alpine vegetation of northernmost Finland." Hult's work was also important for the emergence of the geography discipline in Finland.[261]

Blytt and Hult's findings stimulated two Swedish botanists, Andersson and Sernander to continue their line of research. C.F. Gunnar Andersson (1865–1928) studied at Lund University and in 1909 became Professor of Economic Geography at Stockholm Commercial College.[262] (Johan) Rutger Sernander (1866–1944) studied at Uppsala University and spent his career there. Clements cited 28 of Andersson's publications and 21 of Sernander's—they being among his most cited authors but he did not discuss either one's work in his "Recent Investigations" section of *Plant Succession*.[263] His discussion of their work is under "The Peat Climates."[264] Both were discussed briefly by Fries and in more detail by Söderqvist.[265] Andersson and Selim Berger published (1912) "a vast inventory of the flora of northern Sweden . . . combining the immigration historical problem with extensive discussion of the ecological factors responsible for the geographical distribution of plants."[266] In 1908, Sernander became Professor of Plant Ecology, possibly the first ever, three years before Henry Cowles achieved this title at University of Chicago in 1911.

Sernander became "founder of the Uppsala school of plant geography and plant sociology, the institutional center for Swedish plant ecology from the 1930s to the 1960s."[267] He conducted research on palynology and extended Blytt's chronology back into the last ice age. He also studied Scandinavian plant geography. Clements quoted in English (from German summaries) passages from Sernander's papers of 1890 and 1891.[268]

Danish botanist J. Eugenius B. Warming (1841–1924) wrote the first plant ecology textbook, *Plantesamfund* (1895, German 1896, Polish 1900, Russian 1901, English 1909).[269] He attended Copenhagen University, but interrupted his studies in 1863–66 in order to become secretary to zoologist-paleontologist Peter W. Lund (1801–80) in a tropical savannah-woodland in Brazil.[270] His years in Brazil, with Lund studying fauna and Warming plants, were "most significant for his scientific development."[271] In 1884, he surveyed the physiognomy of North European Spermophyta, "which he arranged in fourteen chief groups with many sub-groups, based upon morphological and biological characters."[272] Clements praised Warming for being the "first to give a consistent account of succession on sand dunes, and his pioneer studies in this field have served as a model for the investigation of dune series in all parts of the world."[273] Warming's discussion of plant communities, although brief, focused plant ecologists attention at this level of study, much as Möbius' discussion did for animal ecologists.

In contrast to Warming's textbook that lacked illustrations was the encyclopedic *Pflanzengeographie auf physiologische Grundlage* (1898) by Strasbourg botanist Andreas Schimper (1846–1901), which (at least in English edition, 1903) contained four maps and 497 illustrations.[274] Like Warming, he had traveled to tropics to study vegetation. Clements summarized Schimper's theoretical ideas and found them less incisive than Warming's.[275] R.J. Goodland accused Schimper of quoting extensively from over 15 of Warming's works and even reproduced Warming's figures without acknowledgement.[276] Yet, Dwight Billings, presumably unaware of Goodland's paper, praised works of both Warming and Schimper and decided Schimer's *Pflanzengeographhie* "was a valid beginning of modern physiological ecology."[277] Malcolm Nicolson discussed both Warming and Schimper and cited Goodland's paper, without mentioning Schimper's plagiarism.[278]

German botanist C.G. Oscar Drude (1852–1933) studied under Grisebach at Göettingen, earning his doctorate in 1873.[279] In 1879, he settled at Dresden Polytechnikum (now Dresden Technical University). He followed Grisebach's example in publishing *Handbuch der Pflanzengeographie* (1890, French 1892) and *Deutschlands Pflanzengeographie* (1896). Ron Tobey discussed the latter book in relation to the

FIGURE 3.5 (a) Henry A. Gleason. Web. (b) John T. Curtis. Hunt Institute for Botanical Documentation. (c) Robert H. Whittaker. Westman & Peet 1982: 97.

rise of American plant ecology.[280] One of those Americans, Clements, cited eight works by Drude.[281]

Nebraskans Roscoe Pound (1870–1964) and Frederic E. Clements (1874–1945), both from Lincoln, studied botany at University of Nebraska and collaborated on *The Phytogeography of Nebraska*, guided at first by Drude's 1896 book and later also by his 1890 book.[282] Their monograph became their doctoral dissertations, for Pound's Ph.D. in 1896 and Clements' in 1897. It was published in 1898 (edition 2, 1900).[283] Pound then left for Harvard Law School, where he eventually became dean.[284] Clements became a leading plant ecologist, with broad influence during 1900–50.[285] He joined University of Nebraska's faculty, 1894–1906, and he published three plant ecology books in 1904, 1905, and 1907! Clements' most important treatise was *Plant Succession: an Analysis of the Development of Vegetation* (1916), which dominated American plant ecology during his remaining life, though it was controversial. In 1926, William Cooper judged: "With Clements' *Plant Succession* ('16) we reach the second point of outstanding achievement."[286] Cooper mentioned several aspects of Clements' achievement and concluded that the greatest contribution of *Plant Succession* was "the extension of study to cover all time from the present moment back to the very beginnings of vegetation upon our planet."

In *Plant Succession*, Clements explained Pound and he had discussed succession in Nebraskan plant formations, but not in the depth that Cowles did, as he could follow succession from the coast inward and see successive stages representing different ages of vegetation.[287] In doing so, Cowles introduced the term *climax* into ecology.[288] His best summation of his studies was in a presidential address at the Association of American Geographers in 1910, "The Causes of Vegetative Cycles:"[289]

> Each climatic cycle has its vegetative cycle; each erosive cycle within the climatic cycle in turn has its vegetative cycle; and biotic factors institute other cycles, quite independently of climatic or topographic change. It is small wonder that within this complex of cycle within cycle, each moving independently of the others and at times in different directions, dynamic plant geography has accomplished so little in unraveling the mysteries of succession.

Overly modest? Clements acknowledged the substantial contribution which Cowles made to understanding succession, without saying so definitively in any one statement.

Just as Drude's treatise provided guidance for Clements and Pound in Nebraska, so did Warming's comparable guide for another American, Connecticut native botanist Henry C. Cowles (1869–1939) at University of Chicago.[290] For his doctoral dissertation, Cowles investigated plant succession.

Two zoologists who earned Ph.D.s from University of Chicago, after studying partly under Cowles were Illinoian Charles C. Adams (1873–1955) and New York State native Victor E. Shelford (1877–1968).[291] Both published noteworthy books in 1913. Adams' *Guide to the Study of Animal Ecology* was presumably the first book to describe this subject in some detail with references. It was organized to accompany a course on animal ecology. One topic was "The Dynamic Relations of Associations and Aggregations, with Special References to Animal Associations."[292] His references included works by Möbius and Stephen Forbes.

Shelford's *Animal Communities in Temperate America, as Illustrated in the Chicago Region* is a substantial work, with 15 chapters, which included six revised articles previously published. He had studied communities in the Indiana sand dunes which Cowles and his plant ecology students had studied. He had 10 chapters on animal communities in nine different environments: large lakes, streams, small lakes, ponds, between land and water, swamps and floodplain forests, dry and mesophytic forests, thickets and forest margins, and prairies. His last chapter, "General Discussion," included "Relations of Communities."

The earliest clear diagram of a food web was probably by USDA entomologist Dwight Pierce and colleagues in 1912 on cotton boll weevil and its enemies.[293] Victor Shelford's 1913 diagrams of terrestrial and aquatic food chains had nothing in common with Pierce et al.'s 1912 diagram and he did not cite their article.[294]

The farm of the Quaker Allee family was near Bloomingdale, Indiana. Warder Allee (1885–1955) earned a B.S. degree at Quaker-supported Earlham College and his M.S. and Ph.D. at University of Chicago (1910 and 1912) under Shelford.[295] He taught at four other institutions before returning to Chicago, 1921–50, where he developed an informal school of animal behavior.[296] He was a very productive ecologist, using extensive laboratory experiments on behavior in social insects. Three books he published during 1930s drew upon experiments during 1920s. His doctoral students included the Park brothers, Orlando (Ph.D. 1929) and Thomas (Ph.D. 1932), both of whom served as ESA presidents. The second issue of 1956 *Ecology* was an Allee memorial number.

Pennsylvanian Edgar Transeau (1875–1960) was a graduate student under Cowles, 1900–01, before earning a Ph.D. at University of Michigan (1904).[297] After teaching at several institutions, he settled at Ohio State University in 1915. Clements cited six of his articles, two in succession (1909 and 1913).[298] Detroit native William Cooper's (1884–1978) mother bought him a copy of Asa Gray's *Manual of Botany*, as he loved flowers as a child. At Alma College, Transeau "influenced my scientific development more than any other person."[299] At graduate school at Johns Hopkins University, he became interested in relationships between vegetation and glacial geology. After transferring to University of Chicago, he realized he had learned Cowles' teachings under Transeau. Chicago geologist Rollin Salisbury "was the greatest teacher I have ever studied under."

Illinoisan Henry Gleason (1882–1975) earned B.S. and M.S. degrees (1901, 1904) at University of Illinois and Ph.D. at Columbia University (1906).[300] In 1908–13, he studied the vegetation of Midwest, mostly in Illinois. Most important of the nine papers he published during that time was "The

Vegetation of the Inland Sand Deposits of Illinois" (1910, 152 pages + photographs).[301] Malcolm Nicolson considered it to be "one of the most intensive studies undertaken by any American ecologist in the first two decades of the twentieth century."[302] It put Gleason in the league with Cowles and Clements. He disagreed with Clements, who claimed there could be only one kind of climax for a given region, which was determined by climate acting upon a specific group of plants. Gleason also claimed a climax process could be reversed, which also challenged Clements' orthodoxy. Clements cited Gleason's 1910 paper in his bibliography[303] but not in his "General Historical Summary," which cited only two papers of 1910, one by English botanist Charles E. Moss and one by himself.[304] Clements did discuss very briefly Gleason's papers of 1907 and 1910, pp. 215–216, but so briefly that he did not mention any theoretical interpretations between Gleason and himself.

A Canadian, who earned his Ph.D. under Cowles (1913), George Fuller (1869–1961),[305] pointed out in "Units of Vegetation and Their Classification" that there was a range of opinions between Clements and Gleason.[306] That was an unsatisfactory situation for Gleason, who in 1926 defended again his "individualistic concept" with more data than he had had in 1910. He still had little immediate influence—Clements continuing to defend his own concepts—but this time Gleason's ideas were not forgotten and became influential after World War II.[307]

A Russian who developed ideas similar to Gleason's was plant ecologist Leonty Ramensky (or Ramenskii, 1884–1953, figure 3.5), who graduated from Petrograd University in 1916 and earned a Ph.D. in 1935.[308] He argued that biotic communities consist of species behaving individualistically. His earliest landmark paper was published in 1924 and the excerpts are now translated into English.[309] There is no evidence of either Gleason or Ramensky being influenced by the other. But like Gleason, his arguments "made little headway against the holistic, superorganismic model of the biogeocenosis (ecological community plus abiotic environment) championed by Sukachëv and his many allies."[310] His influence spread after the publication of a collection of papers, including one by him, in 1956, on ecological evaluation of grazed lands by their vegetation. In his review of this book in Russian, Jack Major found that "This book is a wonderful compendium of information . . . a fitting monument to his career."[311]

Mathematically inclined Russian ecologist Georgii Gause (or Gauze, 1910–86) saw the difficulty of resolving the controversy between Clements' community concept and Gleason's individualistic concept of plant associations verbally and he explained how to make quantitative measurements to resolve the controversy.[312] He apparently was unaware of Ramensky's work.

British plant ecology historian Malcolm Nicolson has compared the work of Clements and Braun-Blanquet (1884–1980) and viewed their differences as representative of American and French plant ecology.[313] His discussion is fine but offering them as examples of national styles is problematic. Nicolson knows that Clements' American colleagues, Cowles and Gleason, significantly disagreed with him and so he did not represent an "American system" of plant ecology.

Swiss plant ecologist Josias Braun (after marriage, Braun-Blanquet, 1884–1980) studied at University of Zurich under Professor Carl Schröter and then earned a Ph.D. at University of Montpellier under Professor Charles-Henri Flahault.[314] Since Braun-Blanquet would become the head of a Zurich-Montpellier school of plant ecology, as Nicolson stated, it seems odd to name him representative of "French" plant ecology, even though he remained at Montpellier.

Yet, differences between Clements' and Braun-Blanquet's ecology were substantial. Braun-Blanquet synthesized his early studies in *Pflanzensociologie: Grundzüge der Vegetationskund* (1928, English 1932). His later work appeared in either German or French.[315] Ecologist Eddy van der Maarel compared his role in history of phytosociology to J.S. Bach's role in history of European classical music.[316] Maarel thought Braun-Blanquet had achieved his three "principles of a systematics of plant communities on a floristic basis." An American comment on Braun-Blanquet: "he developed his methods of community sampling, data reduction, and association nomenclature that dominate much of plant community ecology today."[317]

Danish plant ecologist Christen Raunkiaer (1860–1938) completed his studies at University of Copenhagen in 1885 and taught at Borchs Botanical College, 1885–88. He went on botanical expeditions in Europe and to Virgin Islands. He sought an objective ecological classification of vegetation. In 1987, American plant ecologists stated: "He developed a life form classification that is still widely used . . . and a quantitative method of sampling vegetation whereby data could be treated statistically without bias".[318] His data included *life form*, "basically the position of the plant part that survived the most severe season in any area"[319] He succeeded Warming as professor of plant ecology at the University of Copenhagen in 1911, when Warming retired. Raunkiaer published mainly in Danish and some in French and German, and his writings were little known to English speakers. Five plant ecologists, mostly at Oxford, remedied that by collecting and translating 17 of his papers as *The Life Forms of Plants and Statistical Plant Geography* (1934). There were at least three discussions of his work in English, but no English account of his life.[320]

At ESA's 1916 meeting, Clements spoke on "The Development and Structure of the Biotic Communities" coining *biotic community* and biome:[321]

> The biotic community is regarded as an organic unit comprising all the species of plants and animals at home in a particular habitat . . . . The biotic community, or *biome*, is fundamentally controlled by the habitat . . .

Clements did not focus on the biome in his research, but since Shelford studied animal communities, they had a meeting of minds in their *Bio-Ecology* (1939),[322] the first textbook covering both plant and animal ecology. *Bio-Ecology* synthesized their researches since early 1900s and they cited vast references. Their historical survey of previous studies on both plant and animal ecology began with Möbius (1877). He had coined *biocenose* to describe oyster communities but others used the term to designate plant—animal communities.[323] They mentioned their reports on bio-ecology in the yearbooks of Carnegie Institute of Washington, 1926–34. Clements had continued research that had begun in *Plant Succession* (1916) but Shelford had greatly expanded the scope of his investigations. Sometimes with students, Shelford had explored widely, including Mexico and Panama. Beginning in 1918, he spent alternate summers at Puget Sound Biological Station, teaching and researching. With other colleagues there, he wrote "Some Marine Biotic Communities of the Pacific Coast of North America" (1935), a 105-page synthesis in *Ecological Monograph*. Later, he condensed that study as the last chapter of *Bio-Ecology*.[324]

Plant ecologist E(mma) Lucy Braun (1889–1971), lifelong resident of Cincinnati, Ohio, earned three degrees at University of Cincinnati (B.A. 1910, M.A. 1912, and Ph.D. 1914). She joined the university's faculty in 1910, was professor of botany, 1927–46, and professor of plant ecology, 1946–48.[325] Her vegetation studies in Ohio and Kentucky culminated in her *Deciduous Forests of Eastern North America* (1950). She continued researching and publishing; later works included "The Phytogeography of Unglaciated Eastern United States (1955), *The Woody Plants of Ohio* (1961), and *An Ecological Survey of Fort Hill Memorial* (1969). She served ESA as vice-president in 1938 and president in 1950—the first woman in either office. She was one of the 50 botanists to be awarded the Certificate of Merit by a jubilee committee of the Botanical Society of America (BSA), "for her Contribution to our knowledge of the origin and structure of the Eastern American deciduous forest."[326] BSA's jubilee volume included her "The Development of Association and Climax Concepts" (1958). Her biographer, Ronald Stucky, judged her "One of the most original thinkers for developing field of plant ecology during the first half of the twentieth century in North America."[327] However, she was not a revolutionary.

Clements' concept of plant community dynamics was challenged and largely superseded by Gleason's continuum concept, thanks to Curtis and Whittaker. John Curtis was from Waukesha, Wisconsin—20 miles west of Milwaukee. He attended Carroll College and was attracted to plant physiology.[328] After graduating in 1934, he was accepted as a graduate student by the Botany Department, University of Wisconsin, Madison. He majored in physiology, minored in pathology, and never took a course in ecology. He spent summer 1935 at the university's Trout Lake Biological Station and collaborated with limnologist Chancey Juday on algal research. Curtis was senior author of their paper.[329]

Both Curtis' M.S. and Ph.D. (1937) dissertations were on orchid seed germination. In 1937, he became instructor of botany at U.W.-Madison and in 1938, he began teaching a course with two other botanists on physiology, morphology, and ecology. In 1940, he became Director of Research at University of Wisconsin Arboretum. He joined ESA in 1941 or early 1942. In 1941, he won a Guggenheim Fellowship to

FIGURE 3.6 (a) Arthur G. Tansley. By Wilfrid G. de Glehn, 1940. National Portrait Gallery, London. (b) Raymond L. Lindeman, 1939. College of Biological Sciences, University of Minnesota.

study Michigan, Wisconsin, and Minnesota forests. After two months, World War II began and he became research director of a Rubber Development Corporation experiment station to obtain rubber latex from *Cryptpstegia* vines (commercial failure).

After returning to U.W.-Madison in 1946, Curtis began research on Wisconsin vegetation—his major life work.[330] His school of continuum ecology began with Gleason's *individualistic concept*, in which each species' distribution differed from other associated species. Continuum studies by him and numerous graduate students culminated in Curtis' *The Vegetation of Wisconsin* (1959), 657 pages). His legacy in plant ecology seems comparable to that of Cowles.

Despite Clements and Shelford's example, plant ecologists Henry Oosting (1903–68) and Rexford Daubenmire (1909–95) wrote textbooks on plant communities.[331] Animal ecologists could not slight plants as easily as plant ecologists could slight animals. In their encyclopedic *Principles of Animal Ecology* (1949, xii + 837 pp.), Clyde Allee (1885–1953), Alfred Emerson (1896–1976), Orlando Park (1901–69), Thomas Park (1908–92), and Karl Schmidt (1890–1957) entitled one of five sections "The Community."[332] Five of six chapters in that section were by Orlando Park and the sixth was coauthored by him and Schmidt.

Aquatic ecologists (mostly animal ecologists), whether limnologists or marine ecologists, like land ecologists, considered plants. Even when aquatic animals were more conspicuous than phytoplankton, discussions of animal communities included plants. Franz Ruttner (1882–1961), from Bohemia, now in Czech Republic, became director of the biological station at Lunz am See, Austria. The first edition of his *Fundamentals of Limnology* appeared in German in 1940. The third edition, translated into English in 1963, devoted 101 pages to aquatic environments and 145 pages to biotic communities, subdivided into 70 pages on plankton and 75 pages on vascular plants, animals, and decomposing organisms. George Reid's *Ecology of Inland Waters and Estuaries* (1961) devoted some 200 pages to environments and 130 pages to aquatic life—subdivided into 15 pages on Protista and plants and 113 pages on animals and aquatic communities. Robert Wetzel's *Limnology* (1975) had 280 pages on aquatic environments and 370 pages on plants, animals, and communities.

Georgian Lee R. Dice (1887–1977) earned a B.A. degree at Stanford University (1911) and M.A. and Ph.D. (1914, 1915) at University of California, Berkeley.[333] During summer 1919, he settled permanently at University of Michigan, Ann Arbor. He served as vice-president (1938) and president (1952) of the ESA and was ESA's Eminent Ecologist for 1964. His research focused upon relationships and distributions of species. His main publications were "Life Zones and Mammalian Distribution" (1923), *The Biotic Provinces of North America* (1943), and *Natural Communities* (1952). *Natural Communities* was a well-organized summary of contemporary knowledge, but with no new insights.

Eugene Odum's textbook, *Fundamentals of Ecology* (1953, ed.3, 1971), like Charles Elton's *Animal Ecology* (1927), was a landmark that helped reorient ecology.

Kansan Robert Whittaker (1920–80) was a son of a zoology professor. He majored in zoology at Washburn University in Topeka (B.A. 1942) and then joined U.S. Air Force for 3.5 years.[334] He served as a meteorologist in England, and in his spare time read works by Clements, Weaver, and Shelford. He thought Shelford's book was "rather awful," probably *Animal Communities in Temperate America* (1913, reprinted 1937) rather than his *Laboratory and Field Ecology* (1929).[335] Departing the Air Force, he applied to Department of Botany, University of Illinois for graduate school. He was rejected because he had taken few botany courses as an undergraduate. However, Department of Zoology accepted him and in March 1946, he began studies under Shelford, who retired in June, leaving Whittaker to Charles Kendeigh (1904–86). Whittaker was interested in Merriam's life zone concept, developed in western mountains. So, Kendeigh suggested he study communities in Great Smoky Mountains for a dissertation. He studied there both insect and plant distributions and relationships. He only found time to work up his vegetation data for a dissertation, thus receiving a Ph.D. in zoology (1948) for a plant ecology dissertation—published in 1956 in *Ecological Monographs.* He had promised his doctoral committee to publish his insect data, which he did in 1952 in *Ecological Monographs.* He became one of the few ecologists who studied both plant and animal ecologies.

By 1956, he was sufficiently well known to be invited by the BSA to be one of the 40 authors who provided an article for BSA's Golden Jubilee Volume (1958). Whittaker's "Recent Evolution of Ecological Concepts in Relation to the Eastern Forests of North America" followed Bran's similar article. Her article mildly criticized two of his earlier articles and praised his "Vegetation of the Great Smoky Mountains" (1954). His 1958 article mainly criticized Clements' vegetation units, but he also evaluated Braun's modifications of Clements' system. Whittaker concluded that hers was an improvement over Clements' but did not go far enough (toward Gleason's).

Whittaker had a broad vision and was a critic of ecological theory. He showed his broad vision in his 230-page survey, "Classification of Natural Communities" (1962), beginning with Humboldt and ending in 1950s with Braun, Braun-Blanquet, himself, and others.[336] Yet, he did not champion a particular classification of vegetation. His quest became quantitative in "Gradient Analysis of Vegetation" (1967). He explained that gradient analysis "seeks to understand the structure and variation of the vegetation of a landscape in terms of gradients in space of variables on three levels—environmental factors, species populations, and characteristics of communities."[337] His understanding progressed further in *Communities and Ecosystems* (1970), with chapters on "Community Structure and Composition," "Communities and Environments," "Production," and "Cycling and Pollution," in terrestrial ecosystems. He provided a vegetation map of world lands, with a dozen vegetation formations.[338] That was a century since Grisebach's map. Whittaker subdivided world biomes: two types of temperate forests, two types of thorn woodlands and shrubs, distinguished subarctic and subalpine taiga, four types of deserts, and five types of hydric communities.[339] He thought gradient analysis had "changed the conception of vegetation as much as research on the genetic basis of variation and evolution has changed the concepts of plant species." Shortly before he died, ESA chose Whittaker as the Eminent Ecologist of 1981.[340]

Donald Strong, Jr. et al. edited another noteworthy synthesis, *Ecological Communities: Conceptual Issues and Evidence* (1984), which nevertheless led a reviewer to conclude: "mechanisms underlying the structure of animal communities will not be easily revealed."[341]

Plant ecologist Paul Keddy suggested (2007) that Gleason was not entirely correct and Clements entirely wrong.[342] Vegetation in different regions can exhibit different degrees of interdependency, and therefore the hypotheses of Gleason and Clements can be seen as different positions on a continuum between randomness and structure. His claim fits a pattern seen occasionally in history of science: conflicting hypotheses become reconciled with scientific progress.

Conclusions

The history of studies on succession, community, and continuum was largely the domain of plant ecologists, which is understandable, since the earth's biomass is 99% plants. There were many worthy observers who collected useful data but only a few steered this area of plant ecology in a new direction. Linnaeus was the first theoretician in natural history since Aristotle and Theophrastos; he collected data and developed the concept of *oeconomiae naturae*, which extended the balance of nature to include plants. Alexander von Humboldt collected much more data and constructed a correlational science that explained how environmental factors affected species distributions. He also exerted much influence upon ecological botany during the 1800s. Anton Kerner was influenced by Humboldt but also went his own way in defining communities or formations, in part of Danube Valley (1863). Edward Forbes, Karl Möbius, and Addison Verrill were marine zoologists whose researches focused on near-shore animal communities in which animals were more conspicuous than associated phytoplankton. Eugenius Warming wrote the first comprehensive synthesis of plant ecology (1895). American plant ecologists Henry Cowles and Frederic Clements advanced the knowledge of plant communities and succession. Clements' rigid climate-controlled community succession dogma dominated in America for several decades. Henry Gleason and Leonty Ramensky, on opposite sides of the earth, concluded that plant species requirements are unique which determine their distributions, independent of neighboring species, and cohabiting species did not constitute actual communities. Neither Gleason nor Ramensky gained a substantial following, but John Curtis was more successful and Whittaker also contributed. Keddy suggested that Gleason was not entirely right and Clements entirely wrong but that they defended extremes on a continuum of situations.

D. Ecosystems, Systems, and Productivity

Bibliographical Introduction

Two histories of ecosystem concepts, by Joel Hagen (1992) and Frank Golley (1993), provide our introduction to much of this part of our history. Having the second published history, Golley explained their differences, from his perspective:[343]

> Hagen and I use approximately the same material, but we each tell two different stories about ecosystem ecology. His approach is to embed the ecosystem concept in a history of American and English ecology. My approach has been to focus on the concept and the individuals contributing to concept development and then to consider causal factors that led to pattern we observe. Joel Hagen is a historian and comes to the story from outside . . . I was a participant in the story and approach it from within.

French historian of ecology Jean-Marc Drouin wrote his doctoral dissertation (1984) on *La naissance du concept d'ecosysteme*,[344] which focused on key papers by Stephen Forbes (1880), Arthur Tansley (1935), and Raymond Lindeman (1942). He published a lengthy summary.[345] Hagen began his history with Darwin (1859) and Golley his with Kerner (1863). Robert McIntosh (1985) embedded concepts discussed here in a still broader context.[346] Robert O'Neill et al. discussed (1986) how ecologists have viewed ecosystems in *A Hierarchical Concept of Ecosystems*.[347] Chunglin Kwa surveyed in four publications (1987–93) the aspects of American ecology systems and American participation in the International Biological Program (IBP). Robert Ricklefs and Gary Miller's ecology textbook (2000) had a historical chapter on ecosystems,[348] omitted from their brief version (2007). Christian Lévêque's *Ecology from Ecosystem to Biosphere* (2003) also has historical discussions.[349]

Danish botanist Eugenius Warming (1841–1924) attended University of Copenhagen but took leave from classes for three years, 1863–66, to be secretary to zoologist P.W. Lund in Brazil.[350] While there, Warming studied tropical vegetation, later subject of two substantial publications.

He is most remembered for *Pflanzesamfund: Grundträk af den okologiske Pflantesgeografie* (1895, German 1896, Russian 1901, and English 1909), his first textbook on plant ecology. It lacked illustrations, but had an extensive bibliography. The title of Danish and German editions included "ecological plant geography," replaced in English edition by "An Introduction to Plant Communities." Section 1 discussed climate soils, animals, and effects of vegetation;

Section 2 discussed symbiosis, mutualism, and commensalism; Section 3 discussed adaptations of aquatic and terrestrial species, ecological classification, distinguished between xerophyte and mesophyte terrestrial "communities." Yet, he did not place chapters under these headings; instead, he had 10 "formations" determined by moisture or soil: hydrophytes (aquatic), heliphytes (marsh), oxylophytes (sour soil), hylophytes (saline), lithophytes, psychrophytes (cold soil), psammophytes (sand, gravel), eremophytes (desert, steppe), chersophytes (waste), psilophytes (savannah), two formations based on species types (sclerophyllous and coniferous), and a last formation of mesophytes. His final section was on "Struggle between Plant Communities."

Warming's successor as professor of botany at University of Copenhagen was his former student, Christen Raunkiaer (1860–1938), who studied "life forms," which could be used to compare plant formations of different regions (1903, English 1934). His life forms were based on parts of vascular plant species that survive unfavorable seasons. He published a one-page article 1903) defining five alternatives: Phanerophytes (surviving buds or shoot apices borne on negatively geotrophic shoots projecting in the air); Chamaephytes (surviving buds or shoot apices very close to ground), Hemicryptophytes (surviving buds or shoot apices situated in soil surface); Cryptophytes (surviving buds or shoot apices buried in ground at a distance from surface that varies in different species); and Annuals or Therophytes (plants of summer or of favorable season). He listed sub-types under each heading. Readers could do little with his scheme until he published a 103-page monograph of explanations and examples (1907, English 1934). W.G. Smith explained his concepts in the first issue of *Journal of Ecology* (1913). Americans became interested in what L.A. Kenoyer entitled "A Study of Raunkiaer's Law of Frequence" (1927). Henry Gleason (1929), George Fuller (1935), and Frank Preston (1948) also discussed that topic. Five Oxford and Cambridge plant ecologists published 17 of Raunkiaer's papers in English translation in 1934. So, Fuller and Preston did not have to consult the Danish versions.[351]

American demographic statistician Alfred Lotka (1880–1949), oddly, wrote a book, *Elements of Physical Biology* (1925), for an audience whom he barely knew, biologists, who generally lacked a mathematical background.[352] Its chapters included "The Fundamental Equations of the Kinetics of Evolving Systems" and "Inter-Species Equilibrium." He perhaps first diagramed the natural environmental cycles of elements. An audience developed after his book appeared.

Animal ecologist Charles Elton (1900–91) was educated at Oxford University and remained there.[353] He participated in three Oxford expeditions to Spitzbergen and was head of scientific work on the third one (1924). He was impressed by transfer of nutrients from sea to land when sea birds brought fish to their young, and he drew a diagram showing a complex food web illustrating this, entitled "Food cycle among the Animals on Bear Island."[354] His diagram has been reproduced at least twice since.[355] Not just this diagram won Elton notice in histories of ecosystem concepts[356] but Elton also seemed to have had an implicit awareness of ecosystems without naming and defining them. Chapter titles in his landmark *Animal Ecology* (1927) included: "The Distribution of Animal Communities," "Ecological Succession," "Environmental Factors," and "The Animal Community." The niche concept was prominent in his discussion of animal communities.[357]

Ecosystem Concept

Englishman Arthur Tansley (1871–1955) was the founding editor of BSA's *Journal of Ecology*[358] and he received two articles from South African John Phillips—"The Biotic Community" (1931) and "Succession, Development, the Climax, and the Complex Organism: an Analysis of Concepts" (1934–35)—which he published.[359] Yet, he decided Phillips had followed Clements' misguided superorganism concept. He attacked Phillips' concepts in *Ecology*, "The Use and Abuse of Vegetational Concepts and Terms" (1935). In quest of a term to counter Phillips' *com-plex organism*, Tansley coined the term *ecosystem*, which included inanimate environment, plants, and animals.[360] His idea of ecosystem placed it in a series of systems from atoms to universe. He did not use ecosystem to take plant ecology in a new direction.[361] He merely wanted to replace Phillips' muddled ideas. Nor were contemporary ecologists pushed in a new direction by reading his paper.

The first significant use of the term *ecosystem* was by American "post-doc" at Yale University, Dr. Raymond Lindeman (1915–42)[362] For his doctoral dissertation at the University of Minnesota, he had studied limnology and succession at Cedar Bog Lake for five years. In September 1941, he went to Yale on a fellowship to study with Evelyn Hutchinson. Lindeman was already familiar with some of Hutchinson's publications and he had corresponded with Hutchinson and his student Edward Deevey. Lindeman's goal was to revise and expand the last chapter of his dissertation under Hutchinson for a scientific article.

In "The Trophic-Dynamic Aspect of Ecology" (1942), Lindeman explained: "The trophic-dynamic viewpoint . . . emphasizes the relationship of trophic or 'energy-availing relationships within the community-unit to the process of succession."[363] He briefly discussed Tansley's term *ecosystem* as being "of fundamental importance in interpreting the data of dynamic ecology,"[364] but without realizing he was exploring what would become a major aspect of modern ecology. He died before his paper was published and Hutchinson saw to its appearance in *Ecology*.

Post-War Ecosystem Studies

Eugene Pleasants (EP) Odum (1913–2002) and Howard Thomas (HT) Odum were sons of Howard W. Odum (1884–1954), Professor of Sociology at the University of North Carolina (UNC), Chapel Hill. Their father was the leading sociologist in the South and his publications were a precursor of human ecology. His landmark *Southern Regions of the United States* (1936) had chapters entitled "Geographic Factors and Physical Setting," "Regions of Natural Richness and Abundance," "An Agrarian Country," and "The Tennessee Valley." He was his sons' earliest intellectual influence.[365]

As a youth, EP Odum was interested in wildlife, especially birds, and life in streams.[366] While in high school, he visited the North Carolina State Museum in Raleigh and met its director, Herbert H. Brimley. "H.H. Brimley and his brother C.S. did more than anyone else to encourage me to develop my interest in birds."[367] A close friend of EP Odum was Coit Coker, son of zoology professor Robert Coker. EP and Coit wrote a regular column in *Chapel Hill Weekly* on local birds and encouraged readers to plant shrubs attractive to birds.[368] Upon graduating in 1934, the friends bought a used car and toured scenic parks in the West. After earning an M.A. degree at UNC (1936), EP Odum taught biology at Case Western University in Cleveland, 1936–37, while taking two physiology courses.[369] During summer 1937, he banded birds for U.S. Fish and Wildlife Service and measured the heart rates of small birds for a dissertation at University of Illinois under animal ecologist Charles Kendeigh. EP Odum earned his Ph.D. in 1939 and his dissertation was published in *Ecological Monographs*.[370] In 1940, he became an instructor of zoology at University of Georgia, where he remained. He published the first edition of his outstanding *Fundamentals of Ecology* in 1953, and in 1954 he was promoted to full professor. In 1957, he was given an endowed chair. In 1961, Kendeigh published *Animal Ecology*, which had a very different scope than EP Odum's *Fundamentals*.

His biographer, Betty Jean Craig, noted the synthetic characteristic of EP's *Fundamentals*:[371]

> Odum derived his vision of nature from Clements's theory of the climax formation, or superorganism' Shelford's concept of the biome; Smut's articulation of the characteristics of the whole; Tansley's ecosystem concept; Leopold's definition of land; Lotka's theories of energy transfer; and Lindeman's energetics.

These were his most important published influences, but he also was influenced by his younger brother to whom he offered co-authorship.[372] HT Odum thought that would give him more credit than he deserved but after it appeared without his name on the title page, he felt under-appreciated, and so the title page of the second (1959) and third editions (1971) acknowledged: "In collaboration with Howard T. Odum."

EP Odum's list of intellectual debts included both holists and reductionists. How did he reconcile these different perspectives? Ecologist-philosopher Donato Bergandi complained that Odum did not notice the contradiction (1995). Bergandi compared discussion of ecosystem in the three editions of *Fundamentals* and found: "In the second and third editions, the individual organisms disappear and are replaced by 'functional processes.'"[373] He found it "truly paradoxical that an author as strongly involved epistemological holism as Odum confines himself to understanding the ecosystem essentially in terms of energy."[374] EP Odum may simply have added brother HT Odum's reductionistic approach to his own holistic approach. In 2001, Robert O'Neil complained that EP Odum's ecosystem concept was no longer adequate (without citing Bergandi's critique).[375]

The Atomic Energy Commission (AEC) in 1951 built a plant on Savannah River (SRP), near Aiken, South Carolina, to produce tritium for atomic bombs, and it

FIGURE 3.7 (a) Eugene P. Odum. University of Georgia Archives. (b) Howard T. Odum. Web. (c) Stanley I. Auerbach. *ESAB* 87 (2004) 91.

invited universities of Georgia and South Carolina to submit proposals for biological inventories on its 250,000-acre reservation. EP Odum and colleagues submitted a proposal for both inventory and ecological study of interrelationships.[376] AEC offered each university much smaller grants than expected, but Odum decided he could use graduate students for research and accepted the grant. Thus began a long and important relationship between AEC and Georgia's research team under Odum, which enabled him to increase the number of graduate students his Zoology Department could support and also hire full-time researchers.

Odum's plan was "to use energy flow through biological food chains as a means of linking plants and animals"[377] and radionuclide tracers enabled his teams to quantify movements of matter and energy through ecosystems and determine times involved. They could map food webs and shift focus from an ecological structure to dynamics. For Odum, that was the beginning of modern ecology research, teaching, and institutional growth.[378]

HT Odum grew up in the shadow of brother Gene's shadow, who taught him about birds. HT learned about electricity from *The Boy Electrician*, and in high school in Chapel Hill, he worked for UNC zoology professor Robert Coker, who taught him about fish and philosophy of biology.[379] After two years studying at UNC, HT Odum served three years in the U.S. Air Force during World War II. He was "a tropical meteorologist, where undoubtedly he gained his basic interest in large systems and the energetics behind them."[380] He served in Puerto Rico and Panama Canal Zone and then returned to graduate from UNC in 1947. He went to Yale and worked under Evelyn Hutchinson for a Ph.D. In a 1943 paper, Hutchinson wrote on use of aluminum and other elements in plants and animals. He stated that "The most unfortunate uncertainty involves the position of strontium."[381] He suggested to HT Odum that he conduct dissertation research on the strontium cycle.[382] He provided HT with examples of his own studies on cycles of phosphorus, nitrogen, and carbon; his carbon study included a helpful diagram.[383]

From HT Odum's dissertation, *The Biogeochemistry of Strontium* (1950), he published four brief papers in 1951 and two more longer articles containing his data in 1957.[384] Strontium, the fifteenth most common element on Earth, is not physiologically important in most species. He argued in his dissertation and in two papers that the strontium cycle of the environment and in organisms is stable.[385] Shortly after finishing his dissertation, radioactive strontium was discovered to be a dangerous product of atomic bombs.

Even before obtaining his Ph.D. (1951), HT Odum became assistant professor of zoology at the University of Florida, 1950–54, where he conducted a pioneering study on productivity in Silver Springs, 1951–56.[386] Afterwards, he moved to Duke University, 1954–56 (where he taught me), then to three other institutions before settling back at University of Florida in 1970. Despite these moves, publishing was extraordinary:[387]

Odum published, in addition to the Silver Springs monograph (Odum 1957a): (1) an explication of the theoretical foundations for his career (Odum & Pinkerton 1955), (2) an award-winning (ESA's George Mercer Award) description of a coral reef's ecosystem's trophic structure and energy flow (Odum & Odum 1955), (3) a synthesis of his method for measuring integrated aquatic ecosystem metabolism (Odum 1956), (4) a comparative analysis of ecosystem metabolism in Florida springs (Odum 1957b), and (5) a pioneering microcosm study of stream ecosystems (Odum & Hoskin 1957).

The AEC provided EP Odum with research funds for "Trophic Structure and Productivity of a Windward Coral Reef Community on Eniwentok Atoll." Because HT Odum had developed expertise relevant for this study, EP Odum obtained his collaboration for six weeks during summer 1954.[388] A U.S. military base was on Einwentok Atoll which had conducted atomic explosive tests. EP Odum's biographer explained their situation at Japan Reef research site and their simple equipment.[389] It was already known that coral animals have a symbiotic relationship with single-cell algae in their tissues. Coral metabolism produces a waste, CO_2, needed for algae metabolism and algae produces a waste, O_2, needed in coral metabolism. The Odums' measurements indicated, however, that *zoo- xanthellae* algae were not numerous enough to meet all nutritional needs of coral polyps. The Odums discovered in corals "plant protoplasm exceeds the animal biomass (about 3 to 1) and the filamentous green algae have a greater biomass than the zooxanthellae (about 16 to 1)."[390] Besides nutrients the algae provided, coral obtained other nutrients from ingesting plankton. "Thus, the coral reef is like most known self-sufficient ecosystems in having a much greater weight of plant biomass than animal biomass."[391] After publishing their coral reef study (1955), both brothers continued studying ecosystems, but even so, they followed divergent paths.

EP Odum was the president of Ecological Society of America in 1966, and his presidential address, "The Strategy of Ecosystem Development" (published 1969) was an important synthesis of his thought. He constructed a "tabular model of ecological succession" to explain "trends to be expected in the development of ecosystems," listing 24 attributes under six categories: community energetics, community structure, life history, nutrient cycling, selection pressure, and overall homeostasis. EP's fifth attribute (under community energetics) was on food chains. HT Odum was researching this subject,[392] but EP Odum's citations on this did not include HT Odum's work. EP Odum explained that simple food chains in developing ecosystems become complex as the ecosystem matures.[393]

The pattern which American zoologists Robert MacArthur and Edward O. Wilson found in stages of island colonization paralleled ecological succession on continents.[394] Species with high rates of reproduction and growth (*r*-selection) are more likely to survive in early uncrowded stages of island colonization, but as time goes

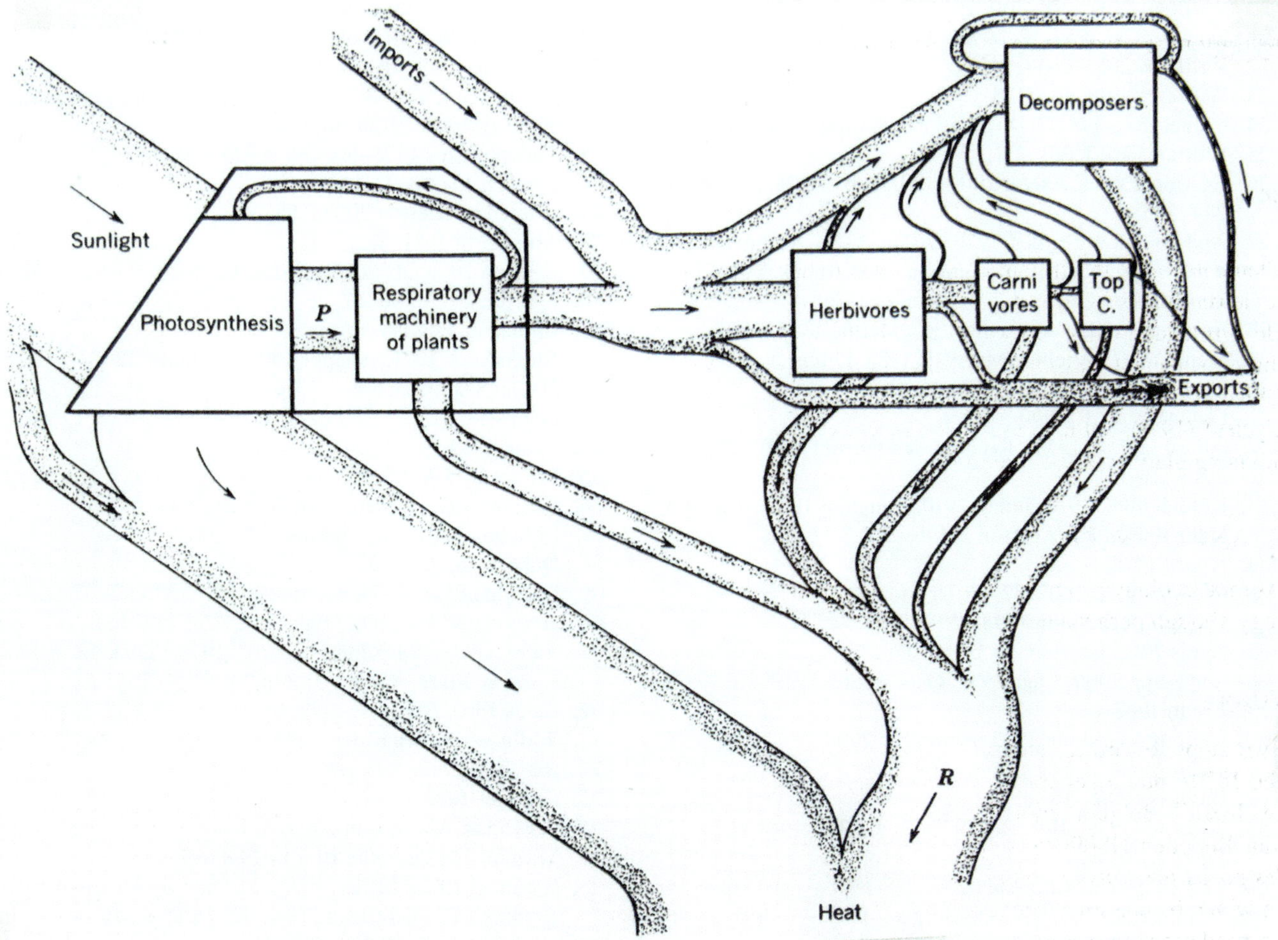

FIGURE 3.8 Energy flow diagram for Silver Springs ecosystem, Florida. HT Odum 1957a: 61.

by, selection pressure favors species with lower growth potential (*K*-selection) but better capabilities for competitive survival under equilibrium densities. Quantity production in young ecosystems yields over time to quality production and feedback control in mature systems. Yet, a few early succession species persist through later stages.

EP Odum acknowledged that not all ecosystems develop from early stages to maturity. Physical perturbations can impose an intermediate development. Examples include fluctuating water level ecosystems—marine estuaries and intertidal zones or freshwater Florida Everglades. Life cycles of many species are adapted to such environments. Wood storks in Everglades breed when water levels are falling and small fish become concentrated in shrinking pools, where easily caught. If water levels do not fall, storks do not breed.

Ecosystem concept has become one of the foundations of ecology. Yet, in 2001, Robert O'Neil rhetorically asked, "Is It Time to Bury the Ecosystem Concept?"[395] Numerous titles in his "Literature cited" show he was not a solitary voice crying in the wilderness. Complaints claim that the ecosystem concept in practice is simplistic and so a distortion.

NOTES

1. Egerton 2012a: 4–7.
2. Stresemann 1975: 7, Scarborough 1977, 1979, Morge 1973: 49–51, Egerton 2012a: 12–13.
3. Wallace 1970, Morge 1973: 65–68, Weisheipl 1980, Egerton 2012: 27–30.
4. De Renzi 2000a, b.
5. Ornstein 1938. McClellan 1985.
6. Stimson 1948, Purver 1967, McClellan 1985: index, Sorrenson 2000.
7. Hahn 1971, McClellan 1985: index, 2000.
8. Bates 1965: 5–9.
9. Bates 1965: 1–27, McClellan 1985.
10. McClellan 1985: 257.
11. Simpkins 1975.
12. Gage 1938, Gage & Stearn 1988.
13. Neave & Griffin 1933: 1.
14. Neave & Griffin 1933: 2.
15. Stroud 1992: 28–58.
16. Foote 1970, Maclean 1972, Gray 1981, Carter 1988, Williams 2013: 73–94, Egerton 2015b: 241–242.
17. Williams 2013: 95–121.
18. Ledyard 2005, Williams 2013: 124–131.
19. Dunmore 1965–69, Williams 2013: 150–178.

20. Williams 2013: 179–200, 206–221.
21. Dodge 1971: 195–212, 1976: index.
22. Williams 2013: 150–164.
23. Williams 2013: 164–178.
24. Pritzel 1871–77: 173, Duyker 2003, Williams 2013: 165–178.
25. Williams 2013: 201–231.
26. Egerton 2006d, 2007d, 2009c, 2011a, b, 2012a, b, c, 2015b: 244–245.
27. Andrist 1967, Snyder 1970, Allen 1975, Ambrose 1996, Ambrose & Abell 1998, Cutright 2003, Tubbs & Jenkinson 2003, Dalton 2008.
28. Bartlette 1962, Goetzmann 1967, Meinig 1886–2004.
29. Stanton 1975, Viola & Margolis 1985, Philbrick 2003.
30. Leonhart 1985, Stann 1985.
31. Fedunkiw, M.P. 2000.
32. Allard 1970, 1978, Rivinus & Youssef 1992, Egerton 2018b:31–32.
33. Kofoid 1898, 1910, Juday 1910, Vaughan 1937, Jack 1945, Hiatt 1954, 1963, Egerton 2014b, c.
34. Allard 1978.
35. Rozwadowski 2002: 42–76, Egerton 2014c: 396–399.
36. Fischedick & Shinn 1993.
37. Ayres 2012, Egerton 2013: 355–358.
38. Sorrenson 2000, Gage 1938, Gage & Stearn 1988.
39. Sheail 1987.
40. Rivinus & Youssef 1992, Fedunkiw 2000.
41. Steere 1958.
42. Burgess 1996: 115.
43. Shelford 1938, Burgess 1977: 2, Croker 1991: 120–121, Egerton 2015b.
44. Croker 1991: 121, Cassidy 2007: 77–80.
45. McIntosh 1976: 255–356, Burgess 1977: 7, 1985: 66–67, Croker 1991: 64–66, 2001: 125, 169–170.
46. Burgess 1977: 3, graph reproduced in Egerton 2015b: 247.
47. Taylor 1938.
48. Moore 1938.
49. Burgess 1977: 16–18.4.
50. Frey 1963a, Hasler 1963, Fry & Legendre 1963: 488–489, Northcote & Larkin 1963: 460, Slack 2010: 115–151.
51. Lauff 1963.
52. Egerton 2014c: 348–374, 2016c: 372.
53. Burstyn 1980, Fye 1980, Schlee 1980.
54. Deacon 1980, Haedrich & Emery 1980, Revelle 1980.
55. Cullen 2005.
56. Barrett & Barrett 2001, Craige 2001: 31.
57. Bocking 1997: 151–202, 2004.
58. Dauben 2000, Egerton 2000, Helvoort 2000, Hoppe 2000, Mills 2000, and more.
59. Leroy 1957.
60. Egerton 2015b: 249–289. See Frank N. Egerton [online].
61. Paracer & Ahmadjian 2000: 6.
62. Stanley 2006: 857.
63. Sapp 1994.
64. Cheng 1971.
65. Quoted in: Clendening 1942: 4, Hoeppli 1959: 5.
66. Herodotos 1926–38, vol. 2, pp. 135–137, bk. 3, section 108–109.
67. Herodotos 1926–38, vol. 1, p. 98, bk. 2, section 28.
68. On balance of nature: Egerton 1973: 326, 2001a: 95–96, 2001b: 151, 2012a: 2–3, 2015a: 81.
69. Balme 1970, Chroust 1973, Souder 2000, Egerton 2001a, b.
70. Aristotle 1965–91: vol. 2, 547b6–17.
71. Townsend & Crane 1973: 387–389.
72. Aristotle 1965–91: vol. 3, 623b15 to 628b22.
73. Aristotle 1965–91: vol. 2, 557a9.
74. McDiarmid, J.B. 1976, Wohrle 1985, Ophuijsen & Realte 1998, Egerton 2012a: 14, n. 49.
75. Theophrastos 1976–90: vol. 2: 333–335.
76. Egerton 1983: vol. 1, 448.
77. Quoted in Egerton 2003a: 88.
78. Ainsworth 1981: 41.
79. Egerton 2008a, b, 2012b, 2013a, b.
80. Egerton 2012a: 80–82.
81. Linnaeus, English edition 3, 1775: 64–65.
82. Chichester 1921, Snapp 1999, Calhoon 2004, Egerton 2012a: 87, 2015a: 84.
83. Dobbs 1750.
84. Grant 1949.
85. Pritzel 1871–77: 303, Egerton 2008b: 170–171, King 1975.
86. Knight 1799, Herbert 1837, Gärtner 1844–49.
87. Limoges 1970: 151, Egerton 1973: 341.
88. Darwin 1859: 92–93.
89. Darwin 1859: 73–74, McAtee 1947, Egerton 2007a: 52–53.
90. Darwin 1860/61, 1862, Egerton 2011c: 359–363.
91. Bates 1863, McKinney 1970, 1976, Dickenson 2004, Egerton 2012b: 42–49, Drouin 2014: 151–157.
92. Bates 1862, Blaisdell 1992: 63–141.
93. Wallace 1865, Blaisdell 1992: 144–156.
94. Blaisdell 1992: 186–197.
95. Dressler 1982: 375.
96. Paracer & Ahmadjian 2000: 3.
97. Mitman 1992: 65–66, Brooks 1998: 97–133.
98. Naumov 1973, Miller 1976, Mitman 1992: 66–67, Sapp 1994: 20–23, Dugatkin 2006: 12–36 2012, Borrello 2010: 30–38, Hale 2014: 227–235.
99. Evans & Evans 1970, Mallis 1971: 362–368, Shor 1976, Burgess 1996: 110–112.
100. Wheeler 1911, 1923, Evans & Evans 1970: 255–256, 263–266, Wilson 1971: 317–318, Borrello 2010: 28–29.
101. Engel 1983: 172–173, Kimler 1990, Mitman 1988, 1992, 1999, Sapp 1994: 1137–139, Burgess 1996: 11, Sterling 1997a.
102. Park 1967, Wilson & Michener 1982, Burgess 1996: 39–40.
103. Allee et al. 1949: chapters 24, 31–35.
104. Grafen 2004, Dugatkin 2006: 86–106.
105. Grafen 2004: 113.
106. Grafen 2004: 116, Cronin 1991, Hamilton 1963, 1964.
107. Wilson 1975: 118.
108. Wilson 1994: 315–327.
109. Ruse 1979, Segerstrâle 2000.
110. Dugatkin 2009: 326–332.
111. Woolfenden & Fitzpatrick 1984, 1996: 16–17, Skutch 1999, Dugatkin 2006: 120–121.
112. Allen 1979: 260–264.
113. De Waal 2014, Stix 2014.
114. Kingsolver & Paine 1991: 313.
115. Allaby 1994: 88.
116. Egerton 2013b.
117. Foster 1965, Philip & Rozeboom 1973, Kean et al. 1978, Penso 1981, Grove 1990.
118. Böhner 1933–35, Ainsworth 1981, Campbell et al. 1999.
119. Thompson 1939, DeBach 1974.
120. Bulloch 1938, Winslow 1943, Brock 1961, Clark 1961, Waterson & Wilkinson 1978, Wilkinson 1992.

121. Howard 1933, Mallis 1971: 79–86, Hatch 1972, Sterling 1997c, Egerton 2013a: 73–74.
122. Howard & Fisk 1911.
123. Schwerdtfeger 1973: 381.
124. Scudo & Ziegler 1978: 396.
125. Ewing 2006.
126. Kiger 1962.
127. Egerton 2015a: 98–100.
128. See above and Egerton 2015a: 85–87, 101–102.
129. Van der Pas 1976.
130. Dammerman 1945: 66.
131. Regal 1982: 498.
132. Ainsworth 1976: 26–197.
133. Jost 1930, Robinson 1971, Paracer & Ahmadjian 2000: 231–233, Egerton 2012b: 314–316.
134. De Bary 1866: 291, relevant passage translated in Mitchell 2002: 195, and quoted in Egerton 2015a: 104.
135. Schwendener 1867, Sanders 1975, Ainsworth 1976: 97–99, Cittadino 1990: 27–31, 52–55, Mitchell 2002: 195, 2007: 110, Kärnefelt et al. 2012: 17–18.
136. Summarized in English by Schneider 1897a: 24–25.
137. Schwendener 1869: 38–39, relevant passage translated in Mitchell 2002: 195.
138. Mitchell 2011: 124–125.
139. Translation in Khakhina 1992: 21.
140. Famintzin & Baranetsky 1867a, b.
141. Baranetsky 1868, translated in Khakhina 1992: 21.
142. De Bary 1868, translated in Mitchell 2011: 126.
143. Reess 1872.
144. Stahl 1874, Mitchell 2006: 153.
145. Kärnefelt et al. 2012: 18.
146. Mitchell 2018: 165.
147. Collander 1967: 23–28, Lawrence 1974, Vitikeinen 2000.
148. McDiarmid 1976, Egerton 2012: 6–7, see above, n. 14.
149. Theophrastos HP 2.1.2, 1916, CP 2.17.5, 1976.
150. Sapp 1994: 6–7, Mitchell 2015: 5–7.
151. Desmond 1977: 649.
152. Mitchell 2011: 128, citing Weddell 1875: 255–256.
153. Mitchell 2011: 130.
154. Mitchell 2011: 140.
155. Cattell & Brimhall 1921: 604, Sapp 1994: 32–34, Mitchell 2011: 129.
156. Plitt 1934, as quoted by Mitchell 2011: 129.
157. Schneider 1897.
158. Schneider 1897: 930–932, outline quoted in Egerton 2015a: 108–109.
159. Galløe 1908, Mitchell 2011: 132.
160. Söderqvist 1986: 46–47.
161. Mitchell 2011: 132.
162. Warming 1909: 224, quoted in Egerton 2015a: 108.
163. Warming 1909: 403.
164. Mitchell 2011: 132, citing Knowles 1913.
165. Desmond 1977: 565, Mitchell 2011: 133.
166. Mitchell 2011: 134, Tobler 1925.
167. Kemna 1897, Florkin 1970.
168. Boucher et al. 1982: 317, Dajoz 1984: 83–84, Sapp 1994: 7–8, 18–20.
169. Beneden 1876: 1.
170. Beneden 1876: 68.
171. Beneden 1876: 56.
172. De Bary 1879a: 21, 1879b: 306–309.
173. Allee et al. 1949: 243.
174. Belt 1985 [1888]: 218–222, Brown 1960, Van Ripper 2004, Egerton 2013a: 40–42.
175. Belt 1985 [1888]: 20.
176. Willis & Oniki 1978: 246.
177. Wheeler 1907: 677.
178. Pound 1893.
179. Monnier 1975.
180. Buchner 1965: 70–71, Sapp 1994: 77.
181. Tétry 1971.
182. Sapp 1994: 142.
183. Temple 1977.
184. Mitchell 1990: 131.
185. Witmer & Cheke 1991, Cheke & Hull 2008.
186. Lanner 1996: 14–15.
187. Lanner 1996: 68.
188. Howe & Smallwood 1982: 202–203, Herrera 2002.
189. Pasteur 1982: 172.
190. Baker 1952, 1976, Egerton 2008c: 416–417.
191. Egerton 2012c: 203–204.
192. Goreau 1963, Muscatine & Porter 1977.
193. Muscatine & Porter 1977: 454–455.
194. McCalman 2014.
195. Egerton 2015a: 117–120, 122–123.
196. Caullery 1952: 16.
197. Breitwisch 1992.
198. Feder 1966.
199. Sterling 1997b.
200. Beebe 1924: 121–122.
201. Beebe 1928: 147.
202. Margulis 1998.
203. Schneider & Boston 1991, Schneider et al. 2004.
204. Möbius 1877, 1880: 721–729.
205. Linnaeus 1749, 1775: 39–129, Egerton 1973: 333–337.
206. Lee 1921, Carley 2004, Egerton 2009a: 44.
207. Quoted in Gorham 1953: 259.
208. Gorham 1953: 259–261, Egerton 2009a: 44.
209. Clements 1916: 8–9, Gorham 1953: 262.
210. King 1685: 950–951.
211. Reproduced in Egerton 2009a: 46.
212. Roger 1970, 1989, 1997, Egerton 2007b, 2012: index.
213. Clements 1916: 9.
214. Buffon 1742: 234, translated in Clements 1916: 9.
215. Buffon 1742: 238, translated in Clements 1916: 9.
216. Pritzel 1871–77: 187–193, Hagberg 1952, Blunt 1971, Stafleu & Cowan 1976–2009: I, 71–111, Morton 1981: 259–276, Koerner 1999, Magnin-Gonze 2004: 120–128, Egerton 2007a, 2012a: index.
217. Egerton 1973: 335–337.
218. Linnaeus 1775: 368, Nicolson 2013: 96.
219. Linnaeus 1775: 369–371.
220. Linnaeus 1781, translated by Blair in Hagberg 1952: 183.
221. Linnaeus 1781: 133.
222. Egerton 1973: 335–357, 2007a.
223. Du Rietz 1957b: 68, Egerton 2015c: 430–431.
224. Mullett 1968, Sweet 1972: 407–410, Mitchison 2004.
225. Anderson 1794: 49, Egerton 2009a: 48.
226. Pritzel 1871–77: 347, Bylebyl 1976, Nicolson 1987: 172–174, 2013: 96, Magnin-Gonze 2004: 155–156.
227. Willdenow 1811: 422, quoted from Buell 1949, Nicolson 2013: 96.
228. Morton 1981: 314.
229. Beckinsale 1971, Carozzi 1987: 207–219.

230. Clements 1916: 10.
231. Drouin 1994: 149–150.
232. Clements 1916: 13–14.
233. Drouin 1998: 15–16.
234. Pritzel 1871–77: 214, Querner 1974, Stafleu & Cowan 1976–2009: III, 438–439, Nicolson 1996: 294–295, 2013: 97.
235. Meyen 1846: 88.
236. Meyen 1846: 50–61.
237. Tobien 1972, Stafleu & Cowan 1976–2009: II, 1123–1127.
238. Becking 1957: 414, Maarel 1975: 215, Nicolson 2013: 97.
239. Christensen 1924–25: I, 464–472, II, 230–231, Spärek 1932, Müller 1976, Egerton 2009a: 49–52, 2012a: 193.
240. Hoeppli 1959: 477.
241. Steenstrup 1842, Clements 1916: 14–16.
242. Clements 1916: 16, Egerton 2009a: 52.
243. Pritzel 1871–77: 178, Maarel 1975: 215, Stafleu & Cowan 1976–2009: II, 804–805.
244. Maarel 1975: 215.
245. Maarel 1975: 215.
246. Lecoq 1854–58: I, 58–90, Nicolson 2013: 97.
247. Clements 1916: 16–17.
248. Reissek 1856.
249. Clements 1916: 17, Christensen 1924–26: I, 472–492, II, 288–292.
250. Vaupell 1857, Clements 1916: 17, Christensen 1924–26: I, 472–492, II, 288–292.
251. Egerton 2011b, 2012a: 151–156, Egerton & Walls 1997, Walls 1995.
252. Thoreau 1980.
253. Spurr 1952.
254. Kerner 1863, Pritzel 1871–77: 163, Stafleu & Cowan 1976–2009: II, 525–530, Nicolson 1996: 297–298, Egerton 2013b: 342–343.
255. Kerner 1951: ix.
256. Clements 1916: 21–22, Pritzel 1871–77: 30, Stafleu & Cowan 1976–2009: I, 241–243.
257. Clements 1916: 21.
258. Clements 1916: 22–23, Collander 1965: 74–77, Stafleu & Cowan 1976–2009: II, 31, Nicolson 2013: 97–98.
259. Fries 1950: 70.
260. Clements 1916: 22.
261. Rikkinen 1988.
262. Fries 1950: 75, Stafleu & Cowan 1976–09: I, 45.
263. Clements 1916: 492–493.
264. Clements 1916: 378–388.
265. Fries 1950: index, Söderqvist 1986: index.
266. Söderqvist 1986: 52.
267. Söderqvist 1986: 44.
268. Clements 1916: 380–382.
269. Christensen 1924–26: I, 617–665, 786–806, II, 367–399, Müller 1976, Nicolson 2013: 98.
270. Jensen 1932.
271. Christensen 1932: 156, also Christensen 1927, Tansley 1927, Goodland 1975.
272. Warming 1909: 4.
273. Clements 1916: 23.
274. Sanders 1975, Cittadino 1990: 97–115, Matagne 2009: index.
275. Clements 1916: 26–27.
276. Goodland 1975: 243.
277. Billings 1985: 5.
278. Nicolson 2013: 98.
279. Stafleu & Cowan 1976–2009: I, 682–683, Magnin-Gonze 2004: 203, Egerton 2013b: 345–346, Nicolson 2013: 98.
280. Tobey 1981: index.
281. Clements 1916: 478–479.
282. Tobey 1981: 51–58.
283. Pound & Clements 1898.
284. Wigdor 1974: chapter 3, White 1981, Burgess 1996: 89.
285. Tansley 1947, Clements 1960, Ewan 1971, Sears 1973, Tobey 1981: 76–87, Hagen 1988, 1992: 20–28, 1999, Burgess 1996: 26–27, Egerton 2013b: 358–360, Nicolson 2013: 99–100.
286. Cooper 1926: 393.
287. Clements 1916: 26.
288. Drouin 2013: 55.
289. Cowles 1911: 182.
290. Sears 1958, Tobey 1981: 104–108, Engel 1983: 137–150, Mitman 1992: 16–19, Cittadino 1993, Burgess 1996: 31–32.
291. Croker 1991, Harmond 1997, Mitman 1997.
292. Adams 1913: 130–141.
293. Pierce et al. 1912, reproduced in Allee et. al. 1949: 53, Egerton 2014a: 68.
294. Shelford 1913: 70, 167, reprinted in Egerton 2007d: 54, 55.
295. Emerson & Park 1955, Schmidt 1957, Kimler 1988, Croker 1991: 31–34, Mitman 1991, 1999, Burgess 1996: 11, Sterling 1997a, Egerton 2014a: 72–73.
296. Banks 1985.
297. Burgess 1996: 106.
298. Clements 1916: 495.
299. Cooper 1913: 130.
300. Nicolson 1990, Burgess 1996: 46.
301. Gleason 1910.
302. Nicolson 1990: 106.
303. Clements 1916: 481.
304. Clements 1916: 30–31.
305. Nicolson 1990: 119, Burgess 1996: 44.
306. Fuller 1918.
307. Nicolson 1990: 130–136.
308. Major 1962, Sobolev & Utekhin 1973, McIntosh 1975, 1983, Ramensky 1983.
309. Ramensky 1983, McIntosh 1985: 137.
310. Weiner 1999: 375.
311. Major 1962.
312. Gause 1936.
313. Nicolson 1989.
314. Acot 2008.
315. Maarel 1975: 218, Nicolson 1989: 101.
316. Maarel 1975: 213.
317. Barbour et al. 1987: 23.
318. Barbour et al. 1987: 22.
319. McIntosh 1985: 135.
320. Smith 1913, Gleason 1929, Fuller 1935.
321. Quoted from Croker 1991: 65.
322. Clements & Shelford 1939.
323. Clements & Shelford 1939: 6–19.
324. Clements & Shelford 1939: 313–353.
325. Stuckey 1973, 1994, 1997, 2001, Braun 1994, Burgess 1996: 18.
326. Steere 1958: 16.
327. Stuckey 1994: 5.
328. Burgess 1993: 1–4, 1996: 33, Fralish et al. 1993.
329. Curtis & Juday 1937.
330. Burgess 1993, Nicolson 2001.
331. Oosting 1948, Daubenmire 1968, Burgess 1996: 35, 83.

332. Allee et al. 1949: 436–597.
333. Evans 1948, Burgess 1996: 36–37.
334. Westman 1982, Westman & Peet 1982, Whittaker 1994, Burgess 1996: 113.
335. Whittaker 1994: 617.
336. Whittaker 1962.
337. Whittaker 1962: 207.
338. Whittaker 1970: 154–155.
339. Whittaker 1970: 52–64.
340. Westman 1982.
341. Southwood 1985: 872.
342. Keddy: 497.
343. Golley 1993: xii.
344. Drouin 1984a.
345. Drouin 1987.
346. McIntosh 1985: 193–241.
347. O'Neill et al. 1986: 20–34.
348. Ricklefs & Miller 2000: 171–184.
349. Lévêque 2003.
350. Christensen 1927, 1932, Müller 1976, Coleman 1986, Egerton 2013b: 346–347, 2017: 300.
351. Smith 1913, Kenoyer 1927, Gleason 1929, Raunkiaer 1934, Fuller 1935, Preston 1948.
352. Gridgeman 1973, Israel 1988, Kingsland 1995: index, Egerton 2014a: 73.
353. Cox 1979, Crowcroft 1991, Kingsland 1995: index, Burgess 1996: 39, Southwood & Clarke 1999.
354. Summerhayes & Elton 1923, reprinted in Elton 1927: 58.
355. Golley 1993: 44, Egerton 2007a: 58.
356. Hagen 1992: index, Golley 1993: index, Ricklefs & Miller 2000: 175–176.
357. Elton 1927: 63–68.
358. Ayres 2012: 95–96.
359. Phillips 1921, 1934–35.
360. Tansley 1935: 299–300, Hagen 1992: 82–87, Golley 1993: 31–34, Ayres 2012: 135–138, Egerton 2013b: 355–357.
361. Sheail 1987.
362. Cook 1977, McIntosh 1985: index, Hagen 1992: 87–94, Golley 1993: 48–56, Burgess 1996: 66, Brady 2008, Slack 2010: 149–151, Sterner 2012, Hoffman 2017.
363. Lindeman 1942: 157.
364. Lindeman 1942: 158.
365. Craige 2001: 3.
366. Craige 2001: 12–13.
367. Odum 1949: vi.
368. Craige 2001: 14.
369. Craige 2001: 22.
370. Odum 1941.
371. Craige 2001: 44.
372. Craige 2001: 40–41.
373. Bergundi 1995: 156.
374. Bergundi 1995: 166.
375. O'Neill 2001: 3275–3276.
376. Craige 2991: 48–50.
377. Odum 1987 in Craige 2001: 52.
378. Barrett & Barrett 2001.
379. Taylor 1988: 224.
380. Brown et al. 2004: 4.
381. Hutchinson 1943: 337.
382. Taylor 1988: 224, Limburg 2004: 31.
383. Hutchinson 1941: 43–57, 1944, 1948.
384. Limburg 2004: 33.
385. Odum 1951a, b.
386. Odum 1956a, 1957a, b, Taylor 1988: 226–228, Kemp & Boynton 2004: 44.
387. Kemp & Boynton 2004: 44.
388. Odum & Odum 1955: 318, Craige 2001: 58–59.
389. Craige 2001: 59.
390. Odum & Odum 1955: 298.
391. Odum & Odum 1955: 301.
392. Egerton 2007a: 63–65.
393. Odum 1969: 263–264.
394. MacArthur & Wilson 1967, Burgess 1996: 69–70, 114.
395. O'Neil 2001: 3275.

4 Aquatic Ecology

A. LIMNOLOGY IN AMERICA, 1930S–90S

In Part 1C we explored the rise of limnology from the 1870s to the 1920s.[1] E.A. Birge and Chancey Juday in Wisconsin and August Thienemann in Germany remained active after 1920s but are not discussed here, though their influence is recognized.[2] Instead, I discuss major aspects of American limnology, 1930–90, by emphasizing Evelyn Hutchinson, Arthur Hasler, and their students. Detailed *Limnology in North America* (1963)[3] is an indispensable resource for half of this period. It includes Canada, Mexico, the West Indies, and Central America, though it does not emphasize history. "Mac" Mcintosh did briefly survey history of limnology in *The Background of Ecology*, four pages of which cover the time period of this discussion.[4]

Three directories of biological stations, compiled by U.S. biologists, are helpful, though they discussed far more salt-water than fresh-water stations: Homer Jack, *Biological Field Stations of the World* (1945) is brief and includes land stations; Robert Hiatt, *Directory of Hydrobiological Laboratories and Personnel in North America* (1954) is more detailed than Hiatt, *World Directory of Hydrobiological and Fisheries Institutions* (1963). American Institute of Biological Sciences reprinted from *BioScience* a booklet of seven articles (ii + 72 pages) of historical relevance: *Topics in Biological Field Stations* (2010). Laura Tydecks et al. compiled worldwide data on biological field stations and published a summary of their findings and a web site for details (2016).[5] Still more recently, Stacy McNulty et al. celebrated "The Organization of Biological Field Stations at Fifty (2017).[6]

Before discussing two American schools of limnology, I summarize some of Alexandra Rizhinashvilli's recent survey of studies by Vladimir Zhadin (1896–1974) during the 1930s and 1940s.[7] He was from Muron, Vladimir Region, USSR. He graduated from Moscow University in 1924 and earned a doctorate in biology (1935). During 1930–50, USSR was under great stress from worldwide depression, then from Hitler's invasion, and, after 1945, from recovery from World War II. Hydrobiologist Zhadin had studied productivity of rivers, and later of reservoirs and lakes, a progression from his initial studies on systematics and variability of freshwater mollusks. Zernov invited him to speak at a 1932 Faunistic Conference at Zoological Institute AS USSR in Leningrad. The title of his talk and subsequent publication was "A Biocoenotic and Faunistic Study of Regions and Their Biotopes in Connection with Economic Operations that Change the Living Conditions of Bodies of Water."[8] It was well received, and he later commented that the conference at 50 had been the "impetus for moving to Leningrad."[9] His suggestions for the future were adopted at the conference. He subsequently worked at the Zoological Institute with five assistants. They collected much data on geographical changes in rivers from one region to another, and compared rivers to ponds behind dams. For a 1949 report, his team collected over 1,000 benthic samples and "several thousand plankton samples."[10] He tied changes in benthic river communities to the accumulation of sediments in water, carried by currents. Suspended organic particles served as food, while inorganic particles could create turbid water that impeded faunal development. In 1965, he won an Einar Naumann Medal from the International Association of Theoretical and Applied Limnology.

English limnologist Jack Talling surveyed the history of limnology in England (2004)—mostly a bibliographical guide—and the World (2008)—a historical perspective.

G. Evelyn Hutchinson (1903–91) was an English ecologist who spent his career at Yale University.[11] Arthur D. Hasler (1908–2001) was a Utah limnologist who spent his career at the University of Wisconsin-Madison. He surveyed the history of limnology at Wisconsin, 1940–61 (1963); a volume celebrating a century of history of limnology at UW-Madison (1987) broadened that survey; and more recently, four biographical essays provide information on his life and career.[12] Hutchinson (1947–48) and Hasler (1949–50) served as presidents of a new American Society of Limnology and Oceanography, and they were the first limnologists elected to the National Academy of Sciences. Hasler was ESA president in 1961.

Hutchinson and Students at Yale University

Hutchinson's father was an outstanding mineralogist at Cambridge University, who taught Evelyn mineralogy, geology, and flora of Cambridgeshire.[13] Evelyn became interested in aquatic insects at age 6, and he won a school prize, a book on insects by Jean Henri Fabre.[14] At age 15, he published a note on a swimming grasshopper (1918). His interest became aquatic insects.[15] At Cambridge University he received a prize in zoology and earned a B.A. degree in 1925. A. M.A. degree was awarded after further work in one's field (his in 1928). The zoologist who influenced him most was next-door neighbor George P. Bidder, who studied sponges.[16] Bidder's sponge course raised ecological questions on environmental interactions. Hutchinson also took J.T. Saunders' course on hydrobiology, which covered both limnology and marine ecology.[17]

After graduating, a Rockefeller Fellowship enabled him to go to Stazione Zoologica in Naples to investigate endocrine glands in octopuses. Next, he applied for and received

DOI: 10.1201/9780429342202-4

FIGURE 4.1 (a) G. Evelyn Hutchinson in 1935. Web. (b) Arthur D. Hasler. Center for Limnology, University of Wisconsin-Madison.

a 2-year lectureship at the University of Witwatersrand in Johannesburg, South Africa. He studied the fauna, not lions or zebras, but water bugs—their temperature tolerance and distribution. Only 14 species were known when he arrived, but 40 were known when he left.[18] His fiancée, Grace Pickford, graduated from Newnham College, Cambridge, and received a Newnham College Traveling Fellowship to travel and study in South Africa. They married, and the substantial article they wrote with Johanna F.M. Schuurman on hydrobiology of pans and other inland waters was Hutchinson's first important contribution to limnology (1932).[19] Their researches were matched at about the same time by Thienemann's first limnological investigation of tropical lakes.[20] In 1929, after reading Thienemann's *Die Binnengewässer Mitteleuropas* (1925), Hutchinson realized his recent investigations had led him to become a limnologist.[21]

Hutchinson obtained a lectureship at Yale University in 1928. His wife, Grace Pickford, worked on and earned a Ph.D.[22] Both joined the Yale Zoology Department; their marriage ended with him divorcing her. His childhood interest in water bugs persisted, and he became the world authority on them. He expanded his interests as he aged, but seldom did he drop an earlier interest. In 1981, he published a lifetime of "Thoughts on Aquatic Insects."[23] His great influence derived from numerous publications and mentoring outstanding graduate students.[24] When he retired in 1971, Marion Kohn drew an impressive Hutchinson Family Tree of Intellectual Descendants.[25] He had supervised 34 completed doctoral dissertations; five more were in preparation, and he had strongly influenced nine other dissertations. Only his graduate students who studied limnology are discussed here.

Missourian Gordon Riley (1911–85) earned his B.S. degree at Drury College (1933), his M.S. (1934) at Washington University, St. Louis.[26] He went to Yale to study embryology, but took Hutchinson's limnology course and switched fields. Riley had a strong background in chemistry, so Hutchinson asked him to write a dissertation on the copper cycle in Connecticut lakes. In March 1937, Riley went on a cruise in the Gulf of Mexico, as a water chemist—his first trip at sea. "The limnological methods he had pioneered worked fine for oceanography."[27] He pursued a productive career in oceanography at Dalhousie University, Halifax, Nova Scotia. He remembered that Hutchinson "had inspired students with an almost euphoric feeling about science, and we were drawn to him with utter fascination."[28]

Later, he reflected that:[29]

> I am deeply indebted to Evelyn for introducing me to his own scientific philosophy about ecology, which was ahead of its time and has permeated all my plankton work. He maintained that populations needed to be studied in terms of dynamic processes—rates of production and consumption and the way these are affected by ecological factors.

Riley's career shows how easily an American ecologist could move between limnology and biological oceanography.[30]

Albany, New Yorker Edward Deevey (1914–88) earned his B.A. degree (1934) in biology at Yale and became

Hutchinson's second graduate student.[31] He studied the biological history of Linsley Pond (= lake), near campus, by distinguishing plant species, using pollen grains embedded in bottom sediments at different depths, and fauna from fossils at parallel depths. That was in the 1930s, before radiocarbon dating was available, but when it became available, he pioneered its use to date field specimens. He earned his Ph.D. in 1938 and published his findings in four articles, 1939–42. Also in 1938 Deevey married fellow graduate student Georgiana Baxter (1914–82), who published studies on invertebrate oceanic plankton. His first academic position was at Rice Institute, Houston, Texas, which ended in 1943 when he began three years of biological research for the U.S. Navy at Woods Hole Oceanographic Institution. After World War II, he joined the Yale faculty, 1946–1968, rising in rank from instructor to professor. In 1951, he received a grant to establish a Yale Geochrometric Laboratory. In 1968–71, he was at Dalhousie University. In 1971, he moved to the University of Florida, Gainesville, as professor and curator of the Florida State Museum of Natural History.

Milwaukeean Thomas (Tommy) Edmondson (1916–2000) became interested in aquatic animals as a child. As a teen, he moved to New Haven with his widowed mother and two brothers.[32] She wanted her sons to go to Yale. In high school he became interested in rotifers, and his biology teacher introduced him to Professor Hutchinson, who became interested in him.[33] Edmondson earned a B.S. in zoology (1936) at Yale, then went, as Hutchinson recommended, to the University of Wisconsin and studied under Juday for a year. There he met limnology student Yvette Hardman (1915–2006), whom he later married. Still later, she edited *Limnology and Oceanography* for 19 years. While in Wisconsin, Edmondson collected sessile rotifers and used data collected by Birge and Juday on Wisconsin lakes "to examine the relative insignificance of calcium concentration, bicarbonate concentration, and pH as measures of alkalinity."[34] That study (1940) was a practice run for his dissertation. After returning to Yale, he wrote it on rotifers (Ph.D. 1942), then published his findings in three articles in *Ecological Monographs* (1944–46).[35] After he taught at Harvard and at Woods Hole, the Edmondsons moved to the University of Washington in 1949, where Tommy and Yvette spent the rest of their careers. He was an active limnologist, most remembered for the restoration of Lake Washington; she became editor of *Limnology and Oceanography*, 1967–1986.

Minnesotan Raymond Lindeman (1915–42) earned his Ph.D. at the University of Minnesota, with a dissertation on *Ecological Dynamics in a Senescent Lake* (1941).[36] Several months later, he went to Yale on a post-doc to study under Hutchinson. He published two papers from his dissertation and at Yale wanted to explore more deeply its concluding chapter. The opening sentence of his new paper explained: "Quantitative productivity data provide a basis for enunciating certain trophic principles, which, when applied to a series of successional stages, shed new light on the dynamics of ecological succession."[37] He had studied succession of shallow lakes into marshes and then into land. This paper is often seen as beginning a "new ecology," more dynamic, quantitative, ecosystemic, theoretical, and less descriptive than the "old." Lindeman died as he completed this paper. Too theoretical for two referees—Juday and Welch—for *Ecology*, but Hutchinson prevailed and had *Ecology* publish it (1942).[38] Its implications extended far beyond a limnological context.

Another innovation: on 21 June 46, Hutchinson obtained a sample of radiophosphorus (P32) that was introduced into Linsley Pond, in 24 doses distributed across the lake. This is the first known experiment in nature using radioisotope tracers.[39] He was assisted by graduate student Vaughn Bowen, Hutchinson's research assistant. They returned a week later and collected water samples at four different depths.[40] Previous studies indicated that "there is a continual liberation of phosphorus from the mud into the free water."[41] They discovered "phosphorus had entered the epilimnion during a period when there can have been practically no vertical mixing," and "Of the 2.8% of the P32 put into the lake and not recovered, the greater part has probably entered the aquatic plants and sediments in contact with the 0–3 m. layer."[42]

North Carolinian H(oward) T(homas) Odum (discussed earlier), like Hutchinson, was the son of a prominent university professor (University of North Carolina, Chapel Hill) and also the younger brother of ecologist E.P. Odum (discussed earlier).[43] During World War II, H.T. served in the Air Force, "as a tropical meteorologist, where undoubtedly he gained his basic interest in large systems and the energetics behind them."[44] After the war, he graduated from the University of North Carolina (1947) and went to Yale for graduate work under Hutchinson. Hutchinson had published an analysis of "Biogeochemistry of Aluminum and Certain Related Elements" (1943), which provided an example for comparative studies on other elements. Hutchinson suggested to Odum dissertation research on strontium's natural cycle.[45] Odum's dissertation was entitled *The Biogeochemistry of Strontium* (1950), and he published two articles from it in *Science*.[46]

Hutchinson proved himself a talented limnologist in a wide variety of investigations, and his *Treatise on Limnology* (4 vols., 1957–93) is undoubtedly the longest ecology treatise by one person.[47] Each volume has a theme: (1) *Geography, physics, chemistry* (1957, xiv + 1015 pages), (2) *Lake biology and limnoplankton* (1967, xi + 1115 pages), (3) *Limnological botany* (1975, xi + 660 pages), (4) *Zoobenthos* (1993, xi+ 944 pages). He died before finishing volume 4, which Yvette Edmondson edited; he had enlisted her assistance in 1988. All four volumes have numerous charts, illustrations, and bibliographies. He had planned a fifth volume on productivity, succession, sediments, and paleolimnology.

Modern science is a collaborative effort, and Hutchinson was a great collaborator, who generously acknowledged the contributions of others. An anthology of his writings devotes less than a quarter of its pages to limnology,[48] though the following section on theory includes limnological subjects.[49]

He retired from Yale faculty in 1971, and in recognition of his scientific and pedagogic contributions, both *Limnology and Oceanography* and *Connecticut Academy of Arts and Sciences Transactions* produced special issues in his honor. March 1971 issue of *L & O* opened with Yvette Edmondson's "Some Components of the Hutchinson Legend," with photographs, bibliography, and lists of 34 doctoral students who wrote dissertations under his supervision, five students who began unfinished dissertations under him, and nine others who wrote dissertations under other faculty, with him exerting strong influence on their dissertation committees. Alan Kohn went a step further in "Phylogeny and Biogeography of *Hutchinsonia*: G.E. Hutchinson's Influence Through His Doctoral Students." Marion Kohn provided two illustrations: "Hutchinson's Phylogenic Tree of Intellectual Descendants" and of 22 "Aquatic or Semi-aquatic Animal Species Named to Honor Hutchinson." *Limnology in North America*'s chapter 4, "New England" (1963), was by two former Hutchinson students, John Brooks (b. 1920) and Deevey.[50]

Hasler and His Students in Wisconsin

Arthur Hasler was born (1908) into a Mormon family in Lehi, Utah, and grew up in Provo.[51] His boyhood interests were fishing, raising livestock, camping, nature study, and Boy Scouts. After graduating from Brigham Young University (B.A. 1932), he expected to follow his father into medicine, but his father was ill, and during the Depression he could not afford medical school. He had majored in zoology, and he attended the University of Wisconsin in Madison because he could earn money while a graduate student.[52] He was one of 13 students who earned Ph.D.s under Chancy Juday. Juday had suggested he write a dissertation on digestion in plankton crustacea. He did and published his findings.[53] In summer 1935 he worked for the U.S. Fish and Wildlife Service studying the effects of sulfate pulp mill wastes on oysters in lower York River, Virginia. He transferred sick oysters from York River into nearby unpolluted Piankatank River and healthy oysters from there into York River. Undisturbed oysters in both rivers served as controls. He found that sick oysters recovered when put into Piankatank and healthy oysters sickened when put into York River.[54] This experiment initiated field experiments as a hallmark of Hasler's limnology.

After two years with U.S. Fish and Wildlife Service, he returned to Madison and completed his Ph.D. (1937) and became an instructor in zoology; he was promoted to assistant professor in 1941. He had become fluent in German during teenage Mormon field service in Germany. When in 1945, he entered Germany with U.S. Strategic Bombing Survey, he became friends with Karl von Frisch in Munich—who later shared a Nobel Prize with Konrad Lorenz and Nikolaas Tinbergen for studies on animal behavior.[55] He had begun studying bees in summer and fish in winter.[56] In 1949, he went on a lecturing tour of the United States and was Hasler's guest when he lectured at the University of Wisconsin.[57] Hasler obtained a Fulbright Research Scholarship for 1953–54, which he spent at Frisch's Munich laboratory, refining his skills in researching fish behavior.

Birge had run a successful limnology program at the University of Wisconsin, with Juday's collaboration, for four decades, with over 400 publications.[58] There was no specific plan. In contrast, Hasler planned a scientific school of limnology, based on experimentation, which lasted over four decades.[59] He supervised M.S. theses of 41 students, Ph.D. dissertations of 53 students, and partially supervised 14 other Ph.D. dissertations. Unlike Birge and Juday, Hasler often collaborated with his graduate students, which often ended with joint publications. He also collaborated with colleagues, resulting in joint publications. He was sole or joint author of 195 publications, 1935–84.[60] He taught a 250 popular limnology course for upper classes and grad students (including me).

Five of his students earned doctorates in 1947: Nelson on fish morphology, Ogawa and Yokoyama on development and seasonal variation of perch blood cells, Walker on fish olfactory discrimination of aquatic plants, and Jones on aquatic plant competition. After an algal bloom in Lake Mendota, Hasler had Sara Elizabeth Jones (who would marry limnologist David G. Frey, a Juday student, Ph.D. 1940) conduct experiments to see if rooted aquatic plants inhibited algal growth. R.N. Pond had conducted experiments that seemed to indicate no impact (1903) since rooted plants received nutrition from soil and phytoplankton from water, so they did not compete.[61] Since the 1920s, fishery managers had expressed doubts about Pond's findings but had not experimented to test his results. Jones' experiments showed that rooted plants inhibited algae. She and Hasler published that in *Ecology*.[62]

After World War II, AEC and the new NSF provided funds enabling ecology to become a "big science" in some fields and some places. David Coleman discussed important aspects of this history in *Big Ecology: the Emergence of Ecosystem Science* (2010). Hasler's limnological school participated in this growth; he used research funds to hire research managers since he could not advise all students simultaneously. In 1962, NSF funds enabled him to build a Laboratory of Limnology on Lake Mendota, within campus, which opened in 1963, with an address by Hutchinson, "The Lacustrine Microcosm Reconsidered."[63]

Another aspect of Hasler's "big limnology" was arranging for the University of Wisconsin to host the Fifteenth Congress of the International Association for Limnology (IAL), August 20–25, 1962—IAL's first meeting in America. To assist with preparations, Hasler hired John Wright, from Montana. Attending were over 600 limnologists and other scientists, from 34 countries.[64] Excursions to American lakes and rivers were arranged for both before and after the meetings. This congress stimulated American limnologists to tell the history of their science in North and Central America and the West Indies. David Frey, University of Indiana, edited a remarkable *Limnology in North America* (University of Wisconsin Press, 1963).

Thirty-two authors wrote 20 regional chapters and five thematic chapters, and Hutchinson wrote a concluding essay, "The Prospect before us."[65] As befitted the host university, two chapters discussed Wisconsin limnology: Frey wrote on the Birge-Juday Era, and Hasler wrote on "Wisconsin, 1940–1961."[66] *Limnology in North America* is a treasure unmatched anywhere else in the world. Only the United States and Canada could readily provide resources to support this venture, and the American Society of Limnology and Oceanography had nurtured a cooperative spirit that led to this achievement.

Hasler's most important investigations were on homing migrations of salmon.[67] Salmon are anadromous—breeding in headwaters of streams, migrating to oceans to mature; years later, returning to home streams to reproduce and die—the reverse migration from catadromous eels, which Johannes Schmidt described.[68] A personal experience initiated Hasler's studies:[69]

> My early research. Together with what I had read about salmon and homing in other animals, prepared me for an experience I had in 1946 while on vacation in Utah's Wasatch Mountains, where I grew up. As my children and I hiked along a path I had frequented as a boy, the familiar fragrances of columbine and mosses growing just ahead but out of sight triggered a whole flood of memories. The names of boyhood chums and visions of the nearby mountain meadow where we played ball vividly flashed across my mind. I thought, 'Well, Art, you're a salmon coming home, and it smells like home.'

If he had been at a western university, he might have investigated salmon, but being in Wisconsin, he wondered if his question about migrating fish being guided by odor could be tested on Wisconsin fish.

The earliest dissertation relating to this hypothesis which he had a graduate student undertake was to test bluntnose minnows to discriminate between odors of plants. The aquarium apparatus developed for these experiments was used in the doctoral research of several students. Theodore Walker first use this apparatus in controlled experiments in 1947, and he showed fish can distinguish odors of different plant species. He explained that aquatic plants "may serve as signposts to guide fish into feeding grounds . . . [and] other natural odors may direct migratory fishes in locating their homing areas."[70] The next test, by Warren Wisby, was to see if bluntnose minnows could distinguish between odors of two streams—also tested in experimental aquaria, using water from different Wisconsin streams. After two months of training, minnows learned to distinguish water leading to food from water leading to an electrical shock. Hasler and Wisby soon published two papers on this, the second being similar to Wisby's dissertation (1952).[71] Next, in 1952, Wisby and Hasler designed experimental apparatus to test odor discrimination in salmon and reported their success in another paper (1954).[72] Hasler collaborated with an editor on "The Homing Salmon," a cover story in *Scientific American* (August 1955).[73]

The next question: how do salmon find their river after several years at sea? Perhaps they depend upon the sun. Hasler and collaborators tested this hypothesis on an immature Pumpkinseed *Lepomis gibbosus* in an experimental tank sitting on a pier in Lake Mendota, with running lake water. They trained the pumpkinseed to enter the "training box in a consistent compass-direction at any time of day."[74] The fish was unable to find its way under an overcast sky. Hasler's last major publication, with coauthor Allan Scholz, was *Olfactory Imprinting and Homing in Salmon: Investigation into the Mechanism of the Imprinting Process* (1983).

In 1964, Canadians introduced kokanee salmon into the Canadian waters of Lake Huron, and in late 1960s coho and chinook salmon were introduced into Great Lakes. The latter two species were introduced to counter the (introduced) alewife die-offs that began plaguing Lake Michigan shores: great masses of them washed ashore, rotting and stinking. Salmon did eat the alewives, and salmon became the basis of a very popular sport fishery.

Most prominent of Hasler's students was Indianan Gene Likens (b. 1935), who earned his B.S. degree (1957) at Manchester College before moving to Madison for his master's and doctoral degrees (1959, 1963).[75] His dissertation, on the movement of radio-sodium in a lake, provided a basis for three articles.[76] In fall 1961, Likens taught at Dartmouth College, New Hampshire, and after receiving his doctorate, he joined the Dartmouth Zoology Department. There, he also became a leader of an ongoing forested ecosystem study at the U.S. Forest Service's Hubbard Brook Experiment Station.[77]

Since the University of Wisconsin and its Laboratory of Limnology are on Lake Mendota, that lake has been convenient for limnological studies, by faculty and graduate students, which made Mendota the most studied lake in the world. Hasler edited an IBP volume, *Coupling of Land and Water Systems* (1975). Bacteriology Professor Thomas Brock synthesized such studies in *A Eutrophic Lake: Lake Mendota* (1985).

Limnology Elsewhere in North America

Hutchinson at Yale and Hasler at Wisconsin were noteworthy examples of American limnology's progress, 1930s to about 1990, but there were very extensive contributions made elsewhere that deserve recognition. My long survey of limnology for this period in the *ESA Bulletin* (2016) devoted slightly more pages to progress elsewhere than to Hutchinson and Hasler. It is impractical to summarize that portion here; it is available, open access, online.[78]

B. North America's Great Lakes

Five Great Lakes between Canada and the United States contain almost a fifth of the world's fresh-water and over 95% of North America's fresh-water. Half of the water is from precipitation, half from rivers and streams. The drainage basin covers 295,000 square miles (767,000 square

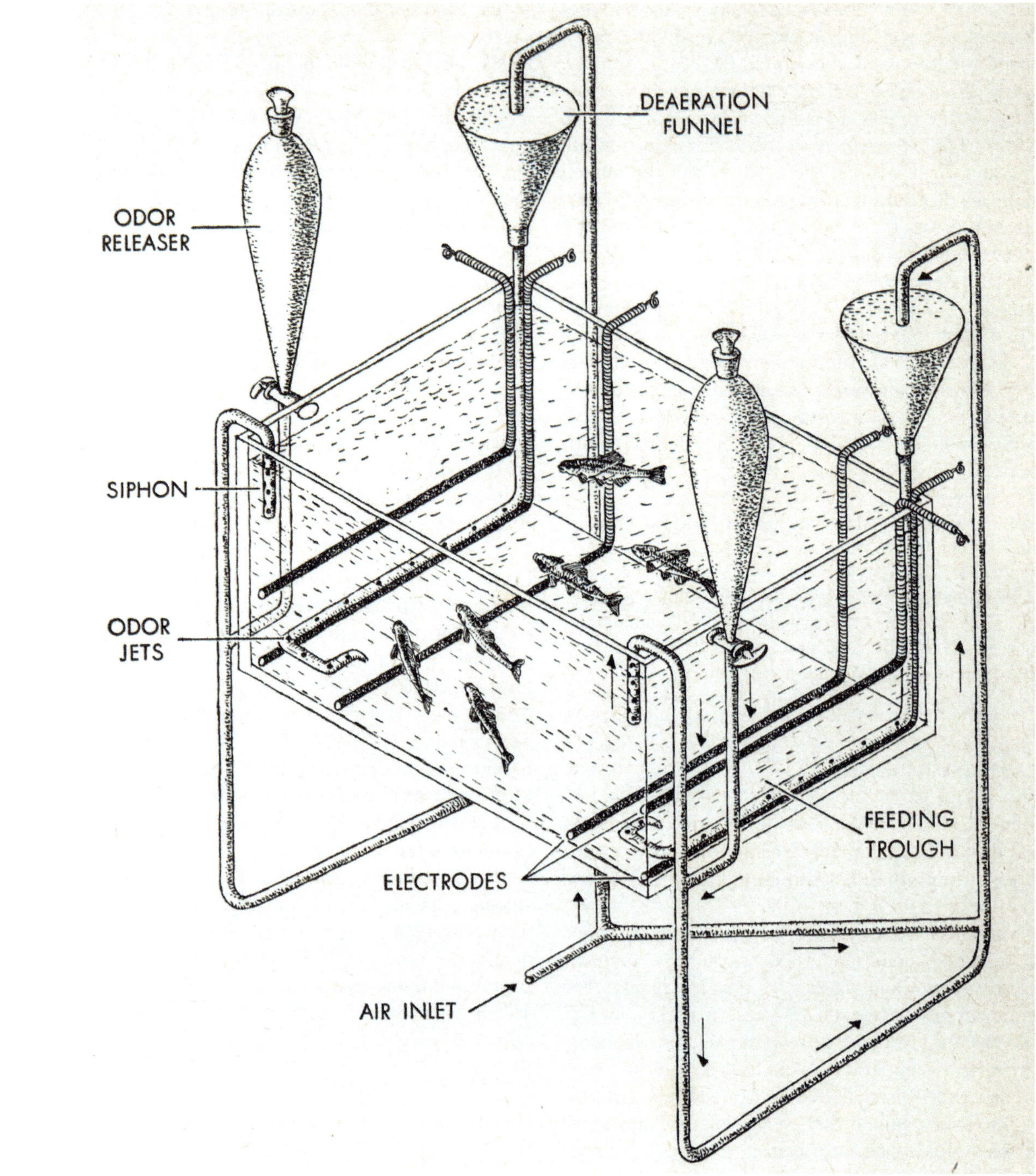

FIGURE 4.2 Fish discrimination of odors. Eric Mose drawing. Hasler & Larsen 1955: 73. Courtesy, *Scientific American.*

km), and the lakes cover 94,250 square miles and hold 5439 square miles of water, shorelines extend 10,210 miles.[79] The distance from Kingston, Ontario, at the east end of Lake Ontario to Duluth, Minnesota, at the west end of Lake Superior is 780 miles and from Gary, Indiana, at the south end of Lake Michigan to Nipigon, Ontario, on the north side of Lake Superior is some 500 miles. Canada possess 31% of the lakes and the United States 69%. Lake basins were gouged from river beds by glaciers, which gradually melted 8,000 years ago.[80] Aquatic and terrestrial endemic species have only a relatively brief association. Before civilization, these lakes had virtually no pollution, and Native Americans fishing had insignificant impacts on fish populations.[81] (A different story unfolded for land animals, which did decline from hunting pressure, and humans may have exterminated megafauna.)[82]

Alida Malkus, of Michigan, wrote a history of peoples who have lived in this region, *Blue-water Boundary: Epic Highway of the Great Lakes and the St. Lawrence* (1960). Her book appeared three years after St. Lawrence Seaway opened, with two-page bibliography. Illinoisan journalist-historian Russell McKee wrote a readable, well-illustrated history, *Great Lakes Country* (1966), with numerous maps and no bibliography. American environmental historian Susan Flader edited for the Forest History Society a symposium volume, *The Great Lakes Forest: an Environmental History* (1983, 18 papers by 22 authors). It emphasized the three western lakes. American environmental author William Ashworth wrote *The Late Great Lakes* (1986, six maps). Canadian author Wayne Grady's *The Great Lakes: the Natural History of a Changing Region* (2007) is a comprehensive, well-illustrated, beautiful book. Michigan environmentalist-journalist Jeff Alexander's *Pandora's Locks: the Opening of the Great Lakes-St. Lawrence Seaway* (2009) is broad in scope, documenting unforeseen catastrophic biological consequences of improving commercial access to these lakes. Canadian ecologist John Riley's *The Once and Future Great Lakes Country: an Ecological History* (2013) is an attractive, well-documented book that emphasized lands around the three eastern lakes. Limnologists Alfred Beeton and David Chandler briefly surveyed the history of limnological studies of these five lakes, 1800s–1961 (1963). U.S. Federal fisheries biologist John Van Oosten compiled a useful bibliography: *Great Lakes Fauna, Flora and Their Environment* (1957), and Elaine Prantner and others compiled *Annotated Bibliography of Limnological and Related Studies concerning Lake Erie and Influent Tributaries* (5 vols., 1974).

Several Americans published their findings from explorations around Great Lakes. Albany, New York native Henry Schoolcraft (1793–1864) studied geology at Middlebury College in Vermont and later joined a Michigan Territory expedition to Lake Superior. He published his observations and became a U.S. Indian Agent.[83] Troy, New York native Douglass Houghton (1809–45) attended Rensselaer School, in Troy, graduated in 1830, then went to Detroit to teach science.[84] After Michigan became a state in 1837, its legislature hired him as state geologist; he conducted extensive surveys, for which he received national recognition. His 1841 report on Michigan's Upper Peninsula indicated deposits of copper and stimulated the first U.S. mining rush. On an expedition to Lake Superior, he drowned. Two mining companies established profitable copper mines where he had explored, lasting until 2015.

Great Lakes forests before the coming of Euro-Americans were spectacular.[85] In the north were conifer forests, and in the south, deciduous forests, with mixed forests in between.[86] Natives adapted their harvesting to the resources of their region. Fishing was most important in the waters of northern portions of Lakes Huron and Michigan, and wild rice was available from Green Bay west to Lake Superior and on around southern Lake Superior. Agriculture was practiced only south of the rice region, south of Lake Superior, in portions of Lakes Huron and Michigan shores south of the main fishing waters, and around Lakes Erie and Ontario.

These lakes became valuable assets for Euro-American economies: first, to transport furs and, later, to transport fishes, lumber, agricultural products, minerals, sand, and people. There were about 300,000 people in the basin by the early 1800s; by 1990, there were 37 million (7.5 million in Canada, 29.5 million in the United States); today, there are 105 million, with 46 million jobs. Great Lakes ecology includes rivers flowing into them, with rivers and lake waters intermingling, as do their inhabitants. River water quality depends upon what washes from adjacent lands, from logging, farming, mining, and manufacturing. Rarely were there early precautions to protect against these hazards.[87]

Logging was the first significant transformation of Great Lakes environments.[88] Logs were floated down rivers to lumber mills; lumber was loaded onto schooners going to urban areas; lumber from around the lakes was cheaply transported by sailing vessels to Chicago, Milwaukee, Detroit, Toledo, Cleveland, Buffalo, Toronto, and elsewhere. The pace of logging increased significantly after the Civil War until the end of the century.[89]

An unintended consequence of logging was the accumulation of limbs, stumps, and sawdust, all inflammable. Destructive forest fires were frequent. Following an unusual summer drought, on 8 October 1871, there were two very destructive fires. The well-remembered Chicago fire destroyed a city and killed 300 people.[90] Less remembered is the Pestigo fire that burned 2,400 square miles of forest on both sides of Green Bay, Wisconsin and killed 2,200 people—America's worse human disaster.[91] Of boreal forests, Wayne Grady stated:[92]

> In Canada, an average of 9000 forest fires are started every year, 85 percent of them by lightning and most of them in the boreal forest. They consume nearly 5 million hectares (74 million acres) of forested areas. By comparison, spruce budworm infestations destroy an average of 500,000 hectares (1.2 million acres) per year, and logging accounts for about 800,000 hectares (2 million acres).

By the early 1900s lumbering had moved west, after logging Great Lakes forests. Rivers were often inadequate for logs or lumber transport; then, railroads were used. Tamarack was used for railroad ties.[93] During the 1890s, larch sawflies killed many tamarack trees. Nevertheless, by 1915, Austin, Nicholson Company, Chapleau, Ontario, produced 4,000 ties a day. In 1920, creosote was introduced, which increased the durability of a tie from 7 to 20 years. During the 1920s, Canada exported from boreal forests more newsprint than any other country, mainly to the United States.

After forests were logged or burnt, farmers came. However, glaciated lands that supported forests were less suited for crops.[94] Only land south of Lake Erie became good farmland. The Great Lakes country also had areas without forests. John Curtis' *Vegetation of Wisconsin* (1959) devoted chapters 5–13 to different types of forests and chapters 14–21 to other vegetation: prairie, sand barrens, bracken-grassland, savanna, tall shrubs, fens, meadows, bogs, aquatic communities, beach and dunes, sand cliffs, and weed communities. His accounts can be supplemented by plant guides to particular groups: water plants, trees, or wildflowers.

Wildlife—mammals, birds, fish, insects, and more—were resources used by natives, and also later by Euro-Americans. There are also guides to particular groups, which are useful for ecologists, hunters, fishermen, and nature hobbyists. Early Europeans were interested in fur trade, and they recruited Natives to assist. Beaver was the first and primary interest, as long as they lasted, but the temptation to trap more than could be sustained meant they gradually disappeared from east to west.[95] In an environmental history of the Toledo region and western Lake Erie, Harold Mayfield learned that bison disappeared by 1812, beaver by 1837, wolverine by 1842, mountain lions by 1845, lynx by 1848, gray wolves and black bears were rare by the 1860s and disappeared during the 1870s, white-tailed deer by 1889. Following a fire or logging in coniferous regions, aspens, willows, and birches grow back faster than conifers, but as they are eaten by moose, conifers can then gain a foothold.[96]

Ospreys and bald eagles prey on fish near shore but are too few to affect fish populations. Gulls are fairly numerous and eat fish, invertebrates, bird eggs, and chicks, but gulls are also scavengers who follow fishing boats and scoop up bycatch and scraps that fishermen toss overboard. Egrets, herons, and cranes snatch small animals from shallows of rivers and streams near where these flow into lakes.[97] Double-crested cormorants have been harassed or killed by fishermen who think they are competitors for fish. Great Lakes zoologists have debated this. David Fielder has seen some merit in fishermen's conclusions (2008, 2010), while Linda Wires disagreed in *The Double-Crested Cormorant: Plight of a Feathered Pariah* (2014).[98] Cormorants migrate south for winter, and southern aquaculturists have also persecuted them.

Water birds have been indirect victims of zebra and quagga mussel invasions into Great Lakes. Mussels live at the bottom of lakes and filter plankton from water. However, their sucking can cause material from the bottom to mix with their food; botulism bacteria live in mud and can be taken in by mussels. When such mussels are eaten by larger animals, botulism can accumulate in their tissue and pass up the food chain to birds.[99]

Before Europeans arrived, there were about 150 fish species in these lakes, and now about 170. Five species became extinct after commercial fishing began: blackfin, longjaw, short-nose cisco, blue pike, and harelip sucker. Great Lakes fish have been studied more than other animals because commercial and sport fishing continue and are economically important. Canadian ichthyologists William Scott and Edwin Crossman produced a monumental *Freshwater Fishes of Canada* (1979, xix + 966 pages), illustrated with drawings and distribution maps. Two American ichthyologists, working separately, have authored remarkable monographs, with more details for smaller regions than Scott and Crossman: Milton Trautman, *The Fishes of Ohio* (1981, xxvi + 782 pages) and George Becker, *Fishes of Wisconsin* (1983, xii + 1052 pages).

When fur trade began declining in the 1830s, the American Fur Company began commercial fishing in Lake Superior, though it lasted only from 1834 to 1841.[100] The earliest decline of a species humans ate was Atlantic salmon in Lake Ontario in the 1840s. The earliest individual entrepreneurs flourished in 1851, when 8,000 barrels of fish, holding 800,000 pounds, went over portage at Sault Ste. Marie, at the eastern end of Lake Superior.[101] The founding of Superior, Wisconsin, in 1854 facilitated this commerce, as was the opening of Paul Point-St. Louis Road in January 1956. Copper prospectors arrived in Lake Superior area in 1856. In 1870, further access for commercial fish was the beginning of the construction of the Northern Pacific Railroad from St. Paul, Minnesota.[102] Railroad builders ate fish, and when they finished, a new avenue for shipping fish became available.

U.S. Congress established the U.S. Lake Survey in 1841, under Topographical Engineers, U.S. Army. Its goal was to chart shorelines, islands, mouths of rivers, and other features of lakes, and to make charts for use by ships.[103] Water levels and surface temperatures were easily measured and recorded in various places.[104] A U.S. exploration of the Great Lakes, in June 1848, was led by Swiss professor at Harvard Louis Agassiz (discussed earlier). Its goal was to explore Lake Superior.[105] Agassiz wrote over half of the volume he edited on the expedition (1850), and Eliot Cabot took almost a third of it in his narrative of the expedition. Agassiz's most important contributions were accounts of fishes collected and an essay, "The Erratic Phenomena about Lake Superior."

Significant adverse human impacts began after the U.S. Civil War ended in 1865. Over-fishing was a problem in the late 1800s in both Great Lakes and North Atlantic.[106] Spencer F. Baird (discussed earlier), assistant secretary of the Smithsonian Institution, convinced Congress to create the U.S. Fish Commission (1871), with him as director (without a salary increase). With attention from the Fish Commission, U.S Great Lakes became the fifth coast. Baird hired James W. Milner (1841–79), a biologist at Waukegan, Illinois, to compile a report on U.S. Great Lakes fishery, which was published in *U.S. Fish and Fishery Commission Report*, 1872–73.[107] U.S. data were far from precise, but Baird liked Milner's work and brought him to Washington as an employee of the U.S. Fish Commission.[108] Canada published comparable data on its own Great Lakes fishery. Milner recommended to states that they establish their own fishery commission, build hatcheries, and restock commercial species.[109] They did, but restocking did not prove to be a cure-all for declining fisheries.[110] In 1879 Milner died and Baird hired Ludwig Kumlein to make a second survey. Charles Smiley compared their data. There had been a slight increase in pounds of fish caught yearly, but fishing

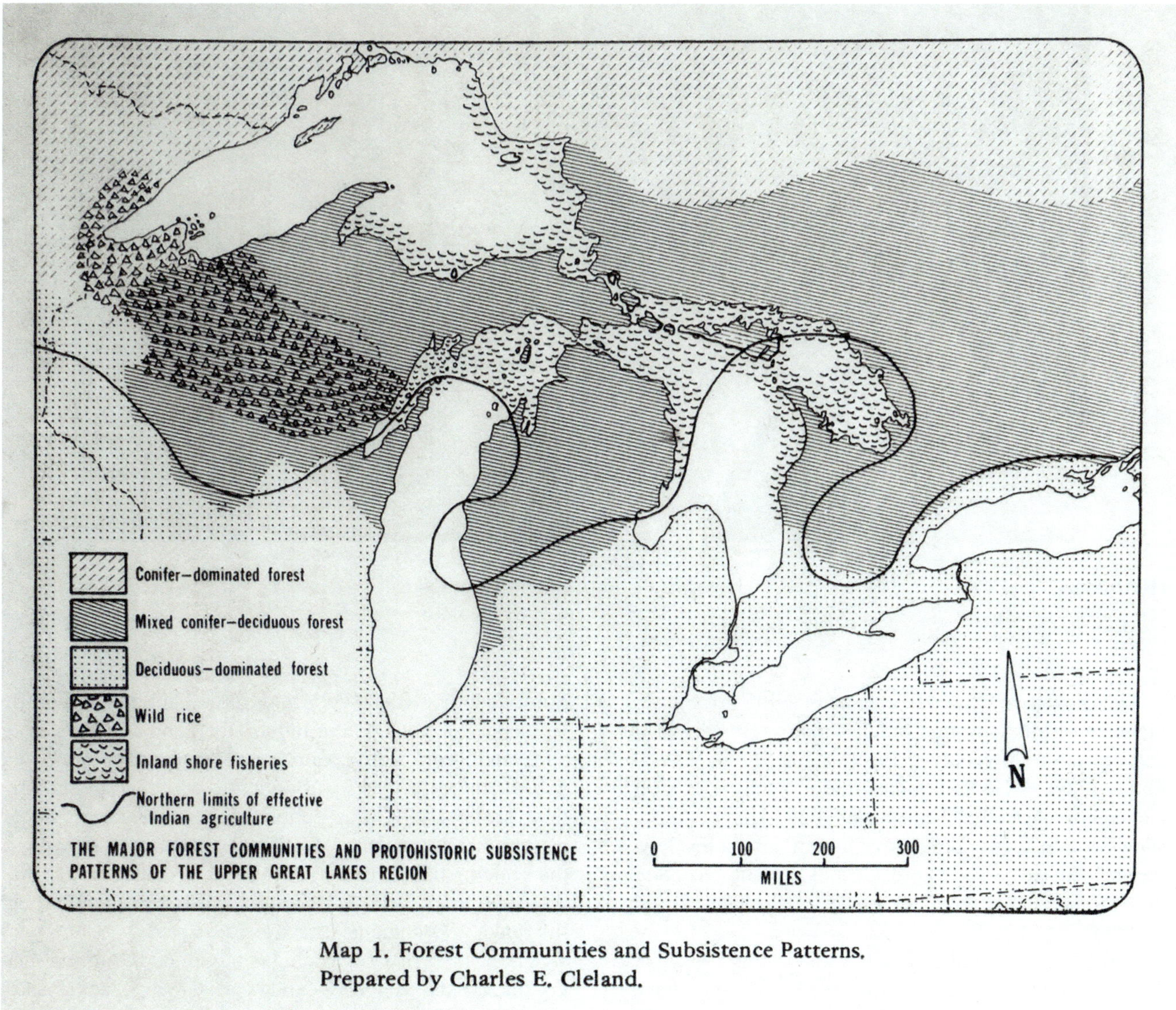

FIGURE 4.3 Great Lakes biota. Map, Charles E. Cleland. Flader 1983: 88.

capacity had increased by 500%, and the average size of fish caught had declined significantly. Smiley interpreted these data as indicating "remarkable diminution, if not complete collapse, is to be anticipated in the coming decade."[111] His recommendation to avoid that was to increase the size of mesh of fishing nets, stop water pollution, and increase fish propagation.

The University of Michigan zoologist Jacob Reighard (1861–1942) in 1889 studied fish embryology to assist Michigan's Fish Commissioner improve fish culture.[112] He also accompanied the commission's lake surveys to collect specimens for the University Museum. He then studied food and parasites of fish collected. In summer 1893, he established the Lakes' first field station, at Lake St. Clair (north of Detroit, between Lakes Huron and Erie), with financial assistance from Michigan's Fish Commission. In summer 1894, the station moved to Traverse Bay, Lake Michigan, directed by zoologist Henry B. Ward, who published a report on his findings.[113] In 1898, the U.S. Fish Commission provided funds and Reighard established a station at Put-in-Bay, western Lake Erie.

Fisheries Research Board of Canada, established in 1901, funded a Georgian Bay Biological Research Station on an island in the mouth of Go Home River, with $1,500 a year, under director R.R. Bensley.[114] That summer, both laboratory and boathouse were built, laboratory equipped, boathouse stocked with 35-foot sailboat, rowboat, flat-bottomed punt, floats, leads, plankton, and other nets. In 1902, Professor C.H.C. Wright, University of Toronto, conducted a hydrographic survey, with two of his students making a map of the region. Bensley began collecting fishes, and others began collecting birds, mammals, and plants. Professor W.J. Loudon collected meteorological data. Its final summer of operation was in 1913; Canada entered World War I in 1914.

FIGURE 4.4 (a) D(onald) S. Rawson. Web. (b) Henry A. Regier. Courtesy H. A. Regier.

Ontario native D(onald) S. Rawson (1905–61, figure 3.5) pioneered limnology on Canada's plains. He earned three degrees at the University of Toronto (B.A. 1926, M.A. 1927, Ph.D. 1929).[115] He had joined the Zoology Department, University of Saskatchewan, in 1928. His bibliography shows he was a very productive limnologist in his 33 years there.[116] He had served as president of Limnological Society of America and director of Fisheries Research Board of Canada.

The United States and Canada established an International Joint Commission in 1912, under the Boundary Waters Treaty (1909), focused upon water management and quality. The binational Great Lakes Fishery Commission was also established in 1894 and was operating in 1946. In 1972 it was strengthened by a Great Lakes Water Quality Agreement, including plans to reduce pollution. Canada's St. Lawrence Seaway Management Corporation and U.S. St. Lawrence Seaway Development Corporation were both established in 1954. Congress had established National Science Foundation in 1950, which became active in 1951. Congress established U.S. Sea Grant Program in 1966.[117]

Reighard encouraged two of his graduate students to study aspects of Great Lakes fisheries, Michigan natives John Van Oosten (1891–1966) and Walter Koelz (1895–1989). Although younger, Koelz (M.A. 1917, Ph.D. 1920) was ahead of Van Oosten (M.A. 1921, Ph.D. 1926). Koelz's research produced two substantial studies: the survey of Great Lakes fishing industry (1926) and *Coregonid Fishes of the Great Lakes* (1929). During his decade studying fishes, he also studied the fishing industry for the U.S. Commissioner of Fishes (1926). On Lake Erie, "considering the immense quantity of netting employed in so small an area as Lake Erie, it is surprising that any fish are left."[118] Subsequently, Koelz collected birds and plants for the University of Michigan in India, Tibet, and elsewhere.[119]

Grand Rapids native Van Oosten remained committed to fish.[120] His M.A. thesis was on aging whitefish by studying their scales (1921), was published; he then studied scales of lake herring or cisco for his Ph.D., plus "a critique of the scale method" (1926).[121] He spent his career at the Ann Arbor office of the U.S. Bureau of Fisheries, rising through the ranks to become its director.

The University of Michigan continued training fishery biologists who had careers at Great Lakes Fishery Laboratory. Two later examples were Stanford Smith (1920–2013, B.S. 1943, M.A. 1951, Ph.D. 1954), son of a professor of entomology at Stanford University in California, and Denver native Alfred Beeton (1927–2019. B.S. 1952, M.S. 1954, Ph.D. 1958). Beeton earned his three degrees at the University of Michigan, and Smith earned his M.A. and Ph.D. there. Smith was the only fishery biologist I knew. He was very happy as an investigative scientist and had no administrative ambitions. He was a team player, and when fishery biologist Ralph Hile changed authorship of one of Smith's papers because James Moffett needed a paper on his vitae for 1957, Smith accepted it.[122]

Beeton was a limnologist, who would have studied under Hutchinson or Hasler if he had gone to Yale or Wisconsin. He joined U.S. Great Lakes Fisheries Laboratory, Ann Arbor, in 1957 and did research there until 1966, when he resigned to become a professor of zoology and Assistant Director of the Center of Great Lakes Studies, University of Wisconsin-Milwaukee, where he stayed also for a decade before joining the Michigan Sea Grant Program, back in Ann Arbor. He published two studies on Lake Erie limnology (1961, 1963) and two on Great Lakes eutrophication

(1965, 1972); he chaired a conference on pollution of the Great Lakes (1980), and later he was a National Oceanic and Atmospheric Administration (NOAA) scientist at the Great Lakes and served on numerous committees.[123]

Lake trout *Salvelinus mamaycush* was a valuable commercial species, and a prime target of sea lamprey. Fishing pressure and sea lamprey predation led to lake trout population decline.[124] That crisis led to a study, *Biology, Population Structure, and Estimated Forage Requirements of Lake Trout in Lake Michigan*,[125] then to a conference on lake trout research.[126] That assessment led to *Strategies for Rehabilitation of Lake Trout in the Great Lakes*.[127] These various assessments publicized existing literature and helped expose problems needing further research.

Around 1960, walleye *Stizostedion viterum* population in Lake Erie collapsed—following the highest catch ever.[128] Two other *Stizostedion* species, blue pike and sauger, had already disappeared from Lake Erie. Three Ontario fishery biologists—Henry Regier, Richard Ryder, Robert Ferguson—and four from Ohio—Vernon Applegate, Jerry Manz, Harry Van Meter, David Wolfert—collaborated to study ecology and management of walleye in western Lake Erie (1969). Western Lake Erie is the shallowest and most productive part of Erie, though also the most polluted part. They concluded that pollution and enrichment of water by agricultural fertilizer had inhibited walleye numbers to some extent, though unregulated fishing was more important.[129] In 1979, fishery biologists J.C. Schneider (Michigan) and J.H. Leach (Ontario) collaborated on a history of "Walleye Stocks in the Great Lakes, 1800–1975: Fluctuations and Possible Causes." They started research at Lake Superior and proceeded eastward to Lake Ontario, discussing the situation in each part of each lake. They found "The stocks were exposed to various sources of stress, including high exploitation, pollution, interaction with foreign species, which, if not important in the decline, may have suppressed recovery."[130] A symposium in 2006 on *Status of Walleye in the Great Lakes* had three fishery biologists edit the findings, including walleye status in each lake, also in Lake St. Clair and its two rivers.[131]

Great Lakes Fishery Commission organized an interagency committee charged with studying bloater population decline in Lake Michigan that began in the mid-1960s. The committee published its findings and conclusions in 1985: *Population Dynamics and Interagency Management of the Bloater* (Coregonus hoyi).[132] The committee recommended cessation of harvesting to allow bloaters to recover their former population. Lake Michigan is within U.S. borders, and the three states on its shores agreed. By the early 1980s, the committee estimated bloater population had returned to that lake's carrying capacity. Since overfishing was an obvious potential cause, why did it take a commission to determine that was the cause? If states had simply been told it was likely caused by overfishing, there would have been controversy over whether to take seriously this recommendation.

Lake sturgeon (*Acipenser fulvescens*) is the largest fish in the Great Lakes, but is not a "top predator." It is a bottom feeder on small animals and organic detritus. It takes years to mature, and so is vulnerable to overfishing.[133] Sadly, European settlers did not appreciate it as a resource, and it was treated as a nuisance, since it damaged fishing nets if caught, and they were often stacked on shore and burned. During the 1880s, its food value was recognized, then overfished! Sturgeon have existed 100 to 200 million years, and it responds when given a chance to grow, in unpolluted waters.

John Hudson and Susy Ziegler recently surveyed the history of Great Lakes commercial fisheries, including scientifically informed management, and found a gradual decline, with no recovery prospects in sight.[134] The future belongs to sport fishing management in the Great Lakes.

Canals, Locks, and Seaway

Canals and locks by 1825 facilitated commerce and exposed lakes to invasive species to adversely impact native species. Erie Canal, connecting Hudson River, at Albany, to Buffalo at the eastern end of Lake Erie, was begun in 1817 and completed in 1825.[135] It is 363 miles long, has 83 locks, and diverted trade from Great Lakes into the United States that otherwise might have gone through Canada. The Erie Canal opened Midwest to European immigrants; population of Ohio, Michigan, Indiana, and Illinois grew 15 times from 273,000 in 1810 to 4,217,000 in 1850.[136] Five shorter canals, built in 1828–37, connected the Erie Canal to ports on the south shore of Lake Ontario.

Canada responded to the economic challenge of the Erie Canal with Welland Canal (mainly financed by New Yorkers), connecting Lakes Ontario and Erie. Welland Canal Company was founded in 1824, admitted ships in 1829, and completed in 1833.[137] The first version had 40 locks and admitted ships 100 feet long. It saved Canada's commerce. Illinois and Michigan Canal, built 1835–48 and enlarged in 1871, connected Lake Michigan at Chicago with the Illinois River.[138] Soo Locks, connecting Lakes Huron and Superior, were built because of stimulus of Erie and Welland Canals. In 1797, Northwest Fur Company built a small Soo lock for canoes and bateau. The first version for ships opened in 1855, but by the 1870s it needed repairs and enlarging, and Michigan turned it over to Federal Government; it was enlarged three times by 1960.[139]

Enlargements were justified by usage. In 1928, its grain passage in its seven ice-free months almost equaled that of the Panama Canal in 12 months. In 1955, tonnage that passed through Soo equaled that of Panama and Suez Canals combined. Canada also built its own Soo canal, 1986–90, but the US canal dominated the passage and Canada's canal usage remained small.[140]

While the St. Lawrence Seaway was under construction and immediately afterwards some called it the greatest engineering project in history.[141] One wonders if claimants considered Egypt's pyramids, China's great wall, Panama's canal, and two bombs dropped on Japan. The seaway was certainly the greatest joint project of Canada and the United

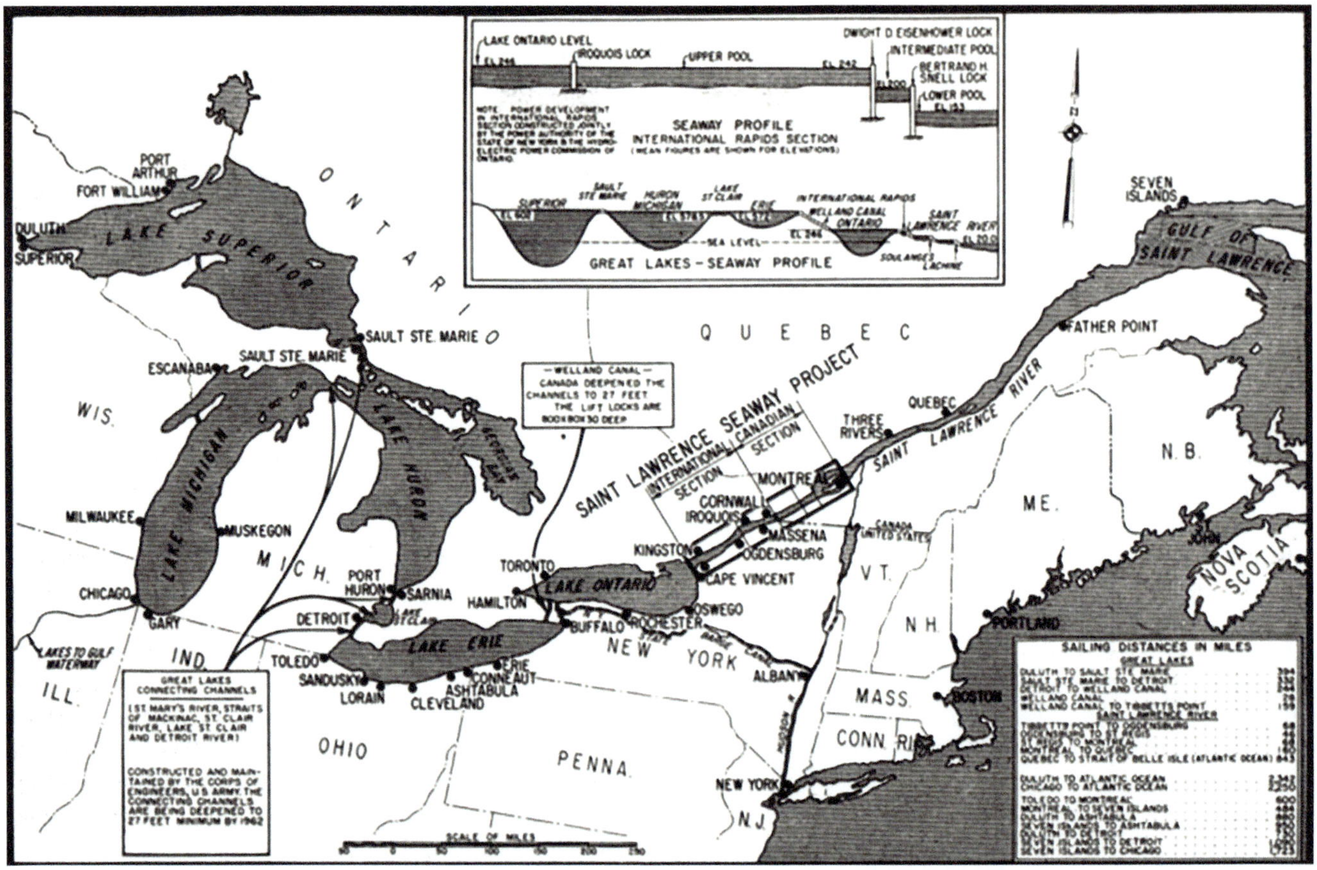

FIGURE 4.5 St. Lawrence Seaway to Great Lakes. Map, web.

States, costing $1 billion.[142] From the eastern end of Lake Ontario to Montreal is 182 miles, including 46 miles of rapids. Railroads and coal companies opposed building the seaway,[143] but Canada began preparing for it in 1951, and in 1954 it finally passed Congress at Senator John F. Kennedy's urging.[144] Construction began that year, and it opened in 1959, allowing ships up to 742 feet long to travel its 2,340 miles to Duluth, Minnesota. Great Britain's Queen Elizabeth and U.S. President Eisenhower were honored guests at opening ceremonies in June.

However, an economist's 1954 projection of the Seaway carrying 50 million metric tons annually never happened.[145] Freight shipped beyond Montreal peaked at 23 million metric tons in 1978 and dropped to 9 million metric tons by 2007. That shortfall, with uncalculated invasive species carried in ships' ballast water dumped in the lakes, creating a biological catastrophe, leading to recalculations to the Seaway's value. Environmental reporter Jeff Alexander decided the Seaway's fiftieth anniversary was a good time to reexamine its pros and cons.[146] His evaluation is a fine guide. Most cargo ships now merely carry cargo among these lakes and ports.[147] Revenue from running the Seaway did not even cover interest costs on loans for building it.[148] "The seaway saga signals the hubris to which twentieth-century governments, experts, and planners were prone when dealing with plans to radically reshape water systems."[149]

Invaders and Responses

The Great Lakes now have catastrophic numbers of invasive species; most, if not all, were assisted by humans—a few intentionally assisted, but most inadvertently. Erie and Welland Canals, Soo Locks, and St. Lawrence Seaway facilitated not only commerce but also invasive animals.[150] These included 34 fish species, 1819–1974, which Lee Emery discussed.[151] Jeff Alexander counted 186 nonindigenous species that have invaded Great Lakes;[152] others may have invaded since 2008.

The first invasive was likely sea lamprey, which was also most devastating to fish. It reached Lake Ontario by 1835.[153] It probably came via Erie Canal, then into one or more side-canals to Lake Ontario. Sea Lamprey did not seem to reach Lake Erie until after Welland Canal was improved in 1919. They were in Lake Erie by 1921, where they did not flourish yet still reached Lake Huron by 1933, Lake Michigan by 1936, and Lake Superior by 1946; its damage to fish in these three largest lakes was catastrophic.[154] Previously, fishermen caught about 15 million pounds of lake trout a year; by the early 1960s, catch had dropped to 300,000 pounds. In three decades, sea lamprey eliminated lake trout from the lakes except Superior. Female sea lampreys usually lay about 60,000 eggs in rivers, but sometimes as many as 100,000. Young remain in river bottoms for up to six years before swimming into the lakes and parasitizing large fish.

The early investigator of sea lamprey in Great Lakes was New Yorker Vernon Applegate (1919–80), who earned three degrees in zoology at the University of Michigan (1943, 1947, 1950). He joined the Institute of Fisheries Research, Michigan Department of Conservation, 1945–50, before moving to the U.S. Fish and Wildlife Service to super-vise sea lamprey research at Hammond Bay Biological Station, which he established.[155] In 1950, the U.S. Fish and Wildlife Service published his *Natural History of the Sea Lamprey*, Petromyzon marinus, which he had mainly researched at the Michigan Institute (as he acknowledged).

Binational concern over sea lamprey depredations led Canada and the United States to establish Great Lakes Fishery Commission (GLFC) in 1954, operational on 1 July 1956.[156] GLFC began publishing its valuable technical reports in 1961. Suppression of sea lampreys had been by adding a lampricide to streams in which they bred. A problem was the four native fresh-water lampreys,[157] two of which were parasitic and two that were not, were significantly smaller than sea lampreys. A sea-lampricide was wanted that did not kill native lampreys. After testing some 6,000 compounds, Applegate's researchers found one—3-trifluoromethyl-4-nitrophenol (TFM). In 1958, the first year of use, GLFC spent $915,000 killing 122, 275 lampreys in tributaries of Lakes Michigan and Superior, or $7.48 each.[158] It is unlikely sea lampreys can be eliminated from Great Lakes, but TFM effectively suppresses their numbers to a tolerable level. However, TFM works in streams, but not in rivers; research continues. Canada pays 31% of U.S. research on alternatives (since Canada controls 31% of Great Lakes).

The second invader was alewife, first discovered in Lake Ontario in 1873, which also likely entered via Erie Canal, and found in Lake Erie in 1931.[159] Predatory fishes had been reduced in numbers by commercial fishermen, and alewife rapidly expanded its range. However, it was poorly adapted to Great Lakes, and in some years there were massive die-offs littering beaches, to the dismay of residents and tourists. The worst year was 1967, with several hundred million pounds of dead fish on Lake Michigan beaches.[160] After salmon were intentionally introduced, attitudes toward alewives changed, because alewives and smelt were major forage species of salmon.[161]

A third invasive was common carp, intentionally introduced from Germany in 1877 as a food fish.[162] However, carp never became a popular human food in Great Lakes country. Omnivorous carp is now a pest species in shallow habitats. Eggs of brown trout *Salmo trutta* were sent from Germany in 1883 and later, and it became established in streams flowing into lakes and along lake shores.[163] Rainbow trout *S. gairdneri* were stocked in a tributary of Lake Huron in 1876, and Michigan stocked in upper Lakes Huron and Michigan nearly two million annually, 1880–1961, and nearly one million annually, 1962–81. In 1912, rainbow smelt was introduced into Crystal Lake, Michigan, which drains into Lake Michigan. It is now abundant in all Great Lakes and is a popular food for humans and salmon.[164]

Amazingly, salmon introductions have transformed Great Lakes fisheries, in ways that have pleased sport fishermen.[165] Atlantic salmon *Salmo salar* were native to Lake Ontario, until fished out by 1890.[166] Lake trout were then the top predator in all five lakes; sea lampreys eliminated it from all but Lake Superior. Three Pacific salmon species were introduced and remain. "Between 1873 and 1933, many serious attempts were made to establish chinook salmon [*Oncorhynchus tshawytscha*] in the lakes, but all introductions failed."[167] Perhaps lake trout competition prevented their survival. In 1956, the Ontario Department of Lands and Forests attempted to establish a pink salmon *O. gorbusccha* fishery in Hudson Bay for Native Americans.[168]

That failed, and 21,000 fingerlings were dumped into Current River, which flows into Lake Superior. They survived there, and pink salmon is the only self-sustaining salmon population in the Great Lakes, but never abundant enough for a sport fishery or to spread to other lakes. Two biologists at Michigan Department of Natural Resources, Drs. Howard Tanner and Wayne Tody reintroduced chinook salmon again, in spring 1966, as a sport fish and in hopes they would eat the alewife.[169] Millions more were introduced, 1967–1984, and both objectives were achieved. In 1966, both Michigan and Ohio stocked coho salmon *O. kisutch* and have done so ever since, as it became a popular sport fish.

Canadian fishery biologist W.J. Christie edited a 1971 symposium on Salmonid Communities in Oligotrophic Lakes, which concluded that "the only defense against further invasions by pest species appeared to be maintenance of sufficiently dense piscivore stocks to assure their suppression."[170] However, what the density was which could both suppress pest species and support fishing pressure required more research. More research occurred, leading to an International Symposium on Stock Assessment and Yield Prediction (1987), including fish community rehabilitation.

Thus far, the only introduced species that created a major crisis was the sea lamprey. There was no realization the Great Lakes-St. Lawrence Seaway would introduce a major new crisis until a newly graduated zoology student from Guelph University, Sonya Santavy, on a one-year fellowship, on 1 June 1988 took bottom samples in Lake St. Clair and discovered an unknown mussel, later identified as a zebra mussel *Dreissena polymorpha*, from the Black and Caspian Seas.[171] Amy Benson reported it was probably discharged by a ship in Lake Erie, 1985/86 (found there a month after Santavy's discovery).[172] Zebra mussels initiated a crisis in 1989, when they clogged the intake pipe at Monroe, Michigan, south of Detroit.[173] Clogging municipal and power plant intake pipes in the Great Lakes was not the only problem they caused. They also absorbed bottom botulism bacteria, and when larger organisms consumed them, botulism bacteria progressed up the food chain to waterfowl, which died by the thousands from botulism, as did poisoned fish not eaten by waterfowl.[174] Within two years of their discovery, zebra mussels had invaded all five lakes. Quagga mussels *Dreissena rostiformis* are larger than zebras, were discovered in Lake Erie in 1989, and spread quickly to Ontario, Huron, and Michigan, and had invaded Superior by 2005. Quaggas come from Dnieper River, Ukraine, and outcompete zebras and often replace them.[175]

Senator John Glenn of Ohio was co-chair of Senate Great Lakes Task Force, and he quickly understood the catastrophic significance of zebra mussel invasion into Great Lakes, and he led bipartisan passage of the Nonindigenous Aquatic Nuisance Prevention and Control Act (1990).[176] Oceanic ships introducing foreign species arose long before mussels precipitated an invasive crisis. Nine species dis-covered 1959–83 were attributed to ocean freighters, but since none were catastrophic, no new steps were taken to eliminate this possibility.[177] A new discovery in 1984, spiny water flea *Byhthotrephes longimanus*, did stimulate comment because thousands of them clung to fishing lines.[178] In 1986 European ruff *Gymnocephalus cernuus* was dis-covered in Duluth Harbor; it competes with native species for food, and its spiny dorsal fin prevents larger fish from eating it.[179] Federal biologists discovered TFM would kill it, but state governments blocked its immediate use while ruffs were localized, and so ruffs spread into other lakes.[180]

Discovery of quagga mussels in 1989 raised the level of urgency. The cause of invasives problem was clear: foreign ships having little or no cargo filled their ballast tanks with water to prevent their ships from floating high in oceans, and when they arrived at a port, they dumped ballast water and took on cargo. Dumped water introduced foreign organisms. A requirement that foreign ships exchange ballast water just before entering St. Lawrence River was passed in Canada and the United States.[181] Ships arriving with cargo were exempt from inspections, creating a heavily used loophole, slowly remedied. Canada did so in 2006 and the United States in 2008.

Round goby *Neoogobius melanostomus* preys on mussels and was discovered in St. Clair River in 1990.[182] However, it does not consume enough to prevent mussels from spreading. It also competes for food with native mottled sculpins, johnny darters, and logperch. Round gobies are too small for human consumption. In the 1990s, two new species of plankton appeared *Echinogammaraus ischnus* in Detroit River and *Cercopagis pengoi* in Lake Ontario.[183] All these species from the Black and Caspian Seas could live together, to the detriment of native species.

Invasive insects have also had destructive impacts in the Great Lakes Basin.[184] When European earwigs arrived in the mid-1900s, they spread from Great Lakes ports. Gary Dunn named seven insects and also six foreign predators or parasites deliberately introduced to combat those pests. Such efforts sometimes worked as hoped, and sometimes it backfired when the deliberate introduction found an alternative food easier to obtain. Dunn provided appendices listing both organizations and periodicals concerned with insect pests in the Great Lakes Basin.[185]

In 2005, Dr. Stephen Brandt, Director of Great Lakes Environmental Laboratory, commented: "The zebra mussel has had perhaps the most profound effect on the Great Lakes ecosystem, second only to human beings."[186] That impact led to change in the U.S. invasive species policy and broadened the scope of responsibilities for the International Joint Commission.[187]

Pollution

Great Lakes region became a region of heavy industry, which disposed of wastes by dumping them into waters they also used to transport much of their commerce.[188] First of pollutions was sawdust from logging, which did not spread far from sawmills at mouths of rivers flowing into lakes. It lasted until these forests were logged and lumber companies move west. Second, cutover lands went into farms where practicable, and wetlands were drained for more farms. Farmland erosion became a serious problem in Green Bay, Saginaw Bay, and western Lake Erie, all being shallow.[189] Third, shipping. During the 1800s, most freight went by sail, because cheapest, and time was not an expensive factor. However, by 1820, steamboats reached Great Lakes, with *Walk-in-the-Water*, 132-feet long and 220 tons, which established regular service between Buffalo and Detroit. When it foundered on a Lake Erie sandbar, its engine was salvaged to run a Saginaw Bay sawmill.[190] Other steamboats appeared, and trees close to lakes became their fuel.

The first lake iron ship for cargo, Onoko, was built and operated at Cleveland in 1881. Gradually, others appeared, yet cargo sailboats were cheaper and persisted until the 1920s. Steam and diesel brought speed and pollution. For most of the 1900s, more cargo was shipped on Great Lakes than from all Atlantic and Pacific ports combined.[191] Shipping increased significantly when the Great Lakes-St. Lawrence Seaway opened.[192]

Rachel Carson's *Silent Spring* (1962) raised awareness of pollution by pesticides, especially DDT, and this became a Great Lakes problem.[193] DDT could pass up a food chain to top predators, including humans, becoming more concentrated at each level, concentrated in tissues. Environmental Defense Fund in 1967 challenged heavy use of pesticides in courts. In winter 1969, Michigan's Department of Agriculture discovered high amounts of dieldrin in coho salmon from Lake Michigan, and it prevented the sale of 500,000 pounds of the lake's fish. Later, U.S. Food and Drug Administration seized 14 tons of Lake Michigan coho because of high levels of DDT. In 1970, high levels of mercury were discovered in fish from St. Clair River, Lake St. Clair, Detroit River, and western Lake Erie. Michigan and Ontario temporarily banned fishing in these waters until the situation was remedied. In 1971, Monsanto's pollution of lakes with polychlorinated biphenyls (PCB) was stopped. The complex pollution problem is never permanently resolved in the Great Lakes.

Accidental transmission of invasives by oceanic ships has been called "biological pollution." NOAA ecologist Gary Fahnenstiel announced in 2004 that the only way to stop invasive pollution is to ban oceanic ships from the Great

Lakes.[194] Since he never again spoke out in public about his claim, his superiors may have forbidden it. Yet others began demanding Seaway closure, including government officials, observing that Great Lakes fisheries were much more valuable than oceanic shipping. To settle the matter, two conservative economists, John Taylor and James Roach, were commissioned to study this. In 2005, they reported banning oceanic shipping would add $55 million to shipping costs, but the fishing industry was losing $200,000 to $500,000 a year due to invasives from oceanic ships. In 2007, Taylor and Roach further reported that closing oceanic shipping would create 1,000 new jobs. Shipping officials replied: closing the Seaway was impossible, which was true with existing laws.

Industrial pollution was an equally serious problem. Environmental historians have identified the most serious polluting companies.[195] On 2 December 1970, President Nixon signed an executive order creating the Environmental Protection Agency (EPA), but it was years before it became active and effective. In 2006, a federal judge ordered EPA to regulate ballast water discharges. EPA and shippers appealed, but a higher court sustained the order. The Federal government was slow to implement this order. A competent retired Coast Guard officer, Eric Reeves, stated that a practical solution was to pour chlorine into ballast tanks. Shippers, governments, and environmentalists objected, and never did. No solution has gained much support.

Since I live in a Wisconsin city on Lake Michigan, I sought and received Wisconsin Sea Grant funds for summer research. I began with "Overfishing or Pollution? Case History of a Controversy on the Great Lakes" (1985), on disagreement between federal fishery biologist John Van Oosten in Michigan (1891–1966) and Ohio State University fish ecologist Thomas (Hux) Langlois (1898–1968) over the cause of decline in commercial fish catch. I pursued this controversy further in "Pollution and Aquatic Life in Lake Erie: Early Scientific Studies" (1987) and in "Missed Opportunities: U.S. Fishery Biologists and Productivity of Fish in Green Bay, Saginaw Bay, and western Lake Erie" (1989a). My Sea Grant support came from Wisconsin Sea Grant Institute, whose own funds were approved by a committee that included federal Great Lakes fishery biologists, whose predecessors I criticized in my third study. The head of the Wisconsin Sea Grant Institute therefore asked me to change my conclusion. I declined and understood I would receive no more research funds from Wisconsin Sea Grant Institute. I had a fourth manuscript for publication by the Great Lakes Fishery Commission on The Role of Sea Lamprey Control in Restoration of Fisheries in Great Lakes (1989b). My manuscript was returned because I had only told the U.S. part of the story. I replied, "Please find a Canadian coauthor to tell Canada's story." A qualified person was found, whose only action was to give a talk at a symposium based on what I had written! The Commission published Stephen Nepszy's *Parasites of Fishes in the Canadian Waters of the Great Lakes* (1988); perhaps if my title had indicated, only for U.S. waters, it might have been accepted? Or might this have been another government agency discipling me for conclusions to my 1989 paper?

Ecosystem Management

In the 1950s, when sea lampreys created a crisis, ecologists were studying ecosystems, in hopes of gaining theoretical understanding from particular studies.[196] In 1972, Stan Smith was pessimistic, and entitled a paper, "Destruction of the Ecosystem in the Great Lakes and Possibilities for Its Reconstruction." He summarized three factors that he thought destroyed that ecosystem: marine invaders, pollution, and overfishing. His plan for reconstruction involved: (1) improvements of land uses in the drainage basin, (2) elimination of sources of physical, chemical, and biological pollution entering these lakes, and (3) restoring favorable and productive fish populations.[197] The second goal was the big challenge.

In 1977, GLFC asked its Science Advisory Committee to find ways to rehabilitate Great Lakes. An international report resulted: "Rehabilitating Great Lakes Ecosystems," which recommended using Green Bay (Wisconsin) and Bay of Quinte (Ontario) as test cases.[198] After further thought and experience, four binational organizations—IJC, GLFC, International Association for Great Lakes Research (IAGLR), and Great Lakes Tomorrow (GLT)—sponsored a workshop in 1983 on "Managing the Great Lakes as Home." Two Canadians, W.J. Christie and J.R. Valentyne, and two Americans, M. Becker and J.W. Cowden, edited the workshop report, which made two recommendations:[199]

1 establish a continuing group to monitor and evaluate progress in implementing an ecosystem approach in the Great Lakes Basin, with annual reports and five-year review in full;
2 establish a public information program with examples and illustrations describing how to practice an ecosystem approach in government, industry, citizen groups, and in personal behavior.

There were subsequent reports on progress in Green Bay and the Bay of Quinte.[200]

In 1988, Iowan political scientist and environmentalist, Lynton K. Caldwell, at the University of Indiana, edited nine papers on "Perspectives on Ecosystem Management for the Great Lakes," eight of which discussed government management.[201] Only Canadian Henry Regier, assisted by Lee Botts and John Gannon, addressed ecological aspects of government management. Also in 1988, GLFC and IJC sponsored a workshop on "An Ecosystem Approach to the Integrity of the Great Lakes in Turbulent Times."[202] There were six papers on policy implications and 11 on theory and testing implications. The most relevant here are one paper by two Americans and one by two Canadians: Americans Thomas Fontaine and Donald Stewart, "Trophic Dynamics and Ecosystem Integrity in the Great Lakes: Past, Present, and Possibilities"; Canadians Robert Steedman and Regier, "Ecological Bases for an Understanding of Ecosystem Integrity in the Great Lakes Basin. Fontaine and Stewart explained the scope of their complex paper: "a comprehensive, yet preliminary, conceptual and mathematical model approach that describes causal relationships among fish food webs, nutrient recycling, and contaminant processes in the southern basin of Lake Michigan.[203] Steedman and Regier explained knowledge they sought: "know how ecological integrity develops, how it responds to turbulent or surprising external influences or stresses, how it may be measured, and how it may be managed."[204] John Magnuson thought "Numerous workshops and publications did lead to informed efforts to restore degraded ecosystems."[205] An eight-member binational Habit Management and Planning Task Force wrote 15 pages of "Guidelines for Fish Habitat Management and Planning," with 19 recommendations.[206]

By 1990, fishery biologists at Great Lakes realized that global warming would likely affect fisheries negatively, and they decided to prepare for that danger.[207] During the 1990s, Dalhousie University marine fishery ecologists addressed an issue that also concerned Great Lakes fishery biologists: impact fishing for top predators exerts on oceanic ecosystems. Ransom Myers and Boris Worm argued that "the global ocean has lost more than 90% of large predatory fishes. Although it is now widely accepted that single populations can be fished to low levels, this is the first analysis to show general, pronounced declines across widely varying ecosystems."[208] Peter Ward and Myers documented for oceans what had already been documented in the Great Lakes: "Shifts in Open-ocean Fish Communities Coinciding with the Commencement of Commercial Fishing."[209]

The ecological history of each of the five Great Lakes contributes to understanding their collective history. Those histories are not given here but are discussed in my more detailed account of Great Lakes ecology (online).[210] Conclusion: humans have subjugated America's Great Lakes to various environmental assaults, some severe, including species extinctions. Yet, limnologists, fishery biologists, and other environmental investigators have worked diligently to understand the dangers in order to find ways to mitigate them.

C. MARINE ECOLOGY, 1920S—90S

In Chapter 2, marine ecology was discussed from the 1870s to the 1920s. Several references can guide this continuing survey: Vaughn et al. included marine ecoogy in *International Aspects of Oceanography* (1937); Jack's briefer survey was only on biology: "Biological Field Stations of the World" (1945); Hiatt et al. wrote a larger, but more restricted *Dictionary of Hydrobiological Laboratories and Personnel in North America* (1954); Hiatt later edited *World Directory of Hydrobiological and Fisheries Institutions* (1963). Schlee's *The Edge of an Unfamiliar World: a History of Oceanography* (1973) discussed marine ecology, with background information from oceanography; Mills' *Biological Oceanography: an Early History, 1870–1960* (1989) concentrated upon plankton studies at Kiel, Germany, Plymouth, England, and Woods Hole, Massachusetts. There have been eight International Congresses on History of Oceanography, including marine ecology, held in different countries, 1966–2013, with volumes of papers published within one or two years. This history for the 1920s to 1990 is told here in two narrations: (1) Beebe, Bigelow, and Ricketts (older ecologists) and (2) Carson, Riley, Cousteau, and Clark (younger ecologists).

A. Beebe, Bigelow, Ricketts

William Beebe

William Beebe (1877–1962), from Brooklyn, was America's best-known and most widely read naturalist, 1900–50.[211] When school age, his family moved to East Orange, New Jersey, largely motivated by his love of nature. He excelled at sports, played musical instruments, and educated himself about natural history. In summers, he and parents vacationed at the Bay of Fundy, where he collected and studied invertebrate animals. After graduating from high school, he went to Columbia University in September 1896. He studied under Professor Henry Fairfield Osborn, but after three years, he left to become assistant curator of birds at the new Bronx Zoo.[212] Beebe published numerous articles on birds after 1906.[213] Zoo director William T. Hornaday decided Beebe needed a scientific journal, and he founded *Zoologica*, which opened with Beebe's "Geographical Variation in Birds with Especial Reference to the Effects of Humidity" (1907).

Beebe returned to his childhood interest in marine life when he led an expedition to Galapagos Islands in 1923. He discovered that "three-fifths of the birds on this island [Indefatigable] were wholly dependent on water life for food, while a number of others included fish and crustaceans as a part of their diet."[214] Six members of the expedition collected specimens by fishing, trolling, scooping up, dredging towing plankton nets, and seining in different marine environments. In three weeks, they collected 60 of 156 known species of near-shore fishes.[215] They discovered two new species and added 20 species not previously known from Galapagos.[216] Beebe's interest in species extended beyond collecting for the Bronx Zoo. His fascination about their behavior is seen in this comment about blennies:

> Two closely related species had diametrically opposite escape reactions; one was found on wet rocks or seaweed out of water, and when alarmed dived to the pool bottoms, the other perched on stones or fronds beneath the surface and instantly leaped clear of the water on being disturbed.[217]

A plankton net pulled through apparently empty waters revealed a "tremendously rich haul" of plankton that could be observed in an aquarium aboard ship. Transparent, hermaphroditic *Sagitta* worms were "very voracious and feed on copepods, diatoms, shrimps, and even larval fish."[218]

He was curious about Galapagos marine iguanas, which graze on sea weed—the world's only marine lizard. While he observed an iguana, a crab crawled onto it and plucked off and presumably ate three ticks.[219] The iguana was undisturbed by the crab, and after it left, Beebe counted 16 remaining ticks on the iguana.

Beebe's expedition was successful, and afterwards a member of New York Zoological Society offered use of another ship for another expedition.[220] He accepted, and this expedition lasted from 11 February to 30 July 1925. In a preface to *The* Arcturus *Adventure*, he explained that this book (1926), like *Galápagos: World's End* (1924), is a "scientifically accurate, popular presentation of the high lights and vivid experiences of the expedition,"[221] and that technically accurate accounts would appear in *Zoologica*, volume 8. The *Arcturus* was well equipped to study oceanic life, and the expedition collected enough specimens to provide a basis for Beebe's speculations. They crossed the Panama Canal on March 28, and they did not find Humboldt Current, as expected. Instead, they made one of the first scientific studies of what became known later as El Niño-Southern Oscillation (ENSO). Another expedition in 1925, led by Robert Cushman Murray, ornithologist at the American Museum of Natural History, also discovered the absence of Humboldt Current. He and Beebe knew each other, but their expeditions did not make contact. Murphy published his discovery in an article (1926) and in his *Oceanic Birds of South America*.[222] However, neither Beebe nor Murphy correlated this periotic condition with the weather. That connection was explained by Sir Gilbert T. Walker, at India's Department, in a series of articles, 1910–36. However, Swedish meteorologist Jacob Bjerknes specifically connected El Niño and Southern Oscillation (1969).[223]

Before sailing on *Arcturus*, Beebe bought a copper diving helmet with hose to study ocean life underwater. He called his descents diving, though he always descended a ladder from a boat. On his first descent, his team lowered a dead crab to use as bait. With that, he was soon surrounded by about 500 fish of 20–25 species. On his second descent, he harpooned "my first *Pomacentrus leucurus*."[224] On another descent, he decided to collect rocks from the bottom. He filled five pails with rocks, hauled aboard his boat. The pails were left out in the sun, and at the end of the day, "There had crept out an amazing array of interesting beings—beautiful sea-worms, starfishes, squillas, hermit crabs, and shrimps of every hue, a number of strange larval fish," including moray eels just like adults, but only 1.5 inches long.[225] He realized these animals could not be obtained by dredging, and that he had opened a new door to exploration.

In 1926, Beebe published two books, one on pheasants and one on *The Arcturus Adventure*, 11 articles (scientific and popular), eight book reviews, and eight brief notes.[226] One article was a scientific account of the *Arcturus* expedition.[227] Another was "The Guianas," coauthored by Henry Gleason, in ESA's *Naturalist's Guide to the Americas*, edited by Shelford. On Thanksgiving Day, a brief notice in *New York Times* announced "Beebe to Explore Ocean Bed in Tank."

A reader, F. Otis Barton Jr. (1899–1992), shared Beebe's ambition to explore the sea. Son of a wealthy family, he had grown up spending summers at Cotuit, Vineyard Sound, Massachusetts.[228] At age 16, he saw in an equipment catalog for underwater salvage a metal helmet, hose, and underwater suit. He sketched a wooden helmet and took it to a Boston cabinet maker who built it. With his brother manning a hose pump, he donned his helmet and climbed down a ladder at Cotuit Pier. Underwater exploration provided entertainment for the remaining summer. He then studied engineering at Columbia and Harvard Universities. After graduating, he spent a year sailing around the world. In the Philippines he met divers who hunted fish and pearls, and he was dazzled by coral reefs colors and associated animals. Someone introduced him to Beebe, who he persuaded to collaborate on under-sea exploration. Beebe had thought of building a steel cylinder, but Barton explained that only a sphere could withstand deep water pressure.

The concept was simple, but its production was complex, took several years to complete, and required much of Barton's fortune. They launched it in 1930, but in 1927, Beebe obtained another ship, the *Lieutenant*, for four months, mid-January to mid-May. Beebe chose to explore Haiti's waters because "There is not a single list of Haitian fresh-water or marine fish" and "Fish are the dominant reason for my choosing this part of the world for a winter of work."[229] That was his last expedition using his underwater helmet. He was better prepared than on the *Acturus* Expedition, having developed a waterproof case for a motion picture camera, and Bell Telephone provided an underwater phone for communicating with his handlers aboard *Lieutenant*. He collected a black sponge growing at 20–30 feet deep and sliced it into chunks, which exposed shrimp, small crabs, and slim gobies living inside.[230] He and his assistant, John Tee-Van, assembled a collection of Haitian fishes, and in an appendix to *Beneath Tropical Seas* (1928) listed families, common names, and a number of species in each family; there were 277 species, not named in that appendix, but given in their scientific report, with illustrations.[231]

Beebe's first wife left him before he was a celebrity, and in 1927, after he was, he married a fashionable novelist half his age. She disliked tropical heat, and so in spring 1929, they and two assistants sailed to balmy Bermuda to see if it was suitable for research. They were entertained by Governor Louis Bols, who took them on a tour of a small nearby island, Nonsuch, with a large vacant building which Bols offered to Beebe as a research station, and he added some motorboats. Beebe gratefully accepted, then returned to New York to raise funds to equip a marine laboratory on Nonsuch.[232] Many patrons contributed, and he returned to Nonsuch with equipment and a staff. It became the

FIGURE 4.6 William Beebe & Otis Barton with Bathysphere. Nonsuch Island, Bermuda. Web.

launching site for Bathysphere, which was sent down on test runs, and minor problems were fixed before Beebe and Barton first descended on 6 June 1930. They had sent it down empty to 2000 feet, but their deepest descent in 1930 was to 1,426 feet.[233] Beebe arranged with the National Geographic Society for an article with his text and photographs and colored illustrations by Else Bostelmann. The article was a success, and three more followed.[234] In August 1932, he arranged with a new National Broadcasting Company for a live broadcast of their descent in Bathysphere, which was broadcast in the United States and Britain to about 14 million listeners. His later book was *A Half Mile Down* (1934), with drawings, his photographs, and eight of Bostelmann's colored paintings.

Was Beebe first to discover symbiotic fish cleaning? It would have been difficult to observe, except from below the surface. He was the first known to H.M. Feder, "Cleaning Symbiosis in the Marine Environment" (1966), who cited Beebe's *Beneath Tropical Seas* (1928), without page number.[235] I did not find an account of it in that book; in 1931, Beebe called fish cleaning "an amazing exchange of courtesy."[236] His 1923 observation of a crab removing three ticks from a marine iguana differed from marine fish cleaning in that the iguana did not solicit cleaning, but fish do. Beebe did not mention his 1923 observation.

After four successful articles in *National Geographic*, why did Beebe stop using the Bathysphere? Diminishing returns. After 1934, he resumed his coastal explorations with two voyages of *Zaca*, in 1936 into Gulf of California—described in *Zaca Adventure* (1938)—and in 1937–38 in Guadalupe Island south to Panama and to Gorgona Island, near Columbia—described in *Book of Bays* (1942).

Henry Bigelow

Bostonian Henry Bigelow (1879–1967) attended Milton Academy, then studied for a year under Alpheus Hyatt at Boston Natural History Museum.[237] He earned three degrees at Harvard University (B.A. 1901, M.A. 1904, Ph.D. 1906). During his senior year, he learned that Alexander Agassiz was planning an expedition to Maldives Islands, in the Indian Ocean. He persuaded him to include himself.[238] They reached Colombo, Sri Lanka on 16 December 1901, and Agassiz took command of *Amra*, with crew, as prearranged.[239] He assigned Bigelow to preserve and describe jellyfish, which Bigelow did in a substantial article.[240] Again, in 1904–05 he accompanied Agassiz in exploring Galapagos and Easter Islands and Mexico's west coast.

Bigelow worked under Agassiz in 1901–10, but after Agassiz died Bigelow lacked direction until 1911, when English oceanographer John Murray visited the MCZ and suggested he explore the Gulf of Maine, extending 200 miles between Cape Cod and Newfoundland. The U.S. Bureau of Fisheries learned in 1912 that commercial fishing was declining there, and so was receptive to Bigelow's request for research funds to study there.[241] In 1912, Bigelow used a small Fisheries vessel, *Grampus*, from Woods Hole to begin. He read ICES publications and knew about area studies being conducted by European oceanographers and marine ecologists in North Sea, in hopes of better understanding fish ecology to improve declining commercial fisheries. Bigelow applied knowledge, skills, and determination to his task, and he produced three large volumes, 1925–27, on fishes, plankton, and physical oceanography. *Fishes of the Gulf of Maine* was coauthored by Dr. William W. Welsh (1850–1924), of the U.S. Bureau of Fisheries, who wrote most of that volume, but died before its completion. *Fishes of the Gulf of Maine* was well-illustrated, mostly with republished drawings. Ecologically important was information on habits and food, well-known for most commercial, but little known for non-commercial species. The plankton volume has numerous maps and micro-photos of plankton. Roger Revelle stated Bigelow's three volumes were "the first genuinely oceanographic study to be completed in North America."[242] They also made the Gulf of Maine "the best-known body of water in the world, oceanographically."[243]

In 1923, Bigelow became member, and often chair, of the American Commission of Fisheries Investigations, until 1932. Those investigations paralleled those of ICES, and he was invited to attend ICES meetings, and did, 1931–40.[244] In 1927, the National Academy of Sciences' Committee on Oceanography asked Bigelow to write a report on U.S. contributions. He spent a year investigating and writing a report, which recommended that America needed a program of environmental and marine biology investigations on the east coast equaling what Scripps Institution did on the west coast.[245] NAS sent his proposal to Rockefeller Institute, which gave $2.5 million to build an oceanographic institute and $50,000 a year to run it, for 10 years. Bigelow became the first director of Woods Hole Oceanographic Institution (WHOI). He found a site for it at Woos Hole, ordered a ship which M.I.T. Professor George Owen designed, and Bigelow had the steel sailing ship built in Copenhagen. Bigelow staffed both WHOI and ship, and he chose former student Columbus Iselin (1904–2012) captain of 142-foot *Atlantis*.[246] When Bigelow retired, Iselin became his successor as head of WHOI. WHOI opened in 1931. Bigelow was elected to the National Academy of Sciences in 1931. Bigelow served as WHOI director for a decade, then president of its board of trustees for a second decade, then chairman of the board for a third decade.

Walter Fisher, Director of Hopkins Marine Station, Monterey Bay, invited him to teach a marine biology course there. He went, and while there, measured temperatures and salinities of Monterey Bay waters from surface down to 600 m. U.S. Navy's Hydrographic Office asked him to write a textbook on waves for naval officers. He agreed, if he had a coauthor to handle aspects beyond his expertise.[247] Bigelow and Edmondson's *Wind Waves at Sea, Breakers and Surf* (1947) was obviously useful: USSR published a Russian translation in 1951. When Gordon Riley asked Bigelow for research time on *Atlantis*, he encountered some resistance:[248]

> Henry and I [became] good friends, but we did not see eye to eye. I was very much under the influence of the oceanographers at the Plymouth Laboratory [England]—Atkins and Harvey and the others—who were trying to develop quantitative measurements of plankton and ecological variables and to analyze seasonal cycles and regional variations in a quantitative way. To Bigelow this was a profound and distasteful oversimplification.

Their clash in opinions resembles the Hensen-Haeckel conflict, with Riley defending a Hensen-like perspective and Bigelow defending a Haeckel-like perspective.

Bigelow advanced at both Harvard and MCZ, becoming a professor of zoology in 1931, and made Alexander Agassiz Professor in 1944. He thought he set a longevity record as an officer at Harvard for 55 years (1905–60).[249] He made professional friends readily, and he won many professional awards. Alfred Redfield considered Bigelow "Shrew, wise, well-informed, a pragmatic realist, widely traveled, and the best-informed naturalist that he had ever met."[250]

Two publications celebrated his contributions: *Papers in Marine Biology and Oceanography Dedicated to Henry Bryant Bigelow by His Former Students and Associates on the Occasion of the Twenty-fifth Anniversary of the Founding of the Woods Hole Oceanographic Institution, 1925* (xx + 498 pp.). No editor listed here under Graham et al. 1955 since Graham was first of 56 authors. No article was on Bigelow, but his bibliography was included. *Oceanus* is a popular magazine WHOI publishes; volume 14, issue 2 (July 1968) was named H.B. Bigelow volume and included remembrances and photographs by 18 authors and some of his bibliography (page 29–32). WHOI became a leading oceanographic center. Retrospective talks were delivered at its fiftieth anniversary: Burstyn (1980), Deacon (1980), Haedrich & Emery (1980), Revelle (1980), Schlee (1980).

Edward Flanders Ricketts

Chicagoan Edward Flanders Ricketts (1897–1948) had an early interest in nature.[251] He attended the University of Chicago and was strongly influenced by Professor Warder Clyde Allee's course on animal ecology. Like Beebe, Ricketts only stayed three years and left without a degree. He married and in 1923 moved wife and son to Monterey,

FIGURE 4.7 (a) Henry B. Bigelow at helm of *Grampus*. Web. (b) Edward F. Ricketts. Ricketts et al. 1968: viii.

near Stanford University's Hopkins Marine Station. Ricketts established Pacific Biological Laboratories, Inc. to sell specimens he and others collected along shore. John Steinbeck novel, *Cannery Row* (1945), was set at Rickett's laboratory and nearby sardine canneries, and in 1958 Monterey renamed that street "Cannery Row." Ricketts was gregarious. One friend was teacher Jack Calvin, attracted to the sea. In summer 1931, Calvin and wife Sasha spent 53 days traveling the Inside Passage in a canoe with sailing from Tacoma to Juneau.[252] For June-July 1932 Calvin had use of a boat, *Grampus*, and took Ricketts, Sasha, her sister Xenia, and mythology authority Joseph Campbell (another Ricketts friend), to Alaska.[253]

Ricketts took notes on animals collected, and Calvin assisted him in turning those notes into his masterpiece, *Between Pacific Tides* (1939).[254] Calvin also photographed many specimens, and local artist Richie Lovejoy drew other specimens. Previous books on animals of shore and adjacent waters organized species accounts according to zoological classification. Ricketts realized no one discovers animals in that arrangement. He organized his accounts according to microenvironments where species lived: rocky shores—subdivided into exposed rocks, protected rocks, and rockweed pools; bays and estuaries—subdivided into rocky shores, sand flats, eelgrass, mud flats; and sandy shores—not subdivided. Illustrations were similarly organized, with an unusual feature: drawings are on a left plate and photographs on a right plate. The book's geographic range included much of Washington, Oregon, and California. It indicated each species environment, food, and other ecological aspects. The first edition had 320 pages, second edition (1948), which Ricketts prepared, but did not live to publish, had 365 pages. Although some of Ricketts' comments made Ralph Buchsbaum uncomfortable, he wrote a friendly review in *Ecology*.[255]

Species accounts were based upon both Ricketts' observations and literature cited. The last organisms discussed were phyto- and zooplankton, with discussion on fluctuations in abundance and important environmental factors, using 17 charts. A plankton bibliography followed those accounts. The rest of the bibliography was organized by classification because that was how most literature was organized by other authors. Finally, a brief bibliography listed works not limited to specific species. Some entries had annotations explaining content more fully than titles did.

Between Pacific Tides became a means for a relationship between Ricketts and British seashore ecologist Alan Stephenson.[256] Stephenson had joined British Great Barrier Reef Expedition, 1928–29, and he later spent a decade leading an ecological survey of South Africa's coast. In 1947, Stephenson and wife Ann visited Ricketts, and he said he so admired *Between Pacific Tides* he had read "every word."[257] Conversely, Ricketts had long admired Stephenson's published papers, and he wrote to Steinbeck that Stephenson was "probably *the* greatest ecologist."[258] Ricketts, of course, could not have judged ecologists in other specialties. Much later, the Stephensons published *Life between Tidemarks on Rocky Shores* (1972), with worldwide coverage.

In 1940, Ricketts and Steinbeck collaborated on a six-week expedition to the Gulf of California, the shores of which are a desert, to collect specimens and data as a basis for a sequel to *Between Pacific Tides*. Afterwards, Steinbeck wrote *The Sea of Cortez* (1941), a 271-page narrative, based on Ricketts' notes. Illustrations were similar to those in *Between Pacific Tides*, plus eight pages of color photographs of seashells collected. There was an annotated bibliography, pages 272–589 pages, and an index to Steinbeck's narrative, pp. 590–598. Excepting a few quibbles, Karl Schmidt wrote a warm review of it in *Ecology*.[259] Ricketts hoped that his findings in the Gulf would have scientific significance, and eventually they did. M.T. Dungan et al. in 1982 reported a catastrophic decline in a top carnivore star-fish *Heliaster kubiniji*, which had been abundant in 1940.[260] However, there was not another full expedition there until 2004. That Sea of Cortez Expedition and Education Project retraced sampling sites Ricketts had used, took similar samples, and compared results from the two expeditions. Human presence in the Gulf was still modest, but many aquatic species had declined and two species had become more abundant.[261]

Between Pacific Tides appeared in (at least) four editions and was widely used. *The Sea of Cortez* has had a much smaller audience, though Steinbeck's contribution to it was reprinted in 1995, and the original book was reprinted in 2009. Richard Grant argues that its significance has been under-appreciated:[262]

> It was the most renowned literary collaboration between a great American novelist and a great American scientist, and it produced a highly original and influential book. *The Log from the Sea of Cortez*, Steinbeck's account of the trip, is a high-spirited blend of adventure travel writing, marine science, and freewheeling philosophy, shot through with a marvelous sense of wonder at the interconnectedness of the world.

There was no sense at the time that the boat used, *Western Flyer*, had historic significance, and as it became older, it was simply abandoned to the elements. Yet, in 2019 geology businessman John Gregg bought it for $1 million and planned to spend $2 million more to repair it for use as an education and research vessel.[263]

Ricketts made three expeditions along Canada's coast to write a volume 3, on Canada's and part of Alaska's Pacific coast. His first trip in 1932 was with Calvin along the Inland Passage to southern Alaska (see earlier). His other two trips (1945, 1946) were along Vancouver Island to Queen Charlotte Island. He accumulated notes and specimens for a book, but died in 1948, soon after his car collided with a train. When Steinbeck received the news, in New York he lamented, "The greatest man in the world is dying and there is nothing I can do." Steinbeck did reach Cannery Row in time for the funeral.

Ricketts saw in1947 how many sardines were being caught and canned and warned in a newspaper article that high amounts could not be sustained.[264] He lacked the authority to be taken seriously, and the fishery collapsed (with partial recovery in 1986).[265] He still worried about a collapse in 1948, about when he died.[266] He left his notes to Hopkins Marine Station, and his notes and writings for a third volume are now published.[267] His lifetime project of documenting Pacific coastal invertebrates was continued by others and culminated in *Pacific Coast* (1985), edited by Bayard and Evelyn Connaughey.[268] Kathrine Rodger began publishing his posthumous works in 2002. Michael Lannoo discussed Ricketts' philosophical writings (2010).

B. Carson, Riley, Cousteau, Clark

Rachel Carson

Pennsylvanian Rachel Carson (1907–64) grew up on a small farm near Allegheny River and Pittsburg.[269] Her family had horses, cows, and chickens, but were not farmers. She had no playmates and spent free time reading and wandering in woods and fields, observing wildlife.[270] Her mother had similar interests and used Anna Comstock's *Handbook of Nature Study* (1911) to encourage Rachel's interests.[271] Rachel decided to become a writer and wrote five brief essays for a children's magazine, 1918–22, which were published. She was an exceptional student and won a scholarship to Pennsylvania College for Women (later Chatham College) in Pittsburg. She became close to her biology teacher, Mary Skinker, who departed after Carson's third year for Johns Hopkins University in Baltimore and summers at Woods Hole. Three of six biology faculty at Johns Hopkins were members of ESA, who published in *Ecology*.[272] Skinker became a parasitologist at the U.S. Department of Agriculture, and she urged Carson to take civil service exams to qualify for a position as a federal zoologist.[273]

Carson did so and was hired at the Bureau of Fisheries part-time to write scripts for 52 radio programs (7 minutes each), entitled "Romance under the Waters."[274] She also wrote articles for *Baltimore Star*'s Sunday magazine on Chesapeake Bay fisheries and related subjects. In April 1936, the Bureau of Fisheries asked her to write an introduction for a brochure on fisheries, but what she submitted was considered suitable for *Atlantic Monthly* and she should write something simpler for a brochure. She did as advised, and it was published as "Undersea."[275] Later, she commented: "From that essay, everything else followed."[276]

She had read ecological literature, but "Undersea" did not follow scientific models. She invited readers to "enter vicariously into a universe of all-pervading water."[277] Her essay reads as if she were swimming through a marine environment, explaining what she saw. Her goal was not to give a warm fuzzy picture of an inspiring scene. Rather, she wrote from a Darwinian perspective. Discussing plankton, she explained:[278]

> Many of the fishes, as well as the bottom-dwelling mollusks and worms and starfish, begin life as temporary members of this roving company, for the ocean cradles their young in its surface waters. The sea is not a solicitous foster mother. The delicate eggs and fragile larvae are buffeted by storms

raging across the open ocean and preyed upon by diminutive monsters, the hungry glassworms and comb jellies of the plankton.

This essay "established her unique voice, scientifically accurate and clear, yet poetic and imaginative, capturing the wonder of nature's cycles and relationships."[279] An editor at Simon and Schuster wrote, asking if she was writing a book. She had not thought of that but liked the idea. She met with him and one of his authors, ocean traveler Hendrik van Loon, and they formulated a plan, which included selling chapters to *Atlantic* or similar magazines, when written. She based *Under the Sea-Wind* (1941) on her own investigations and library research. No bibliography, but she listed in her "Foreword" the main works and authorities on whom she drew.[280] She told three life histories in three "books," sanderlings, mackerel, and eels—subjects of her radio scripts and newspaper articles.

Sea-Wind was a complex "novel," with main characters and plots, but also other characters and subplots. Sanderlings nest near freshwater lakes in the Arctic and later migrate to southern South American beaches, stopping on Atlantic beaches, going both ways. Foxes and jaegers menace egg and young chicks, so parents carry away egg shells after chicks hatch and divert foxes with broken-wing ploys to lure them away.[281] Parent sanderlings migrated south before their young, who migrated instinctively without guidance.

Sanderlings' story ends in chapter 3, with chapters 4–5 of Book I mainly on mullets. Common mullets *Mugil cephalus* inhabited coasts of southern Europe and northern Africa, as well as American coasts from Cape Cod to Brazil.[282] Mullets spend summers in North Carolina sounds and estuaries, living singly, but in fall they enter oceans, gather into schools, and migrate southward to reproduce.[283] Fishermen awaited them with open nets. Ospreys also awaited them, but with negligible impact on populations compared to fishermen. Bald eagles robbed ospreys when possible. Fishermen's nets hauled in many mullets, but also small fish of no commercial value which fishermen discarded on beaches.[284]

Book II, on mackerel *Somber scombrus*, Carson misleadingly entitled "The Gull's Way." Mackerel, in deep water at the edge of the continental shelf, virtually hibernated during four cold months, living on fat stored during eight active months.[285] They lived on both sides of the Atlantic and ranged further north than mullet. In America, they lived from the Straits of Belle Isle south to Cape Hatteras, and in Europe, Norway south to Mediterranean and Adriatic Seas.[286] In spring, they rise near the surface and swim toward shore, in such numbers that they reminded Carson of long-gone flights of passenger pigeons.[287] She observed one migration a mile wide and 20 miles long.[288] Near shore, they released eggs (40 to 50 thousand per female) and milt, then swam to feeding grounds off New England. Since many fertilized eggs are eaten by zooplankton, vast numbers of them must be deposited for the species to survive.[289] When surviving, young gained independent motion, swam into deep waters during day, and up toward surface at night. They ate plankton, and as they grew, they consumed larger prey, but also became vulnerable to larger predators. Growing mackerel cluster into schools, which have some advantages, but also become targets for different predators.[290]

Carson had summarized the life history of eels in 1938 and now elaborated it in "Book III: River and Sea." There is some overlap between breeding grounds of European (*Anguilla anguilla*) and American eels (*Anguilla rostrata*) in Sargasso Sea, but many American eels breed west and south of Sargasso Sea.[291] In 1941, movements of adult eels from freshwater streams to breeding grounds remained unknown, but E.J. Schmidt (1912) had described the movement of billions of larvae to rivers from which their parents had come.[292] It took a year for young to reach American streams from which adults had come, and three years to reach European streams. Females lived 10 to 12 years in streams before swimming downstream to meet males near a river mouth. Whether both sexes swam together to breeding grounds was also unknown. Eels were largely scavengers, but if they encountered live fish caught in a gill net, they ate them, often leaving few for fishermen.

Under the Sea-Wind was published 1 November 1941 and received very positive reviews. However, Carson only enjoyed limelight for 36 days before Japanese bombers diverted everyone's attention by bombing Pearl Harbor. Sales declined and Carson was disappointed. A small consolation was Beebe's reprinting of her two chapters on eels as a final selection in *The Book of Naturalists: An Anthology of the Best Natural History* (1944). Beebe's first selection was from Aristotle.

Some of Carson's writings for the U.S. Fish and Wildlife Service (which had superseded Bureau of Fisheries) were routine, and as a diversion, she suggested she research Federal wildlife refuges and write 12 pamphlets on them. USFWS agreed, which enabled her to get out of Washington and examine refuges and their wildlife. In April 1946, she visited Chincoteague, mostly on Virginia's part of barrier Assateague (9,030 acres versus only 417 acres in Maryland) along the Atlantic flyway, featuring shore birds and waterfowl. Accompanying her was wildlife illustrator Shirley Ann Briggs, whose cartoon of them crossing lobby of Channel Bass Hotel in field clothes contrasted with clothes of elderly vacationers in lobby. In July 1946, Carson and her mother visited coastal Maine, at Boothbay Harbor. It was a working vacation near a Federal hatchery.[293] She fell in love with the area and wanted a summer home there, if affordable.

In February 1947, Carson visited three coastal plains refuges in North Carolina: Mattamuskeet, Swanquarter, and Pea Island. The last is on another barrier island and the other two are west of Pamlico Sound. Pea Island's 5,880 acres sheltered in winter about half of the world's 20 to 30 thousand snow geese. Swanquarter's 43,000 acres sheltered diving ducks—redhead, scaup, canvasback, bufflehead, and ruddy ducks. That refuge is close enough to Mattamuskeet to be administered from the latter. Mattamuskeet is the largest and most popular of the 3. Its 50,177 acres includes 30,000-acre Lake Mattamuskeet, which is mostly knee deep. Yearly, it attracted 212 species of birds, making it attractive to birders, but even more to hunters. On her trip, Carson took illustrator Kay Howe, whose seven illustrations added appeal to the pamphlet. In Lear's edition of Carson's uncollected writings, an abridged version of the Mattamuskeet pamphlet appeared, with none of Howe's illustrations. (I worked at Mattamuskeet Wildlife Refuge in summer 1958, and I obtained a copy of that pamphlet.) Carson's 12 pamphlets on refuges initiated a series of books on U.S. Wildlife Refuges.[294]

On a trip to Parker River Refuge, Massachusetts coast, she conferred with Bigelow, whose books on fishes and plankton of Gulf of Maine (1925, 1926) she had cited in *Sea-Wind*. She discussed her hope of writing another book on oceanography, and he encouraged her. Beebe urged her to go to Bermuda to study marine life underwater.[295] She was unable to do so, but went to Miami Marine Laboratory, took a class, and descended a ladder under an 84-pound diving helmet. Visibility was poor and the current so strong she never let go of the ladder. She did see live coral and was grateful for the experience. She conducted extensive library research in order to explain recent progress. She thought the public was ignorant of many advances made during and after World War II, and *The Sea Around Us* (1951) had a bibliography including recent outstanding works.[296]

Oxford University Press produced a book that is physically unimpressive. Two of Howe's drawings were the only illustrations, neither being especially noteworthy. She had feared the book would resemble a textbook, which it did not, but it was a popularized textbook, though readers might not realize it.[297] It is indeed about the sea (or ocean), and marine life was a minor theme. Scientists thought Earth was about 2.5 billion years old, but she cautioned, correctly, that older rocks might still be found.[298] Life presumably arose in oceans, but no details were known. "Some of the things that floated in the sea must have developed the magic of chlorophyll,"[299] which used CO_2 from air and sea water to "build the organic substances they needed for life." One Howe illustration was Carson's "Chart of the History of the Earth and Its Life," with representative fossils illustrated for each geological period. She explained what fossil records reveal about the evolution of life.

She knew from experience the abundance and variety of life one could collect from a ship, with a fine net. Norwegian ethnologist Thor Hyderdahl and five companions in 1947 had drifted on a raft of balsam logs for 101 days, from South America's coast to Pacific Islands, and Carson asked him to describe sea life observed. He did, and she used his data to construct a food chain.[300] Tropical botanist Ray Fosberg and geologist Harry Ladd advised her on the formation of oceanic islands,[301] on which she wrote an outstanding chapter. She was intrigued with how plants and animals colonized a distant island. *The Sea Around Us* became a best seller, which had happened to none of the books she listed. It won the National Book Award and remained on the best seller list for 86 weeks.

Volcanic Krakatoa provided a natural opportunity to study colonization, since its explosion in 1883 obliterated all life. It was within Dutch East Indies, and Dutch scientists followed its recolonization. For 25 years only a spider and a few blades of grass were seen, but by 1908 it contained birds, mammals, snakes, lizards, insects, mollusks, and earthworms. Ninety percent of species arrived by air. She told a sad story of how unique island species were quickly exterminated by invasive species humans intentionally or unintentionally introduced from continents.[302]

Oxford Press republished *Sea-Wind* in 1952, which led to its translation into foreign languages.[303] Amends for paucity of illustrations was compensated for in a very well-illustrated edition for young readers, with some discussions simplified by Ann Terry White.[304] Even that edition had foreign translations. Royalties, honors, and fame enabled Carson to resign from Fish and Wildlife Service and buy an oceanside lot at Boothbay Harbor, Maine, where she built her summer cottage. She wrote a third book, *The Edge of the Sea* (1955), unlike either previous book. It explained ecology of shore-life in three environments along America's Atlantic coast: rocky shores of New England, beaches of Middle States, and mangrove and coral coasts of Florida Keys. In 1948, she had hired Bob Hines as illustrator for Fish and Wildlife Service, and they developed a good working relationship. He welcomed collaboration on her third book, and he went to shores with her and participated in collecting animals and observing them in their habitats. His numerous illustrations in *Edge of the Sea* were widely praised.[305]

Her book celebrated marine invertebrates, in either tidal (littoral) or neritic (sublittoral) zones. Rocky coasts' first inhabitants were probably barnacles and mussels that strain food from seawater, and spores of green algae, that coated rocks. Grazers that scraped algae off rocks followed. Plankton strainers and grazers attracted predatory whelks, starfish, crabs, and worms.[306] Above tide line, lichens clung to rocks and slowly eroded rock surfaces. Periwinkle snails grazed on lichens and algae; snails lived out of water as much as in it, though some snails made a transition to becoming land snails.[307] Barnacles could also live where submerged at high tide and exposed to air at low tide. After three to five years when barnacles die, their shells become refuges for baby periwinkles, tide-pool insects, young anemones, tube worms, or new barnacles. And so on for other species, and different sets of species below Cape Hatteras and in Florida Keys. *Edge of the Sea* did not have a bibliography and ends with a classification of invertebrates, from protozoa to sea squirts. Hines' illustrations had a more three-dimensional appearance than Frech's outline drawings in *Under the Sea-Wind* (1941). Carson's reputation from these books on seashore regions would later be eclipsed by that of *Silent Spring* (1962).

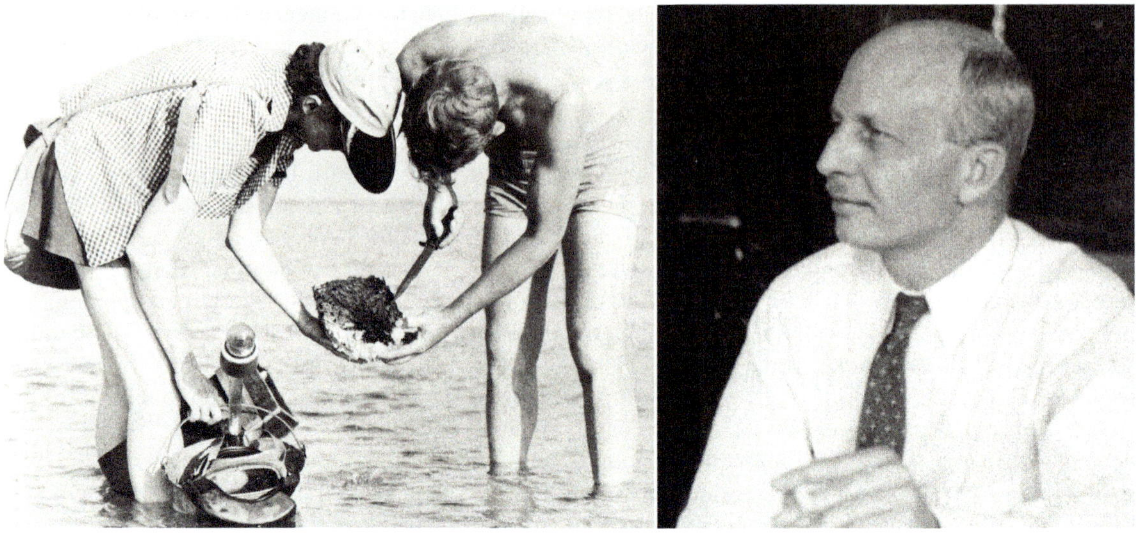

FIGURE 4.8 (a) Rachel Carson & Bob Hines. 1952. Brooks 1972: plt. 5. (b) Gordon A. Riley. Manuscripts & Archives, Yale University Library.

Gordon Riley

Missourian Gordon Riley (1911–85) wrote autobiographical reminiscences in 1982–83, now online.[308] He found elementary school boring and read natural history and science literature on his own. At Drury College, in his hometown, Springfield, he majored in biology (B.S. 1933), planning to teach it in a high school. However, Washington University in St. Louis provided a scholarship, and he earned an M.S. degree (1934) there before going to Yale to study embryology. In 1935, after hearing Hutchinson lecture, he switched over to limnology and became Hutchinson's first graduate student.[309] Hutchinson was a superb mentor. Since Riley had a sound knowledge in chemistry, Hutchinson had him write a dissertation on *The Copper Cycle in Connecticut Lakes and Its Biological Significance* (1937). Hutchinson also introduced him to oceanography.[310] Riley then took Professor Albert Parr's oceanography course, which led to Riley being invited to serve as chemist on a brief cruise of *Atlantis* off the mouth of Mississippi River in March 1937.[311] He enjoyed the cruise, and afterwards more oceanographic than limnological opportunities became available to him.

Riley is most remembered for his mathematical models in marine biology. His plankton studies of 1946–47 built upon previous work by Thomas Park and Hutchinson.[312] Riley was also influenced by two marine biologists at Plymouth Laboratory of the Marine Biological Association, H[ildebrand] W. Harvey and Richard H. Fleming. Riley was interested not only with plankton population dynamics, but also with environmental factors, such as temperature and water chemistry. That was his "synthetic approach," used first in 1946. During the 1950s at Yale, he linked deep-water nutrients with surface production. In 1956, he published three studies he had done at Long Island Sound during 1952–54.[313] He discovered that Long Island Sound productivity was limited by introduction availability. He created a school of oceanographers in the late 1950s and the early 1960s.

Riley's *Reminiscences*, published posthumously, included unflattering assessments of three professional contemporaries, whom he knew when they were all beginning their careers. Martin Burkenroad was "brilliant, erratic, and somewhat paranoid." George L. Clarke: "There was no real warmth of the man, and in some respects, he wasn't a very nice person. He was lazy, always trying to grab a major share of the credit for things in which others . . . had done most of the work." John Ryther had earned three degrees at Harvard, working under Clarke. Ryther published "The Measurement of Primary Production" (1956). Riley's criticism was presumably a prepublication personal communication, but Ryther became angry and changed nothing. Riley published criticism of Ryther's model in "Theory of Food Chain Relations in the Ocean" (1963). In a later conversation, Ryther admitted he had ignored data that did not fit his model, and Riley questioned his honesty in his *Reminiscences*. To evaluate Riley's assessments, one should consider obituaries of these three ecologists.

A turning point in Riley's career and life came in July 1965, when he left Yale's Bingham Oceanographic Laboratory to become Director of Dalhousie University's Institute of Oceanography. His major endeavors went into building and administering staff and programs at the Institute.[314] Only one of 13 papers in his *Selected Works* (1982) was published after he moved to Nova Scotia. Harvey and Riley were the "most influential biological oceanographers of their era."[315] He never published a synthesis of his specialization. His *Reminiscences of an Oceanographer* merits publication, with an editor's annotations.

Jacques-Yes Cousteau

French naval officer Jacques-Yves Cousteau (1910–97) developed an aqualung, which he and others used to explore oceans below the surface.[316] He turned that activity into a post-naval career and became a world-famous author, film star in his own marine documentaries, and leader in marine conservation. His story is told here from his writings and biographies in English.

His parents were from a village, St.-André-de-Cubzac. His father attended law school in Paris, and after they married, they moved there. However, when Elizabeth had her second son, Jacques-Yves, she returned to St.-André. In 1914, when World War I began in France, Daniel Cousteau's millionaire employer returned to America and the Cousteaus returned again to St.-André. After the war, wealthy Americans returned to Paris, and Daniel's new millionaire employer had houses in Paris and New York. In 1920, employer returned to New York and Custeaus went along. Their two sons attended New York schools, and Jacques became "Jack." He was sickly and never enjoyed sports. Instead, he built models of boats, airplanes, and machines. When he went to camp for two weeks in Vermont, he refused to ride a horse, and as punishment he had to clear sticks from a pond, which he enjoyed. In 1923, Daniel's employer returned to live in Paris, and the Cousteau boys returned to French schools, where Jacques "fit in" no better than he had in New York. He became absorbed with models and photography. In 1907, he was fascinated by a movie based on Jules Verne's *Vingt Mille sous les Mers*.

His first triumph was making an acceptable film of a cousin's wedding. He went on to make films like what he saw in theaters, using acquaintances as actors. When he graduated from school at the top of his class in 1929, he joined the French navy to fly airplanes. However, in 1936, he had a car accident that ended his aviation career, though he remained in the navy. At a party in Paris, he met Simone Melchior, who regretted women could not join the navy. In 1937, they married in an impressive ceremony, with fellow officers in formal dress. After a honeymoon in the Alps, they returned to Toulon, where Cousteau was stationed. Another officer suggested he exercise the arm injured in his accident by swimming, and he lent him goggles, fins, and snorkel. When Cousteau saw marine-life he had not known existed, it changed his life. He supplemented his dives with reading in the naval base library. Simone shared his skin-diving until their first son, Jean-Michel, was born in 1938. Cousteau's next venture was underwater photography.

A naval gunsmith designed a waterproof camera casing, and in 1943 an engineer, Émile Gagnan, helped develop Cousteau's underwater breathing equipment. They used three compressed air cylinders, weighing 50 pounds each in air, but weighing nothing under water. After many experiments, they had an apparatus to patent. Cousteau and colleagues made a movie, *Épaves*, which admirals saw and realized aqualung's military potential.

After World War II, the French navy created Groupe de Recherches Sous-marine under Cousteau and a colleague's leadership. Their first task was to clear harbors of wrecks, mines, and unexploded munitions, which took a year. At the first Cannes Film Festival in 1946, *Épaves* won a special prize, which led Cousteau to contemplate a career of underwater photography. Technical problems remained, foremost being nitrogen narcosis, which took the life of a diver in 1947 before they heard of an American remedy, which was to mix helium in the cylinders with air. An American reporter, James Dugan, wrote in 1948 a glowing article on Cousteau, his aqualung, and underwater color photography. His article aroused American interest in the aqualung, being manufactured in Canada.

Cousteau made the most of his rising fame, which grew steadily as he conducted explorations, which he documented on film and in books, the profits from which supported his family and funded new expeditions. Swiss professor-inventor-explorer August Piccard (1884–1962), who had ascended 11 miles above the Earth, enlisted Cousteau's aid in using a "bathyscape" that could descend into the deepest abyss—much deeper than the half-mile Beebe and Barton had reached. In 1949, Cousteau and crew sailed to West Africa to meet Piccard's ship at Dakar, where France had a naval base. On a first voyage, Piccard and Theodor Monod descended 200 feet. A deeper test, with no one aboard, ended with equipment being damaged by a turbulent sea.[317] Cousteau did not continue working with Piccard, because in 1951 he developed his own ideas for a bathyscaphe.[318] On their way back to France, Cousteau's ship stopped at Salvage Islands, between Madeira and Canary Islands for a few weeks of exploration, which he discussed in his first English-language book, *The Silent World* (1953).[319]

When they reached Tunisia, they heard talk of monk seals *Monachus albiventer*, assumed to have been exterminated centuries earlier. The mayor of La Galite, who told them about these seals, was met with skepticism until he showed where they lived.[320] They observed them on land and in the sea, and Cousteau captured a pup for a pet. Back in France, when he discovered it ate $200 worth of fish a month, he gave it to the Marseilles Zoo. Despite his book's title, *The Silent World*, Cousteau explained that hydro-phones had recorded sounds in the sea, mostly from whales and porpoises, but some fish croaked like frogs, and shrimp pincers sometimes made sounds when they snapped shut.[321] America's navy had discovered this during World War II, when it was developing sonar to detect enemy submarines.

Compared to his later books and films, *The Silent World* seems a little superficial. It told the story of his introduction to oceanography down to 1951 and a variety of explorations. Yet, it became a best seller—selling over 486,000 copies in the United States in 1953. In 1955, he released a film with the same title but broader content (which I saw in a theater) that was also a commercial success. The book's illustrations came from an article Cousteau had published in *National Geographic* (1952). The strong impression book and film had seems attributable to their being innovative and opening a new avenue of exploration and research.

Cousteau and his team were able to exploit this new opportunity better than anyone else.

Cousteau's 1952 article in *National Geography* was first of a dozen, and he also received generous grants from the National Geographic Society. Page one in Robert Poole's history of that society (2004) has a photograph of Cousteau and President Gilbert Grosvenor celebrating its centennial in 1988. In 1953, then president, Melville Grosvenor, had introduced Cousteau to M.I.T. Professor Harold Edgerton, who had pioneered extremely fast photography. As Grosvenor expected, they developed a fruitful collaboration. Edgerton had developed camera and flash for depths deeper than Cousteau's team could descend. Edgerton's camera recorded life at extreme depths.[322] In 1953, the French navy built its smallest vessel, a bathyscaph, which Cousteau had envisioned and helped plan. It enabled him and others to descend 2.5 miles.[323] By 1960 Cousteau had developed a more mobile diving saucer.[324] In 1961, the National Geographic Society had President Kennedy present Cousteau with the Society's gold medal for excellence in exploration.[325]

Cousteau continued telling his life story, for 1951–61, in *The Living Sea* (1963), featuring his new ship, *Calypso*, its crew of 30, and guests. In 1983 he published a lavish book on the ship and its expeditions.[326] It had been built in Seattle in 1942 as a minesweeper, and he had found it for sale as surplus at Malta and had it modified for oceanographic uses. Simone became ship master, who maintained harmony, while Jacques-Yves decided research agendas and where to go. They placed their sons in a boarding school and on holidays took them on *Calypso*. In December 1951, Cousteau took *Calypso* and crew on their first voyage to Red Sea, lasting 38 days. Although this expedition occurred over a year before *The Silent World* appeared, discussion of it first appeared in *The Living Sea*. Cousteau thought that "its exuberant coral structures [were] second only to those of the Great Barrier Reef in extent and exceeding it perhaps in splendor."[327] He had worked for the French Navy for years before he formally resigned in 1957, to become Director of the Oceanographic Museum of Monaco.

Cousteau decided to add dolphins in a large pool to the museum exhibits. A clearly depressed dolphin deliberately swam rapidly into the side of the pool and died from the concussion.[328] He decided dolphins might commit suicide if left alone in a pool for several days, but if several were placed there together, they boosted each other's morale and accepted food after five or six days. Cousteau asked his assistant, Albert Falco, to devise a means to capture them humanely. Falco took a year and devised tongs which he used to capture 27 dolphins after 50 attempts. In the museum pool, a veterinarian examined them and administered vitamins and antibiotics. Their keeper taught them to butt a balloon to get a sardine. Sardines were expensive, and eventually the trainer substituted whitings. A disgusted dolphin spurned the whiting and thoroughly splashed water on the keeper. Cousteau eventually decided the target species, *Delphinus delphis*, resented captivity, and he released them. He claimed, however, that another species, bottlenosed dolphin *Tursiops truncates*, enjoyed shows.

The Living Sea summarized many episodes, of which only one more is discussed here. *Calypso* visited Aldabra Islands, part of the Seychelles (though 800 miles away), midway between Madagascar and Africa. Like Galapagos, half a world away, Aldabras have both large sea turtles that nest there and giant tortoises. Cousteau thought only those two groups of islands had giant tortoises, but similar species also live in Seychelles, Madagascar, and Mascarenes.[329] Cousteau learned that the British governor of Seychelles planned to allow development of tourism there. Cousteau proposed a less disruptive alternative, a research institute. The Seychelles governor found a compromise, setting aside the South Island as a nature preserve.

By 1959, Cousteau was concerned over pollution in the Mediterranean Sea, and he persuaded Prince Rainier to have Monaco sponsor at the museum a meeting of the International Atomic Energy Agency. He and Rainier welcomed 450 delegates from countries having atomic energy programs or wanting such a program.[330] Expert delegates included physicists and biologists. Physicists thought nuclear wastes could be safely stored in strata beneath the Mediterranean Sea. Biologists disagreed. France supported the physicists and Cousteau supported the opposition. He alarmed mayors of French Mediterranean cities, who joined him in demanding France abandon its burial plan, which the government did. In 1973, Cousteau founded the Cousteau Society to spread his conservation messages worldwide.

Another Cousteau project was Continental Shelf Station number one, or Conshelf One, a four-room abode for two men on seafloor off Marseilles.[331] Albert Falco and Wade Wesly agreed to live in it for a week, during which they would work outside five hours and stay inside 19 hours a day. They had telephone communication with *Calypso*, and television on ship monitored activity inside Conself One. Falco and Wesly had daily physical exams by a physician who dived down to Conself One. Meals were prepared daily and taken down. They had a radio and TV that received one French channel. Falco became depressed on the third day and requested classical music records, which were taken down and helped his morale. After they returned to *Calypso*, Wesly said he could do it again, but Falco did not volunteer to do so.

A year after *The Living Sea* came *World Without Sun* (1964), book and film, which was about five "Oceanauts" who lived in Conself Two (or Starfish House), a five-room abode erected in the Red Sea in June 1963, about 25 miles north of Port Sudan. Head oceanaut was Algerian marine biologist Raymond Vaissièr, from Monaco Oceanographic Institute. There were 45 men involved, and the French oil industry paid the bill, equivalent to $1.2 million. The five in Starfish House lived at a pleasant temperature, but it was hellishly hot for their support team on *Calypso*. It seemed a successful experiment and the film won an Oscar for best documentary of 1964. Therefore, Cousteau sponsored a Conshelf III in the Mediterranean Sea with 150 technicians,[332] simultaneously with U.S. Navy's Sealab II, near Scripps Institution of Oceanography, cooperating in this endeavor.[333]

Cousteau made 61 films in French and English for TV, some of which were available also as books. Some were regional natural histories, others were on groups of animals. An example of the former was *Three adventures: Galapagos, Titicaca, the Blue Holes* (1973). An example of a book on groups of animals was *Life and Death on a Coral Reef* (1971). Examples on individual kinds of animals were three books on whales, dolphins, and sharks, published in French and English in 1970s, in which he was senior author with a coauthor. Those three books were republished as a handsome boxed set in English in 1987. He also published an *Encyclopedia of the Ocean* (20 vols., 1966–76) and a synthesis volume, *The Ocean World* (1979). He coauthored a book on conservation that appeared posthumously.[334] His *Calypso* enterprises died with him, but his technological achievements for observing marine life, his books and films live on.

Eugenie Clark

New Yorker Eugenie Clark (1922–2015) was raised by her Japanese immigrant mother and grandmother; her father died when she was 2.[335] Her mother worked, and when Eugenie was 9, her mother began leaving her on Saturdays at the Battery Park Aquarium, with a wide variety of fish, which fascinated her. She persuaded her mother to buy a home aquarium, and she began raising fish. A minister introduced her to Queens County Aquarium Society. She read Beebe's books on his marine expeditions. At Hunter College, she majored in zoology (B.A. 1942). She earned M.A. (1946) and Ph.D. (1950) at New York University.

In 1949 she obtained a U.S. Navy grant to go to Micronesia to study South Pacific fish, especially poisonous species. She used rotenone in tidal pools to kill fish. She sent parts of poisonous fish to determine which were safe to eat. Some were, others not. She learned spear fishing from a native fisherman. The American Museum of Natural History, New York City, gratefully received jars of her specimens, labeled and described.

She received a Fulbright Research Fellowship in 1950 to study fish at Ghardaqa Marine Station, on Red Sea in Egypt, south of Sinai Peninsula, which she reached in early 1951. While there, Cousteau passed near-by without them discovering each other. Austrian zoologist Hans Hass (1919–2013) also explored at Red Sea, beginning November 1951, without meeting her or Cousteau.[336] Cousteau, going by in the shipping lane, found little in the water, but she, near shore, had a better experience. She and Hass (separately) were the first ichthyologists to collect there since two Germans did so 70 years earlier.

Her first book, *Lady with a Spear* (1953), was very popular and was translated into seven foreign languages. Why so popular? She was an attractive woman who did not wait for men to show the way. She was a role model for girls—attested to by biographies of her for teens[337]—neither of the two other female ichthyologists were divers. Her book prompted invitations to give talks, and after one in Englewood, Florida, Anne and William H. Vanderbilt asked if she would be interested in founding a biological station on the Gulf coast near Sarasota. Yes, she was, and she established Cape Hazen Marine Laboratory, 1955–66. Her second book, *The Lady and the Sharks* (1969), described her experiences there. She studied fish species in eastern Gulf of Mexico. Sharks were easily caught at night on baited hooks attached to a rope. She and her team gained most attention from studies on sharks. She dissected some and kept others in a pen beside the lab's wharf. In summer 1959, Dr. Lester Aronson, head of Department of Animal Behavior, American Museum of Natural History, came to the Lab, and she asked him if studies had been done on shark behavior. He only knew of studies on sharks smelling. He helped her and her team design experiments to teach sharks to hit a target with their snouts to obtain food. These experiments were successful, and then they thought of new experiments which showed sharks were more intelligent than had been assumed.

In 1966, she became an associate professor at City College of New York for two years. In 1969, she became a popular professor at the University of Maryland, teaching courses on Life in the Oceans, Ecology of Oceans, Ichthyology, and Marine Vertebrate Zoology. *The Lady and the Sharks* (1969), on Cape Hazen experiences, described her shark experiments, her discovery of the first known hermaphroditic fish species, *Serranus subligarius*, and much more.

She returned frequently to Ras Muhammad on Red Sea, publishing four articles on fish there in *National Geographic* in 1970s.[338] In 1983, she convinced Egypt's government to make Ras Muhammad its first national park.

In 1997 she married long-time friend, Henry Y. Kon (her fifth husband), and they lived in Sarasota.

C. Conclusions

These seven marine ecologists illustrate progress in American marine ecology, from the mid-1920s to about 1990. Three popularizers and a scientific collector—Beebe, Carson, Cousteau, Ricketts—received more discussion here than three academics—Bigelow, Riley, and Clark. (Clark was partly a popularizer.) Why? Popularizers sought broad audiences and became public figures and attracted more biographers than did academics like Bigelow and Riley.

Beebe was its most prominent popularizer, 1924–51. He wrote scientific articles and popular books on his explorations and observations. His and Barton's history-making Bathysphere descents provided the earliest observations of live deep-sea animals. Eugenie Clark's testimonial of the influence of Beebe's books on her as a teenager was undoubtedly similar to other teens who went on to study marine life.

Bigelow was a leading academic marine ecologist at a leading university, who wrote very sound, conventional works, widely consulted. His contributions were partly his publications and partly his work as first Director of Woods Hole Oceanographic Institution.

Ricketts was a charismatic intellectual entrepreneur, who was well-enough educated for his chosen profession of supplying marine specimens to academic institutions. His *Between Pacific Tides* was organizationally innovative for explorers of western America's seashores. *Sea of Cortez* was partly a similar book for the northern half of Mexico's seashore, but also partly a novelty since a prominent novelist, John Steinbeck, wrote a narrative of their expedition to collect marine animals to document the seashore survey that constitutes most of the book. Ricketts aspired to write a third volume on seashore animals of Pacific coasts of Canada and Alaska and had made three explorations for this project, which was cut short by his accidental death. Later editors have published some of what he had accomplished on that work.

Carson began her career as a biologist at the U.S. Bureau of Fisheries and traveled U.S. Atlantic coasts from Maine to Florida, and also along the Gulf states. Her contributions to marine ecology were in three popular and original books, 1941–55. One, *The Sea Around Us* became a best seller. It won the National Book Award and remained on the best seller list for 86 weeks.

Riley trained under Hutchinson in limnology, but when opportunities arose in marine ecology, he switched over. He became a popular teacher-researcher-administrator at Dalhousie University. His lasting influence was in mathematical model building of plankton populations.

Cousteau's fame in the third quarter of the 1900s surpassed that of Beebe in the second quarter, thanks to his development of aqualung, underwater motion pictures photography, TV programs on his explorations, and numerous books. His books and TV documentaries of explorations included original observations on marine animals.

Clark trained as an ichthyologist who studied marine species. Her interests soon broadened to include ecology. She published extensively in *National Geographic* and in two autobiographical books on her experiences at two marine research stations.

Bigelow and Clark both established marine research stations; his WHOI at Woods Hole was on a grander scale than hers at Cape Haze.

NOTES

1. Also, Egerton 2014b.
2. Juday & Hasler 1946: 472–480, Thienemann 1959: 78–489, Frey 1963b, Elster 1974: 13–19, Egerton 2008b, 2015e: 37–38.
3. Frey 1963a.
4. McIntosh 1985: 124–127.
5. Tydecks et al. 2016.
6. McNulty et al. 2017.
7. Rizhinashvili 2020.
8. Zhadin 1934.
9. Zhadin 1991: 33, cited in Rizhinashvili 2020: [11].
10. Zhadin 1940: 545.
11. Slack 2010.
12. Hasler 1963a, b, Beckel 1987, Egerton 1987, 2008a. Carpenter, Kitchell et al. 2001, Likens 2002, Magnuson & Quinn 2005.
13. Burgess 1996: 56–57, Slobodkin & Slack 1999, Slack 2010: 17–20, Egerton 2015e: 40–42.
14. Slack 2010: 22, 138.
15. Slack 2010: 34.
16. Slack 2010: 50.
17. Slack 2010: 54.
18. Slack 2010: 74–75.
19. Hutchinson et al. 1932, Slack 2010: 117.
20. Rodhe 1974, Egerton 2008b: 36–37.
21. Hutchinson 1979: 208.
22. Slack 2010: index.
23. Hutchinson 1981.
24. Edmondson 1971, Kohn 1971.
25. In Edmondson 1971: 163, and reprinted in Egerton 2016b: 231.
26. Burgess 1996: 92, Egerton 2015e: 42–43, 2016b: 232.
27. Slack 2010: 126.
28. Riley 1971: 178.
29. Quoted in Slack 2010: 127.
30. Slack 2010: 142.
31. Burgess 1996: 36, Edmondson 1997, Slack 2010: 128–132, Egerton 2015e: 115.
32. Hutchinson 1979: index, Burgess 1996: 38, Hairston 2000, Lehman & Lehman 2009, Slack 2010: index, Egerton 2015e: 49, 2016b: 232.
33. Slack 2010: 115–117.
34. Hutchinson 1988: 1231.
35. Edmondson 1940, 1944, 1945, 1946.
36. Cook 1977, Reif 1986, Burgess 1996: 66, Egerton 2016b: 233–234.
37. Lindeman 1942: 157.
38. Cook 1977: 23–24, Hagen 1992: index, Golley 1993: 48–52.
39. Slack 2010: 159.
40. Hutchinson & Bowen 1947: 149.
41. Hutchinson & Bowen 1947: 148.
42. Hutchinson & Bowen 1947: 152.
43. Hall 1995, Burgess 1996: 82, Brown & Hall 2004, Egerton 2015e: 99–102.
44. Brown et al. 2004: 4.
45. Limburg 2004: 31.
46. Odum 1951a, b.
47. Slack 2010: 1151–155.
48. Hutchinson 2010: 83–157.
49. Hutchinson 2010: 222–230, 236–241.
50. Brooks & Deevey 1963.
51. Burgess 1996: 52, Egerton 2008a, 2015e: 70–73, Likens 2002, Hasler 2016.
52. Egerton 1987a: 94–95, 2015e: 70–73, Likens 2002: 3–4.
53. Hasler 1935, 1937.
54. Galtsoff et al. 1947.
55. Von Frisch 1967.
56. Von Frisch 1941a, b, Egerton 2016a: 66–67.
57. Von Frisch 1967: 138–139, 165–166.
58. Juday & Hasler 1946.
59. Hasler 1964.
60. Hasler 1979, 1984.
61. Pond 1903.
62. Hasler & Jones 1949.

63. Hasler 1963b, Hutchinson 1964.
64. Hasler 1962.
65. Hutchinson 1963.
66. Frey 1963b, Hasler 1963a.
67. Hasler 1960, 1966, Hasler & Scholz 1983.
68. Egerton 2014b: 402.
69. Hasler 1975: 186.
70. Walker & Hasler 1949: 63.
71. Hasler & Wisby 1950, 1952, Wisby 1952.
72. Wisby & Hasler 1954, Hasler 1954.
73. Hasler & Larson 1955.
74. Hasler et al. 1958: 353.
75. Burgess 1996: 66.
76. Likens & Hasler 1960, 1962, Hasler & Likens 1963.
77. Bocking 1997a: 116–147, Likens 2013.
78. Part 57.
79. Bogue 2000: 4, Riley 201: 5–6.
80. Ashworth 1987: 15–26, Grady 2007: 38–61, Riley 2013: 3–7.
81. Bogue 2000: 5–8.
82. Holman 1999: 248–253, Riley 2013: 9.
83. Schoolcraft 1821, Shor 1975, Bremer 1982, 1987.
84. Rezneck 1972, Krause 1989.
85. Bourdo 1983, Tanner 1987: 13–17, Grady 2007: 85–179.
86. Cleland 1983: 83–88, Tanner 1987: 13–17, Riley 2013: 3–28.
87. Hartman 1972, Ashworth 1987: 67–240, Bogue 2000: 113–136, Riley 2013: 55–227.
88. McKee 1966: 186–193, Williams 1976: 28–29, Wells 1978, Twing 1983, Flader 1983, Ashworth 1986: 67–80, Dempsey 2001: 19–36, 2004: 46–48, Kates 2001: 1–14, Dennis 2003: 125–127, Dickmann & Leefers 2003: 118–148, Riley 2013: 154–195.
89. Mahan & Mahan 1991: 30.
90. Longstreet 1973: 121–136.
91. Kouba 1973: 100–104, Lutz 2002.
92. Grady 2007: 99.
93. Grady 2007: 106–109.
94. Kates 2001.
95. McKee 1966: 112–129, Gates et al. 1983: 57–71, Ashworth 1986: 37–44, Grady 2007: 130–136, Riley 2013: 111–148.
96. Grady 2007: 99.
97. Rush et al. 2015.
98. Fielder 2008, 2010a, b, Wires 2014: 24.
99. Alexander 2009: 196–201.
100. Nute 1926.
101. Kaups 1978: 48.
102. Kaups 1978: 55.
103. Blust 1976.
104. Smith 1957.
105. Lurie 1960: 148–152, 1970, Egerton 2011: 165–166, 2012: 150–151.
106. Beeton & Chandler 1963: 543–546, Bogue 2000: 19–109.
107. Cited in Bogue 2000: 347, note 14.
108. Phillips 1880.
109. Bogue 2000: 182–185.
110. Dempsey 2004: 43–46.
111. Smiley 1882: 29.
112. Smith 1957a, b: 412–413, Bocking 1990: 478–495.
113. Smith 1957a, b: 413, Bocking 1990: 482–483.
114. Johnstone 1977: 46–51.
115. Northcote & Larkin 1963, Hammer 2007.
116. Northcote & Larkin 1963: 482–483.
117. Miloy 1983.
118. Koelz 1926: 54.
119. Egerton 1985: 5.
120. Egerton 1985: 5–7.
121. Van Oosten 1923, 1929.
122. Smith's personal communication to me.
123. Beeton 1961, 1963, 1965, Beeton & Edmondson 1972, Beeton et al. 1981.
124. Walters et al. 1980, Wells 1980.
125. Eck & Wells 1983.
126. Eshenroder et al. 1984a.
127. Eshenroder et al. 1984b.
128. Regier et al. 1969: 1–7.
129. Regier & Applegate 1969, 1972.
130. Schneider & Leach 1979: 1.
131. Roseman et al. 2010.
132. Brown et al. 1985.
133. Auer 2013, Baker & Auer 2013, Dempsey 2013.
134. Hudson & Ziegler 2014.
135. McKee 1966: 164–166, Shaw 1966, Bourne 1992, Sheriff 1996, Bernstein 2005, Stagg 2010: 28–35 Egerton 2018a: 99–102.
136. Alexander 2009: 17.
137. Styran & Taylor 2004.
138. Conzen 1988: 7–8.
139. Malkus 1960: 2–12, Dempsey 2004: 137–138, Stagg 2010: 66–67.
140. Stagg 2010: 85–86, 350–251.
141. Malkus 1960: 13, 217, Dempsey 2004: 138–140.
142. Barry 1972: 10–32, Becker 1984, Alexander 2009.
143. Hills 1959: 57–63.
144. Hills 1959: 89–90, Alexander 2009: 42.
145. Becker 1984: 109–110, Alexander 2009: 320.
146. Alexander 2009: xxv.
147. Barry 1972: 32–59.
148. Stagg 2010: 230.
149. Macfarlane 2014: 230.
150. Aron & Smith 1971.
151. Emery 1985.
152. Alexander 2009.
153. Creaser 1932, Applegate 1950, 1951, 1961, Applegate & Smith 1950, 1951, Applegate & Brynildson 1951, Piavis 1961, Aron & Smith 1971, Smith 1971: 210, Lack 1973, Emery 1985: 5–6, Grady 2007: 277–281, Alexander 2009: 25–35.
154. Ashworth 1986: 119, Grady 2007: 277–281.
155. Van Meter 1980.
156. Crowe 1975.
157. Hubbs et al. 2004: 58–59.
158. Alexander 2009: 31.
159. Aron & Smith 1971: 13, Emery 1985: 4–5, Alexander 2009: 55–62.
160. Brown 1968, Colby 1971: 18.
161. Argyle 1982, Wells 1985, Wells et al. 1987.
162. Emery 1985: 8.
163. Emery 1985: 8.
164. Emery 1985: 8–9.
165. Parsons 1973, Crawford 2001.
166. Hubbs & Lagler 1964: 47, Scott & Grossman 1973: 192–197.
167. Emery 1985: 6.
168. Emery 1985: 13–14.
169. Tanner & Tody 2002, Dennis 2003: 190.

170. Christie 1974.
171. Alexander 2009: 108–109, Santavy 2014, Egan 2017: 108–147.
172. Benson 2014: 10.
173. Alexander 2009: 110–111.
174. Alexander 2009: 194.
175. Egan 2017: 122–124.
176. Cangelosi 2014.
177. Alexander 2009: 100.
178. Alexander 2009: 100–102.
179. Alexander 2009: 102–103.
180. Alexander 2009: 129–135.
181. Alexander 2009: 139–150.
182. Alexander 20090: 156–159.
183. Alexander 2009: 160.
184. Dunn 1996: 21, Riley 2013: 259–277.
185. Dunn 1996: 269–271.
186. Quoted in Alexander 2009: 105.
187. Burrows 2014, Reid & Wilkinson 2014.
188. Barry 1972: 146–175, Kuchenberg 1978: 154–207, Mortimer 1981, Ashworth 1986: 154–185, Kehoe 1997: 3–76.
189. Harris et al. 1982: 5–7.
190. Ashworth 1986: 97–101.
191. Dennis 2003: 144.
192. Alexander 2009.
193. Dempsey 2004: 163–169.
194. Alexander 2009: 311–328.
195. Ashworth 1986: 154–172, Dennis 2003: 167–170.
196. Egerton 2015c, 2017.
197. Smith 1972: 41.
198. Francis et al. 1979.
199. Christie et al. 1986: 16.
200. Harris et al. 1982, Minns et al. 1987.
201. Caldwell 1988.
202. Edwards & Regier 1990.
203. Fontaine & Stewart 1990: 153.
204. Steedman & Regier 1990: 257.
205. Magnuson 1980, Magnuson et al. 1990, Harris et al. 1982.
206. Cooley & Pacific 1987.
207. Regier et al. 1989, 1990, Hill & Magnuson 1990, Magnuson et al. 1990, Magnuson 1992, Regier & Goodier 1992.
208. Meyers & Worm 2003: 282.
209. Ward & Myers 2005.
210. Egerton 2018a: 111–121.
211. Welker 1975: 75, Berra 1977, Sterling 1981, 1997, 1999, Gould 2004, Matsen 2003, 2005, Kroll 2008: 65–94.
212. Bridges 1974: 63, Welker 1975: 10, Gould 2004: 54–55.
213. Berra 1977: 27–41.
214. Beebe 1924: 127.
215. Beebe 1924: 128–134.
216. Berra 1977: 141–142 lists new species Beebe discovered.
217. Beebe 1924: 129.
218. Beebe 1924: 136.
219. Beebe 1924: 121–122.
220. Beebe 1926a: v.
221. Beebe 1926a: viii.
222. Murphy 1926, 1936: I, 101–108.
223. Bjerknes 1969, Katz 2002: 98.
224. Beebe 1926a: 87.
225. Beebe 1926c: 93.
226. Berra 1977: 63–65.
227. Beebe 1926b.
228. Matsen 2005: 3–12.
229. Beebe 1928: 11, 7.
230. Beebe 1928: 124–133.
231. Beebe 1928: 225–227, Beebe & Tee-Van 1928.
232. Matsen 2005: 53–54.
233. Welker 1975: 127, Gould 2004: 290.
234. Beebe 1931, 1932a, b, 1934.
235. Feder 1966: 329.
236. Beebe 1931: 676.
237. Bigelow 1964, Redfield 1976, Shor 1988, Brosco 1989, Lewandowski 1999.
238. Bigelow 1964: 9, Schlee 1972: 254–255.
239. Agassiz 1913: 384–385.
240. Bigelow 1904.
241. Brosco 1989: 245.
242. Revelle 1980: 12.
243. Schlee 1978: 78.
244. Bigelow 1964: 30, Lucas 1980: 502–506.
245. Bigelow 1928.
246. Schlee 1978: 16–18, 1980, Rozadowski 2011: 169–170.
247. Bigelow 1964: 37.
248. Riley 1979: 4, quoted from Mills 1989: 272–273.
249. Bigelow 1964: 33, 41.
250. Quoted in Fye 1980: 2, see Redfield 1976.
251. Ward 1974: 195–211, Tamm 2004: 7–9, Lannoo 2010, Egerton 2016b.
252. Calvin 1933.
253. Tamm 2004: 181–187.
254. Ricketts & Calvin 1939.
255. Buchsbaum 1940.
256. Tamm 2004: 246–247.
257. Quoted in Tamm 2004: 247.
258. Letter dated 22 September 1947, quoted in Tamm 2004: 247.
259. Schmidt 1942.
260. Dungan et al. 1982, citing Steinbeck & Ricketts 1941: 378.
261. Sagarin et al. 2008: 372–375, Egerton 2016b: 394–396.
262. Grant 2019: 29.
263. Grant 2019.
264. Ricketts 1947, Tamm 2004: 81–82.
265. Gore 1990: 8, Palumbi & Sotka 2011: 120–121.
266. Tamm 2004: 266–267, 279–280.
267. Hedgpeth 1978, Rodger 2002, 2006.
268. Ricketts et al. 1985.
269. Brooks 1972: 15–17, Ihde 1997, Burgess 1996: 22–23, Lear 1997: 8–12, 1999, Lytle 2007: 15–18.
270. Carson 1922.
271. Lytle 2007: 23–24, Hazlett 2008: 150–151.
272. Dritschilo 2006: 357.
273. Lear 1997: 79.
274. Lear 1997: 78, Souder 2012: 52–53.
275. Carson 1937.
276. Quoted in Lear 1998: 3.
277. Carson 1998: 4.
278. Carson 1998: 7.
279. Lear 1997: 88.
280. Carson 1941: xviii—xix.
281. Carson 1941: 65.
282. Jordan & Evermann 1923: 252–253.
283. Carson 1941: 79–83.
284. Carson 1941: 100.
285. Carson 1941: 105–106.

286. Jordan & Evermann 1923: 273–275.
287. Carson 1941: 111.
288. Lytle 2007: 45.
289. Carson 1941: 116–119.
290. Carson 1941: 146–148.
291. Zim & Shoemaker 1956: 52.
292. Schmidt 1912, Carson 1941: xviii.
293. Lear 1997: 134.
294. Laycock 1965, Murphy 1968, Riley & Riley 1979, Grove 1984, and maybe others.
295. Lear 1997: 162–165.
296. Carson 1951: 217–220.
297. Kroll 2008: 95–123.
298. Carson 1951: 4.
299. Carson 1951: 8.
300. Carson 1951: 19–20.
301. Lear 1997: 155.
302. Carson 1951: 91–97.
303. Brooks 1972: 339–341.
304. Carson 1958.
305. Lytle 2007: 107, Souder 2012: 183–185.
306. Carson 1955: 45.
307. Carson 1955: 50.
308. See also Burgess 1996: 92, Egerton 2015a: 42–43.
309. Hutchinson 1982, Slack 2010: 121–127, Egerton 2016a: 232, 2017b: 124–130.
310. Riley 1982–83: 3.
311. Wüst 1964: 10, Schlee 1978.
312. Riley 1946, 1947, Mills 1989: 266–313, Anderson & Gentleman 2012, Egerton 2014c: 356–357, 2017a: 127.
313. Riley 1956a, b, c, Mills 1989: 299.
314. Riley 1982–83: 124–144.
315. Mills 1989: 307.
316. Almost all of his numerous publications contain autobiographical information; also, Cousteau & Paisner 2010 [2004], Madsen 1986, Munson 1989, Matsen 2009, Egerton 2017a: 130–139.
317. Beach 1969: 242–243, Idyll 1969: 19, Matsen 2009: 96.
318. Cousteau 1960, Cousteau & Dugan 1963: 278–279.
319. Cousteau & Dumas 1953: 194–197.
320. Cousteau & Dumas 1953: 198–202.
321. Cousteau & Dumas 1953: 242–244.
322. Edgerton 1955.
323. Cousteau 1954, Houot 1954.
324. Cousteau 1960.
325. Anonymous 1961.
326. Cousteau & Sivirine 1983.
327. Cousteau & Dugan 1963: 33.
328. Cousteau & Diolé 1975: 31–53, cited from 1987 edition.
329. Crumly 2009: 923, Gerlach 2009: 831.
330. Cousteau & Dugan 1963: 306–312.
331. Cousteau & Dugan 1963: 314–325.
332. Cousteau 1966.
333. Hellwarth 2012: 124.
334. Cousteau & Schiefelbein 2007.
335. Clark 1953, 1969, La Bastille 1980: 159–12, Balon 1994, Yount 1994: 59–71, Eilperin 2015, McFadden 2015.
336. Hass 1953: 23, Rozwadowski 2011: 173–174.
337. McGovern 1978, 1998, McGovern & Clark 1991, Butts & Schwartz 2000, McLoone 2000: 16–23, Ross 2000, Kroll 2008: 124–151.
338. Clark 1972, 1974, 1975, 1978.

5 Aspects of Animal Ecology

A. PHYSIOLOGICAL ANIMAL ECOLOGY

Plant ecologists began studying physiological plant ecology long before ecologists did so for animals. Dwight Billings began "The Historical Development of Physiological Plant Ecology" (1985) with discovery of photosynthesis in 1770s in western Europe.[1] Much later, Frederic Clements had a "Vision of a Unified Physiological Ecology."[2] Malcolm Nicolson found "Physiological ecology constituted a major programme of research with American plant ecologists in the early decades of the twentieth century."[3] Since there were several decades between origins of plant and animal physiological ecologies, influence here of plant ecology on animal ecology is uncertain.

An early use of term *physiological ecology* referring to animals was by Eugene Odum (1913–2002) in his dissertation (1941) at the University of Illinois, under animal ecologist Charles Kendeigh. Odum may have first used the term concerning an animal investigation. A decade later, another doctoral candidate wrote his Ph.D. dissertation at Purdue University on *Physiological Ecology of Salamanders* (1951).[4]

F. John Vernberg (b. 1925) wrote that dissertation under animal ecologist Clarence Goodnight.[5] Vernberg had conducted both field work and laboratory experiments. John and wife Winona Vernberg (1924–2008) had met while both worked at a U.S. Naval hospital, and they married in 1945, before being discharged. They attended DePauw University, Greencastle, Indiana, where John earned his B.A. and M.A. degrees (1949, 1950). They went on to earn Ph.D.s at Perdue University, West Lafayette, Indiana, in 1951, he in physiological ecology and she in parasitology. He became an instructor in zoology at Duke University, Durham, North Carolina, and she received funds to conduct her parasitology research. John rose from instructor to professor at Duke University, 1951–69 (and taught me in fall 1954). He at first continued his salamander research, and he learned that western North Carolina, where he began collecting, had the greatest salamander diversity in the world.

They conducted separate research in summers at Duke University Marine Laboratory, on an island close to Beaufort NC where he also taught a course in physiological ecology of marine animals. In 1956, he was offered new responsibilities as assistant director of the marine lab, which he accepted. They moved to the Beaufort area. John and Winona began collaborating in 1962, when they received a joint grant.[6] In August 1966, he participated in an NSF three-day conference at Williams College on physiological ecology of marine animals. In summer 1967, they attended a meeting of a parasitology society in Tucson.

In 1968, a Duke alumnus, University of South Carolina zoologist Ted Cole, visited the Vernbergs at Duke's Marine Lab and offered John the directorship of a new Baruch Institute for Marine Biology and Littoral Sciences, being cofounded by the Baruch Foundation and the University of South Carolina. It would be located on 17,500 acres of beach, forest, and marshes.[7] In 1969, he became Baruch Professor of Marine Biology and founding Director of the Baruch Coastal Research Institution, University of South Carolina, until retirement in 1996. His new position helped strengthen John's leadership role as physiological ecologist of marine animals. She became Research Professor of Marine Biology, University of South Carolina, 1969–75, then Professor of Public Health and Director of Environmental Health Staff, 1975–77. Later, both were professors with administrative responsibilities. John and Winona published *The Animal and the Environment* in 1970, though their manuscript had been completed at Beaufort, before their move to Columbia, South Carolina. Two years later, they published *Environmental Physiology of Marine Animals* (1972), with Winona as senior author.[8] Since they studied what physiological ecologist's study, and since they stated: "we have tried to highlight some of the physiological mechanisms through which these animals have been able to meet the challenges of their environment. We have not written this book for any one particular scientific discipline,"[9] they were doing physiological ecology. John Vernberg's other publications included *The Animals and the Environment* (1970), *Physiological Ecology of Estuarine Organisms* (1975), *Functional Adaptations of Marine Organisms* (1981), and an autobiography, *It's a Long Story* (2018).[10]

In a forward to Brian K. McNab's *The Physiological Ecology of the Vertebrates* (2002), James H. Brown stated "Physiological ecology emerged as a coherent discipline in the 1950s," but then admitted it "never really had its own theoretical framework and empirical underpinnings." So, how could it have been a coherent discipline in 1950s? Perhaps it only became a common focus during the 1950s. McNab's book provided a coherent discipline. I do agree with Brown that pioneering work was accomplished by "Bartholomew, Bogert, Irving, Kendeigh, Pearson, Schmidt-Nielsen, Scholander, Hobart Smith, and others."

Two "others" would be Danish August Krogh (1874–1949) and his daughter, Bodil Krogh (1918–2015). August Krogh had attended the University of Copenhagen and studied animal physiology (Ph.D. 1899) under Professor Christian Bohr. Krogh's lifetime researches were on respiratory mechanisms and gas exchanges in animals and humans, first expressed in *The Respiratory Exchange in Animals and Man* (1916). His comparative approach continued in his studies. He won

DOI: 10.1201/9780429342202-5

FIGURE 5.1 (a) Winona and John Vernberg. Vernberg 2018: cover. (b) Bernd Henirich. Author.

a Nobel Prize for Physiology and Medicine in 1920. His *Osmotic Regulation in Aquatic Animals* (1937) was surprisingly comprehensive in 242 pages. His final synthesis was *The Comparative Physiology of Respiratory Mechanisms* (1940).

Norwegian Knut Schmidt-Nielsen (1915–2007), as a child, wanted to read Samuel Baker's *Wild Animals and Their Ways*, so his mother bought a copy for his eighth birthday. "It opened a new world" of adventure.[11] In 1934, he entered the University of Oslo and stayed three years. He intended majoring in zoology, and a friend suggested he specialize in physiology, which he did. In summers 1936–38, he and his father fished in Lake Grönningen. Knut measured and examined stomach contents for723 brown trout and Arctic char caught. Trout mostly ate insect larvae and plankton, and char ate mostly bottom organisms, but there was some overlap in kinds of food.[12] That led to his first publication, coauthored by father Sigval (1939). His mother, Signe, confided in Kunt she had worked hard in Sigval's laboratory, but never had a chance to do her own research, and she regretted marrying him.[13] Knut would recall this later.

Sigval Schmidt-Nielsen wrote to Krogh and asked if his son, Kunt, could come to work in his laboratory, and Krogh agreed. Krogh suggested he study osmotic regulation in freshwater crayfish and saltwater crabs and explore their differences. Schmidt-Nielsen found that to be an interesting project. His osmoregulation study became his undergraduate thesis at the University of Copenhagen. Krogh had four children, the youngest, Bodil (1918–2013), still living at home. When Schmidt-Nielsen taught me physiology (1957), I found him strikingly handsome; Bodil agreed.[14] He graduated in February 1941, and a few months later she became first woman in Demark to earn a Doctor of Dental Surgery degree. She began teaching at Copenhagen's School of Dentistry. He began graduate studies and earned his doctorate in June 1946. He was from Trondheim, a central Norway port. His Norwegian father, Sigval, was a chemist, who in 1913 became a professor of organic chemistry at a new Norwegian Institute of Technology at Trondheim. His Swedish mother, Signe, studied physics at Institute of Technology, Stockholm, under Nobel Prize winner Svante Arrhenius. She was third woman to earn a doctorate in Sweden, and first to do so in physics. Knut's paternal grandfather, Olaf, built a railroad from Trondheim to Oslo and then became inspector of fisheries at Trondheim and director of a small fisheries museum. Olaf performed an experiment: in 1903, he released over 1000 flounder hatchlings into a lake to see if they could live in freshwater. Answer: they lived but did not reproduce.

Knut Schmidt-Nielsen bought three textbooks from a departing student; one of these "became the most influential book I have ever owned."[15] Yet, it might seem to have little relevance to his career: Richard Hesse, *Tiergrographie auf oekologische Grundlage* (1924). He explained:[16]

> [I]t discusses the distribution of animals in relation to their physiological characteristics. It covers life in the high mountains and in the depths of the ocean. It examines the importance of high temperatures and lack of water for life in deserts, of the salts in the ocean for osmotic problems, and of the lack of salt for life in fresh water. Hesse's book brought to my attention questions that have occupied much of my scientific life.

This explanation seems to place Richard Hesse (1868–1944) among founders of physiological ecology. Hesse was

FIGURE 5.2 (a) August Krogh. U.S. Library of Congress. (b) Laurence Irving. University of Alaska Archives. (c) Per Scholander. Schmidt-Nielsen 1987: before p. 1.

from Nordhausen and earned his Ph.D. at the University of Tübingen (1892). He taught in several German universities and ended his career, 1926–35, as a professor of zoology and director of Zoological Institute of the University of Berlin. Two American ecologists translated and revised his book,[17] W.C. Allee and A.E. Emerson, who were later joined by three others to write *Principles of Animal Ecology* (1949), that frequently cited Hesse et al. 1937.

In November 1946 the Schmidt-Nielsens crossed the Atlantic and reached Swarthmore. Irving asked him to develop a method to determine chlorine levels for field studies. He did, and Irving decided to use it to test animals in Arizonian deserts. Schmidt-Nielsen had never seen a desert, and he wanted to stay and conduct research there. The U.S. Forest Service had an experimental station south of Tucson which he and Bodil could use. He decided to study kangaroo rats, which they live-trapped and raised in cages with only dry barley to eat. They had 65–66% water content of their bodies, like other mammals, despite having no drinking water for weeks. He discovered they use less water to excrete urea than do other mammals. Also, how did they survive extreme heat of deserts? By staying in burrows during the day and only coming out at night. He and Bodil published their findings in *Physiological Reviews* (1952) and for a general audience in *Scientific American* (1953). Both continued researches based upon findings about kangaroo rats. She did comparative studies on physiological differences between different species of rats, focused on kidney function. She earned a Ph.D. from the University of Copenhagen in 1955, possibly using her physiology publications for a dissertation.

Swiss Max Kleiber (1893–1976), from Zurich, attended the Federal Institute of Technology, located there. He graduated in 1920, and earned his Sc.D. degree in 1924, with a dissertation on "the energy concept in the science of nutrition." In 1929, he joined the Animal Husbandry Department, University of California, Davis, where he remained. He researched energy metabolism in animals, published in "Body Size and Metabolism" (1932), which was still being cited 70 years later.[18] It was the first synthesis for physiological ecology. In 1934 he developed a reliable way to predict basal metabolic rate (BMR) of animals of different sizes to predict nutritional needs—Kleiber's Law. That and later generalizations he synthesized in *The Fire of Life: an Introduction to Animal Energetics* (1961). It also served as a synthesis of physiological ecology.

Bostonian Lawrence Irving (1895–1979) earned a B.A. degree from Bowdoin College, Brunswick, Maine (1916), and an M.A. degree at Harvard University (1917). During 1917–19, he served in the U.S. Army in World War I. Afterwards, he earned a Ph.D. at Stanford University (1924) and joined its faculty for three years. In 1927, Irving joined the Department of Physiology, University of Toronto, and studied physiology of fish, diving mammals, and birds. He received several grants, and he trained some outstanding students. In 1937, he became chair of the Zoology Department, Swarthmore College, Swarthmore, Pennsylvania. While at Toronto and Swarthmore, he spent summers teaching and researching at Woods Hole. To study northern animals, he went on expeditions to Alaska and northern Canada. In 1939, he obtained a Rockefeller grant to enable Krogh's protégé, Per Frederick Scholander (1905–80), to come to Swarthmore and collaborate in studying physiology of diving mammals and also to escape World War II. For Scholander, this was an opportunity of a lifetime. He later wrote: "never had I met a person who so perfectly fitted my ideal" as Irving.[19] They became a very productive research team in polar studies. Scholander's *Experimental Investigations on the Respiratory Function in Diving Mammals and Birds* would appear in 1940. Another product of their research was Irving's monograph, *Arctic Life of Birds and Mammals. Including Man* (1972).

Scholander was from a professional family. His Swedish father was an engineer, and his Norwegian mother was a pianist. His parents divorced, and he moved to Norway with his mother. In the University of Oslo medical school, he found a course on physiology very interesting. He then became interested in likens and went on an expedition to explore for them. He then wrote *Flowering Plants and Ferns of Southeast Greenland* (1933), which Oslo botanists accepted as a Ph.D. dissertation. In Greenland he also became fascinated with diving seals and obtained a grant to study diving seals, whales, and birds. He also taught human anatomy at the University of Oslo.

During World War II, Irving became a major and chief physiologist at Air Force Proving Ground, Elgin, Florida. Scholander became a captain serving under Irving. Afterwards, Irving returned to Swarthmore and also directed a new Naval Arctic Research Laboratory, Anchorage, Alaska. In summer 1946, Irving and Scholander visited Krogh in Copenhagen and met his daughter Bodil and her husband Knut Schmidt-Nielsen, both also physiologists. Irving invited them to come to Swarthmore as research associates. They gladly accepted. In August 1947 Irving and his research team of five from Swarthmore and two from Cornell went to the Navy's research station at Point Barrow, Alaska, to determine the metabolic rate of seven mammal and two bird diving species.

Perhaps Schmidt-Nielsens' example inspired Scholander to marry Irving's daughter, Susan. They had a happy marriage, as she was content to follow him around the world. She undoubtedly edited his posthumous autobiography (1990), though she only put her name on the afterword.[20] In a scientific memoir on Scholander, Knut Schmidt-Nielsen wrote:[21]

> The most impressive aspect of Scholander's scientific life was his versatility as a biologist and his ability to make significant contributions in a broad range of fields. He was first of all a physiologist, but his work always signified a fresh approach to broader biological problems and principles. He had an ability to ask the right questions, to conceive of simple experiments and design the necessary equipment . . . he was warm and enthusiastic, generous and kind, and utterly unconcerned with the hassles of form and bureaucracy.

Ohio animal ecologist S. Charles Kendeigh (1904–86) earned his B.A. and M.A. degrees (1926, 1927) at Oberlin College before moving to the University of Illinois, Champaign, to earn a Ph.D. (1930) under Shelford. He taught at Western Reserve University, Cleveland, 1930–36,

FIGURE 5.3 Bodil M. Krogh Schmidt-Nielsen, Knut Schmidt-Nielsen, Barbara Wagner. Smithsonian Institution Archives.

before returning to the University of Illinois, 1936–73. Of 65 masters and 51 doctoral students he advised, the most prominent was Eugene P. Odum, who summarized Kendeigh's career:[22]

> While still a student he began his monumental long-term study of the physiological ecology of the House Wren at the Baldwin Bird Research Laboratory . . . . He worked there something like 15 summers. This work (reported in numerous papers and three monographs) pioneered ecological energetics.

Kendeigh was ESA president in 1950, and in 1978 he was awarded ESA's Eminent Ecologist Award.

In fall 1952, Schmidt-Nielsen joined the Zoology Department, Duke University, Durham, North Carolina. In 1957, I took his course on vertebrate physiology. He did not convey warmth to students but seemed amused by everything, with a smile. He had an impish streak in his personality. Course grade depended upon three exams, and for our second one, he used the same questions as on our first one, knowing we had studied entirely different topics. In his autobiography (1998), he stated that his marriage had begun to unravel around January 1950. In 1953, she and their children went with him to Saharan Desert for a year to study camel physiology. She was a collaborator on his first paper on camel physiology but not on later papers. In 1962 he became James B. Duke Professor of Zoology. They were raising three children and so remained together until 1964.

Did Scholander's autobiography inspire Knut Schmidt-Nielsen to write his (1998)? Additionally, Duke University Professor of Biology Steven Vogel wrote a biographical memoir on Schmidt-Nielsen (2008).

In summer 1956, Knut returned to a question he had wanted to study in 1939 but was interrupted by World War II: do seabirds drink seawater? Since Bodil had ceased sharing her kidney research with him, he invited two post-doc researchers, a Dane and a Japanese to collaborate. He forced seawater in a tube into a cormorant's stomach.

> Within a minute or two I made the fastest discovery I ever made. I noticed that the bird, with a quick movement of the head, shook off droplets of fluid that appeared at the tip of its beak . . . revealing a high concentration of chloride.[23]

He discovered glands in the head that secreted salt. They published a scientific paper on it, and he published an article on it in *Scientific American*.[24] He stated that this discovery received much attention from physiologists and popular press. Earlier physiology was based on studies of humans and white rats. Hesse, Krogh, Lindeman, and Irving moved beyond these limitations but with little general notice. The Schmidt-Nielsens broadened the scope of physiology even more and attracted attention. They were also notable for integrating studies in field and laboratory.

One reason for Bodil wanting to leave the marriage was her desire to be recognized for her own research, and not merely as being his assistant. He remembered his own mother having the same feelings. Their divorce occurred in 1966. Each of them found a new spouse (he married his daughter's best friend). She became president of American Physiological Society in 1975–76,[25] and he became president of International Union of Physiological Scientists, 1980–86. Although they had lived close to Duke campus, I had admired him for being the only one on campus to commute on a bicycle.

An important reference work was American Physiological Society's *Handbook of Physiology*, with section four on *Adaptation to the Environment*, D.B. Dill et al., eds., 1964. Chapter authors included Irving, Scholander, B. Schmidt-Nielsen, K. Schmidt-Nielsen, Bartholomew, and Dawson; two chapters were coauthored by K. Schmidt-Nielsen and Dawson, and one written by K. Schmidt-Nielsen alone. Topics included terrestrial animals in cold, terrestrial animals in dry heat, terrestrial animals in wet heat, animals in aquatic environments, and animals in high altitudes. All three chapters having K. Schmidt-Nielsen as an author concerned terrestrial animals in dry heat. None of chapter bibliographies cited his *Desert Animals: Physiological Problems of Heat and Water* (1964). In 1992, Tokyo Academy of Sciences awarded K. Schmidt-Nielsen its International Prize for Biology. Since Nobel Prizes do not include biology, this Showa Prize filled that gap.[26]

Missourian George A. (Bart) Bartholomew (1919–2000) moved to Berkeley, California before entering high school.[27] At the University of California, Berkeley, he earned B.A. and M.A. degrees (1940, 1941). His earliest research was on behavior of cormorants in San Francisco Bay, reported in three papers.[28] After military service in World War II, he went to Harvard University and earned a Ph.D. (1947). He then joined the Zoology Department, UCLA, until retirement in 1989.

His doctoral dissertation had been on effect of light intensity and day length on reproduction in English sparrows (published 1949). At UCLA he studied social and reproductive behavior of three species: elephant seal *Mirounga angustirostris*, Alaska fur seal *Callorhinus ursinus*, and California sea lion *Zalophus californianus*.[29] He continued his sea lion study in Galapagos Islands, where heat challenges them.[30] They spend days in tidal pools or channels and are active at night. He also studied physiological and behavioral components of thermoregulation in colonial nesting birds on tropical and subtropical islands.[31] His focus, and that of his graduate students and collaborators, was on physiology and behavior of animals to ensure their activities occurred within a temperature range that their species tolerated.[32] After publishing eight such papers, alone or coauthored, he published a more general article, "Interaction of Physiology and Behavior under Natural Conditions" (1966). Two decades later, Bartholomew published a retrospective essay on his career (1987).

Since K. Schmidt-Nielsen and G. Bartholomew were contemporary professors who made important contributions to teaching and desert physiology research, it is natural to compare them. Joel Hagen did so in "Camels,

Cormorants, and Kangaroo Rats" (2015). His summary: "The differences [are] between Schmidt-Nielsen's engineering approach and Bartholomew's explicitly evolutionary and ecological approach."[33] Another comparison can be drawn between them as professors. Both trained graduate students who earned masters and doctoral degrees, and both were very productive researchers and publishers. Schmidt-Nielsen wrote six books and he collaborated on some papers, though most of his publications have only his name as author. Bartholomew published no books, and about 90% of his papers were collaborations with students or colleagues. Schmidt-Nielsen's recognition was in him being president of International Union of Physiological Scientists, 1980–86. My impression is that extrovert Bartholomew had a warmer personality than introvert Schmidt-Nielsen. Schmidt-Nielsen had a stronger influence from publications, but Bartholomew's influence from training more physiological ecologists balanced Schmidt-Nielsen's publishing influence.

Russian ichthyologist G.G. Winberg (Vinberg) capably synthesized *Rate of Metabolism and Food Requirements of Fishes* (1960, Russian 1956). British editors of *Fish Energetics: New Perspectives* stated in a preface: "It is almost thirty years since Professor G.G. Winberg established the basis for experimental fish energetics with the publication of his monograph."[34] Ichthyologists understood he had established a new direction for their studies. Building upon his work, 14 ichthyologists from British Commonwealth countries provided 12 studies illustrating the fruit of three decades of research.

Oxford University entomologist John Phillipson wrote a pamphlet, *Ecological Energetics* (1966), which summarized physiological ecology at that time. He explained Lindeman's trophic-dynamic concept's relevance: it prompted ecologists to measure rates of transfers within ecosystems. He cited seven papers by himself (1 coauthored) and six by E.P. Odum (2 coauthored), 1955–1963.

Physiological ecologist Brian K. McNab was born in Chicago (1932), and at an early age traveled with his German-born mother and brother, Dennis Gordon, to Portland, Oregon. (His father was possibly Irish.) As an older child, Brian saw a newspaper notice of a bird walk to be held at Portland's Washington Park Zoo.[35] He later decided that that experience was probably the most important event of his life, since it took him from "an uneducated, lower middle-class family to install a strong interest in birds and mammals." After joining Oregon Audubon Society, he gained an appreciation of unique fauna in rainforest around the junction of Willamette and Columbian Rivers and in Cascade Mountains. After graduating from high school in 1950, he served as a temporary game warden with Oregon State Game Commission. In spring 1951, he entered Oregon State University (OSU) and majored in wildlife management until he learned other such students were all hunters. He switched to a zoology major and worked in summers for U.S. Fish and Wildlife Service. During those summers he wondered how wildlife met the challenges of their lives. He decided in his senior year to become an animal ecologist. In that year he took a course in physiology and wondered about its relevance for animal ecology. He searched for relevant papers and found three by Scholander, with Irving as coauthor of one paper. These papers compared birds and mammals in tropical and Arctic environments. McNab doubted that the compared species were closely related enough to ensure soundness of their conclusions.

After graduating from OSU (1955), McNab went on an expedition to Point Barrow, Alaska to study lemming populations. While skinning lemmings, he measured their kidneys and adrenal glands. He earned his graduate degrees at the University of Wisconsin-Madison (M.S. 1958, Ph.D. 1962). His doctoral dissertation was on three species of deer mice *Peromyscus*, which lived respectively in mesic, intermediate, and xeric environments. While he was in Madison, Scholander gave a guest lecture, and afterwards they had a cordial meeting. Later, Scholander invited him to go on an Amazon expedition, which he did. After completing his work in Madison, McNab saw an ad for a one-semester position at the University of Florida. He applied, got it, and it became a permanent position, where he spent his career.

He published alone two substantial papers and a third with his advisor, Peter R. Morrison, as coauthor, all from his dissertation.[36] Decades later, McNab explained:[37]

> My dissertation was the first study to demonstrate an environmental condition requiring a response, in this case, by species living along a mesic-xeric gradient. Much of the residual variation in the basal rate of metabolism-mass curve in deer mice (*Peromyscus*) reflected the gradient. Desert species and populations had lower basal rates than others found in a mesic environment, and intermediate populations had intermediates.

McNab's later work is discussed later.

British biologists Colin Townsend and Peter Calow edited *Physiological Ecology: an Evolutionary Approach to Research Use* (1981). They indicated they began with Oxford animal ecologist John Phillipson's *Ecological Energetics* (1966); he also contributed a chapter to this volume. It has 13 chapters by 13 biologists: 10 Britons and three Americans. Editors provided no indication of why biologists contributed to this collection, with presumably no financial support. Editors state: "The book represents a synthesis of ecology, physiology, and evolution."[38]

In the mid-1980s, some ecologists and physiologists explored body size and scaling. Max Kleiber had initiated such studies in a long chapter in his *Fire and Life* (1961), "Body Size and Metabolic Rate."[39] Over two decades later, five books continued that investigation: Robert Henry Peters, McGill University, Montreal wrote *The Ecological Implications of Body Size* (1983); William A. Calder III, University of Arizona, Tucson, *Size, Function, and Life History* (1984); Knut Schmidt-Nielsen, *Scaling: Why Is Animal Size So Important?* (1984); and William L. Jungers, SUNY Stony Brook, *Size and Scaling in Primate Biology*

(1985). There are drawings of a few animals in first chapters of Schmidt-Nielsen's and Jungers' works, but otherwise, the books have graphs, diagrams, tables, and equations. A fifth book, by two English professors of zoology, R.M. Sibley and Peter Calow, is a slender text-book, *Physiological Ecology of Animals: an Evolutionary Approach* (1986), with a chapter, "Constraints of Size."[40] It covers the animal kingdom.

One of Bartholomew's top students was Bernd Heinrich (b. 1940, figure 5.1), whose family once lived at Borowke, on the northern boundary between Germany and Poland. As Germany's army disintegrated on the eastern front, American soldiers arrived and advised them to flee westward, beyond the boundary between Russian and Allied occupations. They moved into a hut in a forest, which was a paradise for Bernd, of insects and birds. His naturalist father, Gerd Heinrich, gave him for Christmas a book on beetles, and he collected 447 of them, of 135 species. He also collected eggs of 35 bird species. Gerd Heinrich (1896–1984) became world authority on ichneumon wasps, yet never held an academic position; instead, he sold specimens to museums. After they immigrated to New York City, Gerd received a check for $3,000 for bird skins from Celebes, collected before World War II, and for bird and mice skins from Hahnheide, Germany, collected after the war. While collecting for others, Gerd collected wasps for himself.

Bernd earned two degrees at University of Maine (B.A.1964, M.S. 1966), then went to UCLA and earned a Ph.D. under Bartholomew (1970). Bernd also specialized in insects, but in their physiology and ecology, not taxonomy, as his father did. Bernd was introduced to experimentation by his advisor at Maine, Dick Cook, who suggested a question on *Euglena*, which resulted in three coauthored articles in *Cell Biology* and *Protozoology*.[41] Cook had ties to UCLA and wrote a recommendation for Bernd that led to his acceptance, with a research fellowship. He was captivated by deserts and desert animals, which led to his reading Schmidt-Nielsen's works on camels and kangaroo rats.[42] What he learned about physiological ecology appealed to him. When he talked with Bart about a dissertation topic, Bart asked him to make a list of six potential subjects.

Heinrich found that making a list was a meaningful exercise, and he eventually researched all 6. He began researching beetles, and later found a published paper on his topic. He caught a sphinx moth and measured its body temperature and found it comparable to some birds and mammals; it was regulating its temperature. After much experimentation, he had enough data for a dissertation. Then he saw similar research published in *Nature* by two English biologists, to whom he reported his findings, which differed from theirs. Their response was meant to discourage him from further research, but he felt confident about his findings and was not scared away.[43] His thin dissertation of 89 pages was *Flight Energetics and Temperature Regulation by Blood Circulation in the Sphinx Moth* Manduca sexta (1970). His dissertation committee passed it "with distinction." He joined UCLA faculty, 1971–80, then joined the University of Vermont faculty, where he remained. He discussed his career, first in *In a Patch of Fireweed* (1984), and again in *The Snoring Bird: My Family's Journey through a Century of Biology* (2007). Both books contain his impressive drawings.

Heinrich had begun his physiological ecology with insects and continued doing so during the 1970s–80s. He began by getting NSF grants, but eventually he submitted six proposals which were rejected. Then he came upon ravens feasting on a moose carcass and noisily calling other ravens to the feast. Why? He sent NSF a proposal to investigate, which was funded. Later, he summarized his research as beginning in natural history, which "led to physiological ecology, and then to studies of animal cognition and intelligence."[44] After *Fireweed* and before *Snoring* came *The Hotblooded Insects: Strategies and Mechanisms of Thermoregulation* (1993), an impressive survey of about a dozen different kinds, a triumphant end of his insect physiological ecology.

Long Beach, California native, Glen Eric Walsberg (b. 1971) earned his B.S. degree (1971) at a California State University and his Ph.D. at UCLA (1975) in biology, perhaps under Bartholomew. At Arizona State University, Tempe, he advanced from assistant professor in 1978 to professor in 2004. In 2000, he published "Small Mammals in Hot Deserts: Some Generalizations Revisited." Other zoologists since the Schmidt-Nielsens had also studied desert rodents, and Walsberg was both synthesizer and critic of what was known in 2000. He mapped the hottest deserts in the southwest. The Schmidt-Nielsens had conducted their studies slightly east of that region, so their findings were not quite applicable to the entire Southwest.

McNab synthesized *The Physiological Ecology of Vertebrates: a View from Energetics* (2002). It drew upon his own investigations in chapters 3, 5, 11, 13, 14, and 19. Investigations by others included chapters on ectotherms, osmotic regulation, water-salt exchange, gas exchange, and more.[45] Its 80-page bibliography was an important resource for anyone wanting to study topics covered. McNab's culminating work to date is *Extreme Measures: the Ecological Energetics of Birds and Mammals* (2012), which included subjects studied by most physiological ecologist. His evaluation of his own work in a letter is:[46]

> I have undoubtedly made a unique contribution in the amount and diversity of data that I have contributed on over 300 species, many of which are endangered, rare or unusually interesting. And it has been quite an effort to do so (I went to Papua New Guinea 12 times), but it has been a wonderful experience. I have done this alone without depending on grants or teamwork . . . . I do not think that was my most valuable contribution, which I believe to be in the form of the analysis that I have promoted. My view is, at best, a minority view . . . It has been noticed to the degree to which my ideas have been criticized and rejected by some reviewers.

This pessimistic assessment omitted mention of his two very important books, published by Cambridge University

Press (2002) and the University of Chicago Press (2012). He hopes to be vindicated by history, as Mendel was. He continues his work.

Marine zoologist Barbara A. Block (1958–2019), from Springfield, Massachusetts, earned her B.A. degree (1980) at the University of Vermont and her Ph.D. at Duke University (1986), using its marine station at coastal North Carolina. She was assistant professor at the University of Chicago, 1986–94, then spent remainder of her career at Stanford University's Hopkins Marine Station, where she had an endowed chair at the Tuna Research and Conservation Center, which Hopkins Marine Station and Monterey Bay Aquarium cofounded. Her research was on physiological ecology of tuna and other large fish species. She published 200 papers and was senior editor of *Tuna: Physiology, Ecology, and Evolution* (2001).

B. ETHOLOGY

The term *ethology* has a Greek origin = ethos, and was introduced into English in 1656, meaning study of character or ethics. Psychologist Julian Jaynes traced "The Historical Origins of 'Ethology' and 'Comparative Psychology'."[47] In Paris Professor Isidore Geoffroy St. Hilaire's *Histoire general des Règnes organiques* (3 vols., 1854–62) used *ethologie* to mean *ecology*.[48] However, it did not then gain currency.

Animal behavior studies are commonly divided into ethology, a branch of ecology, and behaviorism, a branch of psychology. This account is ethological but does not emphasize its distinction from behaviorism. Ethologist William H. Thorpe published *The Origins and Rise of Ethology* (1979), followed by Richard W. Burkhardt Jr.'s *Patterns of Behavior: Konrad Lorenz, Niko Tinbergen, and the Founding of Ethology* (2005). My more detailed version of this account (2016) began with a bibliographical guide to histories written from different perspectives.[49]

English historian of biology John Durant divided history of ethology into three phases: (1) formative, until about 1930, when pioneering studies were undertaken; (2) classical, 1930–50, when coherent traditions were established by Lorenz and Tinbergen; (3) mature, since 1950, indicated by Tinbergen's *The Study of Instinct* (1951).[50] Biology history Eileen Crist acknowledged dependence of ethology upon earlier natural history studies, but emphasized a difference in perspective: naturalists viewed their subjects as living beings, whereas ethologists viewed their subjects as "natural objects."[51]

Scientific natural history began at Aristotle's Lyceum in Athens in late 300s BC. Its animal treatises are attributed to Aristotle. In them, behavioral discussions occur mainly in *Historia Animaliun*. In it, behavior was discussed for all kinds of animals for which there was information. However, birds were most easily observed, and there was more information on birds than on other animals. Both Jean-Louis Labarrière (1993) and Liliane Bodson (1996) discussed that information, in books 8–9. In particular, she focused upon reproductive behavior, from courtship to raising young and she found information reported was impressively accurate.[52] An example she discussed in detail was European cuckoo brood parasitism.[53] *Historia animalium*'s discussion of bees visiting flowers comes up later:[54]

> On each flight the bee does not go on to flowers different in form; it goes for example from violet to violet, and does not touch any other before it has flown back to the hive. And having arrived at the hive they shake themselves, and three or four bees attend to each. What they pick up is not easy to see, nor has their method of working it been seen.

The dance was discovered, but not its purpose.

Roman authors were scientifically less sophisticated than authors at the Lyceum. The foremost example is Gaius Plinius Secundus (Pliny the Elder, c. 23–79 A.D.), whose encyclopedic *Naturalis Historia* was a compendium of Greek and Roman nature writings.[55] His attempt to separate fact from fiction was unsuccessful, yet it was his work, not those from Lyceum, which came down from antiquity to European Middle Ages. Lyceum's works were preserved in Byzantine civilization and later transmitted to Muslim civilization.[56]

West European emperor Friedrich II (1194–1250) lived mostly on Sicily, where he was uniquely influenced by Latin, Byzantine, and Muslim civilizations.[57] His important contributions to a future ethology were in *De Arte venande cum avibus* (1244–48, pub. 1596), which discussed both his trained hawks and their avian prey. He was, therefore, an amateur ornithologist—before that science existed. His book attracted ornithologists' attention in 1788. Modern authors on falconry offer their own comments on his book.[58]

Latin Middle Ages received some Greek learning from Muslim scholars and later from Byzantine scholars. A few universities were also founded, and from them several encyclopedists emerged. The best of them was German Albert von Lauingen (Albert the Great, c. 1200–80), from a wealthy family, who studied in Paris about 1240, and taught there until 1248, when he settled in Cologne to teach.[59] His students included Thomas Aquinas. He began his *De animalibus libri XXXVI* around 1256–60, drawing from Aristotle, medieval sources, and his own observations in books XX-XXI.[60] His books XXII-XXVI were based, unacknowledged, upon *De natura rerum* by his former student, Thomas of Cantimpré (c. 1186/1210–1276/94), who 391,392 had studied under him in 1245–48, before completing his studies in Paris.[61] A medieval tale Albert accepted was on osprey: "It hunts only fish and has one webbed foot, for swimming, like that of a goose, while the other foot has hooked talons for seizing, like that of an eagle."[62] More credible was his report on geese, which bred in Sclavia in "moist, sandy, marshlike places," and at the beginning of winter they "come back to our land," seeking "food and the more temperate air."[63]

Johann Gutenberg's development of printing press in 1450s gave an enormous impetus to writing and reading of books. Natural history benefitted immediately, because it stimulated writing of new natural history encyclopedias.[64] One example was by Englishman William Turner (c. 1508–68), physician, who wrote first natural history of birds, *Avium praecipuarum quarum apud Plinium et Aristotelem mention* (1544, English 1903).[65] Although organized around Aristotle and Plinius' observations, it also included his own observations. One of his longest and best accounts was on great gray shrike *Lanius excubitor*:[66]

> It has short wings, and flies as if by bounds upwards and downwards. It lives on beetles, butterflies, and biggest insects, and not only these, but also birds after the manner of a Hawk. For it kills Reguli and Finches and (as I once saw) Thrushes; and bird-catchers even report that it from time to time slays certain woodland Pies, and can put Crows to flight. It does not seize the birds it kills with its claws, after a swift flight, as Hawks do, but attacks them stealthily and soon (as I have often had experience) aims at the throat and with its beak squeezes and breaks the skull . . . when prey happens to be more plentiful, it lays by some for future scarcity. For it impales and hangs the bigger flies and insects on the thorns and spines of shrubs.

Fellow English physician Edward Wooton (1492–1555), older than Turner, but whose *De differenti animalium* (1552) appeared after Turner's *Avium*.[67] It was much broader in scope than *Avium*: chapter 5 was on "Quadrupeds that bear live young," six on "Quadrupeds that lay eggs," seven on "Birds, eight on Fish, nine on Insects, and ten on Squid, Crustaceans, Mollusks. It was a more efficient compilation than before, but had few, if any, first-hand observations.

Some naturalists added illustrations to their books. Swiss Conrad Gesner (Gessner, 1516–65) of Zurich, compiled a well-illustrated *Historia animalium*, (5 vols., 1551–87, German 4 vols., 1557–89) and a posthumously published *Historia plantarum* (2 vols., 1751–71), both of which added his own behavioral observations to traditional knowledge.[68] Gesner was a professor at the University of Zurich and a physician. When an epidemic of plague reached Zurich, he died with his patients. Englishman Edward Topsell (1572–1625) published translation of Gesner's first volume on mammals, and a second volume of selections from volumes 2, 4 and 5 (1607–08).[69] Italian naturalist from Bologna Ulisse Aldrovandi (1522–1605) continued in Gesner's footsteps, even to using Gesner's work with scant acknowledgement.[70] He aspired to write more than Gesner, and by living longer, did, but he only published three volumes on birds and one on insects, leaving six for posthumous publication.

English naturalist John Ray wrote three encyclopedias: *Catalogus plantarum circa Cantabridgiam nascen-tium* (1660), *Historia plantarum* (3 vols., 1686–1704), and *Historia insectorum* (1710). Even these included some behavioral observations, as in *Catalogus*, under deadly nightshade *Atropa belladonna*, he commented that snails and slugs nevertheless eat it.[71] Ray also wrote three natural history books, including *The Wisdom of God Manifested in the Works of the Creation* (1691).[72]

Frenchman René Antoine Ferchault de Réaumur (1683–1757) was an outstanding scientist who devoted most of his attention to insects.[73] His *Mémoires pour server à l'histoire des insects* (6 vols., 1734–42) was a monumental achievement.[74] He was not motivated to describe and classify the species found. Instead, he studied behavior of common species. He wrote two memoirs for volume 7, which were published in the 1900s.[75] His "Histoire des Fourmis" was "most important myrmecological document of the eighteenth century," and he also was "founder of ethology,"[76] or better: "founder of insect ethology." He put an ant colony under a bell jar so he could watch inhabitant's behavior, and he fed them.[77] He did not persist long enough to answer his own question, if ant colonies[78]

> are founded like those of the wasps, by a single mother, without the aid of any worker, or whether they are founded by one or several females that are accompanied by several workers for the purpose of taking charge of operations.

Answer: depends on ant species; both alternatives exist.[79] An amateur English naturalist, Arthur Dobbs (1689–1765) published a critique of Réaumur for rejecting Aristotle's claim that a foraging bee only visits one kind of flower on any flight until it returned to its hive. His paper was appreciated at home, but not noticed abroad even though in Royal Society's *Philosophical Transactions* (1750).

As Réaumur's career was winding down in 1740s, a countryman' career was taking off: Georges-Louis Leclerc, Comte de Buffon (1707–88).[80] Buffon became director of Jardin des Plantes in Paris, a very prestigious position, enabling him to publish, with royal sponsorship, a *Histoire naturelle, générale et particulière* (44 vols., 1749–1804), well-illustrated, with 12 volumes on mammals (1753–67) and 9 on birds (1770–86). These 21 volumes provided more details on mammals and birds than before published. Enough evidence, he thought, for him to draw sweeping conclusions about nature. He included some of these conclusions in descriptive accounts. For example, in a long discussion of red deer, he paused while explaining growth of antlers to discuss growth of plants and different kinds of animals. That discussion led to a conclusion that nature was much less energetic in America than in Europe, seen in its fewer species of mammals, which are smaller than comparable European species.[81]

When that appeared in 1783, American ambassador in Paris, Thomas Jefferson, took umbrage at Buffon's claim that American mammals were inferior to Europe's. He responded by writing an extensive refutation, with tables of European and American data, in his *Notes on the State of Virginia*.[82] He backed up his claims by obtaining from America bones and antlers of a moose—larger than any deer in Europe—which he presented to Buffon.[83]

French Lieutenant des chasses des Parcs de Versailles Charles-Georges Le Roy (1723–89), who rose from being only a game keeper through his relationship with encyclopedists Diderot, d'Alembert, and Helvétius, for whom he wrote articles, including one on instinct.[84] Both Gray and Thorpe judged him to have been the most important early contributor studying animal behavior.[85] That achievement was in *Lettres sur les animaux* (1768),[86] mistakenly gave its date as 1764.) Le Roy urged anyone who wants to understand animals to:[87]

> Leave his study, and plunge into the woods, there to follow every wile and turn of these sentient beings, to appreciatethe development and effects of this facility of feeling, and to see how by repeated sensations, combined by memory, they rise from mere instinct to intelligence.

Le Roy stated that Steller and Buffon had provided examples and that he would "like to have the complete biography of every animal."[88] He envisioned a science like ethology and provided insights that an alert game keeper could observe.[89]

German botanist Joseph Gottlieb Koelreuter (1733–1806) conducted hybridization experiments and discovered that some flowers are structured for wind pollination and others for insect pollination, published in three volumes, 1761–66.[90] He attracted one follower, rector at a Lutheran school, Christian Konrad Sprengel (1750–1816), who studied pollination in *Geranium* in 1787. He rediscovered that insects fertilize flowers while collecting nectar.[91]

German Protestant pastor Ernst Spitzer described the ,dance of honeybees (1788):[92]

> Full of joy they twirl in circles about those in the hive, from above downward and from below upward. In a few minutes, after these had made it known to the others, they came in great numbers to the place!

This discovery before von Frisch (1920–22) apparently had no influence on him.

Another German, N. Unhoch, published "an almost perfect description of the round dance" of bees in 1823:[93]

> The dance mistresses twist and turn in something more than a semicircle, now to the right, then to the left, five or six times, and execute what is a genuine round dance.

But Unholch lacked understanding of its function. These isolated observers lacked zoological education that might have led to further investigation.

Possibly the most-read natural history of all time is *The Natural History of Selborne* (1789) by English clergyman-naturalist Gilbert White (1720–93), who studied the natural history of his small town and its vicinity.[94] He earned his B.A. and M.A. degrees (1743, 1746) at Oxford University. He lived at his family home, "The Wakes." He had two regular correspondents, Thomas Pennant and Daines Barrington, and his *Natural History* consisted mainly of his letters to them. It was not illustrated, but some editors of later editions added illustrations from Pennant's *British Zoology* (1768–70) or Thomas Bewick's woodcut illustrations for his *A History of British Birds* (2 vols., 1797–1804). Birds were White's main natural history interest. White's letters to Pennant began in 1767, and White began keeping a journal in 1768. Most of his accounts were on mating, eating, and raising young, with speculations on migrations and life cycles of plants observed. In a letter to Barrington, 8 February 72, he discussed territory during breeding season. Peter Bircham called White "father of British ornithological ecology."[95]

English poet and popularizer of natural history Oliver Goldsmith (1728–74) first used *territory* in its modern biological meaning (1774):[96]

> [S]mall birds mark out a territory to themselves, which they will permit none of their own species to remain in: they guard their dominions with the most watchful resentment: and we seldom find two male tenants in the same hedge together.

Goldsmith published this after White wrote his letter to Barrington but before White published that letter.

English physician-naturalist Erasmus Darwin (1731–1802) wrote on evolution,[97] without much contemporary influence, but Charles Darwin was respectful of his grandfather, read his books, and even wrote his biography.[98] Erasmus Darwin's *Zoonomia* (1794) contained his ideas on evolution:[99]

> [A]nimals undergo perpetual transformations; which are in part produced by their own exertions in consequence of their desires and aversions, of their pleasures and pains, or of irritations, or of associations; and many of these acquired forms or propensities are transmitted to their posterity.

Scots immigrant to the United States Alexander Wilson (1766–1813) settled in Philadelphia and became friends with William Bartram.[100] He decided to describe and illustrate all the birds of America, which meant east of the Mississippi. He emphasized their habits in his own observations and cited relevant works. His *American Ornithology* (9 vols., 1808–14) contained his own illustrations, hand-colored. The last two volumes were completed from Wilson's notes after he died by his friend George Ord. *American Ornithology* was a monumental achievement that earned Wilson the title "Founder of American Ornithology." His opus appeared in 9 or 10 editions by 1878, some including Charles Lucien Bonaparte's supplementary volumes.[101]

The most popular subjects for animal behavior studies were vertebrates, with insects a close second. English entomologists William Kirby (1759–1850) and William Spence (1783–1860) wrote a very popular illustrated *Introduction to Entomology* (4 vols., 1815–26).[102] Five chapters described injuries that insects cause to humans and crops, two chapters were on beneficial insects, and 18 chapters on behavior. The last discussions included care of young, food, habits, societies, defense, locomotion, sounds, illumination, hibernation, and instincts.

French immigrant Jean Jacques Audubon (1785–1851), who Anglicized his first two names after reaching America, was more flamboyant than the serious Wilson, and he aspired to outdo Wilson with his own illustrations in *The Birds of America* (4 vols., 1827–38, 155 plates).[103] However, Audubon was less cautious than Wilson with scientific information in his *Ornithological Biography* (5 vols., 1831–39). Audubon could not find all the birds Wilson portrayed, and he copied five birds from Wilson without acknowledgement.[104] A good number of editions of *The Birds of America* have appeared, some cited by Wood.[105]

Audubon in April 1804 tied silk threads to legs of young pewees along banks of Perkioming Creek before its flowing into Schuylkill River, at his father's Mill Grove farm near Philadelphia. He wanted to know if any of them would return in 1805. He found two did.[106] This was possibly first banding of birds to study natural history, with tangible results. He also tested whether vultures could smell and concluded two species could not.[107]

Like Wilson, Audubon traveled widely to study bird and maybe sell his book, and in Charleston, South Carolina he met amateur naturalist Rev. John Bachman (1790–1874), who had become interested in birds during an acquaintance with Alexander Wilson.[108] Audubon and Bachman liked each other, and Bachman provided him information on local birds. Eventually, Audubon's two sons married Bachman's two daughters. As Audubon was finishing work on American birds, he and Bachman began collaborating on a comparable work on American mammals, with help from Audubon's sons. In 1842, Audubon planned an expedition up the Missouri in quest of mammals; it began in March 1843 and returned in November.[109] He painted over half the plates, and his son John W. Audubon painted the rest. His other son, Victor, handled publication of *Quadrupeds of North America* (30 parts, 5 plates each, collected into 3 volumes, 1845–53). Unfortunately, two editions of 1951 and 1989 omitted Bachman's name on their title pages.

In summer 1832, Audubon traveled to Boston and met English botanist Thomas Nuttall (1786–1859), curator of Harvard University's Botanic Garden and lecturer in natural history.[110] They "clicked" immediately, for Nuttall had explored the West before Audubon, and his biographer, Jeannete Graustein, provided maps of his six western journies. He was interested in birds as well as plants, and he and Audubon exchanged information on birds, on which both wrote comprehensive works. Wilson and Audubon were ornithological explorers who published findings in multi-volume works, as discovered, but Nuttall organized his *Manual of the Ornithology of the United States and of Canada* into two volumes: *The Land Birds* (1832, viii + 603 pp.) and *The Water Birds* (1834, vii + 672 pp.), both with uncolored illustrations. Nuttall integrated his own natural history observations with those of Wilson, Audubon, and others. His *Manual* received positive reviews, and new editions appeared in 1840, 1891, 1894, and 1903.[111]

English physician Jonathan Couch (1789–1870) practiced medicine in his native fishing village, Polperro, Cornwall.[112] In early adult life he had assisted Thomas Bewick. He published articles in Linnean Society *Transactions, Magazine of Natural History*, and *Annals of Natural History*. His greatest natural history achievement was his *A History of Fishes of the British Islands* (4 vols., 1860–65). Of interest here is his *Illustrations of Instinct, Deduced from the Habits of British Animals* (London, 1847, 343 pp.), which Thorpe considered credible.[113] Couch died leaving 12 volumes of notes, from which his *Cornish Birds* (2003) was compiled.

Charles Darwin (1809–82) transcended the natural history tradition with his theory of evolution by natural selection, but he also contributed to natural history. His *Journal of Researches into the Geology and Natural History of the Voyage of* H.M.S. Beagle (1839) was an excellent example of natural exploration.[114] *On the Origin of Species* had a chapter on instinct, which Darwin declined to define:[115]

> I will not attempt any definition of instinct. It would be easy to show that several distinct mental actions are commonly embraced by this term, but every one understands what is meant, when it is said that instinct impels the cuckoo to migrate and to lay eggs in other birds' nests.

He devoted the chapter mostly to social insects, because he recognized a challenge to his theory in existence of non-reproducing worker bees.[116] His next book was *On the Various Contrivances by which British and Foreign Orchids Are Fertilized by Insects* (1862), devoted to co-evolution between orchids and insects for mutual benefit, with nectar exchanged for pollination.

One can claim Darwin as the effective founder of ethology in three works: *Origin* (1859), *Descent of Man* (1871), and *Expression of Emotions in Man and Animals* (1872).[117] Thorpe thought *Expression of Emotions* "had a tremendous influence on the development of the study of animal behaviour, in that here was an outstandingly original comparative study of behaviour, a true pioneering achievement." Despite such qualities, however, his book did not have much immediate impact, which Thorpe attributed to Darwin 's revolution being so broad biologists could not get to all aspects of it simultaneously.

(Jean) Henri Fabre (1823–1915) was from a rural village in southern France and was educated to teach school by 1842.[118] He was not satisfied, however, and continued his studies until he earned a doctorate in science at Paris in 1854. Thereafter, he studied insects and published books on his findings. Darwin cited him in *Origin of Species*:

> M. Fabre has lately shown good reason for believing that although the *Tachytes nigra* generally makes its own burrow and stores it with paralyzed prey for its own larvae to feed on, yet that when this insect finds a burrow already made and stored by another sphinx, it takes advantage of the prize, and becomes for the occasion parasitic.

Fabre was no doubt pleased at Darwin's notice of his work, but he did not accept any theory of evolution. Fabre's major publication was a series of *Souvenirs Entomologiques* (10 vols., 1879–1908), containing 50 years of observations. They were accessible to lay readers, and most were translated into English.

Jean Théodoridès, French historian of zoology, commented: "Fabre remains the very model of the self-taught scientist—solitary, poor, proud, and independent. He was also an attentive and minute observer and a writer of unquestionable talent."[119] How did America's foremost entomologist respond to Fabre's death? Harvard's William M. Wheeler commented: "the world has lost its greatest entomologist, a man who combined in an extraordinary degree the gifts of a virile and penetrating observer and those of a literary artist of high distinction."[120] Yet, Wheeler knew French entomologists had not embraced Fabre, and he knew Fabre was "a crotchety and opinionated recluse."

Swiss psychiatrist-naturalist August H. Forel (1848–1931) had a life-long fascination with ants, besides having a medical career.[121] He focused on ant behavior, but without neglecting systematics. At age 25, he published *Les Fournis de la Suisse* (1873). He sent a copy to Darwin, who wrote a thank-you note, and wrote again after reading it: "seldom in my life have I been more interested by any book."[122] He compared Forel's observations on ants with those of Fritz Müller (1822–97) on termites in volume 1 of his *Beiträge zur Termiten* (1873). Forel's book exerted a strong influence on Espinas' *Des Sociétés Animalis* (1878). In 1893, Forel retired from Zurich Medical School to devote his time to studying ants. He published an early paper on insect senses (1886–88), followed by *Des Senses des Insects* (1908, English 1910. His lifetime of studies culminated in his monumental *Les Monde social des Fourmis compare à celui de l'Homme* (5 vols., 1921–22, English, 1928). Forel also published his autobiography (1931).

Londoner Conway Lloyd Morgan (1852–1936) had a general education and a general career as an educator until in 1884 he went to University College, Bristol, where he remained.[123] In 1884, he gained a new focus when he read Romanes' *Animal Intelligence* (1882), with which he disagreed. Thorpe considered Morgan as having made outstanding contributions to both comparative psychology and ethology.[124] He was an early experimenter in comparative psychology, and his contributions appeared in 14 books, of which six were relevant to ethology: *Introduction to Comparative Psychology* (1884), *Animal Life and Intelligence* (1890–91). *Habit and Instinct* (1896), *Animal Behaviour* (1900), *Instinct and Experience* (1912), and *The Animal Mind* (1930).

Ernest Evan Thompson (later, Ernst Thompson Seton, 1860–1946), from seaside South Shield, England; his family immigrated to Canada when he was six years old.[125] They settled on a farm in western Ontario, but four years later, moved to Toronto. During four farm years, he became fascinated with wildlife, and in Toronto he found like-minded others. He aspired to go to college, but got TB, dropped out of school and never went back. In 1876, he was apprenticed to an artist and studied at Ontario School of Art. He sailed to England and submitted a drawing for a scholarship at Royal Academy of Arts, not granted. He frequented British Museum's library and studied works of Alexander Wilson and Audubon. In 1889 he submitted a drawing again for a scholarship and won it. He began studying there in January 1881, and in October returned to Toronto and joined two brothers homesteading in Manitoba. He was more interested in wildlife than in farming, and he began publishing stories on animal behavior he had observed. After publishing *Fauna of Manitoba* (1886), he became Official Naturalist to the Government of Manitoba.

Most of Seton's writings—31 books and numerous articles—were of a popular nature.[126] They also had a somewhat distinctive perspective: "Where the nature literature of the United States focused on the human experience, this new Canadian approach focused on the animal experience."[127] Noteworthy for ethology was his *Lives of*

FIGURE 5.4 (a) Margaret Morse Nice. Dunlap 1996: frontispiece. (b) Ernst Thompson Seton. Self-portrait. Keller 1984. (c) Jean Henri Fabre. Web.

the Game Animals (4 vols. in 8, 1925–28), on mammals. Illustrations included both his own drawings and photographs. He included chipmunks among game animals. His maps of species geographic distributions were mostly from the U.S. Biological Survey. This work won the John Burroughs Bronze Medal in 1927 and the Daniel Giraud Elliot Medal in 1928.

A prominent American zoologist who studied bird behavior was Professor Charles O(tis) Whitman (1847–1910), University of Chicago, and Director of Marine Biological Laboratory, Woods Hole.[128] Richard Burkhardt considered him by 1898 the most influential biologist in America and an important founder of American ethology, though not using the term. He raised and studied domestic pigeons. His research on them was ethological; he first argued behavior is an aspect of species that can provide evidence for evolution.[129] He seemed not to think of founding a new science. He had many irons in the fire, and most of his pigeon research appeared in his *Posthumous Works*, volume 3 (1919).

Milwaukee entomologist William Morton Wheeler (1865–1937) earned his Ph.D. under Whitman and ended his career at Harvard.[130] In 1901, annual *Zoological Record* began using *ethology* as a subject heading, and in 1902, Wheeler wrote a paper suggesting biology needed a term designating study of habits. He had seen three terms used by researchers—*natural history, ecology, and ethology*.[131] He thought *natural history* and *ecology* were best applied to broad subjects, leaving *ethology* to apply to studies on habits.[132] He began publishing on ant ethology in 1903, which led to a breakthrough:[133]

> A great achievement of Wheeler in relation to animal behavior was his discovery of the phenomenon of trophallaxis, i.e., secretions provided by some individuals and castes in the ant societies which attract and reward the attention of other individuals and castes; thus, binding the colony into a whole. This discovery, which was a forerunner of the present work on pheromones in insects, brought clarity and understanding into the study of insect societies.

He announced this in "A Study of Some Ant Larvae, with a Consideration of the Origin and Meaning of the Social Habit among Insects" (1918).[134]

"In Britain, Darwin's evolutionary ideas provided the inspiration for four key figures, now considered the forefathers of bird behavior research: Edmond Selous, Frederick Kirkman, Eliot Howard, and Julian Huxley."[135] Londoner Edmund Selous (1857–1934) was similar to Fabre in being a loner who disdained the ornithological establishment, though his education was in law, not biology, as Fabre's was.[136] He began full-time bird study in 1898. A summation of his studies was his *Evolution of Habit in Birds* (1933). Frederick Kirkman (1869–1945) edited *British Bird Book* (4 vols., 1910–13), on behavior of all British species. He conducted long-term studies of black-headed gulls, and he later became active in a new Institute for the Study of Animal Behaviour.[137] H. Eliot Howard (1873–1940) was another amateur naturalist and successful businessman, who found time to study territory for many species (1920, 1929).

Another Whitman student had a less happy fate. Wallace Craig (1876–1954) was interested in comparative psychology and earned his doctorate at Chicago in 1908, with a dissertation on behavior of ringdoves, also subject of much of his subsequent research.[138] He joined the University of Maine faculty, 1908–22, teaching psychology and philosophy. His salary was quite modest, and he received no university assistance with research. In 1922–27, he held three temporary positions at Harvard, and afterwards held no permanent position. He continued his research and publication and had some influence, as on Margaret Nice and Konrad Lorenz, and his "Appetites and Aversions as Constituents of Instincts" (1918) was reprinted as a classic paper in *Foundations of Animal Behavior*.[139] He was writing what he thought was an important book, when he died of cancer. His manuscript has disappeared.

Craig, zoologist, became stuck at the boundary between zoology and psychology. A contemporary psychologist, Robert Yerkes (1876–1956), explored the same boundary, but with more security as a psychologist.[140] He earned his M.A. and Ph.D. at Harvard (1899, 1902), then served on Harvard's faculty until 1917. In 1911, he founded the *Journal of Animal Behavior*, the first such journal open to psychologists and zoologists[141]—not a successful gamble, however, lasting only seven years. Today, several journals fill this need. Yerkes was President of the American Psychological Association, 1916–17. In 1924, he joined Yale University's Institute of Psychology to study primates. In 1929, he founded an experimental station near Orange Park, Florida, later named Yerkes Laboratories of Primate Biology. He sponsored three students to study primates in nature—gorillas and chimpanzees in Africa, and howler monkeys on Barro Colorado Island, Panama.[142] African studies were too brief to obtain significant findings, but Clarence Ray Carpenter studied monkeys at Barro Colorado, 1931–35, and achieved important results. He then went on to conduct similar studies on gibbons.[143] Two of Yerkes' students at Yale published monographs on chimpanzees in 1937.[144] He retired from directing the laboratories at Yale in 1941 and from Yale in 1944.

Vermonter Francis Hobart Herrick (1858–1940) graduated from Dartmouth College, Hanover, New Hampshire (B.A. 1881) in zoology, with interest in marine Crustaceans.[145] He earned his Ph.D. at Johns Hopkins University (1888), then spent his career at Western Reserve University, Cleveland, retiring in1929. In 1890, the U.S. Fish Commission asked him to study lobsters, resulting in his *The American Lobster: a Study of Its Habits and Development* (1895). Microscopic studies strained his eyesight, and he switched to ornithology.[146] In 1901, Herrick published *Home Life of Wild Birds: a New Method of the Study and Photography of Birds*. Ernst Mayr argued that book marked "a turning point in the research methods of field ornithologists."[147] The book attracted Erwin Stresmann's attention, and he and

Herrick corresponded. Herrick documented nesting with photographs and judged nesting behavior as purely instinctive. He published seven papers on habits of bald eagles, culminating in *The American eagle: a study in natural and civil history* (1934). He then studied nesting in song birds, gulls, and terns, described in *Wild Birds at Home* (1935), with numerous photographs.

German ornithologist Oskar August Heinroth (1871–1945) used the term *ethology* in title of an article, "Beiträge zur Biologie, namelich Ethologie und Psychologie der Anatiden" (1911), without envisioning a new science,[148] but his disciple, Konrad Lorenz, did. Heinroth earned an M.D. degree in 1895, then studied zoology at Friedrich Wilhelm University, Berlin. He was a pioneer investigator of bird behavior, and he made three major contributions: (1) conducted comparative studies of social signals and other behaviors in waterfowl, and saw that their ritualized displays could serve as taxonomic criteria; (2) with his first wife, Magdalene (d. 1932), published a classic *Die Vögel Mitteleuropus in Allen Lebensund Entwicklungsstuffen* (4 vols., 1924–33), drawing upon their knowledge from raising 1000 individuals of 286 species, documenting and photographing their development; and (3)provided a framework for studying behavior throughout life, partly based upon individuals raised. He also published discovering that when humans raise birds from eggs, they imprint on humans.

Massachusetts native Margaret Morse (1883–1974, figure 5.4) was daughter of a history professor at Amherst College.[149] Her mother had graduated from Mount Holyoke College, and Margaret did also (1906). She worked on a graduate degree at Clark University, Worster, for two years before marrying graduate student L(eonard) Blaine Nice (1882–1974). While raising their four daughters, she also earned an M.A. degree at Clark (1915), with a thesis on feeding habits of bobwhite quail, which she had published in 1910. Blaine was at Harvard Medical School for two years, 1911–13, then he joined the faculty of the University of Oklahoma, 1913–27. In August 1919, Margaret read in *Daily Oklahoman* that a game warden advocated opening state hunting season on mourning doves in August, since young were (supposedly) flying by then.[150] She doubted this, and on August 20 she found three dove nests on the campus with young ones. She notified *Daily Oklahoman*, Oklahoma Game Department, and the U.S. Biological Survey and defeated the proposal. Thus began her career as ornithologist and ethologist. She and Blaine collaborated on *The Birds of Oklahoma* (1924). In 1927–36, the Nices were at Ohio State University, Columbus. There, in "Theory of Territorialism and Its Development" (1933), she cited a previous study by W. Meise (1930) and early discoveries by Johann F. Naumann (1820) and Bernard Altum (1868).[151]

Margaret Nice's main behavioral studies were on song sparrows *Melospiza melodia*, conducted in Columbus.[152] She visited Berlin in 1932, and Ernst Mayr introduced her to Erwin Stresemann, and she commented on difficulty of publishing a lengthy article in American ornithological journals. He offered to publish it in *Journal für Ornithologie*, which he did.[153] Her song sparrow article was translated into German and appeared in two parts.[154] She read in that journal Konrad Lorenz's "Beiträge zur Ethologie sozialer Corviden" (1931), and he read her song sparrow article. They met at 8th International Ornithological Congress, July 2–6, 1934, and became friends. In his "Forward" to her autobiography, he recalled: "I was at once impressed by her deep understanding of ethology, its methods and its approach."[155] She met Niko Tinbergen at 56th annual meeting of American Ornithological Union in Washington, 17–22, October 38.[156] Nice's personality and achievements inspired two well-illustrated books for teens: Julie Dunlap's *Birds in the Bushes* (1996), a biography, and Michael Ross' *Bird Watching with Margaret Morse Nice* (1987), a "how to" book, using her example as a model for one's own studies. Nice's song sparrow studies also inspired others to conduct similar studies.[157]

Indianan Warder Clyde Allee (1885–1955) was from a Quaker community near Bloomington, and he attended Quaker-supported Earlham College, Richmond.[158] He went on to earn M.S. and Ph.D. degrees (1910, 1912) at the University of Chicago under Shelford. He then taught at four universities before returning to Chicago, where he taught 29 years, before forced retirement. He taught a course and graduate seminar on animal behavior, published *Animal Aggregations* (1931), *Animal Life and Social Growth* (1932), and *Cooperation among Animals* (1938, ed. 2, 1951), and became senior author of Allee et al., *Principals of Animal Ecology* (1949), including a bibliography of his articles. This encyclopedia provided a definitive organization for the subject.

English Julian S. Huxley (1887–1975) was grandson of famous Thomas H. Huxley, who had also been interested in animal behavior.[159] Julian's life rivaled his grandfather's, with even more international fame.[160] Julian Huxley had become interested in birds at age 8; at 13, he saw a green woodpecker, which inspired him to learn "the language and habits of birds."[161] That interest coincided with publication of ornithological literature with that same perspective.[162] His best remembered ethological study was on reproduction of great crested grebes *Podiceps cristatus* (1914), in which:[163]

> He established that there were three separate types of sexual ceremonies, namely (1) those that took place at the time of 'pairing up'; (2) those connected with the act of coition; and (3) those performed by both birds of a pair simultaneously, after pairing up, without relation to coition.

Another Englishman, William H. Thorpe (1902–86) entered Cambridge University in 1921 to study agriculture. In 1925, with B.A. degree, he began work at the Department of Agriculture on insect pests. He also continued his academic work at Cambridge and earned a Ph.D. in 1929. He next worked at Imperial Institute of Entomology until 1932, when he became lecturer in entomology at Cambridge. In 1943, he published an extensive work on insect learning, and in 1951 he published a similar work on learning in birds. He had become in 1950, first director of Cambridge Ornithological Field Station, at Madingley. "Madingley would become one of the preeminent centers for animal behavior studies in the world."[164] Graduate students conducted doctoral research there. In 1951–52 Thorpe was Prather Lecturer at Harvard University. He read ethological works in German by Konrad Lorenz and Niko Tinbergen before publishing *Learning and Instinct in Animals* (1956). He investigated bird song, using news technology.[165] In 1959 he became Reader in Zoology at Cambridge, and in 1966 he received a chair in zoology.

Lorenz and Tinbergen, who are discussed here, interacted with English writers. Tinbergen was Dutch, but was brought by David Lack to Oxford University in 1949. In 1950, he began publishing mainly in English. Thorpe, Tinbergen, and others organized a symposium at Cambridge University, Physiological Mechanisms of Animal Behaviour for July 1949—sponsored by British Association for the Study of Animal Behaviour and by Society for Experimental Biology—to which two zoologists each from the Netherlands, Germany, and the United States were invited.[166] Tinbergen's biographer judged this meeting as "a climactic event for Niko," when he was reunited with Lorenz for first time since World War II.[167] Physiologists attending were unenthusiastic about new ethological science. Tinbergen trained ethologists at same university as Lack and Elton, and taught their students. Yet, there was little contact among their three programs.[168] Thorpe's conference set a precedent for later conferences. German ethologist Erich von Holst, at a new Max Planck Institute, Wilhelmshaven, organized one for 30 May-5 June 50. Thorpe and Tinbergen were the only non-Germans among a dozen participants.[169] The First International Symposium of Comparative Behavior Researchers met at Buldern, Germany, in 1952. Other symposia followed.

Was it coincidental that the ethology of someone, Lorenz, who had dealt with Nazis,[170] was attacked by Jewish American Professor of Psychology at Rutgers University Daniel S. Lehrman (1919–72)?[171] Lehrman was founder and director of Rutgers' Institute of Animal Behavior. His 26-page 1953 article critiqued 10 papers by Lorenz, 1931–50, and Lehrman's conclusion listed six "serious flaws" in Lorenz's papers.[172] Much later, in an autobiographical essay, Lorenz offered a rather neutral comment on Lehrman's attack:[173]

> Lehrman dismissed, in principle, the existence of innate movement patterns and, in so doing, supported his argument substantially by using a thesis of D.O. Hebb, who had maintained that innate behavior is defined only through the exclusion of what is learned and, thus, as a concept was "non-valid," that is, unusable.

Tinbergen published *The Study of Instinct* in 1951, though completed in 1948.[174] Retrospectively, he saw this book as containing his early discussion of "why" in science. That question continued developing in his thoughts until its ultimate expression in "On aims and Methods of Ethology" (1963). His four "why" questions were: causation, survival value, ontogeny, and evolution.

Royal Entomological Society of London held its Third Symposium in September 1965 on insect behavior.[175] Seven topics were addressed: orientation, rhythms, flight, feeding, sexual, communication, and social insects. Insect physiologist John S. Kennedy (1912–93), Cambridge, UK, spoke on "Some Outstanding Questions in Insect Behaviour" (1966).

Illinoisan entomologist Richard D. Alexander (1929–2018) earned his B.S. degree at Illinois State University (1950), and his M.S. and Ph.D. (1951, 1956) at Ohio State University. He spent his career at the University of Michigan, where he advised over 40 ecology students earning M.S. and Ph.D. degrees, and published over 140 papers and books. At an International Conference on Systematics (1967), he spoke on "Comparative Animal Behavior and Systematics," arguing that behavior is an important means to distinguish between some species.[176] He began publishing on human behavior in a review of Lorenz, *On Aggression* (1966) and Robert Ardrey, *The Territorial Imperative* (1966).[177] In a later collection of Alexander's papers, Bobbi Low explained that review's larger significance:[178]

> [T]his short piece foreshadows Alexander's subtle and encompassing theory of cooperation, and the interplay between cooperation and aggression, as well as laying out, with remarkable sophistication for the time, the interactions between natural selection [and] cultural selection.

Viennese Karl von Frisch (1886–1982) was from an academic family, with university professors in both parents' families. He and three brothers continued that tradition.[179] His mother allowed him to keep in their house his pets: 9 mammals, 16 birds, 26 cold-blooded terrestrial vertebrates, 27 fish, and 45 invertebrates! His father, a professor of medicine, wanted him to study medicine as a safer way to support himself than zoology. He did, for two years, then his father agreed he could switch to zoology. Ethologists agree that his discovery of language of bees was the most important contribution to understanding animal behavior during the 1900s.[180] He did not immediately zero in on honeybees. He first studied ability of some fish to change color to blend in with background. In 1911 he read a claim by a professor of ophthalmology that fish and all invertebrates were color blind.[181] That challenged his experience with fish, and he assumed brightly colored flowers were to attract collectors of pollen and nectar, that pollinated flowers. He began experimenting on honeybees to see if they could distinguish colors. (He continued experiments on fish during winters.)[182] During the next nine years, he gained insights into bee behavior, and finally understood bee "speech."[183] He spent 26 years studying intricacies of honeybee waggling.[184] Lectures he delivered at Cornell University were published as *Bees: Their Vision, Chemical Senses, and Language* (1950). His ethological investigations were not limited to fish and bees. Assisted by his son, Otto, he wrote *Animal Architecture* (1974), in which they surveyed the animal kingdom on building abodes. However, honeybee language was his major contribution, and others have continued that research.[185]

Another Viennese, like Frisch, Konrad Lorenz (1903–88) was son of a professor of medicine, and his parents also allowed him to own many pets.[186] The wide variety of animals Lorenz raised is indicated in his popular *King Solomon's Ring: New Light on Animal Ways* (1952). His ethological studies evolved out of his early experiences raising pets. His father wanted him to study medicine, which he did and earned an M.D. Then, he earned a Ph.D. in zoology. He was especially attracted to ducks and geese; he discovered that when he raised then from hatchlings, they became imprinted on him as their parent. He independently discovered imprinting. He had ethological methodology in common with Heinroth, whose disciple he became. Their correspondence during the 1930s is published.[187] Lorenz published five crucial papers, 1931–41, which are now translated in his *Studies in Animal and Human Behaviour* (2 vols., 1970–71), volume 1.

Deutsch Gesellschaft für Tierpsychologie was organized in 1936, and in 1937 it began its *Zeitschrift für Tierschrift*, with Lorenz as an editor. In Britain, Institute of Animal Behaviour also organized in 1936 and began its *Bulletin of Animal Behaviour*, 1938–51. In 1948, Tinbergen founded *Behaviour* as an ethological journal to replace *Zeitschrift für Tierpsychologie*, which had closed during the war.[188] Lorenz published a paper on imprinting (1935) that impressed Tinbergen, who contacted him. Tinbergen accepted Lorenz's invitation to spend 1938 at Lorenz's home in Altenburg, Austria, which was fruitful for both. Their correspondence and interactions led them to formally establish a science, ethology.[189]

In March 1940, Germany invaded and annexed Austria, and Lorenz was enthusiastic about this Anschluss and joined the Nazi party. He believed its biologically tinged ideology was compatible with his research.[190] Also in 1940 he became a professor of psychology at the University of Konigsberg, but was drafted into German army in 1941. As a member of Nazi Party, he was sent to Poznan to study "German-Polish half-breed" to find if they were fit to reproduce superior German citizens.[191] He became a Russian prisoner of war, 1944–48. In 1950, Max Planck Society established Lorenz Institute for Behavioral Psychology, which employed him until he retired in 1973, though he continued his research, lecturing, and publishing. Lorenz was an ethologists,who made his early papers outdated. His *Studies in Animal and Human Behaviour*, volume 2, contained his publications 1941–63. Additionally, Richard I. Evans republished four papers in English (1975) that had appeared in 1966–70.[192] Lorenz synthesized his and others' findings in *The Foundations of Ethology: the Principal Ideas and Discoveries in Animal Behavior* (1981, German, 1978).

Nicolaas ("Niko") Tinbergen (1907–88) was Dutch, but is placed here among German speakers because his early papers were in German and his early ethological interactions were with Lorenz.[193] His father, with a Ph.D. in Medieval Dutch literature, taught high school in the Hague. The Tinbergen home was intellectually oriented, and Nikko's older Jan, won a Nobel Prize in economics (1969). Being uncertain about a career, after graduating from high school in 1925, his parents sent him to a field station in Germany for two months, where he learned bird banding and nature photography.[194] That experience convinced him to study ethology at Leiden University.

Only, Leiden University did not teach it. He studied biology; in the fourth of a five-year program, he could specialize.[195] He conducted his own nature studies. In summer 1929, he wandered in a coastal forest, wondering about a dissertation topic. He watched a bee-wolf wasp *Philanthus triangulum* dig a hole in the sand, fly away, and return with a honeybee, which it dragged to the end of its hole, and laid an egg beside it, then covered the hole. He decided this wasp would be a good topic. In 1932, he finished his dissertation, received his Ph.D., published first (of 3) paper from his dissertation, and married a friend's sister. He and wife Lies then joined a Dutch government expedition to Greenland during International Polar Year, 1932–33. That experience had an enormous influence upon these flat-landers, as Niko learned diet and habits of native Inuits. He also studied life histories of snow buntings and red-necked phalaropes. He published a memoir entitled *Eskimoland* (1935), and that experience transformed him from a "sentimental naturalist to a hunter who saw his animals as objects."[196]

In May 1940, Germany occupied the Netherlands, yet Tinbergen's life did not change much, except he and Lies turned their house into a hospital.[197] Leiden University refused to expel Jews and was closed. Trains were sabotaged and German soldiers shot. Germans arrested 1900 public figures and confined 1400 of them in a commandeered Catholic college. Inmates organized activities, and Tinbergen lectured there on Greenland and ethology.

His studies were methodologically different from Lorenz's. Lorenz raised free-ranging waterfowl and studied their habits. Tinbergen observed wild birds and experimented on them without interfering with their freedom.[198] In 1949, Tinbergen settled at Oxford University and became part of Britain's biologist community, without disrupting his relationships with Continental ethologists.

1973 Nobel Prize

Before 1973, no ecological science had received any notice to rival other biological sciences, such as physiology or molecular biology. Nobel laureate Peter Medawar had befriended and admired Tinbergen, and "he was probably the main protagonist recommending the honour for the three ethologists."[199] The award required a broadening of the scope of the prize for physiology and medicine. In 1973, Frisch, 86, was unable to go to Stockholm for the ceremony; Lorenz was 69, and Tinbergen 66. Frisch received the award for his discovery of language of hive bees, Lorenz for being "acknowledged founder of the science of ethology," and Tinbergen for having "brought the new science of ethology into full flower."[200]

Just five years later, however, Tinbergen commented: "'Ethology' is the name for a phase in the evolution of the behavioural sciences."[201] Other ethologists thought this judgement was too hasty and pessimistic. One of his former students, and his biographer, Hans Kruuk, stated: "At the time of Niko's retirement [1974], *behavioural ecology* was born."[202]

Anthropoid Apes

Kenyan anthropologist-paleontologist-zoologist Louis Leakey played an unusual and important role as facilitator for three young women who wanted to study behavior of great apes: Jane Goodall, Dian Fossey, and Birutė Galdikas.

English Jane Goodall (b. 1932) & Fossey (1932-85, cover) became most famous living student of animal behavior.[203] Leakey helped arrange for her to study chimpanzees in Gombe Stream National Park in 1960. In 1962, she participated in a symposium sponsored by Zoological Society of London, reporting on two years of chimpanzee observations, emphasizing her discoveries of predation and tool use.[204] Solly Zuckerman, who chaired the symposium, saw she contradicted his 1932 claim that all primates are vegetarians. He rejected her claim.[205] She earned a Ph.D. in ethology (1966) at Cambridge University, under Robert Hinde, with a dissertation on *Behaviour of Free-Living Chimpanzees*. She became a diligent observer and publisher.[206] She had a steady stream of student assistants, who came for varying periods. Their observations enabled her to leave Gombe at times. Her research was funded during the 1960s mainly by the National Geographic Society, Washington, DC, and later also by other sources. She celebrated 30 years of her findings with a volume describing them (1990), and after 50 years, a celebratory volume by other ethologists appeared.[207]

American Dian Fossey (1932–85, cover) was two years older than Goodall and began her research on gorillas in 1966; she established her Karisoke Research Centre in Rwanda in 1967.[208] She studied three groups of mountain gorillas and became obsessed with protecting them from poachers. She followed Goodall's example and in 1970 enrolled at Cambridge University to study for a Ph.D. under Hinde. She completed it in 1976, with dissertation on *The Behaviour of Mountain Gorillas*. The National Geographic Society funded her research, and she wrote a well-illustrated chapter, "Living with Mountain Gorillas" for National Geographic's *The Marvels of Animal Behavior* (1972). She published several academic papers and a book, *Gorillas in the Mist* (1983), two years before she was murdered by poachers, tired of her interference.

Lithuanian parents of Birutė M.F. Galdikas (b. 1946) moved to Weisbaden, Germany after World War II before her birth, and when she was 2, they moved to Canada, where her father worked in a copper mine for a year before they moved to Toronto. The first book she borrowed from Toronto Public Library was on a little monkey, Curious George. Growing up, she read about adventures of Goodall and Fosse. In 1962, Galdikas' family moved to California.

Birutė earned her B.A. degree (1966) in psychology and zoology jointly from UCLA and the University of British Columbia. She earned her M.A. degree (1969) at UCLA in anthropology. There she met visiting Louis Leakey and asked him about studying orangutans in nature. He and the National Geographic Society established a research station in Borneo, where she conducted dissertation research. She completed her Ph.D. in 1978, with dissertation on *Orangutan Adaptation at Tanjung Puting Reserve, Central Borneo*. Little was known about orangutans before she studied their behavior, habitat, and diet. She also had to contend with poachers and devastation of their habitat. She persisted for over 30 years, one of the longest continuing studies of a single species. She established Orangutan Foundation International to support orphans and save habitat. She is now a professor at Simon Fraser University, Burnaby, B.C. Her publications include a memoir, *Reflections of Eden: My Years with the Orangutans of Borneo* (1995) and *Great Ape Odyssey* (2005). There is a book for young teens: Anita Silvey, *Undaunted: the Wild Life of Birutė Mary Galdikas and Her Fearless Quest to Save Orangutans* (2019).

C. Animal Population Ecology

An important early contributor was Belgian mathematician Adolphe Ouetelet (1796–1874), both for his own work and as first teacher, and subsequently collaborator of fellow Belgian Pierre-François Verhulst (1804–49).[209] Although Hutchinson devoted a brief account to Quetelet, his main focus was on Verhulst, who developed the logistical curve in three papers (1838–47).[210] Verhulst would surely have contributed more had he not died 2 years later.

The demographic significance of Verhulst's contributions was not appreciated in his lifetime, and there was a longer delay than with Mendelian genetics for appreciation of Verhulst's achievement for demography to be recognized. Logistic curve was independently discovered in 1911 in the context of bacterial growth, and in 1913 applied to yeast.[211] Finally, in 1920, Raymond Pearl (1879–1940) and mathematician Lowell J. Reed published "On the Rate of Growth of the Population of the United States since 1790 and Its Mathematical Representation." A census being mandated every decade by the U.S. Constitution, they had more data available than for any other country. In a follow-up article in 1921, Pearl cited Verhulst's work—of which he was unaware in 1920—in a footnote.[212] He featured the logistic curve in three books: *The Biology of Death* (1921), *Studies in Human Biology* (1924), and *The Biology of Population Growth* (1925). Pearl was founding editor of *Quarterly Review of Biology* in 1926.

Meanwhile, the Industrial Revolution was transforming agriculture, which began operating on a much greater scale than before. In the short run, that favored insect pests, leading to establishment of U.S. Department of Agriculture (USDA) in 1862, with comparable organizations arising in west European countries. In 1887, a meeting of Fruit Growers Convention in Riverside, California asked USDA entomologist Charles Riley (1843–95) for help against cottony-cushion scale *Icerya purchase* infesting California citrus orchards. He replied, if they got Congress to provide funds, he would send an entomologist to Australia, home of *Icerya*, to discover its predator.[213] The Convention did, and he sent Albert Koebele to Australia, where he discovered a beetle predator and a fly parasite of *Icerya*. By releasing these enemies in infested orchards, this pest was controlled. That was beginner's luck—later pest control attempts were seldom this simple. Other cases are discussed here.

Research on insect pests also became large-scale, and in 1897 it became mathematical and theoretical. Three significant studies, two French and one American, appeared in 1897 on insect populations. Most mathematically ambitious was a monograph by Ad. Bellevoye and J. Lorent, "Les plantations pins dans la Marne et les parasites qui les attaquent." They provided hypothetical calculations on population dynamics of a bee and its parasite, for three years.[214] Appearing in a regional science journal, it would have attracted mainly French-speaking Europeans.

Professor Paul-Alfred Marchal (1862–1942), Institut National d'Agronomie, Paris, was France's leading entomologist. American entomologist Leland Howard offered a testimonial (1930):[215]

> It is impossible for me adequately to express my admiration for Marchal. Following a correspondence beginning in 1894, I have known him personally since 1902, have often visited him in his laboratory and in his home, and spent the better part of three months with him traveling in the United States.

Marchal's brief note, "L'équilibre numérique des espèce set ses relations avec les parasites chez les insects" (187), argued that oscillations in crop pests—Hessian fly, oat midge, army worm—is due to increase in parasite population following increase in host population. Crash of host population is followed by crash of parasite population, which allows hosts to increase again. In 1907, he summarized Bellevoye and Lorent's 1897 paper, and Marchal's paper was translated into English.[216]

A third 1897 paper was by USDA chief entomologist Howard (1857–1950), who had earned his B.S. and M.S. degrees (1877, 1883) under John Comstock at Cornell University.[217] He undertook a "Study in Insect Parasitism: a Consideration of the Parasites of the White Tussock Moth" in 1895 because Washington, DC, "suffered from an extraordinary outbreak" of *Orgyia leucostigma*.[218] Previously, the tussock moth had mainly attacked fruit trees, but after English Sparrows spread throughout American cities, it had rid shade trees of canker worms, and tussock moths apparently filled the void. Howard studied in the city where he worked, on known and potential parasites of tussock moths. Lotka later cited Howard 1897 and provided a graph illustrating his data.[219]

Howard undertook another investigation, with W.K. Fiske, head of USDA's Gypsy Moth Parasite Laboratory, that combatted both gypsy and browntail moths. Their monograph (1911) was detailed, well-written, with many photographs and drawings, and historical introduction. The earliest selection in Robert H. Tamarin's *Population Regulation* (1978) is from Howard and Fisk (1911: 105–109). Tamarin's own comment was: "They believed in the balance of nature, a view prevalent among those who believe in density-dependent population regulation."[220]

USDA entomologist Dwight Pierce (1881–1967), in Dallas,[221] led a team of three to investigate boll weevil

FIGURE 5.5 (a) Tinbergen & Lorenz. (b) von Frisch.

parasites. Their report resembled Howard and Fisk's monograph, but went further with a notable diagram, "The Boll Weevil Complex," starting at cotton plant, and indicating both parasites and hyper-parasites (figure 5.6).[222]

North Atlantic fisheries were declining due to overfishing from both sides of that ocean. Spencer Baird, head of the Smithsonian Institution, understood the situation and proposed to Congress in 1871 that it established a U.S. Fish Commission under his direction, with no increase in his salary. Congress did, and this new agency monitored fish stock in all American waters, including the Great Lakes, to detect species in crisis, so governments could take remedial actions.[223] European nations established an International Council for the Exploration of the Sea in 1902.[224] It monitored fisheries of Europe.

Norway's annual cod catch was very important to both its fishermen and its people, and government worry over fluctuations in catch motivated government to build a research vessel, *Michael Sars* in 1900, and appoint fisheries biologist John Hjort (1869–1948) as founding director of Norway Fisheries, 1900–16, to investigate fluctuations, which he did, with no significant discoveries.[225] Hjort became Norway's representative to International Council for Exploration of the Sea, 1902–48. He served as its vice-president in 1920–38, and president from 1938 until he died in 1948. In 1904, German fishery biologist Friedrich Heincke (Heinche) published his study on age groups of cod and plaice, which alerted Hjort to use the same techniques to census cod and monitor age classes. Hjort became collaborator with John Murray of *The Depths of the Sea* (1912), a very popular reference work.

In 1920s, animal ecologists were not highly trained in mathematics, when two science-oriented mathematicians, Italian Vito Volterra (1860–1940), and American Alfred Lotka (1880–1949), published guides to using mathematics to calculate animal population dynamics.[226] Their procedures are valid for conditions they specify. "However, the assumptions are not drawn from biological reality."[227] Pearl recruited Lotka to Johns Hopkins University, where he wrote *Elements of Physical Biology* (1925). Volterra became interested in population dynamics by his soon-to-be son-in-law, Umberto d'Ancona, who studied Adriatic fishery fish statistics.[228] In 1926, without knowledge of Lotka's book, Volterra published his first population study, with separate English summary.[229] Lotka saw Volterra's summary in *Nature* and wrote its editor, 29 October 1926, claiming priority. *Nature* published Lotka's letter and Volterra's response in its first 1927 issue.[230] Lotka also wrote Volterra directly, and their correspondence was entirely cordial.[231] Lotka's last letter thanked Volterra for sending him a copy of his *Lecons sur la théorie mathématique de la lute pour vie* (1931).

Italian mathematician Giorgio Israel contrasted Lotka's and Voltrerra's contributions:[232]

> Volterra's approach was strictly adherent to the classical physico-mathematical paradigm, while Lotka's point of view was more eclectic and open-minded as regards the new developments of physics, and somewhat skeptical about formal statements . . . while Volterra followed the mechanical analogy. Lotka was receptive for the 'thermodynamic analogy' and had a great interest for the energetic problems in population dynamics.

In 1926, Volterra was already a leading mathematician, whose work attracted more attention than Lotka's. Francesco Scudo suggested Lotka's book was too broad in scope: "Possibly by having thus overwhelmed most readers it was on the whole a flop, as Lotka bitterly lamented shortly later in letters to Volterra."[233] Kingsland thought Lotka had published too early and lacked a biological audience.[234] Another interpretation: Lotka's book was needed to motivate ecologists to learn enough mathematics to be able to use his book. Entomologists William Thompson, William C. Cook, and Royal Chapman soon appreciated the value of Lotka's work (see later). A major concern for Volterra was reconciling mathematical calculations with real biological data. He began with a real fisheries question from D'Ancoa, and he persisted in comparing calculations and data.[235] Ecologists who wrote to him wrote about real problems, but there were only three who wrote.

Russian zoologist Vladimir Alpatov (1898–1979), who first taught ecology in Moscow,[236] read papers by Pearl and wrote him to ask if he could come to Johns Hopkins to study under him.[237] Pearl obtained a Rockefeller grant

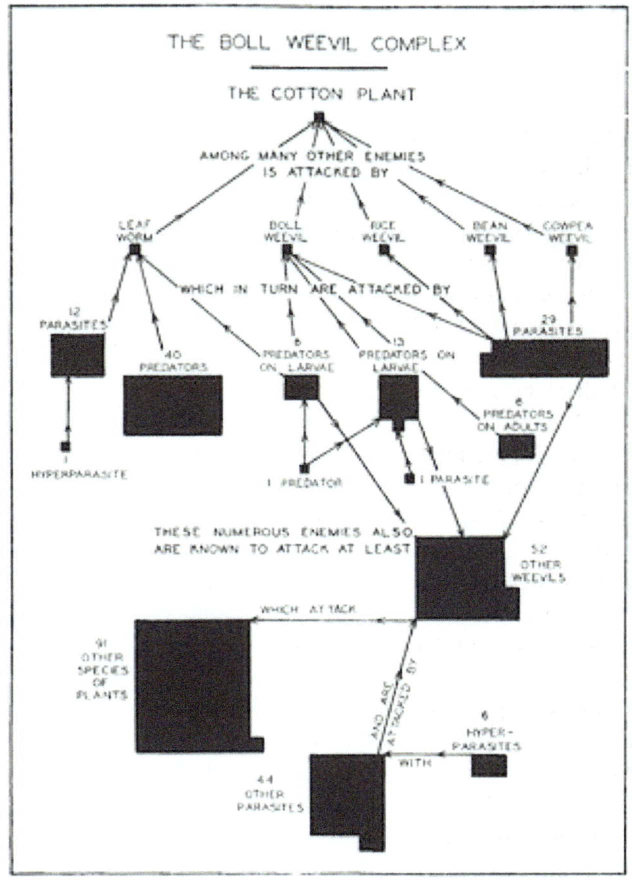

FIGURE 5.6 Boll weevil complex. Pierce et al. 1912.

for him to come, 1927–29. Returning to Moscow, Alpatov discussed Pearl's work with graduate student Georgii Gause (Gauze, 1910–86), who had already surveyed grasshoppers in the Caucasus Mountains and developed a mathematical formula relating species abundance and environmental factors, which appeared in *Ecology*.[238]

Gause began laboratory experiments, with populations of microorganisms, on competition and on predation. On 12 November 32, he wrote to Volterra, thanking him for a letter and a copy of a recent paper and mentioning his current research. He then acknowledged limits to his mathematical ability and invited collaboration:[239]

> I am working in the field of experimental biology, and I am not qualified enough to analyze more closely the mathematical part of the problem and it would be very interesting if it would be possible for you to investigate these equations in the future, and to find their solution in the general form.

Volterra collaborated with a few colleagues, but not with Gause. In a letter of 12 October 33, Gause wrote telling Volterra of his pending book, *The Struggle for Existence*, and that he had confirmed experimentally some of Volterra's mathematical projections, and requesting any papers he had published since 1931.[240] Gause had confirmed the results of competition between two similar species for the same resource led to local extinction of one species.[241] The competitive exclusion principle is also called "Gause's principle."

Not mentioned in his letter to Volterra was his finding on predation:[242]

> We expected at the beginning of this chapter to find "classical" oscillations in numbers arising in consequence of the continuous interactions between predators and prey as was assumed by Lotka and Volterra. But it immediately became apparent that such oscillations are impossible in the population studied, and that this holds true for more than our special case.

Before his book appeared, he published a note in *Science* of his experiments confirming Volterra's prediction on competitive exclusion, but not on predator-prey interactions.[243]

His book became, according to Hutchinson, a "foundation stone of ecology."[244] Fortunately, Gause was publishing in America, because in January 1934, at Ecological Conference of the Academy of Sciences' Botanical Institute, a fanatical Stalinist, I.I. Prezent, denounced Alpatov and Gause for using mathematics in their "formalist, mechanist school."[245] Gause decided it would be expedient for him to switch his research to microbiology.

Two ecologically oriented entomologists, Canadian William Thompson (1887–1972) and American Royal Chapman (1889–1939), were receptive to mathematical assistance from Lotka and Volterra.[246] London, Ontario native Thompson earned a B.A. degree at the University of Toronto (1909).[247] In summer 1908 he worked at USDA Gypsy Moth Parasite Laboratory, and in 1909 USDA sent him to Cornell University to earn an M.S. degree (1911) under Professor Comstock. In 1913, he resigned from USDA and went to Paris to study under Professor Maurice Caullery. In 1914, he studied bacteriology and protozoology at Cambridge University. During World War I, he joined Royal Naval Medical Service. In 1918 he returned to Paris and earned a Ph.D. (1921). Leland Howard, USDA, reemployed him as head of a European Parasite Laboratory. In 1929, he went to England to run Farnham House Laboratory and study insect parasites and predators. During World War II, he returned to Canada to direct Imperial Parasite Service, Canadian Department of Agriculture. In 1948–58, he established biological control laboratories in California, Trinidad, Uruguay, Switzerland, India, and Pakistan.

Culminating Thompson's publications was "Biological Control and the Theories of the Interactions of Populations" (1939), with a 27-page historical introduction, beginning with Darwin and including his own researches. He discussed Lotka and Volterra but was wary of zoologists who substituted mathematical calculation for field data: mathematical construction is so elegant and impressive that there is already a tendency to overestimate its value and to offer it as a description of what really happens in nature.[248] He had already published a polemic, *Science and Common Sense: An Aristotelian Excursion* (1937), which made this point and attacked Darwin's theory of evolution.[249]

Minnesotan Royal Chapman earned B.A. and M.A. degrees (1914, 1915) at his state university and his Ph.D. (1917) at Cornell University under Comstock.[250] He returned to the University of Minnesota as a faculty member and developed a course on ecology—first at that university. His course emphasized "controlled experiment and quantitative field studies."[251] In his *Ecology* paper, "The Quantitative Analysis of Environmental Factors" (April 1928), he cited two of Volterra's papers, showing ecologists what they needed to master. He and Volterra exchanged papers.[252] Chapman also wrote of a student, John Stanley, "who is taking work in mathematics and who is setting up a series of experiments in which we will have *Tribolium confusum* competing with a similar species, *Tribolium ferruginium*" and "we hope from this that we can get the substantiation of your calculations for the cases where two species compete for the same food."[253] Chapman's textbook, *Animal Ecology with Especial Reference to Insects* (1931), had as an appendix an English version of a 40-page article by Volterra, "Variations and Fluctuations of the Number of Individuals in Animal Species Living Together." Since Chapman's textbook never used mathematics, Kingsland suggested he reprinted the article to add an aura of exactness to an insecure ecology.[254]

A(lexander) J(ohn) Nicholson (1895–1969, figure 5.7) was born in Ireland to English parents, who soon moved to Birmingham, where he grew up and attended its university and earned B.Sc. and M.Sc. degrees (1915, 1920).[255] In World War I, he was in Royal Field Artillery. In 1920, he was appointed lecturer in entomology at the University of Sydney. In his course, he itemized factors limiting insect populations, one of which was food.[256] In 1929, the University of Sydney awarded him D.Sc. for work on mimicry. In 1930, he joined Entomology Division of what would become Commonwealth Scientific and Industrial Research Organization. Howard and Fisk (1910) had begun an informal "biotic school of population dynamics," and Nicholson became its chief defender. "Nicholson's 'Balance of Animal Populations' (1933) is probably the most important single paper on population regulation, having sparked the many controversies in this field."[257] That 47-page paper was quantitative, with 11 graphs, but lacked mathematical calculations. In 1935 he collaborated with mathematician V.A. Bailey and published a mathematical version of the 1933 paper, keeping the same title (48 pages). They acknowledged the contributions by Lotka and Volterra, unknown to them when they began.

English animal ecologist Charles Elton (1900–91) was son of an English literature professor. He attended Oxford University and never left. He joined three expeditions to Spitzbergen, 1921–24, and was junior author of expedition reports.[258] He wrote Britain's first textbook on animal ecology (1927), with two chapters on animal populations. In 1931, he attended an American conference and obtained a grant that enabled him to establish a Bureau of Animal Population (BAP) in 1932. He also obtained BSA to initiate *Journal of Animal Ecology*, which he edited for 35 years. Research by him, students, and colleagues resulted in his 500-page *Voles, Mice, and Lemmings: Problems in Population Dynamics* (1942).

Londoner David Lack (1910–73) was son of a physician and an actress.[259] His school and Royal Society for the Protection of Birds encouraged his early interest in birds. He was educated at Cambridge University, yet in 1945, he became head of Edward Grey Institute for Field Ornithology at Oxford. Since BAP emphasized mammals, there was moderate overlap in focus. He led an expedition to Galapagos Islands in 1938, leading to his *Darwin's Finches* (1947). That was followed by *The Natural Regulation of Animal Numbers* (1954) and *Population Studies of Birds* (1966).

Illinoisan brothers Orlando (1901–69) and Thomas Park (1908–92) earned B.S. and Ph.D. degrees under W.C. Allee at the University of Chicago.[260] Both returned to Illinois after teaching a few years elsewhere. Orlando settled at Northwestern University and Thomas went back to the University of Chicago. Orlando studied ecology of nocturnal animals and Thomas studied social relations of insects, in laboratory experiments. Thomas Park taught the first course in population ecology and coined the term. Both became coauthors of Allee et al., *Principals of Animal Ecology* (1949, xii + 837 pages).

Albany native Edward Deevey (1914–88) graduated from Yale University and became a graduate student under Hutchinson.[261] His dissertation was on paleolimnology, yet he shared Hutchinson's interest in quantitative studies. He explained "Life Tables for Natural Populations of Animals," and how to construct them (1947). He began with life tables for humans, compiled for setting insurance rates. There was no rush to follow his example since few ecologists had mathematical training. However, R.F. Morros and C.A. Miller used his instructions for "The Development of Life Tables for the Spruce Budworm."[262]

Robert Tamarin suggested William Thompson was first of a "compromise school of population regulation" (1978).[263] This school saw some truth in both biotic and abiotic schools of population regulation. Thompson also responded to Nicholson and Bailey's criticism of his earlier work.[264] K.E.F. Watt had a different critique: Volterra's conclusions were valid within the context which he defined, but his "assumptions are not drawn from biological reality."[265] The same was true of Nicolson and Bailey (1935). Mark Hixon, Stephen Pacala, and Stuart Sandin found the same

FIGURE 5.7 (a) Charles Elton. (b) David Lack. (c) A.J. Nicholson.

compromise school (2002) that Tamarin identified, citing his 1995 paper.[266] This case was well known among population dynamics researchers: Andrewartha and Birch (1954) defending abiotic factors versus Nicholson (1954) defending competition.

1954

London-born entomologist-ecologist Peter W. Price (b. 1938) detected "a tremendous wave of activity in population ecology building in the 1950s and cresting in the 1960s," but viewed most studies as descriptive.[267] He singled out three as important: Andrewartha and Birch (1954), Lack (1954) and Morris and Miller (1954), to which list I add two by Nicholson (1954a, b) and two by Cole (1954a, b). There was probably little, if any, communication among these ecologists. Yet, they were all at the advancing edge of a common science.

Lack studied controls on bird populations, published in 11 papers, 1946–52, leading to his *Natural Regulation of Animal Numbers* (1954). He commented, "some parts of the population have been studied intensively, others rather little, and some scarcely at all."[268] By banding nestling song sparrows, Margaret Nice found that first-year mortality was 79%, and in individuals older than one year, 45%.[269] Lack and others obtained similar mortality rates in Europe.[270] His main conclusion was: "The reproductive rate of each species, evolved through natural selection, is that which normally results in the greatest number of young surviving to independence." That claim was not controversial, but his further claim, "that in most animals density-dependent variations affect the reproductive rate less than the death-rate," was somewhat controversial. He did agree that too many deer in an area could consume all the food and then starve, and that in some insect species, if their density in a region is high, some will go elsewhere to breed. Lack summarized *Natural Regulation* as an appendix in his *Population Studies of Birds* (1966).[271]

Lack acknowledged Nicholson's influence on his thinking.[272] Nicholson had conducted experiments, consulted Gause's *Struggle for Existence*, and concluded: "Populations do not conform to any single pattern, such as the logistic curve: but there is a single dominant controlling mechanism, namely, density-dependent reaction."[273] "An Outline of the Dynamics of Animal Populations" (1954) was a new lengthy synthesis updating his 1930s papers.[274] He showed that significant work by himself and others supported his earlier conclusions, and he responded to criticisms by Thompson and others to his 1930s papers. He also published two more papers of limited scope in 1954.[275]

Australian H(erbert) G(eorge) Andrewartha (1907–92) collaborated with former student L(ouis) C(harles) Birch (1918–2009) to write *The Distribution and Abundance of Animals* (1954), which Price regards as "a very important and highly regarded treatise, which emphasized weather and shortage of time favorable for reproduction. Their book was encyclopedic (xv + 782 pages) compared to Lack's (viii + 343 pages). Andrewartha and Birch's data and theory were largely in chapters 13 (91 pages) and 14 (18 pages). In chapter 14, they argued that "generalizations about 'density-dependent factors' and competition in so far as they refer to natural populations are neither theory nor hypothesis but dogma."[276] They cited Nicholson as a perpetrator of this dogma.[277]

Indianan animal ecologist La Mont Cole (1916–98) earned his B.S. and Ph.D. degrees (1938, 1944) at the University of Chicago under Allee, and moved to Cornell University before publishing "The Population Consequences of Life History Phenomena" (1954a). He, like Lack, related reproductive potential to natural selection, but Cole was not discussing birds. He saw some utility in species investing less in reproduction than birds do by having a higher degree of fecundity and dispersal, such as ground pine *Lycopodium*, "whose light wind-borne spores may be scattered literally over the whole face of the earth."[278] He viewed life history of an animal as a strategy to maximize fitness.[279] He noted that small mammal population cycles of three or four years were "generally identical with those obtained from a long series of tree rings, presumably because climate favored both groups."[280] We now know that small mammals succeed in raising many offspring in masting years, when oaks produce abundant acorns, and small mammals are less successful when acorns are scarce.[281] Cole had learned from Lotka and Volterra and devoted 7.5 pages to "Computational Methods."

R(obert) F(ranklin) Morris (b. 1916) and C.A. Miller wrote "The Development of Life Tables for the Spruce Budworm" (1954) to show utility of such tables on insect forest pests. They cited Deevey's "Life Tables for Natural Populations" (1947); they adapted what he did for vertebrates to insects. It required much research, and Price attested to their paper's importance for forest entomology: "The reason why Morris and Miller's methodology became so widely adopted rests on the apparent rigor of the approach, and its conceptual simplicity, even though laborious field sampling was inevitable."[282]

All these 1954 studies displayed the breadth of population ecology, without resolving controversies.

1957–58

Ray(mond) Beverton (1922–95) and Sidney Holt (b. 1926–2019), at Britain's Fishery Laboratory, Loswestoft, on East Anglia's coast,[283] published first synthesis of mathematical studies on fish populations: *Dynamics of Exploited Fish Populations* (1957, 533 pages). They used their own earlier studies, but in 1954, when they finished this encyclopedic treatise, there was a substantial literature from other fishery biologists. Their scope was North Sea fisheries, which several European countries exploited. It was important for those countries to share data and agree on harvesting limits. That did not occur.[284]

Since 1933, Biological Laboratory, Cold Spring Harbor, Long Island, had hosted Symposia on Quantitative Biology.

Its 1957 symposium, Population Studies: Animal Ecology and Demography, was funded by three federal agencies and three foundations, and hosted about 150 scientists. The goal was to see what animal population ecologists and human demographers might learn from each other. (Answer: not much.) La Mont Cole's opening talk, "Sketches of General and Comparative Demography," has a history of both demography and animal population studies down to Darwin's *Origin of Species*, mentions Lotka and Volterra's work, and discussed comparative life histories, population regulation, fluctuations, and age structure.

A dispute developed between Australians Andrewartha and Nicholson and became "highly charged and personal," to the dismay of Americans. Andrewartha spoke on "The Use of Conceptual Models in Population Ecology," Birch on "The Role of Weather in Determining the Distribution of Animals," and Nicholson on "The Self-Adjustment of Populations to Change." Discussions following talks were published following talks (1958). Neither Andrewartha nor Birch commented on Nicholson's talk, but he commented on both of theirs.

A. Milne, Agricultural Research Council Unit of Insect Physiology, Durham University, England, cautioned, "Theories of natural control of mammal and bird populations are not likely to be very useful for insects," since warm-blooded animals have "more efficient water-conserving mechanisms, are far less affected by the irregular vagaries of weather, and they may also exhibit 'territorial' behaviour which is important in limiting density."[285] Years later, J.P. Dempster commented:[286]

> Milne's ideas form a halfway house between the views of Nicholson and of Andrewartha and Birch.[287] He believes that the latter underestimate the part played by density-related processes in the determination of animal numbers, but thinks that Nicholson overestimates the frequency with which intra-specific competition occurs in nature.

Hutchinson had the last word at the symposium, but he devoted little more than a page to the talks in his "Concluding Remarks," before explaining "The Formalization of the Niche," maybe the most cited paper in the symposium volume (1958).

Elton, Lack, and Bodenheimer were not at the symposium, but Elton and Bodenheimer and published relevant books in 1958. Bodenheimer had gone to Australia in 1956, where conversation with Nicholson changed his views from skep-tic to acceptance of many of Nicholson's arguments.[288] Nicholson was one of four animal ecologists to whom Bodenheimer dedicated *Animal Ecology To-day* (1958). It had two long chapters on population ecology, devoting a brief paragraph to Andrewartha and Birch's arguments, and seven pages to the progress Nicholson had made in mathematical calculations on animal populations.[289] Elton's *The Ecology of Invasions by Animals and Plants* (1958) broke new ground.[290] Its relevance to population ecology was that local increase in numbers enabled species to invade new territory. His book had no statistics or mathematical calculations; most illustrations were maps showing ranges of invasive species. That it initiated a new line of research was challenged by David Simberloff, who claimed that distinction for his 1981 paper.[291]

Father of Robert H. MacArthur (1930–72) was teaching in Toronto when he was born, but soon he moved the family to Marlboro, Vermont.[292] Natural history was Robert's dominant interest growing up, but he majored in mathematics at Marlboro College (B.A. 1951) and at Brown University (M.S. 1953), but at Yale he studied animal ecology under Hutchinson (Ph.D. 1958). He published his doctoral dissertation as "Population Ecology of Some Warblers of Northeastern Coniferous Forests" (1958). He had studied five species of warblers *Dendroica*, which had overlapping territories. They partitioned their environment in subtle ways, and so they did not violate Gause's law. His paper won ESA's Mercer Award. That related species could partition overlapping ranges to avoid competition had been noted before by Charles Kendeigh: "warblers are so prominent in the mixed evergreen-deciduous forest because they fill so many diverse niches."[293]

Since 1958

MacArthur was a theoretician like Darwin, who synthesized his own and others' field observations to support his theoretical arguments. He is best remembered for general mathematical theories to systematize and explain natural history information of distribution, abundance, and ecology, and creating unified population biology. Most remembered is MacArthur and Wilson's *The Theory of Island Biogeography* (1967).[294] His immediate influence was expressed in a memorial symposium volume, M. Cody and J. Diamond, eds., *The Ecology and Evolution of Communities* (1975).

Appearance of Andrewartha and Birch's *Distribution and Abundance of Animals* (1954) enabled Andrewartha to move to the Department of Zoology, University of Adelaide, South Australia, where he taught population ecology, based upon their book. It was too lengthy for a textbook, so he published *Introduction to the Study of Animal Populations* (1961, xvii + 281 pages). It appeared about simultaneously with Slobodkin's textbook (see later); these two seem to be first textbooks on population ecology. Gordon Orians reviewed Andrewartha's in *Ecology*, stating it was a brief version of his 1954 book, which might be a supplement to a general ecology textbook.[295] He explained that the nonevolutionary thinking in both Andrewartha books reflected the present-mindedness of agricultural ecologists focused upon insect pests.

Lack's *Population Studies on Birds* (1966, appendix) argued against Andrewartha and Birch's weather theory (1954).[296] He did not consider the 1961 textbook. He listed other ecologists who also disagreed with their arguments. They studied insects and Lack studied birds, but Nicholson was an insect ecologist who disagreed with Andrewartha and Birch. They did use mathematics in their books, in 1961 more than in 1954.

New Yorker Lawrence Slobodkin (1928–2009), at age 9, had a canary, rabbit, tortoise, horned toad, baby turtles, newts, and tropical fish. He earned a B.S. degree (1947) at Bethany College, West Virginia, then went to Yale and earned a Ph.D. (1951) under Hutchinson.[297] His dissertation was "a detailed study of the role of age structure in the growth of experimental populations of the microcrustacean *Daphina* epitomized his approach—a quantitative experimental test of a mathematical theory that was intended to apply broadly."[298] He worked for the U.S. Fish and Wildlife Service in Florida and solved the "mystery of the red tide." In 1953 he moved to the Zoology Department, University of Michigan and became a third author of a notable theoretical paper, with Nelson Hairston and Frederick Smith: "Community Structure, Population Control, and Competition." (1960).

Slobodkin's first book, *Growth and Regulation of Animal Populations* (1961, viii + 184 pages), mentioned earlier, was one of first two textbooks on animal population dynamics, along with Andrewartha's. Kingsland commented:[299]

> [Slobodkin's] book covered problems in niche theory, demography, logistic theory, competition, and predator-prey relations; and ended with a projection of a unified ecological theory which would provide answers to practical problems as well as unite both ecology and genetics into a predictive evolutionary theory. It did a great deal to spread the mathematical message of the Hutchinson school, even though Slobodkin himself later became more skeptical of mathematics.

Another Hutchinson student, MacArthur, reviewed it in *Ecology* (summer 1962) and stated that ecologists are either observers or theoretician, and that Slobodkin's textbook was the best yet by a theoretician. MacArthur might not have seen Andrewartha's, which was reviewed in the fall 1962 issue of *Ecology*.

Milne continued his earlier discussion:[300]

> The mechanism of natural control of population is undoubtedly the most controversial problem in animal ecology today. In 1956 at Montreal, I pointed out the shortcomings (as they seem to me) of previous theories of the mechanism and then proposed a new theory for insects.[301] The time has come to consider the inevitable criticism which my theory has aroused.

Milne argued that the old density-dependent versus density independent controversy was "inadequate." His alternative subdivided density-dependent category into imperfectly density-dependent and perfectly density-dependent factors.[302] The imperfectly density-dependent category included a mix of dependent and independent factors, placing Milne intermediately between these two schools of thought.

Lack agreed with Milne that "even the best field studies of the problem suffer from ecological deficiency," but whereas Milne thought a remedy of these inadequacies would resolve controversies, Lack saw differences in theories as due to "difference in basic assumptions concerning the extent to which competition is held to be important or unimportant in natural populations."[303]

Milwaukeean Gordan Orians (b. 1932) earned a B.S. degree (1954) at the University of Wisconsin, then Ph.D. (1960) at the University of California, Berkeley.[304] He joined the Zoology Department, University of Washington, Seattle, where he remained. He wrote a critique of population theory, "Natural Selection and Ecological Theory" (1962) that contrasted Andrewartha and Birch's 1954 book with Lack's 1954 book to end their controversy.[305] He thought ecology has descriptive generalizations, such as competitive exclusion, but its only theory is natural selection. K. Bakker, Zoölogisch Laboratorium, Leiden, wrote "Backgrounds of Controversies about Population Theories

FIGURE 5.8 (a) Wynne-Edwards. (b) Slobodkin. (c) MacArthur.

and Their Terminologies" (1964), praising Orians' paper, but suggesting a better starting point was Nicholson (1933) versus Thompson (1929, 1939). Bakker explained the distinctions in detail, and their significance.

English zoologist Vero Wynne-Edwards ("Wynne," 1906–97) had broad ecological interests, which made him a good teacher and administrator.[306] His father, who taught mathematics and studied Britain's flora, was headmaster at Leeds Grammar School. At age 13, Wynne was sent to Rugby's boarding school, where he flourished. In 1924, he entered Oxford and studied under Huxley and Elton. He was active in Oxford Ornithological Society and annual Oxford Bird Census. He spent parts of three summers at Port Erin Marine Biological Station, Isle of Man. In 1930, he married and became assistant professor at McGill University, Montreal. On their voyage to Canada, he recorded sea birds observed and "discovered the basic pattern of inshore (coastal), offshore (to edge of continental shelf), and pelagic (deep water) marine zones, with different species in each zone."[307] His discovery applied worldwide, and ornithologists accepted it.[308]

On an expedition to North Labrador, he studied fulmars and saw some birds of reproductive age not breeding, "which he thought was a form of population control."[309] During World War II, he taught radar mechanics to members of Canada's Air Force. In 1946, he became a professor of zoology, University of Aberdeen, Scotland, where he remained and became active in academic affairs. His landmark achievement was *Animal Dispersion in Relation to Social Behaviour* (1962, x + 653 pages), summarized in *Scientific American* and in *Science*.[310] Territoriality was a population control. His book was controversial. John Maynard Smith acknowledged territorial breeding behavior in many kinds of birds would tend to limit the number of nesting pairs in a given region but argued that behavior could have evolved by natural selection at an individual level without need to postulate group selection.[311] Wynne-Edwards responded that it was difficult for a field naturalist like himself to have meeting of minds with a laboratory zoologist like Maynard Smith, which was a mischaracterization. Maynard Smith was a theoretician.

Tim Birkhead, Jo Wimpenny, and Bob Montgomerie wondered if Elton's *Animal Ecology* (1927) had misled Wynne-Edwards.[312] Lack became an opponent who devoted a third part of "Theoretical Controversies concerning Animal Populations" to refuting Wynne-Edwards's theory.[313] The controversy was over his discounting importance of natural selection compared to group selection.[314] Ed Wilson's blockbuster, *Sociobiology* (1975), had a chapter on population (4) and another on group selection and altruism (5). His conclusion to chapter 5 was: "although the theory of group selection is still rudimentary, it already has provided insights into some of the least understood and most disturbing qualities of social behavior."[315] Group selection remained a controversial concept, but it was not to be dismissed as a quirk of Wynne-Edwards' imagination.[316] More evidence might help Wynne-Edwards win his case; he published *Evolution through Group Selection* (1986). However, instead of more converts, that book attracted more criticism.[317] His swan song was "A Rationale for Group Selection" (1993).

Slobodkin wrote one of the first textbooks on animal populations, and another Hutchinson student, Robert MacArthur, was the senior author, with Joseph Connell, of the third, *The Biology of Populations* (1966). Joseph H. Connell (b. 1923), from Gary, Indiana, earned a B.S. degree (1946) at the University of Chicago, M.A. degree (1953) at the University of California, and Ph.D. at Glasgow University, Scotland. In 1956, he joined the Zoology Department, University of California, Santa Barbara. He received ESA's Mercer Award in 1963 and Eminent Ecologist Award in 1985. Were MacArthur and Connell dissatisfied with Slobodkin's book? They ignored his book but did cite a paper by him. Instead, they explained their book had been planned in 1960 as third of a series, with the other two being on cells and on organisms. They did not cite any of Milne's papers. MacArthur was at the University of Pennsylvania, 1958–65, then went to Princeton, 1965–72. He became one of the most influential ecologists of his time.

MacArthur and E. O. Wilson's famous *Theory of Island Biogeography* (1967) is, in a sense, a study on population dynamics: species populations expand to new islands as populations increase (or for other reasons); local extinctions occur when island populations cannot maintain their numbers (for various reasons). This dynamic MacArthur brought to his last book, *Geographical Ecology: Patterns in the Distribution of Species* (1972), another noteworthy book. English marine ecologist Arthur S. Boughey (b.1913), University of California, Irvine, published an attractive, concise *Ecology of Populations* (1968, ed. 2, 1973).

Population Ecology

Meanwhile, in England, Royal Entomologic al Society of London held a symposium (1967), with 13 papers on insect abundance. Three speakers were from London, with others from elsewhere in Europe, North America, Africa, and Australia. In introducing the resulting volume, T.R.E. Southwood and O.W. Richards summarized history of insect population studies, 1888–1967.[318] Symposium participants discussed both population theories and methods of investigation.

Numerous publications from 1960s and '70s indicated rising interest in population ecology, including: Lloyd Keith, *Wildlife's Ten-Year Cycle* (1963); William Hazen compiled a large volume of *Readings in Population and Community Ecology* (1964, edition 2, 1970); Maurice Solomon, a brief *Population Dynamics* (1969, x + 60 pages); Lowell Adams compiled a small *Population Ecology* (1970, x + 160 pages); Anthony Alison edited 12 original papers, *Population Control* (1970, 240 pages) on both animal and human demography; Ian McLaren reprinted 10 articles, *Natural Regulation of Animal Populations* (19771, v + 195 pages); P.J. den Boer and G.R. Gradwell edited first of two volumes of a symposium, *Dynamics of Populations* (1971, 611 pages), sponsored by Advanced Study Institution on Dynamics of Numbers in Populations, Oosterbeek, the Netherlands, which included 37 invited papers, opening address, and three conclusions; M.S. Bartlett and R.W. Hiorns edited 19 papers from a symposium sponsored by Institute of Mathematics and Its Applications and by Institute of Biology, Oxford University, *The Mathematical Theory of Biological Populations* (1973, xii + 347 pages).

Hutchinson, in retirement, wrote his only textbook, *An Introduction to Population Ecology* (1978, xi + 260 pages), after two of his former students, Slobodkin and MacArthur, had published textbooks on this. His preface recommended Slobodkin's as more concise than his own, and he warmly acknowledged the influence of his deceased former student, MacArthur. Hutchinson could also have recommended English animal ecologist J. P. Dempster's *Animal Population Ecology* (1975) as being as concise as Slobodkin's, but Dempster was not his former student. Hutchinson's *Introduction* included many historical discussions and footnotes, which may make it entertaining to population ecologists, but possibly distracting to students. It also included numerous drawings, graphs, and mathematical equations. Hutchinson was among the foremost ecologists in the world, and so his book presumably had a wide audience.

Balance in Nature?

Predator-prey relations had been discussed theoretically and studied in laboratory experiments, but what about in nature? Nature conveniently ran an experiment at U.S. Isle Royale National Park, in northern Lake Superior, closer to Canadian than to American shores. There is no physical evidence of moose on Isle Royale before they swam there shortly before 1904.[319] They flourished without predators. National Park Service biologist Adolph Murie (1899–1974) visited there in 1929–30 and estimated there were 1000–3000 moose on its 210 square miles (544 km2), and their browsing was degrading the environment at Isle Royale.[320] A population crash occurred in 1935, reducing population to 500–600 moose. Fire in 1936 burned 20% of island vegetation.[321] Isle Royale became a national park in 1940. Wolf tracks were reported in 1948, but since coyotes were also present, there was uncertainty until in 1951 a park ranger made a plaster cast of a wolf print.

Durward Allen (1910–97), biologist in Washington office of U.S. Fish and Wildlife Service, realized a natural experiment was about to unfold. He applied for funds to study it, but Eisenhower Administration was uninterested. Allen resigned and joined Purdue University's faculty, where he could establish a study group of graduate students in wildlife management.[322] In 1958–75, he had four doctoral students and two postdocs conduct research there. His first student was David Mech (b. 1937), who later studied wolves in Minnesota and at the Arctic Circle.[323] Allen's last student, Rolf Peterson (b. 1949), earned his Ph.D. in wildlife ecology (1974). He remained at Purdue in 1974–75 as research associate, then joined faculty at Michigan Technological University (MTU), in Houghton. When Allen retired in 1975, Peterson assumed supervision of Isle Royale research. Books by Allen (1979) and Peterson (1995) describe their researches.

One of Peterson's students, John Vucetich (b. 1971), earned a B.S. degree (1995) in biology and Ph.D. (1999) in wildlife ecology, with dissertation on "Demographic and Genetic Components of Extinction Risk." Vucetich joined Peterson on MTU faculty.[324] In 2004, they coauthored two papers: "The Influence of Top-Down and Bottom-Up and Abiotic Factors on the Moose (*Alces alces*) Population of Isle Royale" and "The Influence of Prey Consumption and Demographic Stochasticity on Population Growth Rate of Ile Royale Wolves (*Canis lupus*)."[325] For moose, top-down influence is the wolf, bottom-up is vegetation, and abiotic factors are environmental. They found "wolf predation, the dominant cause of death for Isle Royale moose, appears to be a dominant predictor of the moose equilibrium, but has much less influence on variation around that equilibrium."[326] Furthermore, "abiotic factors are an important and complex influence of the dynamics of terrestrial populations." For wolves, "ratio-dependent kill-rate depends on the ratio of prey to predator) may be the important stabilizing force of the predator-prey dynamics on Isle Royale."[327] Also, "Per capita growth of the Isle Royale wolf population appears to increase with the rate at which wolves kill moose." A spring 2014 assessment showed only eight or nine wolves and about 1,100 moose.[328] By 2018, there were only one or two wolves there. On 7 June 18, National Park Service decided to release 20–30 wolves there over three years. Otherwise, the moose would deplete the vegetation.

Fisheries

English fisheries biologist D(avid) H(enry) Cushing (1920–2008), during World War II served in the British Army, afterwards he graduated from Oxford University, and eventually joined Lowestoft Fisheries Laboratory, and continued work of Beverton and Holt. In 1963, Professor Hasler arranged for Cushing to go to the University of Wisconsin to deliver lectures, which became the basis of his *Fisheries Biology: A Study of Population Dynamics* (University of Wisconsin Press, 1968, xii + 200 pages). He evaluated fish stocks by region:[329]

> In the last ten years, the world's catch of fish (about 50 tons in 1965) in freshwater and in the sea has doubled. Some stocks, such as the sardine-like fishes off African coasts, are underexploited. Others, like the bottom-living fish stocks in the northeastern and northwestern Atlantic, are exploited at about the right rate. Other stocks are over fished. For example, Antarctic blue whales have been fished nearly to extinction.

A large conference volume appeared: *The Future of the Fishing Industry of the United States*, D. Gilbert, editor

(1968, 346 pages). The volume included more than population dynamics, but that was a crucial part of conference concerns. Nor would such concerns ever end. J.A. Gulland edited *Fishery Population Dynamics* (1977, xi + 372 pages), to which 15 fishery biologists contributed chapters. W.E. Ricker's "The Historical Development" emphasized methods of assessing size of particular populations of commercial species. The earliest aging of wild fish Fricker knew was by Swedish clergyman H. Hederström (1759). He had counted rings on fish vertebrae. Since no one saw a use for this information, end of story. That same fate befell Dutch Antoni van Leeuwenhoek, who counted annual growth ridges on fish scales in 1684 to determine age; his discovery was translated into English and published in Royal Society's *Philosophical Transactions*.[330]

The University of Toronto Professor Henry A. Regier (b. 1930) explained in his chapter that there were four major types of population problems fishery biologists encountered, and each should be handled differently: (1) large stable resource species; (2) large fluctuating resource; (3) stable systems of small resources; and (4) systems of interacting resources that fluctuate erratically. He explained how to identify each type and how to calculate their populations. Fishery biologist Kenneth Johnstone wrote *The Aquatic Explorers: A History of the Fisheries Research Board of Canada* (1977, xv + 342 pages) to celebrate the Board's first 75 years.

With such fishery investigations ongoing, one might have expected all nations with shores to prudently control their fisheries to avoid depleting them. However, ecologist Robert May and four associates studied "Management of Multispecies Fisheries" (1979) and found the common goal of maximum sustained yield (MSY) was an academic exercise which only worked when fishermen were the only factor impacting targeted species.[331] In the real world, complex factors and interactions meant MSY calculations were invariably too optimistic, with MSY never achieved.

Institute for Marine Studies, University of Washington, Seattle, organized a large multi-authored *The Management of Marine Regions: North Pacific*, E.S. Miles, et al. (1982, xxxiv+ 656 pages), funded by Rockefeller Foundation. This report attempted to achieve an understanding and consensus among coastal nations having North Pacific commercial fisheries.

Two decades after Cushing published his *Fisheries Biology* (1968), he reported "a method was devised in 1965 from which annual catch quotas (total allowable catch, TAC) could be readily estimated."[332] Did responsible management then solve all problems?

> Many pelagic stocks collapsed. . . . A particular difficulty is the effect of fishing when poor year classes succeed each other due to a persistence of environmental factors. The anchoveta stock failed off Peru because of 3 poor year classes (two associated with El Niño) succeeded each other when fishing was intense.

Tim D. Smith complained that fishery science had been driven by economics and politics and admitted: "The ecological effects of fishing can be significant, reducing the number of fish in the sea and changing marine ecosystems in ways we are unable to anticipate even after more than a century of study."[333] He wrote that after a spectacular collapse of western Atlantic cod fishery.[334]

Anne Platt McGinn made clear in *Rocking the Boat: Conserving Fisheries and Protecting Jobs* (1998) that, in practice, for both fishery enterprises and countries, short term advantages trumped long-term sustainability. Carl Safina, *Song for the Blue Ocean* (1997), agreed with her diagnosis but chose to find some hope in rise of many organizations during the 1990s dedicated to preserving the oceans' living resources. Those organizations have substantial virtues but have not reversed degradation of Earth's oceans. Political magazine *Mother Jones* published three articles on over-fishing and pollution (April 2006).[335] Jennifer Hubbard explained difficulties of monitoring fish populations in *A Science of Scales: the Rise of Canadian Atlantic Fisheries Biology, 1898–1939* (2006).

Legacy

By outliving Nicholson, Andrewartha and Birch seemingly had the last word in their dispute, in *The Ecological Web* (1984, xiv + 506 pages), which updated their arguments after 30 years. They hoped "this book will be accepted as a new contribution to population ecology—not merely a revision of our previous work."[336] Reviewer Charles Krebs praised their first book, but not this one. He complained that "Population ecology is exciting now because it is rigorous, quantitative, and experimental, all the virtues espoused by Andrewartha and Birch in 1954 but curiously lacking in this book."[337] He thought they had not kept up with a movement they had initiated, and so their critique was outdated.

Gone but not forgotten: in 1995, centennial of Nicholson's birth, Division of Entomology, Commonwealth Scientific and Industrial Research Organization organized an impressive conference in Canberra of over 200 ecologists from a dozen countries. Andrewartha was dead and Birch did not contribute a published paper. The centennial volume, *Frontiers of Population Ecology*, R.B. Floyd et al., editors (1996, xii + 639 pages) is impressive. In his Preface, Paul Wellings stated:

> John Nicholson is recognized as one of the leading ecologists of this century and is familiar to ecologists throughout the world because of his research on the factors causing population limitation and the concept of regulation through density-dependence. Debate about his research on populations dominated ecology during the middle third of this century and those debates reached their zenith at the Cold Spring Harbor Symposium on Quantitative Biology in 1957.

That debate, Wellings claimed, "did change markedly after 1957 as the focus shifted to the development of a methodology of density dependence (eg. Varley and Gradwell 1960) and the consequences of nonlinear relationships on popular dynamic (e.g., Cook 1965)."

In 1983, Jane Goodall published "Population Dynamics during a 15-year Period in One Community of Free-living Chimpanzees in the Gombe National Park, Tanzania," using data from 1965 to 1980.[338] She had collected data on individuals that sometimes extended from birth to death. She published it in an animal psychology journal, and few population ecologists may have seen it.

UK Midlands ornithologist Ian Newton (b. 1940), from Derbyshire, became interested in birds as a child.[339] He was first in his family to attend a university. At Bristol University he studied zoology. An important influence on him was his attendance (for 40 years) at Lack's annual Edward Gray Institute undergraduate conferences on ornithology. At his first one, as a freshman, he spoke on bullfinch *Pyrrula pyrrula* feeding. He accepted Lack's offer to help him prepare it for publication (appearing in *Bird Study*). He spoke at later conferences on other finches, and he earned a D.Phil. degree from Bristol in 1962. Lack accepted him as a D.Sc. student, with a dissertation on finches (1982). On a postdoc he studied factors influencing bullfinch populations.

After five years at Lack's Institute, Newton joined UK's Nature Conservancy, a government agency, and first studied waterfowl. However, during DDT pesticide crisis during the 1950s and 1960s, raptor populations had declined drastically, and he switched research to them, leading to his *Population Ecology of Raptors* (1979), which attracted attention because of the crisis. He edited *Lifetime Reproduction of Birds* (1989), with 26 chapters by 37 ornithologists from all over the world. By then, he was at Monks Wood Experimental Station, Cambridgeshire, UK. He drew upon much of his research in *Population Limitation in Birds* (1998) and upon his other research in *The Migration Ecology of Birds* (2008) and *Bird Migration* (2010). His online bibliography in 2015 listed: 13 books (written, coauthored or edited), 61 book chapters, and 167 papers, written or coauthored.

Predator-prey relations had been discussed theoretically and studied in laboratory experiments, but what about in nature? Nature conveniently ran an experiment at U.S. Isle Royale National Park, in northern Lake Superior, closer to Canadian than U.S. shores. There is no evidence of moose in Isle Royale before they swam there shortly before 1904.[340] They flourished without predators. National Park Service biologist Adolph Murie visited there in 1929–30 and estimated there were 1000–3000 moose on its 210 square miles (544 km^2), and their browsing was degrading the environment.[341] A population crash occurred in 1935, reducing population to 500–600 moose. Fire in 1936 burned 20% of island vegetation.[342] Isle Royale became a National Park in 1940. Wolf tracks were reported in 1948, but since coyotes were also present, there was uncertainty until in 1951 a park ranger made a plaster cast of a wolf print.

Durward Allen (1910–97), biologist in Washington office of U.S. Fish and Wildlife Service, realized a natural experiment was about to unfold. He applied for funds to study it, but Eisenhower Administration was uninterested. Allen resigned and joined Purdue University's faculty, where he could establish a study group of graduate students in wild-life management. In 1958–75, he had four doctoral students and two postdocs conduct research there. His first student was David Mech (b. 1937), who later studied wolves in Minnesota and at the Arctic Circle.[343] Allen's last student, Rolf Peterson (b. 1949), earned a Ph.D. (1974) in wildlife ecology. He remained at Purdue in 1974–75 as research associate, then joined Michigan Technological University (MTU) faculty, Houghton. When Allen retired in 1975, Peterson Assumed supervision of Isle Royale research. Books by Allen (1979) and Peterson (1995) describe their researches.

One of Peterson's students, John Vucetich (b. 1973), earned a B.S. degree (1995) in biology and Ph.D. in wildlife ecology, with dissertation on "Demographic and Genetic Components of Extinction Risk." Vucetich joined Peterson on MTU faculty. In 2004, they coauthored two papers, one on moose, and one on wolves. A 2014 assessment found only eight or nine wolves and about 1,100 moose. By 2018, there were only one or two wolves. National Park Service decided to release 20–30 wolves there over three years. Otherwise, moose would deplete vegetation, as happened in 1935.

Insect ecologist Naomi Cappuccino, Carleton University, Ottawa, highlighted aspects of history of population ecology, then commented that earlier proponents of population regulation by density-dependent versus density-independent factors often argued at cross-purpose, since these factors are not mutually exclusive: "One of the most refreshing aspects of the renaissance of population dynamics is the absence of polarizing debate," replaced by a "new pluralist synthesis." (1995)[344]

Following her paper in the same journal, Russian expatriate Peter Turchin (b. 1957) claimed that the debate over density-dependent versus density-independent regulation "continues to this very day."[345] Since she was senior editor of that volume, she could have tried to find middle ground between their claims. However, since the scope of his research was much broader than hers, that may have seemed unrealistic. When Turchin published *Complex Population Dynamics: a Theoretical Empirical Synthesis* (2003), he agreed with her that population ecology was a mature science that had outgrown some long-standing debates.

D. SAVING HABITATS AND MANAGING WILDLIFE IN NORTH AMERICA

Informed wildlife and fisheries management, forest and range management, are branches of applied ecology. This account is limited to wildlife.

Prehistoric Extinctions

The earliest human migrants and settlers came from Asia, before 1000 BC. Those who came over land were likely already big game hunters. In Eurasia, large mammals adjusted to increasing efficiency of human hunters as hunting technology evolved. However, Asian big game hunters confronted large mammals in America when their lethal technology had already evolved. They exterminated large mammals before those species developed fear of them.[346]

Early Historical Records of Extinctions

German naturalist Georg Wilhelm Steller (1709–46), on a Russian expedition in 1741, discovered on Bering Island in the North Pacific, Steller's sea cow *Hydrodamalis gigas*, which was exterminated some years later by Russian fur traders, for food.[347] Vertebrate species on islands were vulnerable to extinction when humans arrived. Fossil trilobites, belemnites, and ammonites were known in Europe, but since they were oceanic species, no one could say they no longer lived in the depths.[348] Mammoth bones were known from Siberia, but Europeans had never explored vast areas where mammoths maybe lived. The same situation existed in North America

The great auk *Pinguinus impennis* in the North Atlantic was superficially similar to penguins, though they lived a world apart.[349] They lived along coasts, and loss ability to fly. It was valued for meat, fat, and feathers. George Cartwright, in Newfoundland, warned in 1785 that the high rate of slaughter would diminish them "to almost nothing," unless restrained.[350] The last individuals were killed in 1844 and sold to museums.

United States Policy

Bison/buffalo *Bison bison* lived in gigantic herds on prairies and plains.[351] They may have numbered 60 million in North America before they encountered Euro-Americans with guns. There were 40 million by 1800, 20 million in 1850, and about four million in 1870. Before they obtained European horses, and later guns, native Americans killed some, but their economies did not depend on them. They resisted Americans' invasion, but only slowed it down. Their notable victory was against Custer's outnumbered U.S. troops at Little Bighorn, Montana in 1876.[352] Soon after, they were confined to reservations.[353]

When the Union Pacific transcontinental railroad was being built after the Civil War, Will Cody was hired to supply fresh meat to construction crews.[354] He later claimed to have slaughtered 4,280 bison, 1867–68, and he became known as "Buffalo Bill." When western railroads were completed, eastern upper-class hunters took excursions west and shot bison from railroad cars, leaving the bodies for wolves and coyotes. Commercial hunters followed, who sold hides to eastern processors. U.S. Government did not discourage this, since bison scarcity helped keep Indians on reservations. When Congress passed a hunting limit, President Ulysses Grant vetoed it.[355] In 1886, U.S. National Museum, in Washington, sent taxidermist William T. Hornaday (1854–1937) west to collect bison for an exhibit.[356] He shot 28 and selected six of them for the exhibit. That experience awakened his interest in saving bison from extinction.[357] A few westerners took home calves orphaned by the hunt and raised them.[358] Small herds remained in the United States and Canada. In 1905, Hornaday and others founded the American Bison Society to breed them to save them in national and state parks and on U.S. refuges.[359] Donald Goddard published a collection of Hornaday's conservation writings.[360] In 1907, Canadian Government bought a private herd of bison and moved it to Buffalo National Park, Alberta.[361]

Four American birds became rare, then extinct in the late 1800s to the early 1900s: Labrador duck, passenger pigeon, Carolina parakeet, and heath hen. All four could have been raised in captivity. Labrador duck and heath hen had limited ranges and were never widely known.[362] Passenger pigeon and Carolina parakeet were well known and could have been raised in captivity, but neither generated such concern that saved bison. Passenger pigeons were killed for meat by market hunters, and the last one lived until 1914 in Cincinnati Zoo.[363] Market hunters, railroads, and logging deciduous forests contributed to its disappearance. Carolina parakeets were killed as pests for raiding fruit trees.[364] There were also species of fishes, amphibians, reptiles, insects, and plants that became threatened or actually extinguished.[365]

Saving Habitats and Some Mammals

Congress passed the Hatch Act in 1862, enabling citizens to buy federal lands at an affordable price, after being surveyed by the General Land Office.

Vermonter George Perkins Marsh (1801–82) was son of a Woodstock lawyer.[366] As a boy, he appreciated nature and was upset by clear-cutting forests, followed by wildfires, erosion, and flooding, then scant restoration. He studied law and entered a partnership in 1825, which went well until his partner died in 1832. He struggled alone for a decade, then ran for Congress and served in 1843–49, strongly opposing slavery. He also helped turn James Smithson's bequest into the U.S. Smithsonian Institution. He then joined the U.S. Diplomatic Service as ambassador to Turkey, and later to Italy, where he died. In both countries he saw environments humans had impacted for 2000 years. He joined that vision with his childhood memories of turning Vermont forests into wastelands. He then wrote the founding treatise of the environmental movement: *Man and Nature: or, Physical Geography as Modified by Human Action* (1864). Americans were then preoccupied by the Civil War, but a year later, they began to consider how to build a better nation, and Marsh's book influenced those plans.

Physician-forester-statistician Franklin B. Hough (1822–85), from Martinsburg, New York, was in charge of 1870 U.S. Census. He found evidence of declining availability of timber. In 1875 the American Forestry Association (AFA) was organized, and Hough was an active member. Members collected data on forests, timber, production, reforestation, and AFA published accounts of commercially important trees. In 1876, AFA encouraged Congress to establish an office to oversee U.S. forests. Congress created a special agent in USDA to which Hough was appointed. He traveled widely and published *Report on Forestry* (1877, 659 pages). In 1881, a Division of Forestry was established in USDA, with Hough as head, 1881–83. He published *Elements of Forestry* (1882), first American textbook, with chapters of ecological relevance on soils, climate, reproduction, fires, and insects.

Hough left a legacy worthy of a father of U.S. Forestry, but his successor, Nathaniel Egleston, was a time-server.[367] *100 Years of Forestry* (1976), by William Bergoffen, had appropriate words for Hough and no mention of Egleston. Egleston was replaced in 1886 by professional forester Bernhard Fernow (1851–1923), who had attended Prussia's Forest Academy, then worked in Prussia's Forest Service for seven years.[368] He came to America in 1876 to attend a meeting of AFA, an excuse to rejoin his American fiancée, whom he married and became an American citizen.[369] As America's first professional forester, he became AFA's executive secretary, 1883–95. He was head of the U.S. Forest Service, 1886–98. In 1898 he became the first dean of New York State College of Forestry at Cornell University. That lasted until 1903, when New York's governor vetoed funds for the College because he disagreed with one of Fernow's programs. In 1907, Fernow became the first head of Faculty of Forestry, University of Toronto.

Officers at Halifax Garrison, Nova Scotia, encouraged urban sportsmen to organize a Game and Inland Fishery Society in 1852.[370] Their efforts to prevent unrestrained hunting and fishing were so successful that local government authorized its members to enforce wildlife laws. Even earlier, Hudson's Bay Company had persuaded its trappers not to trap all the beaver, as had happened in eastern USA.[371] Canada's transcontinental railroads were built following America's example, with adverse impact on wildlife, as in the United States. American Progressives preached reform, with influence in Canada. Provincial governments saw that game laws were ineffectual without game wardens to enforce them, and they hired some.

Three successive wealthy Roosevelts in New York State became concerned for habitat reform (before usage of that term). They were not fathers and sons, though their lives overlapped. First was Robert Barnwell Roosevelt (1829–1906), who enjoyed hunting and fishing and witnessed abuses of that privilege, and habitat decline.[372] He expressed these concerns in three books: *Game Fish of the Northern States of America and British Provinces* (1862); *Superior Fishing: or the Striped Bass, Trout, and Black Bass of the Northern States* (1865); and *Game Birds of the Coasts and Lakes of Northern States of America* (1866). He was elected to Congress for New York City, 1871–73, and he became an enthusiastic supporter of the new U.S. Commissioner of Fish and Fisheries.[373] In 1877 he became president of the New York Association for the Protection of Game, which lobbied the New York State Legislature on behalf of wildlife. Two Roosevelt presidents are discussed here.

Scottish John Muir (1838–1914), his father, and two siblings sailed to America in March 1849, leaving behind his mother and four other siblings until they found a place to live. They bought a farm in south-central Wisconsin, on a small lake.[374] The farm was isolated, and John educated himself. He attended the University of Wisconsin for a year, taking courses in botany, geology, and learning to write. He also attended lectures by Ezra Carr on chemistry and natural history. Muir became friends with Carr and his wife, and after they moved to California, they corresponded with him. In 1868, he sailed to California, then walked 100 miles to Yosemite Valley, and felt he had reached home. He herded and sheared sheep, but disapproved of them there because their hooves damaged soil structure and they overgrazed low vegetation. (When I visited Yosemite Valley in August 1957, sheep were still there!)

Muir began publishing influential articles expressing his environmental concerns. By 1864, Yosemite Valley received enough visitors to require some supervision. California Senator John Conness proposed to Congress that it be given to California for a state park, a novel idea.[375] While the Civil War continued, that was a convenient solution to the problem, and Congress agreed, giving about 60 square miles for the park.

By 1870, a semi-volcanic area at Yellowstone River had attracted explorers, and a partly government expedition traveled there to learn what stories were authentic.[376] One expedition leader, Nathaniel Langford, wanted a second transcontinental railroad, in Montana.[377] He publicized geysers and other unique features, which attracted the curiosity of Ferdinand Hayden, head of Geological and Geographical Survey of U.S. Territories, who led an expedition to Yellowstone and verified Langford's claims.[378] As Hayden wrote his report, a lobbyist for a Northern Pacific Railroad asked him to urge Congress to designate Great Geyser Basin as a second park, comparable to Yosemite State Park. Hayden did, pointing out that the area was unsuitable for farming or ranching. Since Yellowstone was not in a state, Congress made it a national park, in 1872, with over two million acres of mountainous terrain, which became Yellowstone National Park.

In 1889, Robert Underwood Johnson, who had edited Muir's articles in *Century Magazine*, met him in San Francisco, and they went to Yosemite Valley, where they decided to campaign for a large Yosemite National Park. In 1890, Muir published two articles on this in *Century Magazine*, and Johnson lobbied for it in Washington.[379] They succeeded, benefitting the nation.[380] Muir encouraged the founding of other national parks, and he wrote the first book on *Our National Parks* (1901).

New Yorker George Bird Grinnell (1849–1938) grew up playing with Audubon's grandsons and attended a school run by his widow, Lucy Audubon.[381] He enjoyed shooting birds and mammals in his neighborhood—then, a common sport. His higher education was at Yale University (B.A. 1870, Ph.D. in mineralogy, 1880). In 1870, he volunteered to assist Yale paleontologist Othniel Marsh (1831–99) in exploring fossil bones out west.[382] That started academic exploration of the west.[383] They rode Union-Pacific Railroad to where Marsh wanted to explore.[384] In 1874, Grinnell was naturalist-mineralogist on Lt. Col. George A. Custer's Black Hills Expedition, then on Col. William Ludlow's 1875 reconnaissance to Yellowstone National Park.[385] In 1876–80. Grinnell was editor of *Forest and Stream Weekly*, then became owner and chief editor of *Field and Stream*, 1880–1911. Conserving wildlife became a major concern. He joined a Harriman Alaska Expedition in 1899.[386]

New Yorker Theodore Roosevelt (1858–1919), at age 8, wandered down Broadway and saw a dead seal from the harbor on display. Later, he identified that moment as beginning his lifelong interest in nature.[387] He rushed home and get notebook and ruler to describe and measure it. As his interest in mammals and birds deepened, he began a personal museum for species he studied. That justified his hunting—in the name of science. He prepared study skins, and when upper-class men made hunting fashionable, he joined Grinnell in founding Boone and Crocket Club in 1877, to promote manly hunting sport and conservation.[388]

As a young man, Roosevelt served three years in New York State assembly, 1881–83, and then as governor, 1898–1900. Fellow Republicans felt he neglected industry and the wealthy, and they stuck him into a "dead-end:" Vice-Presidency under President William McKinley. However, McKinley was assassinated in 1901, and he became President. His popularity among Republicans declined, but increased among American people, ensuring his reelection in 1904.

He became one of America's earliest environmental leaders. His interactions with U.S. head forester Gifford Pinchot were important for developing his environmental outlook.[389] During a controversy over killing birds to decorate women's hats, 14 March 1903, he established America's first wildlife refuge, at Pelican Island, Florida. On April 1, he began a tour of the West—66 days, 14,000 miles. His party included prominent naturalist John Burrows (1838–1914), as they went to Yellowstone National Park for two weeks.[390] Burroughs, a prolific author, wrote two books on his experience: *Camping with President Roosevelt* (1906) and *Camping and Tramping with Roosevelt* (1907). On May 15, Muir joined Roosevelt's train going to Yosemite National Park. He and Roosevelt camped for three days and two nights.[391] Burrows was not troubled by Roosevelt's hunting, but Muir was and urged him to outgrow his killing of animals. Roosevelt agreed to consider that. By 1909, when Roosevelt left office, he had created 50 more refuges and 18 national monuments.[392] After leaving office, Roosevelt and son Kermit went on an African safari, and the Smithsonian Institution, having few African animals, agreed to display what the Roosevelts shot (for science!).

Europeans came to America with a bias against wolves, which killed any unguarded livestock. Some governments paid bounties for killing them.[393] Wolf pelts were also of value. They were hunted from the 1600s through the 1800s. A changing attitude began under the influence of Canadian naturalist-author-artist Ernest Thompson (later, Seton, 1860–1946), who lived mostly in the United States.[394] In 1882, on a train from North Dakota, he saw dogs surrounding a wolf, which was fighting them. His sympathies were with the wolf, and a sketch he later drew of that scene appeared in his *Trail of an Artist-Naturalist* (1940). In 1892, he met a wealthy businessman, with a ranch in New Mexico where wolf predation on livestock was a serious problem. He paid Seton's way to his ranch and offered him bounties on wolves killed. Wolves, unlike coyotes, hunt at night, when guns were useless, and poisoned meat was used to kill them.[395] Seton tried poison and found wolves avoided poisoned meat. He saw a coyote that ate it, then shaken by that and much later asked, "What right . . . has man to inflict such horrible agony on fellow beings, merely because they do a little damage to his material interests?"[396]

At the time, however, he switched to steel traps, using, as ranchers did, traps without teeth, in case a dog was caught. These traps held a leg without damaging it, allowing trappers to free dogs and shoot wolves. After capturing a known wolf, Lobos, he put it in a cage, but it died a day later. In 1898, he published *Wild Animals I Have Known,* with the lead story on Lobos. My father had read it as a boy, and he read it to me when a child. It had little influence upon my father, but it reinforced my developing interest in nature. Seton also wrote later books on animals.

He aspired to be a professional naturalist and had written for his native Manitoba, *List of Mammals of Manitoba* (1886) and *Birds of Manitoba* (1891)—pioneering works, but not very impressive.[397] Next came *Life Histories of Northern Animals* (2 vols., 1909), on 59 mammals, with 68 distribution maps, over 100 colored drawings, and almost 350 black-and-white drawings. It was highly praised by reviewers.[398] With such encouragement, he undertook his climatic *Lives of Game Animals* (4 vols. in 8, 1925–28), covering in volume 1: cats, wolves, foxes; in 2: bears, raccoons, badgers, weasels; in 3: deer, antelope, buffalo, sheep, peccary; and in 4: squirrels, rabbits, armadillo, opossum. This major work was praised by W.J. Hamilton Jr. in a survey of mammalogy, 1853–1953.[399]

Another Grinnell, on the other side of the continent from George, was Joseph Grinnell (1877–1939), who grew up in Pasadena.[400] He earned a B.A. degree in zoology at Throop (later, California) Institute of Technology, Pasadena. He went to Alaska in summer before his senior year, and again after graduating, for almost two years, studying and collecting birds. He earned his M.A. (1901) and Ph.D. (1913) at Stanford University. In 1907, he met Annie M. Alexander (1867–1950), a wealthy collector of vertebrate specimens.[401] After returning from her expedition to Alaska, they discussed her idea of her founding a Museum of Vertebrate Zoology at the University of California, Berkeley; for science only, no exhibits. He was its director, 1908–39, and also on faculty in the university's Zoology Department.[402] His publications were mostly on California birds, with some on California mammals.

Appalachian Mountain Club was organized in 1876; one organizer was Harvard astronomy professor Edward C. Pickering, who wanted to explore and preserve the

White Mountains of New Hampshire. In 1894, it began publishing *Appalachia*. It has now expanded from Maine to Washington, D.C., has 12 chapters, with headquarters in Boston, and about 275,000 members.

In August 1883, three members of the Nuttall Ornithological Club, presumably influenced by the existence of British Ornithologists' Union, decided America should have an ornithological organization; they wrote 48 ornithologists and suggested it.[403] On 26 September, 21 ornithologists met at the American Museum of Natural History, New York, and spent three days discussing details for an American Ornithologists' Union.[404] It would publish a scientific journal, *The Auk*.

Sierra Club was organized in 1892. Muir is remembered as its founder, but others were also involved.[405] In 1889, Muir met with artist William Keith, lawyer Warren Olney, and professors Joseph Le Conte, David Starr Jordan, and others. Editor Robert Underwood Johnson may have suggested a mountaineering club and a goal, to merge Yosemite State Park with surrounding federal land into a Yosemite National Park, and to publish information on it to encourage the public to appreciate it. In 1893, Sierra Club began publishing a *Bulletin*. Today it has 64 chapters nationwide, is headquartered in San Francisco, and has three million members.

In 1906, San Francisco experienced two catastrophes, an earthquake, followed by a fire.[406] When rebuilding, the city decided it needed an additional source of water; it wanted to build a dam in Hetch Hetchy Valley, within Yosemite National Park. Sierra Club fought the proposal, but its members living in San Francisco sided with the city.[407] Washington and Sacramento endorsed San Francisco's request, so Congress passed it. Although Muir and Sierra Club lost then, they started a tradition of strong resistance to attempts to build non-park structures in national parks.

Even so, some such projects succeeded. In summer 1959, I was a ranger-naturalist at Rocky Mountain National Park, and part of orientation for new summer employees was being taken to a visitor center on the western side of the park to a public talk on the Colorado Big-Thompson Project, completed in 1944, for which Alva B. Adams Tunnel was dug within the park, under Continental Divide, to carry westside water to the east side for irrigation. Eighty percent of park rainfall fell on western slopes, where population was thin, so it was sent to eastern slopes with significant population. The tunnel is barely noticeable within the park.

In 1929, Congress appropriated $1 million to add new wildlife refuges to the 51 Roosevelt had established. The U.S. Biological Survey appointed three teams to canvas different regions for favorable purchases.[408] In 1937, land was bought for Aransas National Wildlife Refuge, near Corpus Christi, Texas, to protect wintering grounds for whooping cranes. By 1955, there were 282 Federal refuges, with 18,500,000 acres.[409] Presently, there are at least 562 U.S. Fish and Wildlife Refuges, and there is a map of them for distribution. Books on these refuges include maps of refuges existing when published.[410]

In summer 1958, I worked at U.S. Lake Mattamuskeet Wildlife Refuge, to prepare for tens of thousands of swans, geese, and ducks which arrive via the Atlantic flyway, for the winter. That lake is the largest natural one in North Carolina, apparently dug by a meteor. It is 18 miles (29 km) long, 7 miles (11 km) wide, and 3 to 9 feet deep. In 1913, it was drained for farming, which proved impractical. In 1934, U.S. Wildlife Refuge System bought about half its current 50,000 acres, with more added later. The waterfowl spend nights on the lake, and during the day they fly into nearby fields to feed, with hunters waiting to shoot them. Delta National Refuge, in Louisiana, on the Mississippi flyway, and in the Mississippi Delta, is about the size of Mattamuskeet Refuge, was acquired during the same period, and has as many or more wintering waterfowl.

Gifford Pinchot (1865–1946), like Theodore Roosevelt, was born into a wealthy family.[411] His grandfather had gained wealth by clear-cutting forests in New York State.

His father, James Pinchot, became a wealthy merchant, with a large estate. He owned forests, which led to an interest in forestry. He and like-minded estate owners organized American Forestry Association (discussed earlier). He interested son Gifford in forestry, and after graduating from Yale (1889), he spent a year at France's National Forestry School, and after returning home, became forest manager for George W. Vanderbilt in western North Carolina (where Vanderbilt built the largest private residence in America). Pinchot developed timber management plans and conducted forest surveys. Pinchot wrote about his activities there.[412] He and Fernow (discussed earlier) were founders of American forestry.[413] Upper-class New Yorkers Pinchot and Roosevelt were acquainted but did not discuss forestry until 1899, when Pinchot and friend visited the Governor's Mansion and Roosevelt invited them to spend the night.[414] They had a meeting of minds that continued for the rest of Roosevelt's life. Pinchot became a father figure to U.S. Forest Rangers.

Some western U.S senators were timber barons who did not want rangers interfering with their practices, and during Republican President Taft's administration, 1909–13, they got revenge by cutting Forest Service's budget.[415] America's greatest fire then tore through national forests in Montana, Idaho, and Washington on 20 August 1910, and Forest Service lacked some equipment needed to fight it. Many lives were lost and 3.2 million acres burned. Firefighters had to help pay for medical bills from the fire. Pinchot, who had been dismissed by Taft, blamed short-sightedness of Congress for such a catastrophe.

In 1887, Canada's Parliament passed a Rocky Mountains Park Act, creating the first Dominion Park at Banff Hot Springs.[416] In 1911, Parliament created Yoho and Waterton Lakes National Parks (and the United States added Glacier National Park).[417] A month later, Canada's Cabinet approved the creation of a Dominion Parks Branch in the Department of Interior—the first such agency in the history of the world. James B. Harkin (1875–1955) became

Commissioner of Dominion Parks and guided their further development. A century later, Canada celebrated its innovation and Harkin as Father of Canada's National Parks with a centennial volume.[418]

In America, some national parks were established before any national forests were, yet a National Forest System was established before a National Parks System was. In 1872, the U.S. Army was put in charge of Yellowstone National Park, with Nathaniel Langford as superintendent.[419] John Muir's *Our National Parks* (1901) discussed small parks and forest reservations, Yellowstone, Yosemite, Sequoia and General Grant National Parks—all in the West—with 11 plates, a map, and 11 photographs. He encouraged Americans to value and protect their parks.

Californian Stephen Tyng Mather (1867–1930) was the son of a wealthy businessman. He graduated from the University of California, Berkeley, and joined a company in which his father was involved, which mined and processed borax; he stayed there long enough to become wealthy himself.[420] On vacation in 1914, Mather drove around National Parks and disapproved of their maintenance. He complained to a friend, Secretary of Interior Franklin K. Lane, who had already decided he needed an administrator to organize and run parks. He replied: "If you don't like the way the national parks are being run, come on down to Washington and run them yourself."[421] Mather was a Sierra Club member and occasional mountaineer, and Lane persuaded him to go to Washington, and Lane introduced him to a potential assistant, Horace Albright (1890–1987).[422] They "clicked," and Mather agreed. (Albright succeeded Mather, 1929–33).

They achieved a fast start by publishing *The National Parks Portfolio* (1916), in June, helping Congress create a National Park Service in August.[423] States soon realized that national parks attracted tourists, so states lobbied Congress to create more of them. In 1988, national seashores were added to the System, on all coasts (including Great Lakes). Apostle Island National Seashore, in Lake Superior, was created in 1970. Important for preserving marine habitats was creation of National Marine Sanctuaries and National Estuarine Research Reserves. Charles Seaborn described 12 sanctuaries and 21 reserves in *Underworld Wilderness: Life in America's National Marine Sanctuaries and Reserves* (1996).[424]

Mather, in 1921, called a meeting of state representatives, to encourage states to create state parks. There were places nominated for national parks, but not accepted, that might become state parks. Freeman Tilden, who had published *The National Parks* (1951), also published, 40 years after Mather's meeting, *The State Parks: Their Meaning in American Life* (1962). Since 1921, every state had created parks by 1962, but the number per state did not correlate with size of state. Arizona, a large state with Grand Canyon National Park, had only four state parks. Tilden's 500-page book devoted a half page to Arizona! Oregon, which had Crater Lake National Park, also had 180 state parks! Stung by Tilden's 1962 report, Arizona did better afterwards, enabling Jay Price to write *Gateways to the Southwest: The Story of Arizona State Parks* (2004, 242 pages).

The third Roosevelt who preserved habitats was Franklin Delano Roosevelt (FDR, 1882–1945), who served in New York State Legislature, as governor and the longest-serving U.S. President, 1933–45.[425] Theodore Roosevelt was his inspiration and guide; yet, circumstances enabled Franklin to achieve more than Theodore had, with four terms in office and a Depression requiring new ideas.[426] For the environment, he had Chicago lawyer Harold Ickes (1874–1952) as Secretary of Interior. When Roosevelt ran in 1932, Ickes had organized Republicans for Roosevelt.[427] The Civilian Conservation Corps (CCC) was Roosevelt's idea for putting unemployed young men to build facilities, which in normal times Congress would be reluctant to fund, passed, to counter the Depression.[428] Some funds went to the Department of Agriculture for Forest Service projects, but much more went to the Interior for Fish and Wildlife for refuge projects and to U.S. Parks and Monuments.[429]

A Republican newspaper cartoonist with a strong interest in conservation, Jay ("Ding") Darling (1876–1962), was appointed by FDR to a Committee on Wildlife Restoration, where he suggested a duck stamp program for hunters, to fund federal game wardens, and he also designed the first stamp in 1934. FDR then appointed him head of the U.S. Biological Survey (which would become U.S. Fish & Wildlife Service). Darling became a founder of National Wildlife Federation in 1936.

Bureau of Land Management (BLM) was created in 1946 by merging General Land Office and U.S. Grazing Service, both of which existed in the West.[430] BLM became Federal agency controlling the greatest amount of land. Much change occurred after the war, and BLM attracted little attention. Three books have informed Americans about BLM: James Conway, *The Kingdom in the Country* (1989), with map but no illustrations; Leslie Allen, *Wild Lands of the West: the Story of the Bureau of Land Management* (2002), mostly a picture book; and James Skillen, *The Nation's Largest Landlord: the Bureau of Land Management* (2009), with three maps. Americans who visit BLM lands are about the same people who visit National Forests, to hunt, hike, and camp.[431]

Although American Nature Conservancy history began in 1945,[432] Noel Grove's history of it only starts in 1978, when it bought 18,000 acres for grizzly bears—Pine Butte Swamp Preserve—on eastern slope of Rocky Mountains, Montana.[433] U.S. Nature Conservancy has chapters in every state. Wisconsin chapter published *The Places We Save: a Guide to the Nature Conservancy's Preserve in Wisconsin* (1997), with accounts of 23 preserves and 11 cooperative projects, with one or two photographs for each place, and almost half had a map; Cathy D. Harrington was editor.

Canadians founded their Nature Conservancy in 1962.[434] Nature Conservancy of Canada (NCC) is most active in southern provinces, where most Canadians live and where nature is most threatened. NCC raises funds to buy and maintain wild lands. NCC has collaborated with other nonprofit organizations or with governments. Canada has about a tenth the population of the United States and much lower commercial production, and NCC is a tenth the size of U.S. Nature Conservancy. Bill Freeman wrote an excellent 50-year history of NCC.

There are also private preserves. An early Canadian example was Ohio-born self-educated craftsman (John) "Jack" Miner (1865–1944), who had a waterfowl refuge at Kingsville, Ontario in 1904, with hospital for injured birds.[435] He began banding waterfowl in 1909. An American example: Don Schueler was a professor of English at the University of New Orleans, who with partner Willie Brown, bought 80 acres of run-down farmland north of New Orleans, to turn most of it into a wilderness, and a small part into their home, The Place, where they raised chickens and had a garden and orchard. He wrote about it in *A Handmade Wilderness* (1996). Christopher M. Klyza edited *Wilderness Comes Home: Rewilding the Northeast* (2001), with 12 chapters by different authors who explored such themes as "Ecological Reserve Design," "Stewardship and Sustainability," and "Species Reintroduction and Recovery." Wilderness appreciation is also important for saving reserves. Dick Frankenberg edited 38 essays describing tours on government lands: *Exploring North Carolina's Natural Areas: Parks, Nature Preserves, and Hiking Trails* (2000).

Wildlife Management

Governments conduct most wildlife management, but individuals and private organizations need to monitor government policies and practices. In 1852, California legislature passed a law limiting deer hunting to six months a year.[436] Since it did not also create game wardens to enforce it, the law had little or no influence on hunters. Other states did likewise, down to around 1900. Conservation only began when game species became scarce. Federal government developed roles for agencies in Departments of Agriculture and Interior.[437] Governments in both Canada and the United States began selling hunting licenses to fund game wardens by the early 1900s.[438]

New York Stater C. Hart Merriam (1855–1942) had an M.D. degree, but was more interested in birds and mammals than in medicine.[439] In 1884, AOU asked Congress to establish an office to study birds and mammals.[440] It did, and in 1885 Merriam became head of a division of USDA to study birds and mammals in relation to agriculture. He focused his division's attention upon species having great to slight relevance for agriculture. In 1889, he led an expedition to San Francisco Peak and Little Colorado River, Arizona, and he is most remembered for his 1890 report of findings. Most important was his introduction of his concept of *life zones*—as one ascends a western mountain, one crosses changing climatic zones, progressively colder, and each zone has its own vegetation and fauna.[441] His division became a pest control agency for western farmers and ranchers, especially after he retired in 1910.[442]

Iowan John Lacey (1841–1913) fought in the Union Army during the Civil War, and afterwards he passed law exam and in 1870 was elected to Iowa Assembly.[443] He later served in Congress, 1888–90, 1892–06. He was first in Congress to advocate wildlife conservation.[444] In 1894, he introduced a bill to give Yellowstone National Park authority to arrest and punish poachers of wildlife. He is most remembered for the Lacey Game and Wild Birds Preservation and Dispositions Act (1900), to end killings to sell meat or other purposes.[445] However, enforcement provisions were inadequate, and illegal commerce in wildlife continued. In 1904, he sponsored a bill for enforcement, which finally passed in 1913: Weeks-McLean Bill, protecting migratory birds. Congress asked President Wilson to negotiate treaties with Canada and Mexico to extend this protection to all North America. Those treaties were delayed by World War I but passed in 1918.[446] Canada only restricted selling of shot game, without outlawing it.[447]

In Europe, there were several early examples of bird banding that did not start an ongoing tradition. The effective founder of Europe's ongoing bird-banding tradition was Dutch teacher in Viborg, Hans Mortensen, who read of another Dutchman around 1830 having put brass collars on young ducks and goslings, and one was later reported found near Danzig.[448] Mortensen in 1899 put aluminum bands on adult starlings, but never had any reported re-captured. In 1900, he did so again and received reports of one shot in Norway and another in Holland. In 1903, a German ornithologist tried it, with positive results. In 1909, Hungarian ornithologists did so with positive results.

Audubon reported having banded at Mill Grove, Pennsylvania in 1804 phoebe nestlings with silver strings, and that the following spring two of them returned.[449] But no tradition developed. Allegany, New York, native Leon J. Cole (1877–1948), University of Michigan graduate (B.A. 1901, Harvard Ph.D. 1906), suggested bird banding at a meeting of Michigan Academy of Sciences in 1901. He got the idea from the U.S. Fish Commission tagging fish to track their movements.[450] In winters 1907–08 and 1908–09, New Haven Bird Club sponsored a banding program by him and two others. He and others founded an American Bird Banding Association at annual AOU meeting, December 1909, with him elected president. In 1920, banding records were given to the U.S. Biological Survey, which hired Denver native Frederick C. Lincoln (1892–1960) to manage a Federal banding program; he did so for 40 years, until his death.[451]

Concern for monitoring wildlife motivated Canadian entomologist Gordon Hewitt (1885–1920) to write *The Conservation of Wild Life of Canada* (1921).[452] Comprehensive, with illustrations and maps, it became a benchmark against which to judge future progress. Chapter 12, "the Periodic Fluctuations of Our Fur-bearing Animals," drew upon Hudson's Bay Company (HBC) data on numbers of furs collected, 1820–1914, which indicated periodic fluctuations in smaller mammalian species; he constructed graphs for 13 species. He thought causes were complex. Cambridge ecologist Charles Elton wanted more insight from that data than Hewitt provided, for his *Animal*

Ecology (1927) and his much longer *Voles, Mice, and Lemmings* (1942, 500 pages). One factor was possibly sun cycles of three or four years affecting vole cycles.[453]

Canadian Wildlife Service (CWS) was founded in 1947, with less than 30 staff, who came from other wildlife positions.[454] Increase in personnel occurred gradually. There was no provision for research, but a lack of precise mandate allowed it to arise. CWS was first Canada-wide agency. The Canada-U.S. border is long, and there were mutual inclinations to cooperate across borders.

In 1937 the Wildlife Society was organized, and it published a *Journal of Wildlife Management*. Midwestern universities began adding wildlife personnel to their faculty, who taught Leopold's *Game Management* (1933). Also in 1937, U.S. Senator Key Pittman, Nevada, and Representative Absalom Robertson, Virginia, introduced Federal Aid to Wildlife Act ("Pittman-Robertson Act"), which imposed a 11% tax on hunting guns and ammunition. Funds raised were appropriated to states for wildlife restoration. That included acquisition of lands for habitat restoration or reintroduction of wildlife into areas it formerly occupied.[455] It was the most important wildlife law since Lacey Act of 1913. Pittman-Robertson also set a precedent for subsequent federal-state cooperation.[456]

Indianan Waldo Lee McAtee (1883–1962) earned two degrees at Indiana University, B.A. (1904) and M.A. (1906).[457] In summer 1903, he worked for USBS as assistant biologist and after graduating as in charge of economic ornithology. He was founding editor of *Wildlife Review* and wrote 6,500 abstracts for it. He was also a founding member of Wildlife Society and edited its *Journal of Wildlife Management*, 1937–42. His *Wildlife Food Plants* (1939) was a valuable guide. Ira Gabrielson, retired director of U.S. Fish and Wildlife Service, stated that McAtee was the most versatile biologist he had ever known.[458]

Illinoisan Herbert Lee Stoddard (1889–1970) was self-educated on wildlife and began work as taxidermist for Milwaukee Public Museum, 1910–13, 1920–24.[459] He joined AOU in 1912 and aspired to do more than taxidermy. USBS hired him in 1924 to study bobwhite quail at Red Hills, Georgia, near Tallahassee, Florida. He was main author and editor of *The Bobwhite Quail: Its Habits, Preservation, and Increase* (1931, xxix + 559 pages, 69 plates) and concluded that hunting and habitat loss were important limiting factors to population increase. That was controversial, because its southern habitat was longleaf pine-wiregrass, which needed fire for its maintenance. His book appeared when U.S. Forest Service policy was to suppress all forests fires.

The Bobwhite Quail was an amazing book in 1931, likely the most detailed study of a bird species. It was also "the first major landmark in species management." Its 20 chapters included "Food and Feeding Habits," "Internal Parasites and Parasitic Diseases," "Quail Preserve Development and Management," and "Artificial Propagation." In 1958, he and four colleagues founded Tall Timbers Research Station, which specialized in fire ecology.[460]

One would never expect a comparable book on bobwhites. Yet, nearly 40 years later, "New techniques for measuring numbers, turnover rates, and other features of wildlife were waiting to be applied to the bobwhite."[461] USF&W biologist Walter Rosene (b. 1912) used new advantages and wrote *The Bobwhite Quail: Its Life and Management* (1969, xxv + 418 pages + unnumbered photos).

Iowan R. Aldo Leopold (1887–1948) grew up reading Seton's books.[462] He was the son of a hunter, who passed on that tradition to him.[463] While at an eastern school, he heard of hunters who had broken the law by killing 150 ducks apiece. He assured his mother that "When my turn comes to have something to say and do against it and related matters, I am sure that nothing in my power will be lacking to the good cause."[464] He studied forestry at Yale University (B.S. 1908, M.F. 1909) and then joined the U.S. Forest Service.[465] He never studied wildlife management—there were no such courses. Given a choice of locations, he chose Southwest District.

There, he took an initiative and managed wildlife for hunting. His goals in 1915 included "managing the wild species scientifically," "set aside land for game refuges where hunting permits would be sold," using fund to "pay rangers to work on predator control."[466] An official in Washington opposed setting aside wildlife refuges in National Forests, selling hunting licenses, and paying rangers to do what should be their regular duties. Leopold's immediate supervisor wanted to help him find a way to carry out his plan and created a public relations position for him without specific duties. He hunted wolves to improve deer hunting, then discovered he had too much of a good thing. Arid vegetation could not feed an increased deer population; he realized wolves were essential in nature. In 1944, he wrote an essay about the moment he realized that "Thinking Like a Mountain."[467] In 1922, he had 750,000 acres at headwaters of Gila River surveyed, and in 1924, he persuaded the Forest Service to designate the area Gila Wilderness Area, the first such designation on Federal land.[468]

Then he became assistant director of Forest Products Laboratory in Madison, Wisconsin, a desk job.[469] He had been publishing papers on wildlife and conservation since 1916.[470] In 1928, the Sporting Arms and Ammunition Manufacturers' Institute (SAAMI) hired him to conduct a national survey of game conditions. Hunting had become popular after World War I, as veterans used their knowledge of guns for sport. Farmers increased their acreage of crops with new tractors and equipment. Leopold developed new procedures, as this was the first such survey. He began in Iowa, Michigan, and Minnesota. The University of Wisconsin offered him an office on Madison campus.[471] In March 1929, he presented his findings to date in lecture at Madison on game management. His published *Report on a Game Survey of the North Central States* (1931) included data he next used in *Game Management* (1933), the first textbook on this subject. These two works qualified him for appointment as first professor of wildlife management—in the world—at U.W.-Madison. His professorship was funded by a modest grant from Wisconsin Alumni Research Foundation, for five years.

Leopold became leader of a new field, and his textbook enabled faculty in any American university to teach this new course.[472] (When I studied wildlife management in fall 1958, his *Game Management* was the textbook.) In 1935, he bought a run-down farm beside Wisconsin River as a weekend retreat for his family. Their only crop was trees, whose numbers they increased periodically. In 1948, at the farm, they saw smoke from a neighboring farm, and he fought the fire to stop its spread to his farm, and he died of a heart attack. That was before the U.S. Surgeon General warned that cigarettes are dangerous to one's health, and he smoked. He was an active author, and his most famous book appeared posthumously: *A Sand County Almanac* (1949).

ESA was organized in 1915. By then three Midwestern universities were homes to informal schools of ecology: University of Chicago, terrestrial plants and animals; University of Wisconsin, limnology; and University of Nebraska, prairie plant ecology.[473] ESA organizers might have been influenced by British Ecological Society, organized in 1913, with its *Journal of Ecology*. Most ESA members since 1915 have been concerned about managing natural resources. Ecological research yielded knowledge helpful for resource management. However, in 1945, a majority of members voted not to involve ESA in saving natural areas. Former president Victor Shelford then organized for conservation an Ecologists Union, that later changed its name to The Nature Conservancy.[474]

Minnesotan Murie brothers devoted their careers to wildlife studies, meant to guide management. Olaus (1889–1963), the older, provided Adolph (1899–1974) with guidance in his career.[475] They, like Leopold, grew up reading Seton's books. Olaus' father died when he was 9, and his mother remarried and had another son, Adolph. In 1912, Olaus graduated from Pacific University, Forest Grove, Oregon. He worked two years for Oregon Fish and Game Commission, then joined staff of Carnegie Museum of Natural History, Pittsburgh, and went on expeditions into northern Canada for specimens of birds and mammals, in 1914–15 and 1917.

Olaus Murie was hired in 1916 to "hunt predators in the peninsula of Washington."[476] He killed two bobcats and a cougar, then published an article on killing the cougar and feeling uncomfortable for doing so.[477] In 1920, he went to Alaska for USBS to study caribou. He met an Alaskan resident (since age 9), Margaret (Mardy) Thomas (1902–2003!), who was enthusiastic about Alaskan wildlife and wilderness.[478] They married in 1924, after she graduated from Alaska Agricultural College. (She had earlier attended college three years in the United States.) She became an author, who wrote two books: *Two in the Far North* (1962) and, with Olaus as coauthor, *Wapiti Wilderness* (1966). (Wapiti is the Indian word for elk.) Olaus illustrated both books with his excellent drawings, and he illustrated books by others, including Adolph Murie's *A Naturalist in Alaska* (1961; drawings not in 1963 paperback).

In July 1923, Olaus wrote to his superior: "I have a theory that a certain amount of preying on caribou by wolves is beneficial to the herd, that the best animals survive and the vigor of the herd is maintained."[479] In 1926, he submitted his 93-page caribou report, published by USBS, which stated that neither wolves nor sport hunting limited the population.[480] Instead, he blamed competition with reindeer for forage. In 1927, he and Mardy moved to Jackson, Wyoming, where he studied elk in Teton National Park, leading to his *The Elk of North America* (1951), with his drawings. They moved on to Moose, near Jackson, and raised three children. Later, he wrote *Journeys to the Far North*, which begins in 1914, and posthumously edited by wife, Mardy (1973). Charles Craighead and Bonnie Kreps celebrated Mardy's life in *Arctic Dance: the Mardy Murie Story* (2002).

FIGURE 5.9 (a) Mardy & Olaus Murie. (b) Adolph Murie. Web.

Adolph followed Olaus to Alaska and also condemned USBS for its predator control policy. He studied grizzly bears and wolves in Mount McKinley (now Denali) National Park for National Park Service during 25 summers between 1922 and 1970, and wrote a book on each.

He wrote a memoir, *A Naturalist in Alaska* (1961) in which he discussed the species he studied. He was writing his tenth book in 1974, when he died, and his son, Jan, completed it.

An embarrassment of riches! In a longer essay on this subject,[481] I also discussed Ira Gabrielson, Paul Errington, Frederick Hamerstrom and wife Frances, Irvin Rasmussen, Albert Day, Alexander Wetmore, and wildlife organizations—too much to discuss concisely here. I end with citation of a volume that linked habitat preservation and wildlife management: William McShea and William Healy, editors, *Oak Forest Ecosystems: Ecology and Management for Wildlife* (2002), which has 22 chapters by 34 authors.

NOTES

1. Billings 1985.
2. Hagen 1988.
3. Nicolson 1990: 109.
4. Vernberg 2018: 49–50.
5. Vernberg 2018: 50.
6. Vernberg 2018: 78.
7. Vernberg 2018: 109–110.
8. Vernberg & Vernberg 1972.
9. Vernberg & Vernberg 1972: vii.
10. Vernberg 2018 was self-published.
11. Schmidt-Nielsen 1998: 16.
12. Schmidt-Nielsen 1998.
13. Schmidt-Nielsen 1998: 36.
14. Schmidt-Nielsen 1995.
15. Schmidt-Nielsen 1998: 38.
16. Schmidt-Nielsen 1998: 38–39.
17. Hesse et al. 1937.
18. Kleiber 1932, McNab 2002: 517.
19. Scholander 1990: 36.
20. Scholander 1990: 191–193.
21. Schmidt-Nielsen 1987: 387.
22. Odum 1987: 509.
23. Schmidt-Nielsen 1998: 151.
24. Schmidt-Nielsen et al. 1958, Schmidt-Nielsen 1959a.
25. Dantzler 2015.
26. Schmidt-Nielsen 1998: 320–323.
27. Dawson 2011: 3.
28. Bartholomew 1942, 1943a, b.
29. Bartholomew 1950, 1951, 1952, Bartholomew & Hoel 1953, Bartholomew & Peterson 1967, Dawson 2011.
30. Bartholomew 1966.
31. Bartholomew & Dawson 1954.
32. Dawson 2005, 2011.
33. Hagen 2015: 193.
34. Tytler & Calow 1985.
35. McNab, unpublished manuscript.
36. McNab 1963a, b, McNab & Morrison 1963.
37. McNab to Egerton, 19 December 2018.
38. Townsend & Calow 1981: ix.
39. Kleiber 1961: 177–216.
40. Sibly & Calow 1986: 111–126.
41. Heinrich 2007: 336.
42. Heinrich 2007: 342.
43. Heinrich 2007: 343–358.
44. Heinrich 2007: 427.
45. McNab to Egerton, 4 January 2019.
46. McNab to Egerton, 22 December 2018.
47. Jaynes 1969.
48. Allee et al. 1949: 786.
49. Egerton 2016: 31–34.
50. Durant 1981: 159.
51. Crist 1998.
52. Bodson 1996.
53. Aristotle, *HA* 618a: 26–31.
54. Aristotle, *HA* 1991 transl., p. 343.
55. Singer 1980: 261, Egerton 2001b: 261, 2012a: 12–13.
56. Singer 1980: 262, Egerton 2002b, c, 2012a: 17–24.
57. Allen 1951: 398–400, Stresemann 1951: 8–11 + 2 plates, 1975: 9–12, McVaugh 1972, Egerton 2003a, 2012a: 25–27, Birkhead 2008: 137–142.
58. Michell 1959: 2, 7, Glasier 1978: 12, Gallagher 2008: 147–157.
59. Wallace 1970.
60. Allen 1951: 400–401, Stresemann 1951, 11–12, 1975: 12–13, Egerton 2003b, 2012a: 27–30, Birkhead 2008: index.
61. Aiken 1947, Kibre 1976.
62. Albertus 1999: 599.
63. Albertus 1999: 613.
64. Sarton 1953: 52–132, 1957: 128–171.
65. Raven 1942: 48–137, Stresemann 1951: 13–15, 1975: 13–15, Webster 1976, Bircham 2007: 23–33.
66. Turner 1903: 119–121.
67. Bodenheimer 19228–29: I, 230–233, Raven 1947: 40–42, Wheeler 1976.
68. Allen 1951: 402–403, Stresemann 1951: 18–22, 1975: 18–21, Greene 1983: II, 747–797, Wellish 1984, Egerton 2003: 206–208, 2012a: 34–40.
69. Raven 1947: 217–226, Bircham 2007: 71–73, Birkhead 2008: 36–37.
70. Bodenheimer 1928–29: I, 247–276, Allen 1951: 403–405, Stresemann 1951: 22–24, 1975: index, Castellani 1970, Beider 1973: 85–86, Singer 1982: 263, Aguilar 2006: 29–30.
71. Ray 1975: 114.
72. Raven 1942, Keynes 1961, Bircham 2007: 58–71, Birkhead 2008: index.
73. Tuxen 1973: 97–99, Gough 1975, Egerton 2006b, 2012a: 74–77.
74. Bodenheimer 1928–29: I, 415–448, II, 379–399.
75. Réaumur 1926, English 1926, 1977, 1955.
76. Wheeler 1926: xvi, 36.
77. Réaumur 1926: French, 54–55, English: 142–143.
78. Réaumur 1926: French, 88–89, English, 177.
79. Wheeler 1926: 247, n.
80. Buffon 1954, Roger 1970, 1997, Egerton 2007a, 2012a: 84–86.
81. Buffon 1780: IV, 98–100.
82. Jefferson, French ed., 1785, English, 1787. Cited from 1984 ed.: 169–185.
83. Dugatkin 2009a.
84. Anderson 1994.
85. Gray 1968: 377–379, Thorpe 1979: 10–13.
86. Le Roy 1768, 1994, English 1870.

87. Le Roy 1870: 10.
88. Le Roy 1870: 11.
89. Birkhead 2008: 110–112.
90. Lorch 1966: 219–220, Olby 1973, Morton 1981: 316–321, Mayr 1986, Magnin-Gonze 2004: 103–104.
91. King 1975, Morton 1981: 326–328, Magnin-Gonze 2004: 148.
92. Spitzer 1788, quoted from Lindauer 1985: 129.
93. Quoted from Lindauer 1985: 129.
94. Fitter 1959: 12–38, Groves 1976, Mabey Dadswell 2003, Foster 2004, Burkhardt 2005: 71–72, Bircham 2007: 100–11, Egerton 2007a, b, 2012a: 89–92.
95. Bircham 2007: 100.
96. Goldsmith 1774: V, 301, cited from Nice 1941: 443.
97. McNeill 2004, Fara 2012: index.
98. Darwin 2003.
99. Darwin 1968: 85.
100. Allen 1951: 552–569, Hunter 1983, Egerton 2009: 454–459, 2012a: 127–128, Burtt & Davis 2013.
101. Wood 1931: 630–631.
102. Cameron 1973: 286, Lindroth 1973: 123, Clark 2006, 2009.
103. Herrick 1917, Ford 1964, Streshinsky 1998, Rhodes 2004, Souder 2004, Vedder 2006.
104. Welker 1955: 48–58 + figs. 7–12, Hunter 1983: 93–97 + figs. 15A-17B.
105. Wood 1931: 207–209.
106. Audubon 1831–39: II, Herrick 1917: II, 107–108, Rhodes 2004: 37.
107. Audubon 1826–27.
108. Shuler 1995, Steffen 1999, 2000.
109. Rhodes 2004: 417–430.
110. McKelvey 1955: 139–149, 164–187, 586–626, Beidleman 1960, Graustein 1967: 242–253, Thomas 1974, Desmond 1977: 469, Savage 1979: 174–188.
111. Wood 1931: 496.
112. Bettany 1887.
113. Thorpe 1979: 25.
114. Browne 1995, Egerton 2010b, 2012a: 143–146.
115. Darwin 1859: 207.
116. Richards 1987: 142–152, Prete 1990, Borrello 2010: 7–14.
117. Thorpe 1979: index, Burkhardt 1985, Richards 1987: 185–242, Dugatkin 2006: 1–10.
118. Peattie 1936: 326–347, Fitter 1959: 125–142, Klopfer & Hailman 1967: 23–224, Théodorides 1971.
119. Théodoridès 1971: 508.
120. Wheeler 1916: 74.
121. Forel 1937, Pilet 1972, Tort 1996, Lustig 2004: 285–291, Sleigh 2007: 21–37.
122. Darwin, letters of 28 September 74, 15 October 74; 2015: 480, 498.
123. Field 1949, Klopfer & Hailman 1967: 14–19, Gray 1968 381–383, Clark 1974, Thorpe 1979: 24–30, Singer 1982: 265–268, Boakes 1984: index, Burkhardt 2005: index, Bircham 2007: 336–339, Birkhead 2008: index, Birkhead et al. 2014: index.
124. Thorpe 1979: 26.
125. Wiley 1954, Keller 1984, Anderson 1986, Witt 2010.
126. Seton 1954.
127. Lutts 2991: 34.
128. Mayr 1976, Durant 1981: 167–170, Singer 1982: 269, Pauly 1988: 1219–132, 1994, Burkhardt 1990b, 2005: 19–33, Mitman 1992: index, 1999, Maienshein 1999, Birkhead et al. 2014: 266–267.
129. Lorenz 1985: 267.
130. Evans & Evans 1970, Mallis 1971: 262–268, Shor 1976, Durant 1981: 165–167, Lustig 2004: 298–307, Sleigh 2007: index.
131. Wheeler 1902.
132. Evans & Evans 1970: 213–215.
133. Thorpe 1979: 45.
134. Wheeler 1918.
135. Birkhead et al. 2q014: 248.
136. Thorpe 1979: 30–33, Durant 1981: 178–180, Burkhardt 1990c, 2005: index, Bircham 2007: 336–339, Birkhead 2008: index, Birkhead et al. 2014: index.
137. Burkhardt 2005: 98–103, Birkhead et al. 2014: index.
138. Durant 1981: 185–189, Burkhardt 2005: 33–60.
139. Houck & Drickamer 1996: 276–292.
140. Hilgard 1965, Burnham 1976, Boakes 1984: 148–156, 196–197.
141. Burkhardt 1987, 2005: index.
142. Peterson 2006: 163–166.
143. Carpenter 1934, 1940.
144. Cowles 1937, Crawford 1937.
145. Leutner 1940.
146. Barrow 1998: 187, 197.
147. Mayr in Stresemann 1975: 342.
148. Stresemann 1951: index, 1975: index, Heinroth 1971, Durant 1981: 163–165, 170–174, Burkhardt 1990b, 2005: index., Birkhead 2008: 116–117.
149. Trautman 1977, Nice 1979, Rossiter 1982: index, Ainley 1987, Bonta 1991: 222–231, Mitman & Burkhardt 1991: 180–186, Barrow 1998: 195–198, Burkhardt 1999, 2005: index, Birkhead 2008: index, Birkhead et al. 2014: index, Egerton 2016a: 59.
150. Nice 1979: 42.
151. Nice 1933: 89. On Altum, see Mayr 1935.
152. Nice 1937, 1967.
153. Nice 1979: 115–116.
154. Nice 1933–34.
155. In Nice 1979: ix.
156. Nice 1979: 197.
157. Searcy & Nowicki 1999.
158. Schmidt 1957, Banks 1985, Mitman 1992: index, Dugatkin 2006: 37–60, Egerton 2015a: 46, 2016a: 59–60.
159. Boakes 1984: index.
160. Baker 1976, Burkhardt 1992, Durant 1992, Olby 2004.
161. Huxley 1970–73: I, 26, 40.
162. Klopfer & Hailman 1967: 36–38, Durant 1981: 181–184, Burkhardt 1992: 129–131, 205: index.
163. Baker 1976: 213.
164. Burkhardt 2005: 342, Bircham 2007: 382–386.
165. Thorpe 1961.
166. Thorpe 1979: 81–82.
167. Kruuk 2003: 147.
168. Crowcroft 1991: 61, 135, Kruuk 2003: 7.
169. Burkhardt 2005: 374.
170. Evans 1975: 45–46, Nisbett 1946: index.
171. Lehrman 1953.
172. Lehrman 1953: 358–359.
173. Lorenz 1985: 278–279.
174. Tinbergen 1985: 450–452.
175. Haskell 1966.
176. Alexander 1969.
177. Alexander & Tinkel 1968.
178. Low 2013: 125.

179. Von Frisch 1967: 1–20.
180. Burkhardt 1990a: 312, Raffles 2010: 171–200.
181. Von Frisch 1967: 48.
182. Von Frisch 1941.
183. Von Frisch 1920–22.
184. Von Frisch 1950, Lindauer 1985: 129.
185. Michelsen 1999.
186. Evans 1975, Nisbett 1976, Thorpe 1979: index, Durant 1981: 174–177, Lorenz 19985, Bateson 1990, Chavot 2000, Taschwer & Föger 2003, Burkhardt 2005: index, Birkhead 2008: index, Birkhead et al. 2014: index.
187. Heinroth & Lorenz 1988.
188. Burkhardt 2005: 284–285.
189. Burkhardt 2005: 199.
190. Kalikow 1983, Burkhardt 2005: 231–280.
191. Nisbett 1976: 77–91, Taschwer & Föger 2003: 112–131.
192. Evans 1975: 129–269.
193. Tinbergen 1958, 1985, Thorpe 1979: index, Dewsbury 1990, Kruuk 2003, Burkhardt 2005: index, Bircham 2007: 386–394, Birkhead 2008: index, Bolhuis & Verhulst 2009, Birkhead et al. 2014: index.
194. Kruuk 2003: 33–35.
195. Kruuk 2003: 38–39.
196. Kruuk 2003: 69.
197. Kruuk 2003: 114.
198. Tinbergen 1953a, 1960, 1972.
199. Kruuk 2003: 268.
200. Marler & Griffin 1973: 464, Nisbett 1976: 220–223, Feldman 2000: 280–283, Kruuk 2003: 268–273, Taschwer & Föger 2003: 225–240, Burkhardt 2005: 446–449.
201. Letter to R. Hinde 1978, quoted in Kruuk 2003: 329.
202. Kruuk 2003: 330.
203. Montgomery 1991: 24–454, 193–213, Peterson 2006.
204. Peterson 2006: 290–292.
205. Zuckerman 1962, 1988. On him as primatologist, Burt 2006.
206. Goodall 1965, 1967, 1971, 1986.
207. Goodall 1990, 2010.
208. Montgomery 1991: 46–66, 214–233, Mowat 1987, Schafer 1999.
209. Freudenthal 1975, Pelseneer 1976, Kingsland 1985: 65–66.
210. Hutchinson 1978: 18–21.
211. Hutchinson 1978: 21.
212. Pearl & Reed 1920, Pearl 1921: n.2 on 206, Kingsland 1985: 64–76.
213. Egerton 2012a: 190–192, 2013: 69–70.
214. Bellevoye & Laurent 1897: 101–103.
215. Howard 1930: 240.
216. Marchal 1907, 1908: 353–354.
217. Howard 1933, Graf & Graf 1959, Mallis 1971: 79–86, Hatch 1972, Sawyer 1999, Egerton 2013: 73–74, 2015: 563–564.
218. Howard 1897: 6.
219. Lotka 1925: 90–91.
220. Tamarin 1978: 29.
221. Mallis 1971: index.
222. Egerton 2007: 54, 2014a: 68–71.
223. Allard 1967, 1970, 1997.
224. Schlee 1973: 206–216, 223–225, Cushing 1988: 203–212, Mills 1989: 75–91, Smith 1994: 113–124, Rozwadowski 2002, 2004, Smed 2004a, b, Schwach 2013, Egerton 2014: 396–397.
225. Solhaug & Saetersdal 1971–72: 405–406, Merriman 1972a, Mills 1989: index.
226. Whkittaker 1941, Watt 1962: 245–247, Scudo 1971, 1984, Gridgeman 1973, Volterra 1976, Hutchinson 1978: 221–226, Scudo & Ziegler 1978, Kingsland 1982, 1985: 25–49, 102–116, 1991: 7–10, Burgess 1996: 67–68, Fuchsman 1999.
227. Watt 1962: 247.
228. Scudo 1971: 2, 1984: 23–26, Israel 1993: 487–492.
229. Volterra 1926.
230. Lotka 1927, Volterra 1927, both reprinted by Israel 1993: 496–497.
231. Israel & Millán Gasca 2002: 280–288.
232. Israel 1988: 37.
233. Scudo 1984: 20.
234. Kingsland 1895: 211.
235. Millán Gasca 1996: 5–6.
236. Weiner 1988: 66.
237. Kingsland 1985: 146–148.
238. Gause 1930.
239. Israel & Millán Gasca 2002: 212–213.
240. Israel & Millán Gasca 2002: 213–214.
241. Gause 1934b: 89, 113.
242. Gause 1934b: 140.
243. Gause 1934a.
244. Hutchinson 1978: 120.
245. Kingsland 1985: 160–161, Weiner 1988: 222.
246. Graham 1941, Thorpe 1973, Kingsland 1985: 123, 127–131.
247. Thorpe 1973, Kingsland 1985: index.
248. Thompson 1939: 381.
249. Kingsland 1985: 135–141.
250. Burgess 1996: 25.
251. Graham 1941: 522.
252. Israel & Millán Casca 2002: 123.
253. Israel & Millán Casca 2002: 125.
254. Kingsland 1985: 128.
255. Mackerras 1970.
256. Egerton 2014c: 59.
257. Tamarin 1978: 29, Kimler 1986: 223–227.
258. Summerhayes & Elton 1923, 1928, Cox 1979: 17–28, Crowcroft 1991, Southwood & Clarke 1999, Paviour-Smith 2004.
259. Anderson 2013.
260. Burgess 1996: 85.
261. Burgess 1996: 36.
262. Morris & Miller 1954, Price 2003b: 19.
263. Tamarin 1978: 92–93.
264. Thompson 1939: 319–321.
265. Watt 1962: 246–247.
266. Hixon et al. 2002: 1492, Tamarin 1995.
267. Price 2003b: 19.
268. Lack 1954: 275.
269. Nice 1937, 1979, Trautman 1977, Burkhardt 1980, Birkhead et al. 2014: index.
270. Lack 1954: 85.
271. Lack 1966: 281–287.
272. Nicholson 1933, Nicholson & Bailey 1935.
273. Nicholson 1950.
274. Nicholson 1954a.
275. Nicholson 1954b, c.
276. Andrewartha & Birch 1954: 469.
277. Andrewartha & Birch 1954: 19–20.
278. Cole 1954a: 239.
279. Kingsland 1985: 174.
280. Cole 1954b: 3.
281. McShea 2000.

282. Price 2003: 23.
283. Cushing & Edwards 1996, Ramster 1996.
284. Clover 2006: 106–110.
285. Milne 1958: 254.
286. Dempster 1975: 123.
287. Milne 1957, 1958, 1962.
288. Bodenheimer 1959: 164–167.
289. Bodenheimer 1958: 118, 120–126.
290. Chew 2006: 209–286.
291. Simberloff 2011, 1981.
292. Cody & Diamond 1975: vii—ix, Fretwell 1975, Kingsland 1985: 180–205, Wilson & Hutchinson 1989, Burgess 1996: 69–70, Brown 1999, Pianka & Horn 2005, Kaspari 2008, Birkhead et al. 2014: index.
293. Kendeigh 1945: 433.
294. MacArthur & Wilson 1967: 149.
295. Orians 1962b.
296. Lack 1966: 290–299.
297. Slobodkin 2009.
298. Colwell & Futuyma 2011: 19.
299. Kingsland 1985: 180.
300. Milne 1962: 19.
301. Milne 1957a, b.
302. Milne 1962: 21.
303. Lack 1966: 296.
304. Orians 2010, Birkhead et al. 2014: index.
305. Orians 1962b.
306. Wynne-Edwards 1985, Kimler 1986: 227–231, Newton 1998a, Bircham 2007: 378–380, Birkhead 2008: 257–260, Birkhead et al. 2014: 356–366.
307. Wynne-Edwards 1930, Newton 1998: 476.
308. Saunders 1973: 10.
309. Wynne-Edwards 1939, Newton 1998: 477.
310. Wynne-Edwards 1964a, 1965.
311. Maynard Smith 1964.
312. Birkhead et al. 2014: 357–359.
313. Lack 1966: 299–312.
314. Birkhead 2008: 357–360, Birkhead et al. 2014: index.
315. Wilson 1975: 129.
316. Wilson 1983, Leigh 2010.
317. Personal communication, R. Montgomerie 2015.
318. Southwood 1968.
319. Peterson 1995: 22.
320. Murie 1934.
321. Peterson 1995: 22–23.
322. Allen 1979: xviii—xx.
323. Mech 1970, 1988, 1991, 2000, Mech et al. 2015.
324. Scientia.global/issue-109/.
325. Vucetich & Peterson 2004a, b.
326. Vucetich & Peterson 2004a: 188.
327. Vucetuch & Peterson 2004b: 318.
328. Bergquist 2014.
329. Cushing 1968: 3.
330. Leeuwenhoek letter of 25 July 84; *Phil. Trans.* 15 (1685) 883–895. Egerton 1968: 9.
331. May et al. 1979.
332. Cushing 1988: 259.
333. Smith 1994: 1.
334. Kurlansky 1997, Clover 2006: 121–130.
335. Franklin 2006, Robbins 2006, Whitty 2006.
336. Andrewartha & Birch 1984: xiii.
337. Krebs 1985: 873.
338. Goodall 1983.
339. Newton 2014: see box 10.1, p. 383 in Birkhead et al. 2014.
340. Peterson 1995: 22.
341. Murie 1934.
342. Peterson 1995: 22–23.
343. Mech 1970, 1988, 1991, 2001, Mech et al. 2015.
344. Cappuccino 1995: 12.
345. Turchin 1995: 19.
346. Martin & Wright 1967, Martin & Klein 1984, MacPhee & Schouten 2018.
347. Stejneger 1887, 1936, Hultén 1940: 295–296, Trefethen 1975: 44, Egerton 2008: 43–50, 2012a: 94–96, 2019b: 2.
348. Kolbert 2014: 24.
349. Greenway 1967: 271–291, Halliday 1978: 67–76 + plate 3, Mowat 1996: 17–39, Fuller 1999: 60–85, Barrow 2009: 61–66, Kolbert 2014: 55–66.
350. Quoted in Burnett 2003: 4.
351. Haines 1970: 156–164, 188–200, Roe 1970: 334–520, Lott 2q002, Barrow 2009: 93–96.
352. Connell 1984, Welch 1994.
353. Gibson 1980: 463–512, Wilson 1998: 247–285.
354. Carter 2000: 93–94.
355. Haines 1970: 205.
356. Haines 1970: 206–207.
357. Barrow 2009: 108–114, Bechtel 2012, Dehle 2013.
358. Stewart 1978: 31–33.
359. Barrow 2009: index.
360. Goddard 1995: 124–157.
361. Now Canadian Forces Base Wainwright. Loo 2006.
362. Greenway 1967: 173–175, Halliday 1978: 103e-104 + plate 7, Cokinos 2000: 131–193, 279–304, Barrow 2009: 131, 263–268.
363. Schorger 1955, Greenway 1967: 304–311, Halliday 1978: 87–95 + plate 5, Cpkinos 2000: 195–278, Barrow 2009: 96–100, Fuller 2014, Greenberg 2014.
364. Greenway 1967: 322–327, Hallidat 1978: 86–100 + plate 6, Cokinos 2000: 15–58, Snyder 2004, Barrow 2009: 127–130.
365. Matthiessen 1987: 285–302.
366. Curtis et al. 1982, Lowenthal 2000, Trombulak 2001.
367. Steen 1976: 20–21.
368. Rodgers 1951, Belonzi 1997, West 1999, Mazel 2009.
369. Steen 1976: 23–25.
370. Loo 2006: 14.
371. Loo 2006: 15.
372. Harmond 1997, 2009b.
373. Regier 2001: indez.
374. Melham 1976: 24–37, Clarke 1980: 2–27, Turner 1985: 36–50, Leshuk 1988, Strong 1988, Cammarata & Sterling 1997, Dillon 1999, Worster 2008: 13–66, Zuzworsky 2009.
375. Duncan 2009: 8.
376. Duncan 2009: 24–28.
377. Black 2012: index.
378. Duncan 2009: 29–35.
379. Strong 1988: 96.
380. Jones 1965, Huth 1948, Brower 1990.
381. Reiger 1972: 6–25, 1997, 1999, 2009, Dunlap 1988: index, Burgess 1996: 48–49, Brinkley 2009: index, Lunde 2016, Merchant 2016: 3–12.
382. Reiger 1972: 29, Shor 1974.
383. Bartlett 1962, Penick 1971.
384. Goetzmann 1976: index.
385. Goetzmann 1967: index, Reiger 1972: index.

386. Grinnell 1995.
387. Roosevelt 1926: xx, Dunlap 1988: index, Brinkley 2009: 29–30, Pak 2009, Lunde 2016: 9–10.
388. Gable 1997, Reiger 2001: index, Brinkley 2009: index, Lunde 2016: 138–139, Merchant 2016: 234–235.
389. Roosevelt 1926: 359, 384–392, Pinchot 1947: index, Egan 2009: index.
390. Huth 1957: 180, Renehan 1992: 246–249, Brinkley 2009: 509–510, Lunde 2016: 169–170.
391. Worster 2008: 366–369, Brinkley 2009: 536–546.
392. Brinkley 2009: 818–830.
393. Coleman 2006.
394. Keller 1984: 95–96, Wadland 1997, Sterling 1999b, 2009.
395. Keller 1984: 131–134.
396. Seton 1940: 162.
397. Keller 1984: 181.
398. Keller 1984: 186.
399. Hamilton 1955: 672–673.
400. Grinnell 1940, Linsdale 1942, Shor 1972, Burgess 1996: 49, Chambers 1997, Stein 1997.
401. Sterling 1997a, Stein 2001: 64–65.
402. Chambers 1997: 330.
403. Palmer 1933: 7.
404. Chapman & Palmer 1933.
405. Cohen 1988: 8–10, Turner 1991: 44–49.
406. Thomas & Witts 1971, Beck & Williams 1972: 332–335, Lavender 1972: 339–340, Muscadine 1975: 428–436.
407. Jones 1965, Melham 1976: 177, 183–186, Clarke 1979: 298–299, Dillon 1999b, Worster 2008: index.
408. Perry 1984: 114.
409. Leopold 1955: 799.
410. Laycock 1965, Murphy 1968, Greenwalt 1978, Riley & Riley 1981.
411. Pinchot 1947: 2–6, Fausold 1961, Steen 1976: index, Strong 1988: 61–84, Osborne 1997, Penick 1999, Egan 2009: index, Myers 2009, Lannoo 2018: index.
412. Pinchot 1893.
413. Steen 1976: 47–103.
414. Pinchot 1947: 144–145.
415. Egan 2009: 81–242.
416. Burnett 2003: 6.
417. Campbell 2011a, b.
418. Hart 2010.
419. Dillon 1999a, Black 2012.
420. Shankland 1951: 13–18, Brattain 1999.
421. Shankland 1951: 7.
422. Thorsen 1997.
423. Yard 1916.
424. Seaborn 1996, Sellars 1997.
425. Sterling 1997, Brands 2008, Harmond 1999a.
426. Brinkley 2016.
427. Strong 1888: 152–176, Watkins 1990, Sterling 1997a, Kuhne 2009.
428. Owen 1983: index, Black 2003: index, Brands 2008: 333–334, 389–393.
429. White & Maze 1985: 118, Watkins 1890: index.
430. Skillen 2009.
431. Hirt 1994.
432. Dexter 1978, Birchard 2005, Egerton 2015a: 54–55.
433. Grove 1992: 20.
434. Freedman 2013: 39.
435. Linton & Moore 1984, McNicholl 1997, Loo 2006: 63–92, Boros 2009.
436. Leopold 1955: 795.
437. Dingell & Potter 1978.
438. Loo 2006: 23.
439. Shor 1974b, Sterling 1977: index, 1997g, 1999b, Hermann 2009, Lannoo 2018: index.
440. Dunlap 1988: 35.
441. Merriam 1890.
442. Cain 1978.
443. Sterling 1997e, Gallagher 1981, Pegram 1999, Dehler 2009.
444. Hornaday 1915: 582.
445. Day 1959: 42–43, Bean 1978: 286–287.
446. Day 1959: 47, Cioc 2009.
447. Loo 2006: 25–26.
448. Lincoln 1933: 66, Stresemann 1951: 341–343, 1975: 336–338.
449. Rhodes 2004: 37.
450. Barrow 1998: 171, Egerton 2019a: 30.
451. Reeves 1984a, Henson 1997, Barrow 1998: index.
452. Riegert 1997, Burnett 2003: 9–14, Loo 2006: 11–12.
453. Elton 1942: 190–191.
454. Burnett 2003: 22–24.
455. Day 1959: 222.
456. Gottschalk 1978.
457. Cattell & Cattell 1938: 887, Reeves 1984b, Sterling 1997b, Smith 2005 et al., Lannoo 2018: 142–144, 221–222.
458. Reeves 1984c.
459. Stoddard 1969, Gromme 1973, B urges 1996: 102, Way 1915, Lannoo 2018: index.
460. Stoddard 1969: 283 + photo before 189.
461. Allen 1969: xviii.
462. Flader 1974, Meine 1988: 16, 1997, Lorbiecki 1996, Warren 2006, Lannoo 2018: index.
463. Aley 2009.
464. Letter of 25 May 1904, quoted in Meine 1988: 38.
465. Meine 1988: 80, Strong 1988: 134–151.
466. Lorbiecki 1996: 65.
467. Meine 1988: 458–459, Lorbiecki 1996: 167, Leopold 2013: 114–115.
468. Meine 1988: 226, Lorfbiecki 1886: 90–96, Aley 2009: 248.
469. Warren 2006a: 101–103.
470. Meine 1988: 604–620.
471. Dunlap 1988: 2, Meine 1988: 261–266.
472. Trefethen 1975: 214–216, Kadlec 1978, Schoenfeld 1978.
473. Egerton 2015e: 1–2.
474. Egerton 2015e: 54–55.
475. Murie 1962, Murie & Murie 1966, Burgess 1996: 79–80, Murie & Sterling 1997, Burnett 2009.
476. Glover 1989: 31.
477. Murie 1917.
478. Craighead & Kreps 2006, Brinkley 2011: 263–271.
479. Quoted in Glover 1989: 33.
480. Murie 1973.
481. Egerton 2019a.

6 Biogeography

A. 1700–1840

A general bibliographical introduction of almost three pages opened part 61A (2018) of my essay on terrestrial biogeography and paleobiogeography, 1700–1840, briefly supplemented in parts 61B and 61C.[1] Parts A, B, and C of this survey are condensed revisions of 61A, B, and C. Part D is a new discussion. Elsewhere, I have provided earlier bibliographies on natural history voyages and stations.[2]

Historian Steve Nicholls identified Spanish Jesuit José de Acosta (1539/40–1600) as the first to suggest, in *Historia natural Moral de las Indias* (1590), that terrestrial animals may have crossed a land bridge to reach America.[3] Dutch Protestant minister Abraham van der Mijle (Myl, 1563–1637) wrote *De Origine Animalium et Migration Populorum* (1667) to explain the Americas and Australia having unique species of plants and animals.[4] His answer: there were different centers of creation. His solution to this puzzle was not discredited for almost two centuries.

British Protestant minister John Ray (1627–1705) was the main founder of natural history in Britain.[5] He began his publishing with *Catalogus plantarum circa Cantabrigiam nascentium* (1660), one of the earliest local floras.[6] He first stressed the importance of distinguishing between similar species. His *Catalogus* provided data relevant for biogeography, without theoretical speculations. The same was true for his subsequent books on all known species of plants (3 vols., 1686–1704), birds (1676), fish (1686), and insects (1710). However, he also published two theoretical books: *The Wisdom of God Manifested in the Works of Creation* (1691) and *Miscellaneous Discourses Concerning the Dissolution and Changes of the World* (1692). Although he opposed the idea of spontaneous generation,[7] I am unaware of his discussing why different continents have unique species.

John Lawson (about 1659–1711) was the first Englishman to write extensively on American natural history.[8] He arrived in Carolina in 1700, and the Lord Proprietors hired him to survey the colony's interior, which he did in almost two months, assisted by five Englishmen and four Indians. After he returned to England, he published *A New Voyage to Carolina, Containing the Exact Description and Natural History of that Country* (1709). His survey enabled him to include a map of the colony. He wrote before Linnaeus provided scientific names for species, so some of his identifications now seem imprecise.[9] He took almost 25 pages to describe trees, shrubs, and vines.[10] He identified three kinds of honeysuckle—living in moist grounds, dry land, and swampy woods. He included seven illustrations of mammals, three of snakes, and a turtle shell. He discussed four kinds of cats—panther, cat-a-mount, wild cat, and tiger—presumably cougar, lynx, bobcat, jaguar. Other examples could also be listed. He discussed 27 "beasts," 129 birds, 42 saltwater fish (including dolphins and whales), 20 freshwater fish, 20 shellfish, and 22 "insects," which included reptiles, frogs, and worms but no real insects. He returned to Carolina in 1710 and was killed by Tuscarora Indians in 1711. He had made "the first major attempt at a natural history of the New World."[11]

French botanist Joseph Pitton de Tournefort (1656–1708) became Europe's leading botanist, focusing mainly on taxonomy and systematics.[12] He explored for plants in western and southern Europe, and in 1700–02 in the Middle East. Later, back in Paris he wrote *Relation d'un Voyage du Levant*, but died in a traffic accident before its publication in 1717 (English, 1718). In volume 3, letter 19, he described his visit to Mount Ararat (where Noah's Arc supposedly landed after the flood). He found plants on its side could be divided into elevation zones:[13]

> [A]t the foot of Mount Ararat those plants which were common in Armenia, a little further those which I had before seen in Italy; when I had ascended somewhat higher such vegetables as were common about Paris; the plants of Sweden possessed a more elevated region; but the highest tracts of the mountain, next the very summit, was occupied by the natives of the Swiss Alps and Lapland Alps.

Linnaeus claimed that this kind of vegetation-elevation correlation began with Andrea Cesalpino (1519–1603), but without citation. He had in mind Cesalpino's *De Plantis* (1583).

Englishman Mark Catesby (1682–1749) became interested in natural history under influence of a neighbor, John Ray.[14] Lacking university education, Catesby still became the greatest British natural history explorer of his time. A married sister lived in Virginia, and he went to visit her, 1712–19, and collect plants he could sell to English patrons.[15] He made drawings of plants and collected specimens but did not record his findings. At a meeting of the Royal Society in 1720, a newly appointed governor of South Carolina offered to pay him to make observations for the Royal Society. He did so, May 1722–January 1725, and then spent almost a year in Bahama Islands before returning home in 1726. He had been interested in the differences between nature in Virginia and South Carolina. In writing his *Natural History of Carolina, Florida, and the Bahamas*, his introduction discussed geography, climate, soil, rivers, and Indians of Carolinas, and his accounts of plants and animals were accompanied by 220 plates illustrating 109

DOI: 10.1201/9780429342202-6

birds, 13 amphibians and reptiles, 46 fishes, 9 quadrupeds, 31 insects, and 171 plants. It took him 20 years to produce his book, drawing upon Lawson, engraving and coloring his plates. He published it in parts of 20 plates with text, 1729–47. Modern specialists have identified and commented on his birds and plants.[16] He displayed his plates at the Royal Society and was then voted a member. He lacked illustrations for all the animals he wanted to discuss and borrowed from Sir Hans Sloane's illustrations by John White in the early 1600s, which Catesby plagiarized. Catesby became a skillful artist, and some of his plates are attractive, with plants and animals on the same plate. He conveyed action in his first plate, showing a bald eagle stealing a fish from an osprey in midair. He also wrote for Royal Society an article on bird migration.[17]

Other botanical explorers during the early 1700s also published accounts of their findings, which became data for phytogeography. Peter the Great authorized Danish explorer, Vitus Bering (1681–1741), in Russia's navy, to explore the north Pacific to see if Siberia and Alaska were connected.[18] After almost three years, he reached the Kamchatka Peninsula, where he built a ship and discovered no connection existed between Siberia and Alaska. Bering returned to St. Petersburg in 1730. His valuable collected information led Russia to fund a second expedition, 1731–42, broad in scope and involving hundreds of people, including two capable naturalists.

One, Johann Georg Gmelin (1700–55) was the son of an apothecary (pharmacist) who taught at the University of Tübingen.[19] He attended university lectures since age 14 and earned a medical degree in 1727. Two of his former professors had joined the Russian Academy of Sciences in St. Petersburg in 1725, and in 1727 they invited him to join them. He did, and in 1731 he became an academician. In 1733 the Academy sponsored an expedition to Siberia headed by Vitus Berling, which Gmelin joined—for 9.5 years![20] He returned to St. Petersburg in 1743 and worked on his *Flora Sibirica* (4 vols., 1747–69), which described 1,178 species.[21] He listed species found in six geographical regions of Siberia. In 1747, he suggested there were different centers of species creation—80 years after van der Mijle had done so.

German zoologist Georg Steller (1709–46) studied medicine, became a physician in Russia's army in 1734, and traveled to St. Petersburg.[22] At the Academy of Sciences he learned of Bering's Second Kamchatka Expedition and volunteered to join it, already departed. He was accepted and began his nine-year expedition by crossing Siberia to the Pacific, where he met and sailed with Being. His travels provided many and varied natural history and geographical observations.[23] When they landed briefly on an island near Alaska, he discovered a jay, now Steller's Jay, which was soon supplemented by the discovery of two more bird and two mammal species. The return voyage was catastrophic; their ship began sinking, and they landed on an island, where Bering died, now Bering Island. Survivors built a smaller vessel from remains of their ship, with which they reached Kamchatka. They had survived by eating several Steller's sea cows, and his account was the only one recorded before that species was exterminated by later Russian fur trappers.[24] In 1744, Steller left for St. Petersburg and died on the way. However, his books—*De Bestiis Marinis, Journal of a Voyage with Bering*, and *History of Kamchatka*—survived and were published.

In western Europe, one leader in natural history was Swedish Carl Linnaeus (1707–78), who studied medicine but was more interested in botany.[25] He founded modern nomenclature and systematics, and first conceptualized ecology, under the name *economy of nature* in 1749.[26] In 1732, Uppsala Society of Science provided funds for his natural history exploration of Lapland, which was "the most productive exploratory expedition ever undertaken in Sweden."[27] He later undertook three similar expeditions to Öland and Gotland 1741, Västergötland 1746, and Skåne 1749.[28] He wrote phytogeographical observations on these trips, and Swedish ecologist Einar du Rietz listed six distinct discoveries he published:[29]

1. Linnaeus (1937) gave an excellent summary of the vegetation belt of Scandinavian mountains. Under the name "latera alpium" he described subalpine birch-forest belt, conifer belt "desertum lapponieum" and alpine belt "montes" or "alpes propie" . . . with its dwarf shrub, heath and its "convalles" with willow shrub, of middle alpine belt, "alpes altiores" with light and scanty vegetation . . .
2. Important vegetation differences between ordinary wet mountains of Lapland and dry lichen heath mountains of Dalarne described (1737).
3. Notes on habitats of alpine plants (1737) were often remarkably accurate and detailed.
4. Vivid picture of contrast between different florae of Sweden.
5. Emphasized differences between different alpine florae (1745a, b, 1756). He believed these differences were due to incomplete dispersal.
6. Frequent occurrences of alpine plants along rivers in the conifer forest belt were discovered (1737, 1744); he advanced this as plant dispersal by running water.

Linnaeus also introduced terms *station* and *habitation*, indicating a species' environment and range, in *Stationes Plantarum* (1754), but without complete clarity.

A French contemporary of Linnaeus, Comte de Buffon (1707–88), head of the *Jardin de Roi* in Paris, developed ideas challenging those of Linnaeus.[30] One of Buffon's original ideas was that as species spread from Europe to the rest of the world, new environments modified them into different species. He also speculated that humans may have exterminated mammoths and mastodons, known from fossil bones. He believed that no species found on one continent also existed on any other continent.[31] This idea may have been lifted from Aristotle's *Historia Animalium*.[32]

Linnaeus published in a wide variety of publications, but Buffon published primarily in his elegant *Historia naturelle, générale et particulière* (15 vols., 1749–67 + 7 supplements, 1774–89) and *Histoire des oiseaux* (9 vols., 1770–83). Both Linnaeus and Buffon took steps toward a dynamical science of biogeography (unnamed).

One of Linnaeus' former Swedish students, Pehr Kalm (1716–79), received funds from the Royal Swedish Academy of Sciences in 1748 to collect useful North American plants that could survive in Sweden.[33] His sponsors expected him to search mainly in Canada, but he preferred Philadelphia and vicinity, where most American naturalists lived. This trip was the great adventure of his life, and after returning to become a professor at the University of Åbo in Finland, he devoted much time to writing his *En Resa til Norra America* (3 vols., 1753–61, translated into German, English, Dutch and Italian). He added 60 plants to Linnaeus' *Species Plantarum* (1753), and mountain laurel was named *Kalmia*. He observed American animals as well as plants.

In 1761, Denmark sent an exploring expedition to Arabia. Five professional men included Finish native Petrus Forsskål (1732–63), a Linnean botany student.[34] The expedition first spent a year in Egypt, where he compiled an Egyptian flora. He died from malaria on his way to Arabia. "His death was one of the greatest disasters to research in Swedish natural science."[35] He was "one of the most highly talented men ever produced by Finland."[36] His flora was later published in Copenhagen.

Swedish botanist Carl Thunberg (1743–1828) was a disciple of Linnaeus who earned a medical degree in 1770 and went to Holland, where he became a plant collector, to South Africa in 1772, and in 1775 to Japan, where he pretended he was Dutch.[37] In 1776 he traveled to Dutch colonies, then to London, and finally back to Uppsala in 1779, where he taught botany. His major contributions were *Flora Japanica* (1784), *Travels in Europe, Africa and Asia* (London, 4 vols, 1793–95, Swedish, 1788–93), *Prodromus plantarum capensium* (Uppsala, 2 vols., 1794–1800), and *Flora capensis* (Uppsala, 2 vols., 1807–23). He found and described more new plant species than any other botanist during his lifetime.

German professor of mathematics and physics in Brunswick Eberhard Zimmermann (1743–1815) explored animal geography in two works that discounted some of Buffon and Linnaeus' discussions as too speculative, and he provided extensive data as a basis for his science.[38] Bodenheimer judged his *Specimen Zoologiae Geographiae Quadrupedum* (1777, 685 pages + map) to be "a remarkable zoogeographical publication,"[39] and he summarized its features, including a world map, with major regions indicated in color and mammalian names added where they occur.[40] Larson discussed in more detail Zimmermann's expanded German version, *Geographische Geschichte des Menschen und der vierfüssigen Thiere* (3 vols., 1778–83), which drew upon additional sources and updated his map.[41] Ebach, citing Feuerstein-Herz's doctoral dissertation (2004), saw Zimmermann's two works as innovational and influential despite using only published literature, without any field work. Zimmermann's innovation was to synthesize knowledge from natural history and exploration accounts. He still believed, like Mijle, in multiple centers of creation (excepting humans). Nature as a whole would have been created in each center.[42]

Berliner Karl Willdenow (1765–1812) learned plants from his apothecary father and from botanist Johann Gleditsch.[43] Willdenow published *Florae Berloinensis* (1787) before earning an M.D. degree at the University of Halle. He took over his father's shop and taught botany informally. His *Grundriss der Kräuterkunde* (1792, ed. 4, 1805, English 1805) became the standard textbook and contained a substantial account of plant geography—the first by a German—which explained how studies of physical conditions, stations, and habitations provided a framework for evaluating limits of phytogeographic regions.[44] He distinguished five European floras. His textbook was so successful he was admitted to the Berlin Academy of Sciences in 1794, and in 1798 he became a professor of natural history at Berlin Medical-Surgical College. In 1810, he became a professor of botany, University of Berlin.

During the 1800s, phytogeography progressed faster than zoogeography,[45] beginning with Alexander von Humboldt (1769–1859) and his collaborator Aimé Bonpland (1773–1859). Wealthy, upper-class Humboldt grew up on his family's estate near Berlin, with a mother of French heritage.[46] He attended two universities and a mining academy, 1787–92, and received a broad science education. At Göttingen University, he became friends with Georg Forster, who undoubtedly told him much about his voyage with Captain Cook. In 1798, Humboldt went to Paris to join an expedition around the world, which never occurred. He became friends with French botanist Bonpland, who had also hoped to join that expedition.[47] They traveled to Spain and received permission to explore Spain's American colonies.

They spent over five years (1799–1804) exploring Spanish America.[48] Humboldt's numerous publications afterwards began with his and Bonpland's *Essai sur la Géographie des Plantes* (1807). Humboldt became the leading natural scientist of his time, and his influence on other naturalists was immense. Beyond collecting specimens, Humboldt's goal was to "never lose sight of the harmony of concurrent forces, the influence of the inanimate world on the animal and vegetable kingdom."[49] He wanted to study an unnamed science, ecology. Their expedition explored northern South America, Cuba, and Mexico, collecting specimens and observations.[50] Humboldt's later publications constituted the greatest single contribution to phytogeography.[51] They included: his and Bonpland's *Essai sur la géographie des plantes* (French & German 1807 [not 1805], Spanish 1809, English 2009), *Personal Narrative of Travels to the Equinoctial Regions of the New Continent* (translated title, 1818–29), "On the laws observed in the Distribution of Vegetables" (1816), "Sur les lignes isothermes et de la distribution de la chaleur sur le globe" (1817, English 1820–21, reprinted in Egerton 1977c), De distribution geographica plantarum

(1817), and *Schriften zur Geographie der Pflanzen* (1989). He saw similarities between coastal floras of Africa and South America, and suggested an ancient connection. He organized plant geography by contrasting forms of vegetation in different regions, rather than contrasting regions by identifying differences in ranges of species. Ebach judged this his main innovation.[52] His approach, however, would not work when studying regions for the first time.

Humboldt wanted to explore Asia, to compare his Asian and South American findings. Tsar Nicholas saw a benefit in allowing him to explore Siberia and evaluate its mineral wealth.[53] Humboldt left Berlin 12 April 1829, was gone almost six months, and traveled 9700 miles.[54] However, he mainly was able to compare Altai Mountains and Andes, and Siberian steppes and Venezuelan llanos. He published two works on his Asian expedition: *Fragmens de géologie et de climatologie asiatiques* (2 vols., Paris,1831) and *Asie centrale: Recherches sur les chaines de montagnes et la climatologie comprée* (3 vols., Paris, 1843). He had given 61 public lectures in Berlin in 1827–28 (free and open to all), and their publication became his final grand work, *Kosmos* (5 vols., 1845–62, English, 5 vols., 1849–58).[55] His discussion of plant and animal geography was in volume 1 and was a synthesis of knowledge then available.

German naturalist Pyotr Pallas (1741–1811) studied the geography of both plants and animals. He was the son of a professor at Berlin's medical college.[56] He earned his own medical degree in Leiden, with a dissertation revising Linnaeus' classification of worms.[57] He then studied collections of marine animals in Holland and England. He became friends with English naturalist Thomas Pennant, and his letters to Pennant are published.[58] In 1767, Pallas accepted Catherine the Great's offer of a position at St. Petersburg Academy and remained there over 40 years. In 1768–74, he joined an Academy expedition to Lake Baikal, Ural Mountains, and eastern regions. He published an account of his travels in 1771–86 (3 vols. in German, translated into Russian, French, English, and Italian). These volumes were important for biogeography.[59] Drawing upon Gmelin's *Flora Siberica* and his own studies, Pallas wrote a *Flora Rossica* (2 vols., 1784–1815). After another expedition, 1797–1801, he settled in Crimea and wrote his travel accounts (2 vols., 1797–1801) and *Zoographia Russo-Asiatica* (3 vols., 1811–31).

A medical student with a strong interest in botany, Friedrich Stromeyer (1776–1835), studied under botanist Gmelin at Göttingen University, writing a dissertation on what he called historical plant geography (1800). He wanted to do for plants what Zimmerman had done for animals. Malte Ebach translated Stromeyer's introduction at the end of his *Origins of Biogeography*.[60] The editor of *Journal für die Botanik* wrote a glowing nine-page review of Stromeyer's dissertation, but it still faded from further consideration.[61] Stromeyer became an academic chemist, who watched Humboldt become a famous phytogeographer. In 1820, an anonymous 50-page essay attacked Humboldt's works and praised Stromeyer's work.[62] No one had difficulty guessing who wrote it.

Swedish botanist Göran Wahlenberg (1780–1851) studied medicine and botany at Uppsala University and spent his career there.[63] His botanical explorations began in 1799 before Humboldt influenced him. In 1811 he began publishing floras of central European regions, with plant geography discussed in their introductions. A difficult personality limited his influence.

Genevan Augustin-Pyramus de Candolle (1778–1841) decided early in life to become a botanist, but followed his father's wishes and also studied medicine at the Geneva Academy.[64] Simultaneously, he became friends with nearby botanists, who added to his education. In 1798 he went to Paris to continue his education, but also began his publishing career, which led to his becoming one of Europe's leading botanists. Although Humboldt was an influence, de Candolle's main ecological concepts came from Linnaeus. De Candolle's most important phytogeographicalecological publication was an encyclopedia article (1820, reproduced in Egerton 1977c).[65] He recognized the importance of competition in determining the extent to which a species increases.[66] Not only Linnaeus but perhaps Robert Malthus also influenced de Candolle's statement that plants of a given country, of a given place "are at war one with another" since he had met Malthus in England and he knew the translator into French of *An Essay on the Principle of Population* (1798).[67]

Danish lawyer-botanist Joachim Schouw (1789–1852) was a Humboldt disciple.[68] In 1812, he accompanied Norwegian botanist Christian Smith into Norway's mountains, where Schouw was impressed by vegetation zones up the sides of mountains. He wanted to know the causes, and in 1816 he earned a doctorate in botany at Copenhagen University, where he remained. In 1822 he began naming plant associations by adding *etum* to end of genus names of dominant species.[69] He explored Italy's vegetation and climate on two expeditions, 1817–19 and 1829–30. After his first expedition, he published principles of phytogeography in Danish: *Grundtraek till en aimindelig Pflantegeographie* (1822, German 1823, French 1824). In 1829 he published *Tableau du Climat et de la Végetation de l'Italie*, vol. 1. His work on a second volume appeared much later as a series of papers on Italian plant geography.[70]

British botanist William Hooker (1785–1865) was director of Royal Botanic Gardens, Kew (1841–65).[71] He wrote two articles on plant geography (1835); one closely following de Candolle's 1820 article, as Hooker acknowledged. Hooker distinguished range from habitat: "Botanical geography is constituted by considering plants in relation to their *habitation*, region, of the country in which they grow, and in regard to their locality or particular *station*."[72] Hooker's other article was on the distribution of plants within England and based upon firsthand knowledge[73] (both reprinted in Egerton 1977c). He had a son who became a leading explorer.

FIGURE 6.1 (a) Johann C. Fabricius. (b) Pierre-André Latreille. (c) Jean T. Lacordaire. Web.

Danish entomologist Johan Fabricius (1745–1808) studied under Linnaeus (1762–64), and traveled widely in Europe.[74] He served as a professor of natural history at the Universities of Copenhagen and Kiel. His best work was *Philosophhia Entomologiae* (1778), in which he proposed eight climatic zones for insects.[75] As more data accumulated, Fabricius' zones were refined by Frenchmen Pierre-André Latreille (1762–1833) in *Introduction à la Géographie générale des Arachnids et des Insects, ou des climats propres à ces animaux* (1815)[76] and Théodore Lacordaire (1801–70), who helped found Entomologique de France (1830).[77] He had part of his *Introduction à Entomologie* (3 vols., 1834–38) published in English as "On the Geographical Distribution of Insects" (1838–40). An innovation by Lacordaire was use of statistics to quantify characteristics of species, such as the number of species found in different regions of the Earth.[78]

German zoologist Gottfried Treviranus (1776–1837), from Bremen, studied medicine at the University of Göttingen and returned to Bremen to practice medicine and teach at Bremen Lyceum.[79] His younger brother Ludolph Treviranus (1779–1864) also studied medicine and later became a botanist.[80] Gottfried Treviranus was a very productive author, whose greatest work was *Biologie, oder Philosophie deer lebenden Natur für Naturfirscher und Aerzte* (6 vols., 1802–22), first use of term *biology*. Volume 2 (1803) discussed organization of life, distribution of plants and animals, and influence of external conditions.[81]

There is a vast literature on history of natural history explorations in America.[82] American Thomas Jefferson (1743–1826, president 1801–09) had a strong interest in science.[83] He took the opportunity of his Louisiana Purchase from France (1803) to organize an exploring expedition, headed by his secretary, Meriweather Lewis (1774–1809), who invited U.S. Army veteran William Clark (1770–1838) to be co-leader of what is called the Lewis and Clark Expedition (1804–06), which traveled up the Missouri River and on to the Pacific Ocean and back.[84] America's first expedition became its best remembered one.[85] Two bird specimens brought back to Philadelphia were named and illustrated by ornithologist Alexander Wilson: Lewis' woodpecker *Melanerpes lewis* and Clark's crow (now nut-cracker) *Nucifraga columbiana*. Since that expedition was an early example of American (Jefferson's) government sponsorship of overland success—excepting a plan for maintaining collections afterwards—led to a series of similar land explorations.[86] Two other government expeditions occurred while Lewis and Clark were out west, led by U.S. Army officer Zebulon Pike (1779–1813) and by federal surveyor-astronomer Thomas Freeman (d. 1821). Pike was sent to explore the headwaters of the Mississippi by General James Wilkinson (who became a Spanish spy). Pike led his expedition from St. Louis on 9 August 1805. He and his men were interested in deer, elk, and buffalo, as they depended on them for food. They reached Leech Lake, one source of the Mississippi, on 1 February 1806 and returned to St. Louis on April 30. Pike's journal and maps went to Jefferson, who had them published.[87]

Soon, Wilkinson sent Pike on another expedition, across the prairie to the Rocky Mountains. That expedition left St. Louis on July 15, sailed up the Missouri River to Osage River, then followed it west. He was interested in "prairie-squirrels" (prairie-dogs) and warned of going into their "towns" because rattlesnakes lived in their holes.[88] In November, Pike saw, across the plains, 100 miles away, a mountain later named "Pike's Peak."[89] They decided to climb it and reached its base on November 25. They expected to climb it and return the same day, but they had to spend the night in a cave, and they reached the top the next day, only to discover they had climbed an adjacent, lower peak. Pike decided no one could climb the higher peak, which was covered with snow.[90]

Jefferson sent Thomas Freeman to explore the Louisiana Territory's southern boundary along Red River. His expedition, including naturalist Peter Custis (1781–1842), reached the Red River from the Mississippi on 3 May 1806.[91] They ascended Red River about 615 miles, according to Freeman's estimate, about half its length, before a Spanish army forced them to return downriver. Both Freeman and Custis kept daily journals. Custis could not collect as many specimens as he wished, but he identified 22 mammal species, 36 birds, 17 reptiles, amphibians and fish, 4 insects, 58 trees, and 130 herbaceous plants. He discovered 22 species of plants and animals new to science, though he inadequately published their names and descriptions.[92] Freeman explained prairies they saw as due to Indians burning grass for hunting.[93]

Scotch-American Alexander Wilson (1766–1813) was an independent explorer, going from Philadelphia north into New England (1808) and also west to Pittsburg, then down the Mississippi to New Orleans (1810), continuing eastward to the Atlantic Ocean, and back to Philadelphia, studying birds and selling first two volumes of his *American Ornithology* (9 vols., 1808–14).[94] In his accounts, he stated where he had collected a species and what habits he had observed. He provided 320 bird portraits of 264 species, on 76 plates. His achievement was so impressive that later naturalists named Wilson "father of American ornithology."[95] There were eight birds named for him (six names still valid), and the Wilson Ornithological Society was founded in 1888.

Englishman Thomas Nuttall (1786–1859) was one of the most widely traveled naturalists in North America during the 1800s.[96] His main interest was in plants,[97] leading to his *The Genera of North American Plants, and a Catalog of the Species to the Year 1817* (2 vols, 1818). He gave only a brief characterization of familiar genera, and merely listed its already-known species, listing 15 species under the familiar genus *Viburnum*.[98] For unfamiliar species, he commented on environment; for *Iris verna*: "Habitat. On the gravelly shores of the calcareous islands of Lake Huron near Michilmakinak."[99] (This was an early use of the term *habitat* as we use the term, instead of *habitation* or *station*.)

Nuttall wrote two travel journals. The first written on a western trip that began 12 April 1810 with a stagecoach ride to Pittsburg, then on foot to Lake Erie, by boat to Detroit, and then on through Lakes Huron and Michigan to Green Bay, up the Fox and down the Wisconsin River to the Mississippi, and down it to St. Louis, which he reached in mid-September. Aside from telling his travel experiences, much of his journal describes new plants he found. When he found an interesting insect, he described its anatomy with technical terms comparable to terms he used to describe plants.[100] He also commented on birds, mammals, reptiles, and geology observed. When he reached Sandusky Bay in Lake Erie, in southwest Ohio, he first saw a prairie, and he was quite impressed when told it stretched for 150 miles: "like an ocean of land to an immeasurable distance up the lake's N.W.!"[101] He described its characteristic plants and birds. On 13 March 1811, he began traveling up the Missouri River with fur traders to the Yellowstone River, and later down the Missouri and Mississippi Rivers to New Orleans, reached in December 1811.[102]

With the War of 1812 brewing, Nuttall sailed back to England. He returned to America in May 1815 and resumed his botanical studies. Some of his observations appeared in *Genera of North American Plants* (2 vols., 1818). On 17 October 1817, he was elected a member of American Philosophical Society, and in 1818, four other members donated $50 each toward his expenses up the Arkansas River lasting 18 months. He left Philadelphia on 2 October 1819 and reached Pittsburg on the fifteenth. He then descended the Ohio and Mississippi Rivers to the Arkansas River. He recounted his discoveries in *A Journey of Travels into the Arkansa Territory during the year 1819* (1821), with a map of the river based on his data, five landscape drawings, and a.m. and p.m. temperatures from 20 January to 28 September. Much of his book is travel narration, geological observations, and discussions of Indians, but some observations of ecological interest are included.[103] For example, he observed:[104] "Nearly all the trees throughout this country possessing a smooth bark, are loaded with mistletoe (*Viscum verticllatum*)." In February 1820, he sailed from New Orleans with 300 new plant specimens, which he described in a series of papers.[105] He gave animal specimens to zoologists to describe.[106] In 1823 he became Curator of Botanic Garden and lecturer in botany at Harvard College.

A Philadelphia friend of Nuttall was Thomas Say (1787–1834), a descendent of John Bartram and son of a prominent physician.[107] Thomas and younger brother Benjamin were childhood friends with Titian Ramsay Peale (1799–1885), who interested Say in insects.[108] Say attended the University of Pennsylvania Medical School. On 25 January 1825, six science-oriented friends founded the Academy of Natural Science of Philadelphia. Say was absent but sent a letter of support, and he joined at the next meeting and became its most active member. Say aspired to do for American insects what Wilson had done for American birds.[109] He succeeded: *American Entomology* (3 vols., 1824–28, 54 plates), and so deserves to be acknowledged as "father of American entomology."[110] He also published a similar monograph on American mollusks: *American Conchology* (6 vols., 1830–34).

In June, wealthy Scottish businessman-geologist William Maclure (1763–1840)[111] was elected an Academy member and helped Say become a leading American naturalist.[112] The Academy also helped Say gain prominence. Say began in summer 1813 reading essays on entomology at meetings, and continued doing so in winter 1813–14. He also contributed insects, shells, and other specimens to the Academy's scientific collections, then he arranged and studied them. Say's travels rivaled those of Nuttall, though Say never went further west than the Rocky Mountains. However, he also traveled into Canada and Mexico.[113] Say published volume 1 of his *American Entomology* in 1817, with illustrations, but he could not continue at that time.

One diversion was the Academy's *Journal*, which began publication in May 1817, with Say playing a major part, as author and editor's assistant. He published 5 papers in 12 parts, describing 94 species of crustaceans and 7 new genera (these papers were reprinted in 1969).[114] Another diversion was a trip to Georgian and Floridian islands, which Maclure organized in late 1817 and invited Say and Peale to join.[115] Peale was already drawing insects for Say and on this trip he was Say's assistant. They collected specimens of insects new to science, which Say described in the Academy's *Journal*. Since Say and Peale worked well together on this trip, they were receptive to joining a more ambitious expedition.

In 1819, the U.S. Army sponsored an ambitious expedition led by Major Stephen H. Long (1784–1864), a graduate of Dartmouth College, who had taught at West Point and led two previous expeditions.[116] This was the Army's first expedition to include professional naturalists, and both Long and Secretary of War John C. Calhoun wanted outstanding ones. Long chose Say as zoologist,[117] Peale as assistant naturalist and artist,[118] and William Baldwin (1779–1819), a capable botanist, who left the expedition due to declining health, and died 31 August of tuberculosis.[119] The expedition spent the winter 5 miles south of Council Bluffs, on the Missouri. Long spent the winter in Washington, and when he rejoined the expedition in June, he brought along Edwin James (1797–1861), as botanist, geologist, and surgeon.[120] James had graduated from Middlebury College, Vermont, in 1816. He then studied medicine under his brother and botany and geology under John Torrey and Amos Eaton. All naturalists and Long kept journals, and after the expedition James compiled from the journals *Account of an Expedition from Pittsburg to the Rocky Mountains, Performed in the Years 1819, 1820* (3 vols., 1823), with maps and landscape illustrations by expedition artist Samuel Seymour, and two engravings "after Peale."

James' book was a worthy contribution for the time, but not for the 1900s. He had not identified authors of passages quoted. Reuben Thwaites annotated a version of James' *Account* (1904) that identified authors of passages, from their journals, with map and illustrations. Maxine Benson (1988) published a condensed edition of James' text with valuable introduction, bibliography, and both black-and-white and color illustrations, though only her name is on the title page.

Another solo immigrant naturalist, like Wilson, was Constantine Rafinesque (1783–1840), an immigrant from Constantinople (Istanbul) who did not follow, as Wilson had, natural history traditions.[121] His father was a French merchant, and his mother was German. He was raised in Greece and tutored in geography, geometry, history, and English.[122] He first reached America on 18 April 02. In 1803–04 he took walking field trips and discovered that searching in the wilderness for plants and other things was his favorite activity.[123] On 1 January 1805 he sailed to Italy and spent a decade on Sicily before returning permanently to America. He claimed to discover and name 2700 plant genera, 320 subgenera, 6700 species, and 900 varieties. In 1950, botanists accepted only 18 of his genera, 84 species, and 13 varieties.[124] Rafinesque lacked credibility because he was hypercritical of published evidence by others while being careless about evidence in his own works.[125] In 1818, he sailed down the Ohio River, collecting fish which became the basis of his *Ichthyologia Ohioensis* (1820). In late summer he stopped at Henderson, Kentucky and spent three weeks with immigrant John James Audubon, whom he trusted. However, Audubon was a trickster, and he showed him drawings of 11 fake fish which Rafinesque mistook as real and cited Audubon as his authority.[126] Apparently Audubon never realized that this joke reflected badly on his own reputation as well as on Rafinesque's. Temporary immigrant Charles Bonaparte appreciated Rafinesque:[127]

> The writings of Rafinesque had been most valuable to him and he regretted the recent loss in America of that distinguished naturalist, who had contributed so much to the science of ichthyology with his accurate list of the fishes of Sicily.

Ichthyologist David Starr Jordon provided a later evaluation of his contributions.[128] Despite the Rafinesque controversy, Charles Boewe recently edited 13 essays on him (2003) and Leonard Warren published his biography (2004).

French-American ornithologist Jean Jacques (changed to John James) Audubon (1785–1851) was son of a ship captain in what is now Haiti, but at age 6 was sent with a sister to be raised in France by his father's wife.[129] At 18, his father sent him to a farm he owned in Pennsylvania, where John could practice drawing birds. In 1808, he married a neighbor's daughter, Lucy Bakewell; they moved to Louisville, Kentucky, where Audubon became a partner in running a store. He had had only a brief formal education, and no education in art. In 1810, Alexander Wilson entered the store in hopes of selling a subscription to his *American Ornithology*, showing the first volumes. Audubon could not afford to buy it, but he decided he could outdo Wilson as an artist. In 1820, the store became bankrupt and Audubon decided to collect his own bird portraits for publication. He kept a journal of daily events and bird notes. In 1824 he sought, unsuccessfully, a publisher in Philadelphia and New York, and in 1826 he sailed for London. He obtained a publisher for his elaborate *The Birds of America*, with 435 large illustrations in four volumes, 1827–38, and a separate *Ornithological Biography* (5 vols., 1831–39). He was unable to draw all bird portraits he wanted, and he copied a few of Wilson's to fill in his gaps, without saying so.[130] Despite that and academic complaints on other aspects of his illustrations, Audubon is now the most famous naturalist in America, based upon his life and bird portraits. There are two fine anthologies of his writings.[131]

Another French naturalist who came to America (1823) was Napoleon's nephew, Charles-Lucien Bonaparte (1803–57), with his wife of a year, to live with her father, near Philadelphia, 1822–28.[132] Although born and died in Paris, he lived more of his life in Italy than in France—beyond dominance by his uncle. His interests in nature were broad, but he decided in America, he wanted to continue Wilson's studies on American birds. Wilson had discovered and named some species named earlier by Europeans.[133] Bonaparte wrote four volumes on birds not described by Wilson (1825–33), yet these two naturalists' contributions were quite different.

Wilson collected field observations and birds and drew his own illustrations; Bonaparte studied museum specimens, especially at Academy of Natural Sciences and hired natural history artists—including Peale—to illustrate his volumes. Wilson had never been to Florida, but instead of leading an expedition there, Bonaparte hired Peale to go alone![134] Bonaparte and Say became close friends until Say moved to New Harmony, Indiana in 1825.[135] Bonaparte and Audubon were at times friends, but were also competitors. When Bonaparte's *American Ornithology* volumes began appearing in 1825, Audubon saw he had been advised by Say and not by himself.[136] Later, when they met in London, they revived their friendship.[137] Bonaparte had not remained in America until he completed his volumes, but returned to Italy about five years before completion. He named 20 species new to science.

Bonaparte made friends with British naturalists, and after Audubon published volume 1 of his *Ornithological Biography* in 1831, English ornithologist William Swainson wrote Bonaparte on 1 September 31:[138]

> the errors in nomenclature, as you may suppose, are numerous and two thirds of his *new* species are known even to *me* as *old* ones. So far as technical science is concerned, it is, in short, a complete failure.

Maybe Swainson's letter influenced Bonaparte, but he liked to pick Audubon's brain for useful information on birds and was not interested in abandoning his on-and-off relationship with Audubon. Audubon discovered that Bonaparte's account of the turkey was mostly what Audubon had told him, but was not acknowledged.[139] However, although Bonaparte used Audubon's bird portraits as his American reference in *A Geographical and Comparative List of the Birds of Europe and North America* (1838), he used John Gould's portraits as reference for Europe, and he praised Gould's as more beautiful.[140] Audubon responded by not acknowledging Bonaparte's help when he published his *Synopsis of the Birds of North America* (1839).[141]

An ever-busy Bonaparte began publishing a work on Italian vertebrates in 1832, before his American volume 4 appeared in 1833, but he published it in 30 parts, each appearing when completed, 1832–41, to be bound into three volumes.[142] This was perhaps his most useful work. He also became absorbed by systematics of Europe's fishes.

B. 1840–1940S

Historian of vertebrate zoology Keir Sterling compiled *Selections from the Literature of Biogeography* (1974), which has 28 papers in English, almost all published in the United States, 1830–1948. Ecologists Mark Lomolino, D. Sax, and J.H. Brown compiled *Foundations of Biogeography: Classic Papers with Commentaries* (2004, xx + 1291 pp.), with 72 papers, 1761–1782; some selections are from books rather than from periodicals. A few of these selections are considered here.

After completing his bird volumes, Audubon partnered with Charleston minister John Bachman (1790–1874) to produce *Viviparous Quadrupeds of North America* (3 vols., 1845–48 of plates and 3 vols. text, 1846–54).[143] Audubon's last expedition was up the Missouri River to study and collect mammals.[144] Audubon did not live to complete this work, but he had trained his two sons—John Woodhouse Audubon and Victor Gifford Audubon—to be naturalistartists, and they completed it, with Bachman. They had married Bachman's two daughters, so it became a family project.

German geographer Heinrich Berghaus (1797–1884) went to Paris in 1815 to meet Humboldt; they became regular correspondents, and their correspondence of 1825–58 survived and was published (3 vols., 1863). Berghaus was not a biogeographer but figures in our story. Dutch ornithologist Coenraad Temmick (1778–1858) established the Rijksmuseum in Leiden in 1823. Dutch government began sending out naturalists to its Dutch East Indies in 1820, and of seven sent during the 1820s, only one survived to return home, Salomon Müller (1804–64), who took specimens and notes to Temmick's Rijksmuseum, where he wrote an 82-page paper, "Geographie der Thiere: Ueber die geographiche Verbreitung der Säugethiere im Indischen Archipelagus" (1842), which contributed to both terrestrial and aquatic biogeography.[145] Müller's information was later used by geographer Berghaus for a map showing what is now called "Wallace's Line" (see later) which more fairly could be called the "Müller-Wallace Line."[146]

School teacher Jules Thurmann (1804–55) was from a town in Alsace, France, but his widowed mother raised him in Porrentruy, Switzerland, which remained his home town. He attended Porrentruy College, then earned his B.A. degree from the University of Strasbourg. He gained scientific recognition from his publications on botany and geology. In 1847, he published *Système de géographie botanique*, and in 1849, he distinguished between flora and vegetation (Nicolson's translation): "A land's flora and its vegetation are two quite different things . . . the first means the numbers of the distinct plant which one observes, the second their proportions and associations."[147]

Two Germans, both named Wagner, deserve our notice. Nurnberg native (Johann) Andreas Wagner (1797–1861) earned a doctorate in 1826, and in 1832 he joined the University of Munich faculty and also served as curator of the university's Zoologische Staatsammlung.[148] He was promoted to professor in 1836. He published *Die Grographische Verbreitung der Säugethiere Dargestellt* (3 parts, 1844–46), seeking to do for mammals what Schouw had done for plants. Charles Kendeigh outlined Wagner's book and explained he tried to correlate mammal distribution with types of vegetation.[149]

Bayreuth native F.J.C. Moritz Wagner (1813–87) studied at the University of Augsburg and elsewhere. He became an explorer: in Algeria (1836–39), Armenia (1842–46), and North and Central America.[150] In 1862, he became a professor of geography and ethnography, at the University of

Munich. He studied continents, not islands. Like Wallace in Amazonia, he found that some species' range was limited by rivers, but also by mountain ranges and deserts. He concluded that species only form when a species that formerly had a wide range became divided geographically into different populations that no longer interbreed. Darwin decided this claim was too dogmatic and there are exceptions to this generalization.

Austrian zoologist Ludwig Karl Schmarda (1819–1908) studied at the University of Vienna, then held several academic positions before returning to his alma mater as a professor of zoology.[151] He was a productive researcher on invertebrates and published seven books, the third on geographical distribution of animals (1853). Karl Schmidt found this book discussed the same topics one would expect in a modern book on zoogeography. Schmarda defined 21 terrestrial regions on Earth and 10 marine regions. Both Schmidt and Ebach reproduced his world map and listed names of his 31 regions.[152] He traveled around the world in 1853–57, and published a book on his discoveries (1861). In 1862, he returned to his alma mater as professor. The *Geographisches Jahrbuch* for 1866 had articles by Schmarda on animal geography and by Grisebach on plant geography.[153]

Two animal geographers at that time in London who studied vertebrates were English Philip Sclater (1829–1913), who studied birds, and German Albert Günther (1830–1914), who studied reptiles. Sclater attended Oxford University and practiced law.[154] His main interests were ornithology and zoological organizations in London. In 1856 he traveled around the United States. His lasting influence came from his paper, "On the General Geographical Distribution of the Members of the Class Aves" (1858). His scheme for six world regions of bird distribution is reprinted by James Brown and Mark Lomolimo.[155] Sclater believed all species were created in place. Still, "In this paper Sclater delineated and named the zoological regions that we know today."[156] Günther earned an M.D. at the University of Tubingen (1858), then visited his mother near London.[157] He prepared a catalog of the British Museum's reptiles and published "On the Geographical Distribution of Reptiles" (1858). At the time, their geographical forays led no further. Sclater's map of localized marine mammal distributions (1897) is also reproduced by Brown and Lomolimo.[158] The species listed have not only localized, but extensive, distributions. This map identified six world regions as well!

In 1875, Sclater extended the scope of his survey to all vertebrates; he mentioned leading references for each order and indicated gaps where research was needed. Later, he collaborated with his son on *The Geography of Mammals* (1899). His regions had the virtues of simplicity and plausibility,[159] but their book received an unfavorable review (by R.L.) in *Nature*.[160]

Joseph Hooker (1817–1911) joined an Antarctica expedition, 1839–43, which is discussed later, under aquatic biogeography. One expedition did not cure his itch to travel, however. His next expedition was to India, December 1847–February 1851, where he explored vegetation of Himalayan Mountains.[161] That choice came after discussions with Darwin over relationships between elevation and vegetation.[162] His reports to his father while there appeared as *Rhododendrons of the Sikkim Himalaya* (London, 1849–51), and after returning with numerous plant specimens and records, he published *Himalayan Journals* (1854). His superb *Himalayan Journals* (1854) provided valuable information. For example, his discussion

FIGURE 6.2 (a) Hewett C. Watson, 1839. (b) Edward Forbes, 1849. By T.H. Maguire. Egerton 2003: 45, 124.

of vegetation and climate was quite complex and informative.[163] He coauthored a book on India's flora in 1855, and later he published *Flora of British India* (7 vols., 1872–97). One botanist felt Hooker's separate contributions to phytogeography were so important he collected and republished them.[164] One historian thinks Hooker's *Journals* is "one of the few and certainly the most accomplished examples for India of a scientific travel narrative comparable to those of Humboldt, Darwin, and Wallace."[165] He succeeded his father as Director of Kew Gardens, 1865–85.

Travel books by Darwin and Joseph Hooker encouraged other English naturalists to undertake their own expeditions. School teacher turned botanist Richard Spruce (1817–93) was particularly interested in bryophytes.[166] He first traveled into the Pyrenees, April 1845– April 1846, to study these inconspicuous species in those mountains, and he published "The Musci and Hepaticae of the Pyrenees" (1850). William Hooker suggested he next explore the Amazon Valley, to which several botanists sponsored his travel, in 1849, where he collected 7000 plant species, many unknown. Britain in 1860 sent him into the Andes to collect cinchona plants to grow in India; they arrived alive. Afterwards he collected plants along coasts of Ecuador and Peru before sailing back to England in 1864. He published monographs on "Palmae Amazonicae" (1869) and "Hepaticae of the Amazon and of the Andes of Peru and Ecuador (1885), and wrote *Notes of a Botanist in the Amazon*, edited posthumously by Alfred Wallace (2 vols., 1908). A selection from it was reprinted in 1948.[167]

After a minimal rural education, Alfred Russel Wallace (1823–1913) moved to London in 1837 to live with an older brother and attend public science lectures.[168] In 1838 he apprenticed to another older brother, a surveyor, until 1843, when he became a school teacher, who read science books by Humboldt, Darwin, Lyell, and William Edwards' *A Voyage up the River Amazon* (1847). Edwards collected birds and butterflies.

Henry Walter Bates (1825–92) had been an excellent student, yet in summer 1838 his father apprenticed him to a hosiery manufacturer.[169] These two nature-travel dreamers, Bates and Wallace, met at Leicester Public Library and became friends. Soon, Wallace suggested they travel to tropics and collect specimens to sell to pay expenses and also study the origin of species.[170] William Edwards advised them to go to the Amazon. They left Liverpool 26 April 48 and reached Pará (Belem), Brazil on 28 May. After two months collecting, they sent to London 553 species of Lepidoptera, 4540 beetles, and 400 of other species, some undescribed.[171] They traveled together for about five months, then explored separately. Wallace stayed four years, and his *Travels* (1853a) is much briefer than Bates' *Naturalist on the River Amazons* (2 vols., 1863), who stayed 11 years.

Wallace devoted 13 chapters (279 pages) to a narrative of travels and four chapters (82 pages) to geography, geology, vegetation, zoology, and aborigines. By July 1852, Wallace had a collection of live animals that needed attention to carry to England. He loaded his cargo, alive and preserved, on a brig *Helen*, and they left Pará 12 July. On 6 August, *Helen* burned in the Atlantic, and Wallace only saved his papers, including drawings. On August 15, they were rescued, but on an old, leaky ship that reached England 1 October. His agent had insured his cargo, and he received enough money to live on, and in 1853 he published both his *Narrative of Travels* and *Palm Trees of the Amazon and Their Uses*—that being 16 years before Spruce published his palm monograph. Wallace thought 13 species were new, though eight were not.[172]

Wallace was clearly the main founder of zoogeography. Brown and Lomolino listed 17 biogeographic principles which he advocated.[173] They also reproduced his two-page world map of biogeographic regions, based largely upon Sclater's six regions for birds. Later, Brown and Lomolino commented that during 2 [1.5?] past centuries, we have not added to Wallace's list, just modified which principles were emphasized.[174]

Bates' main interest was insects, and Brazil has a spectacular insect fauna. He could not have survived selling only insects, and he also collected birds, mammals, and a few reptiles and fish. Although his insect interests were broad, three species of ants attracted his attention. Leaf-carrying saüba ants *Oecodoma cephalotes* were a "great scourge to Brazilians" because they damaged foliage of cultivated trees.[175] He thought ants used leaves to thatch domes to cover entrance tunnels to nests, but English engineer Thomas Belt argued they used leaves to grow underground a fungus, *Rhozites gonglyophora*, which they ate.[176] Fire ants *Solenopsis* sp. had limited ranges, but they were formidable enough to drive Aveyros villagers from their homes for several years.[177] Other scourges were army ants *Eciton burchellii* (and nine related species), which had no permanent nest.[178]

Bates most notable discovery he called "mimetic analogies," now called "Batesian mimicry" (a species mimics another). He also discovered "imitative resemblances," now called "protective coloration."[179] He presented evidence for both in an article on Amazonian Heliconidae butterflies. He wrote it (1862) after reading Darwin's *Origin* and becoming a convert. He cited three previous articles with examples of mimetic and coloration had been described, but he was first to see such phenomena as evolutionary strategies.[180] Heliconidae were numerous and seemed unpalatable to predators. It was thus advantageous for other species to imitate them. Darwin was delighted and wrote Bates on 20 November 62: "I have just finished, after several reads, your Paper. In my opinion it is one of the most remarkable & admirable papers I have ever read."[181] Darwin was mostly referring to mimicry, as he had discussed protective coloration in his *Origin*.[182] To be sure Bates' discoveries were not overlooked. Darwin published a five-page review of his paper, and cited it in edition 4 of the *Origin* (1864: 506).[183] Bates explained that similarities between hummingbirds and hummingbird moths were a case of parallel evolution; no mimicry was involved.[184] Darwin urged him to write on his travels, and when he did, Darwin rated it "the best book of Natural History Travels ever published in England."[185]

In an article on animal migrations, Spruce commented (1867) that probably no other large portion of the Earth had as high a portion of its animals that were vegetarian as did Amazon and Orinoco Basin, with carnivores occasionally eating vegetation.[186] He thought only animals which could inhabit a continent from sea to sea were general feeders, with parasites, larger predators, and scavengers, such as vultures and termites. Lepidoptera ranges rarely correspond to main features of vegetation because they feed on understory vegetation. Many trees and lianas exclude caterpillars with resinous or acrid or poisonous juices, and others have leathery leaves, unpalliative to many caterpillars. He listed orders of plants on which caterpillars feed.

Like Hooker, one expedition was not enough for Wallace. He had returned from the Amazon in 1852, after a catastrophe at sea, and yet had published two books in 1853. In March 1854, he traveled to Malay Archipelago on a British Naval ship (no charge). He spent eight years there, 1854–62, and he traveled about 14,000 miles, mostly by ship, while suffering periodic tropical fevers (malaria).[187] He later considered his Malay experience "the central and controlling incident of my life."[188]

Wallace could have had more than one reason for thinking this, but certainly his main reason was a theoretical breakthrough, on species, which began taking shape in 1855, expressed in his "On the law which has regulated the introduction of new species" (1855). As I explained in 2012, this "law" was implicit in Darwin's *Journal of Researches* (1839, ed 2, 1845), though Darwin considered it a generalization, not a "law."[189] It stated that fossils in a place resemble species now living there. Wallace used evidence from Galapagos Islands to support his claim.

Darwin's theory of origin of species provided the key to the puzzle of species distribution.[190] Wallace, who had read both editions of Darwin's *Journal of Researches*, reached the same insights as Darwin on species origins in 1858, and after Wallace sent Darwin his theory, their concepts were published together (1858). A year later, Darwin laid out his evidence and reasoning in *Origin*, which included two chapters on biogeography (11, 12), which amounted to a synthesis of biogeography for that time. Biogeography became a prominent topic in his correspondence, indicated by indexes of volumes to volumes 6–7 of his *Correspondence*.[191] Asa Gray's biographer explained the importance of Darwin's theory for botanists:[192]

> Hooker, Gray, and [Alphonse] de Candolle were all firstrate scientists, capable of establishing a new science of plant geography at the same time they contributed mightily to taxonomy. But Charles Darwin's horizons were already so much broader than theirs that even a single letter of his, written in Haste, rearranged the facts of natural history into new and striking patterns. Hooker did not exaggerate when he confessed, he would never have taken up the subjects of distribution and variation of species but for the advantages he 'had derived from [Darwin's] friendship and encouragement.'

One attraction of Malay Archipelago for Wallace was a chance to observe and collect orangutans, which few Europeans had done. Much of chapter 4 of *Malay Archipelago* (1869) is about shooting them (14) to sell skins in Britain. He expressed surprise its range was limited to

FIGURE 6.3 (a) Philip L. Slater. Web. (b) Albert C.L.G. Günther. Web.

Borneo and Sumatra, and even on those islands its habitat requirements were limiting:[193]

> It seems at first sight inexplicable that the Mias [orang-utan] should be quite unknown in the Sardwak valley, while it is abundant in Sambas, on the west, and Sàdong, on the east. But when we know the habits and mode of life of the animal, we see a sufficient reason for this apparent anomaly on the physical features of the Sardwak district. In the Sàdong . . . Mias is only found where the country is low, level, and swampy, and at the same time covered with a lofty virgin forest. They rarely descended to the ground, and natives said they only had 2 occasional enemies, crocodiles and pythons, and they were strong enough to kill both.

In *Amazon Travels*, he had discussed geography after his travelogue, but *Malay* opened with geography, because of "Wallace's Line." On 15 June 56, he sailed from Bali to Lombok, only 15–20 miles away. Both islands have volcanoes, about 8000 feet high. Yet, Lombok had a strikingly different flora and fauna than Bali. In his *Malay* map, he showed Bali in an Indo-Malayan Region and Lombok in an Austro-Oriental Region.[194] Before publishing *Malay* in 1869, he learned George W. Earl had pointed out each island was separated by a shallow sea from different continents. During ice ages, each would have connected to a nearby continent.

Malay Archipelago was a treasurer-trove of interesting natural histories, and:[195]

> This is the most famous of all books on the Malay Archipelago. It is the greatest travel book on the region and, in its analysis of the geographical distribution of animals, it ranks as one of the most important natural history books of the nineteenth century.

When Wallace published *The Malay Archipelago* (London, 2 vols., New York, 1 vol., 1869), he was much more sophisticated than when he published *Amazon Travels* (1853), after staying twice as long as he had on his first expedition. John van Wyhe though it important enough for him to publish an annotated edition (2015).

Wallace's grand synthesis, *The Geographical Distribution of Animals* (2 vols., 1876) drew upon his knowledge of Amazon Basin, Malay Archipelago, museum researches, and vast reading. Both Darwin and Cambridge University ornithologist Alfred Newton, had urged him to write it about six years earlier.[196] It was well organized, had 20 helpful illustrations, and seven maps of zoogeographical regions. His only guide to literature was mention of zoologists' names and occasionally brief mention of titles, such as: Jerdon's *Birds of India*, Blyth's *Catalogue*, Bonaparte's *Conspectus*, and *Proceedings of the Zoological Society of London* down to 1865. He had mild criticism for Andrew Murray's *The Geographical Distribution of Mammals* (1866); Edward Blyth's *Catalogue* may have been useful, but he dismissed Blyth's "Suggested New Division of the Earth into Zoological Regions" (1871) as unnatural and inconvenient.[197]

Wallace began *Geographical Distribution* with a survey of means of dispersal and migration, then discussed "Distribution as affected by the conditions and changes of the Earth's surface." He accepted Philip Sclater's six zoological regions of the Earth (1858), determined for birds.[198] His book, Part I: each region had four sub-regions—three more than Augustin-Pyramus de Candolle had established for plants;[199] Part II discussed distribution of extinct species; Part III discussed in detail the six geographic regions; Part IV was "a systematic sketch of the chief families of land animals in their geographical relations." Darwin expressed hope (5 June 1876) others would publish on geography of insects, pulmonated mollusks, freshwater fishes, and plants, as Wallace had done for land vertebrates. However, when Darwin read volume 2, he saw that Wallace had discussed fish and invertebrates. His inclusion of plants on his wish list indicates he had forgotten the works of de Candolle (1855) and Grisebach (1872), discussed next. But Wallace had one more contribution, *Island Life* (1880), which helped make him the most important contributor to zoogeography during the 1800s.[200] Karl Schmidt thought *Island Life* was more successful than *Geographical Distribution of Animals*.[201]

Swiss mammal paleontologist Karl Ludwig Rütimeyer (1825–95) "in 1867, makes the suggestion of the former existence of a vast Antarctic continent, connecting all of the southern continents and New Zealand, and this idea received support from T.H. Huxley in 1870."[202] Austrian geologist Eduard Suess (1831–1914), who had studied similar fossil ferns from South America, Africa, and India, coined its name, Gonwandaland.[203]

Suess began his career as a paleontologist, though the scope of his research expanded at the University of Vienna, and his grand synthesis was *Das Antlitz der Erde* (3 vols., 1883–1909, French, 4 vols., 1897–1918, English, 5 vols., 1904–24). His last chapter was on distribution of life on Earth. Sea levels fluctuated, and land bridges seem to have existed at Beringia, Panama, and between Britain and France, and elsewhere.[204]

Finnish botanist Ragnar Hult (1857–99) studied at Uppsala University before earning his doctorate at Helsingfors (now Helsinki) University.[205] He studied physiognomy of vegetation and distinguished seven layers—ground, lower, middle, upper field, shrub, lower and upper forest—which he defined. In summer 1886, he taught a field course in plant geography, a first. Swedish student Rutger Sermander "adopted Hult's ideas with great enthusiasm."[206] When Warming's book appeared (1895), it had a strong influence in Sweden.[207]

English naturalist Henry Dresser (1838–1915) was friends with Wallace.[208] He inherited his father's lumber business and began processing iron ore. Another bird student, Richard Sharpe (1847–1909), proposed to him that they collaborate on a history of European birds. Dresser agreed. Their plan was to publish their *History of European Birds* in parts,[209] and they published 13 parts.[210] In September

1872, British Museum hired Sharpe to work on its bird collection and he resigned from collaboration with Dresser. Dresser continued alone, and by 1882, he had published parts 14–84, including 624 species from western Russia to Britain and south to North Africa and Mediterranean islands. Dresser examined 10,637 bird skins from 102 collections. His monograph was highly praised by reviewers. He then supplemented it with *Eggs of the Birds of Europe* 24 parts, 1905–10. Including photographs of eggs and nests.

Ecology historian Patrick Matagne sees four phytogeographers as most important in the late 1800s: Alphonse de Candolle, Kerner, Grisebach, and Flahault, and he reprinted selections from their works.[211] (Why not add Drude, Warming, Schimper?)

Like his father, Swiss botanist Alphonse Louis Pierre de Candolle (1806–93) studied for a time in Paris,[212] where he earned a doctorate in law (1829), which perhaps was useful when he published *Lois de nomenclature botanique* (1867). In 1835 he succeeded his father as professor of natural history at Academie de Genèva. In 1850, he retired from teaching to focus upon research. His important work for phytogeography was *Géographie botanique raisonnée* (2 vols., 1855, 1365 pages, 2 maps). It was the most detailed synthesis by then (and maybe, ever). He judged heat and moisture as dominant factors affecting plant distribution. Joseph Hooker wrote a detailed review in seven issues of his father's *J. of Botany* (1856). De Candolle believed in special creation to enhance what he could not explain. Hooker was friends with Darwin, whose *Origin* was not yet published, but Hooker offered some of Darwin's arguments instead of de Candolle's. Malcom Nicolson thinks Candolle was more of a floristic phytogeographer than a follower of Humboldt, and Patrick Matagne thinks Candolle's development of a classification based on plants' mode of life contributed to the emergence of a new ecological approach, centered on plant-environmental interactions.[213] De Candolle became an evolutionist after reading the *Origin*.[214]

Austrian botanist Anton Kerner von Marlaün (1831–98) studied medicine at the University of Vienna, only to give up his practice to become a botanist; he was strongly influenced by Humboldt.[215] He became a botany professor at Budapest, then Innsbruck, finally, Vienna. While in Hungary, he had studied and collected plants in botanically little-known eastern Hungary and Transylvania and published *Das Pflanzenleben der Danaulaender* (1863). Kerner sought to identify all plant formations and explain changes over time in them. Kerner's American translator, Henry Conard, claimed it is the "immediate and direct parent of all later works on Plant Ecology."[216] R.J. Goodland reserved that parental honor for Warming (1895), and Dwight Billings and Malcolm Nicolson agreed with Conard.[217]

Meyen (discussed, line 4) taught botany to his young nephew, August Grisebach (1814–79), who later studied medicine and natural history at Göttingen and Berlin.[218] *Humboldt interested him in botanical exploration. He explored in the Alps (1833), and in 1838 he published "Uber den Einfluss des Climas auf Begränzung der natürlichen Floren," which is summarized in English.*[219] *In 1839–40 he took the most important journey of his life, through the Balkans and Asia Minor, where there had been* little botanical exploration. Afterwards, he published both a travel journal (1841) and a two-volume flora (1843–44). He would later undertake three more botanical explorations within Europe (3 of those papers are reprinted in Egerton 1977c). He used *formation* to distinguish different groups of vegetation;[220] Humboldt had used "association," but without establishing it as a technical term, though Schouw had done so. Grisebach became a leader of phytogeography and wrote several progress reports on phytogeography for particular years, two of which are also published—for 1842–43 and 1844–45—in English translation.[221] His *magnum opus* was *Die Vegetation der Erde nach ihrer Klimatischen Anoronang* (2 vols., 1872), which included the first color map of world vegetation.

What changes occurred after Grisebach (1872) published on world phytogeography and Wallace (1876) on world zoogeography—each in two volumes? These were encyclopedic achievements, helpful for organizing emerging sciences, but not having theoretical breakthroughs. If one studied the geography of plants or animals of a particular region, then one could place that study within a broader context by consulting one of these comprehensive works. In America, James Starr Lippincott wrote a 60-page summary on "Geography of Plants" for a USDA annual report for 1863, summarizing for the public progress in this science (reprinted in Egerton 1977c).

French botanist Charles-Henri-Marie Flahault (1852–1935) earned his doctorate at Paris in 1878, then traveled with botanist Gaston Bonnier (1853–1922) to Sweden to study relationships of plants and climate.[222] Their "Observations sur les modifications des végétaux suivant les conditions physiques du milieu" (1878) compared similarities and differences between alpine and polar climates and plants, rather similar to Kerner's comparisons (1863) of treeless alpine and prairie regions and plants. Flahault returned to Sweden four times and became friends with Warming in Denmark. He accepted the botany chair at Montpellier and remained 47 years. In 1891 he started a school of Mediterranean phytogeography, and in 1897 he sent six maps on phytogeography of French coast to Académie des Sciences, which finally published them in 1937.[223]

By the 1890s phytogeography seemed to be a coherent science, and it became a stepping stone to a plant ecology. Three northwest Europeans guided that transition, two Germans, Drude and Schimper, and Danish Warming. The earliest to publish a significant contribution was Oscar Drude (1852–1933), who had studied under Grisebach, receiving his Ph.D. from Göttingen in 1873.[224] He published his *Handbuch der Pflanzengeographie* in 1890, attesting to the maturity of that science, and followed with *Deutschlands Pfanzengeographie* (1896).

Even more significant was *Pflantesamfund* (1895, German 1896, English 1909) by Danish Eugenius Warming

FIGURE 6.4 (a) Charles R. Darwin, age 46, 1855. Web. (b) Joseph D. Hooker. By Charles H. Jeens. Web. (c) Asa Gray in 1886. Gray 1894: II, frontispiece.

(1841–1924), whose plant ecology evolved from phytogeography. German plant ecologist Andreas Schimper (1856–1901) wrote an encyclopedic that (*Pflanzengeographie auf physi-ologischer Gtundlage* 1898, English 1903, 870 pages + 4 maps) was less influential than Warming's unillustrated textbook but was far more impressive visually.

Drude's second book came to the attention of botanists at the University of Nebraska before his first book did. When they obtained his *Handbuch*, they found his vegetational boundaries inaccurate for Great Plains.[225]

Two contemporary Harvard University professors were interested in biogeography: zoologist Louis Agassiz (1807–73) and botanist Asa Gray (1810–88). Gray was from upstate New York and earned an M.D. degree there.[226] He joined Harvard's faculty in 1842 and became America's leading botanist. His prominence came from his publications (requiring a 64-page bibliography),[227] particularly his *Manual of the Botany of the Northern United States* (1848, lxxii + 710 pp.). In 1851 he went to England and at Kew he met Darwin. In 1855, Darwin became interested in Gray and Hooker's correspondence on plant geography and so he began corresponding with Gray himself. In 1857, Darwin explained to Gray his theory on origin of species, which helped Gray interpret the similarities between plants of eastern Asia and eastern North America. He became the important reviewer of Darwin's *Origin* when published in 1859.[228] That put him at odds with Agassiz.

Agassiz was Swiss and earned his Ph.D. degree in 1829 and his M.D. degree in 1830.[229] He then studied under Cuvier in Paris before accepting a professorship at College of Neuchâtel. He published scientific works on fish, paleontology, and glaciology. In 1846 he went to Boston to lecture at the Lowell Institute, which led to Harvard offering him a professorship in 1847. In June 1848, he led a private expedition of students and naturalists to the north shore of Lake Superior, in quest of biological and geological specimens and observations. They later wrote a good collective *Lake Superior* (1850): Eliot Cabot wrote a 124-page narrative, drew eight landscapes, and discussed birds (3 pages); Agassiz, editor, compared shore plants with those of Switzerland (53 pages) and described fish (131 pages), reptiles (4 pages), geology and glaciology (33 pages); two entomologists described insects (49 pages); and there were nine illustrations of fish, reptiles, and insects. *Lake Superior* was well received, and Darwin praised a gift copy, though he would have disagreed with Agassiz on pages 375–376 if he read them: "The fishes and all other freshwater animals of the region of the Great Lakes, must have been created where they live."

Pennsylvanian Spencer Baird (1823–87) attended Dickinson College (B.A. 1840, M.A. 1843) and decided to become a zoologist.[230] He taught at Dickinson College before becoming, in 1850, assistant to Joseph Henry as head of new Smithsonian Institution.[231] In 1878 he succeeded Henry as Secretary (head) of the Smithsonian. He sought naturalists who were either private collectors or on government expeditions who would collect birds and mammals for Smithsonian collections.[232] These specimens had tags indicating where collect—important for zoogeographical studies. Baird himself synthesized reports on mammals of Pacific railroad surveys (1857) and was senior author of *The Birds of North America* (1860). From 1871, he was also head of the U.S. Fish Commission (discussed later).

New Yorker James G. Cooper (1830–1902) James G. Cooper was son of naturalist William Cooper.[233] James earned a medical degree at College of Physicians and Surgeons (1851) and practiced medicine for two years, then joined Isaac Stevens Expedition of the northernmost route for a transcontinental railroad. He served as physician-naturalist, with interests in both plants and animals. Afterwards, he published "On the Distribution of Forests and Trees of North America, with Notes on its Physical Geography" (1859, 35 pages + map). His map described six Natural Provinces, containing 27 Regions.

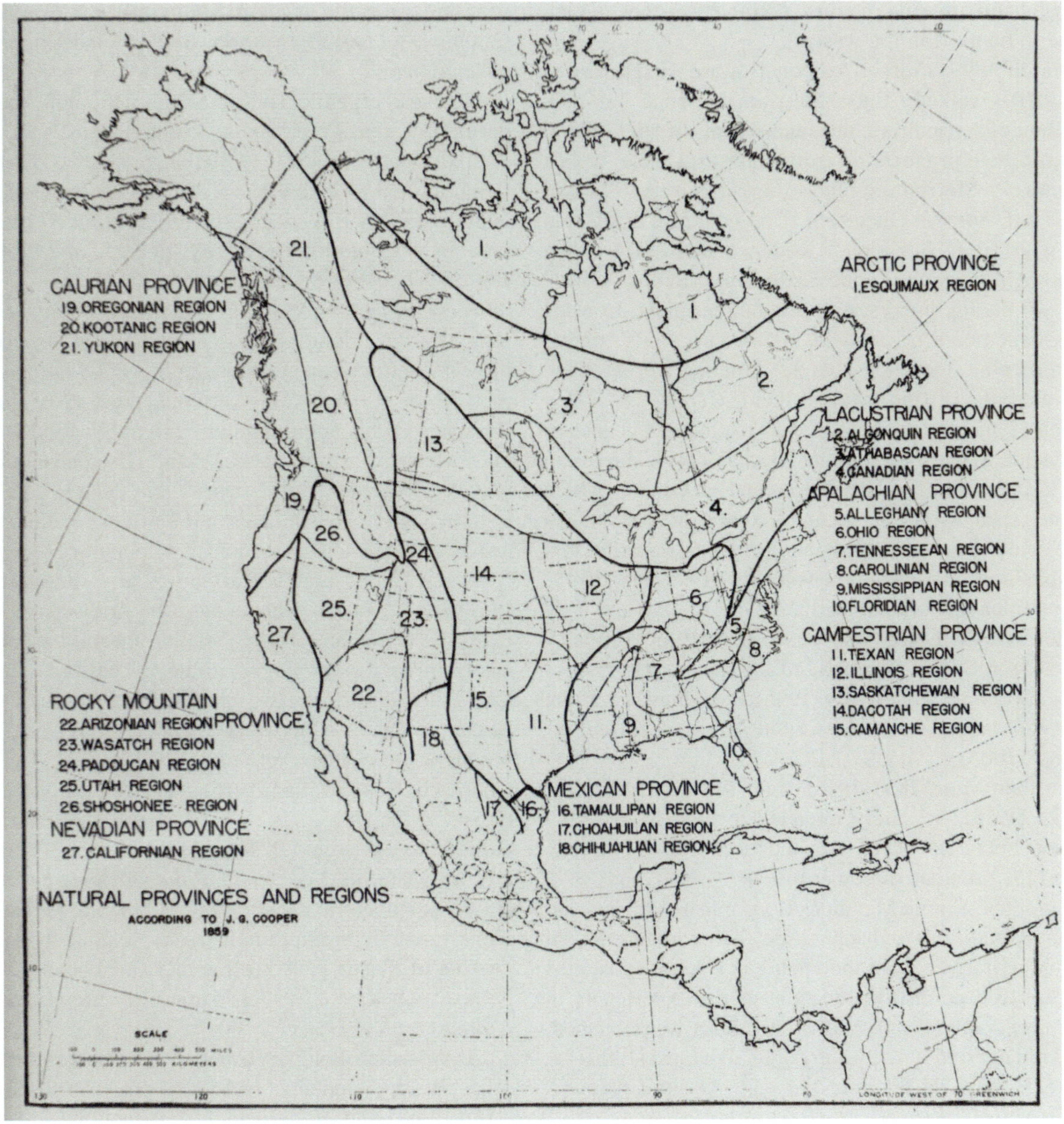

FIGURE 6.5 Natural provinces of North America. By James G. Cooper (1859). Kendeigh 1954: 155–157.

C.Hart Merriam (1855–1942), from a prosperous New York State family, which spent winters in New York City.[234] He had a strong interest in wildlife, and in 1871, his Congressman father took him to see Baird at the Smithsonian. Baird saw Merriam's stuffed bird and mammal specimens, and later he arranged for Baird to accompany Ferdinand Hayden's U.S. Geological Survey of Yellowstone in summer 1872. Baird published a *Report on the Mammals and Birds of the Expedition* (1873) as part of the Survey's report. He studied three years at Yale, then earned an M.D. degree at New York College of Physicians and Surgeons (1879). He practiced medicine for six years, yet in 1884 he published *The Mammals of Adirondack Region*. Its introduction discussed soil, topography, climate, vegetation, and other animals.[235] In 1885, he became head of a new division of USDA to study birds and mammals, and remained until 1910. In 1889 he led an expedition to survey San Francisco Peak and nearby Little Colorado River, Arizona—about 12,000 square miles.[236] He measured elevations and recorded species ranges, but not temperatures, as Humboldt had for Chimborazo and Cotopaxi, published in 1807.[237] Merriam used his data to construct a Provisional Biological Map of North America Showing Principal Areas.[238] His comparison of mountain and continent areas is illustrated in Brown and Lomolino.[239] His life zone scheme was/is widely used and provoked more published discussion than contributions to American biogeography. Victor Shelford, and Karl Schmidt criticized it as simplistic, but Rexford Daubenmire praised Merriam for using data on both plants and animals in constructing life zones, which continue to be used in the west.[240]

In March 1899, railroad tycoon Edward H. Harriman (1848–1909) walked into Merriam's office at the Department of Agriculture, introduced himself, and invited Merriam to help organize an expedition to Alaska, to last two months.[241] Merriam agreed to organize scientists and support staff, but was uncertain if he could be away that long. A word from Harriman to leaders at the USDA ended that worry. Merriam obtained 23 leading scientists, three artists, two photographers, and two taxidermists. In May, east-coast members traveled in luxury cars on a special Harriman train, sightseeing across America to Seattle.[242] A renovated steam ship awaited them, *George W. Elder*. Harriman's family were aboard for a vacation, but he let scientists decide where to stop along Alaska's coast. Harriman wanted useful scientific observations and specimens, which was done, then Merriam spent 12 years, at Harriman's expense, sending specimens to authorities to describe and then publishing 13 substantial volumes of reports (1901–14). Expedition historians William Goetzmann and Kay Sloan devoted their final chapter to discussing results.[243] See also George Bird Grinnell, ed., *Alaska 1899: Essays from the Harriman Expedition* (1995) and Thomas Litwin, ed., *The Harriman Alaska Expedition Retraced: a Century of Change* (2005).

When John Muir (1838–1914) began writing his *Travels in Alaska* (1915), he surely intended to include his 1899 expedition, but he ran out of time and died, so the book only discussed his previous three trips in 1879, 1880, and 1890. One biographer, Linnie Marsh Wolfe, published his journal from that expedition, and later used that evidence for an expedition narrative.[244] On the other hand, John Burroughs (1837–1921) did publish his account of the experience, in his *Far and Near* (1904), one of 29 books he published, though I did not find it in my copy of his *Writings* (1905–22, 23 vols.). He was most impressed with Alaska's glaciers, including Muir Glacier.[245]

Michigan native Conway MacMillan (1867–1929) attended the University of Nebraska, where leading American botanist Charles E. Bessey taught.[246] MacMillan earned bachelor and master degrees (1885, 1886) and studied further at Harvard and Johns Hopkins Universities and became instructor in botany at the University of Minnesota (1887), and later professor. He was active at research and publishing, 1890–1901. His *Metaspermae of the Minnesota Valley* (1892, 826 pages) has an ecological survey of the region. Also notable was his "Observations on the Distribution of Plants along Shore at Lake of the Woods" (1897). He chaired Minnesota's Botany Department until 1906 and headed a state botanical survey, 1900–06.[247] He became dissatisfied with the modest salaries of professors and left academia for business.

Two references by ecologists on history of biogeography during the 1900s merit mention here: James Brown and Mark Lomolino, *Biogeography* (ed. 2, 1998, pp. 32–35) and Mark Lomolino, Dov Sax, and James Brown, eds., *Foundations of Biogeography: Classic Papers and Commentaries* (2004, xx + 1291). Both works are far broader in scope than covered here.

Chicago herpetologist Karl Schmidt thought "the modern period in zoogeography must be dated from 1915," because William Matthew (1871–1930) then published "Climate and Evolution."[248] Matthew was from New Brunswick, Canada, and earned his B.A. degree (1889) at its university, before moving to Columbia University for M.S. and Ph.D. (1894, 1895). When his dissertation advisor, Henry F. Osborn, moved to American Museum of Natural History in 1907, he took Matthew along as his assistant. When Osborn retired in 1911, Matthew replaced him.[249] In 1927, Matthew became a professor of paleontology at the University of California, Berkeley. Of his approximately 240 publications, "Climate and Evolution" was most important. His contributions were widely appreciated at the time,[250] but decades later he was seen by Gareth Nelson, Norman Platnick, and Donn Rosen as having an obsolete "dispersalist" perspective.[251]

In the United States, German plant ecology influenced Midwestern pioneer plant ecologists. Midwestern universities arose later than eastern ones, and around 1900 they more easily hired new ecologists than did eastern universities. The following three have already been discussed earlier. Michigan native Albert S. Hitchcock (1865–1935) studied botany at Iowa State Agricultural College, at Ames, under Charles E. Bessey (1845–1915), a leading U.S. botanist who moved to the University of Nebraska in 1884.[252] Hitchcock earned three degrees (B.S.A. 1884, M.S. 1888, Sc.D. 1920). Did he write his *Ecological Plant Geography of Kansas* (1898) as a doctoral dissertation? Hitchcock joined the U.S. Department of Agriculture in 1901. He published guides to grasses of places in Americas and Caribbean Islands. He is most remembered for his *Manual of the Grasses of the United States* (1935).

About when Hitchcock began Kansas researches, two other Bessey students, Nebraska natives Roscoe Pound (1870–1964) and Frederic E. Clements (1874–1945) began similar, but joint, research on phytogeography of Nebraska. For Pound, plant ecology was a challenge, but his choice for a career was law, at Harvard. For Clements, that research was the beginning of the most influential American plant ecology career during his lifetime.[253]

There was also progress in Chicago. Connecticut native Henry Chandler Cowles (1869–1939) studied botany and geology at Oberlin College, 1889–1893, graduating with high honors.[254] He taught science for a year, then was admitted to the University of Chicago, where he spent his career. He began with geology, but in 1896, John Merle Coulter came to Chicago as head of Botany Department and Cowles switched to botany. Coulter taught ecology in spring 1896. He was also founder and editor of *Botanical Gazette*, which received a review copy of Warming's *Lehrbuch der ökologischeb Pflanzengeographie* (German ed., 1896). Cowles wrote a synopsis of it for Coulter, who reviewed it in *Botanical Gazette*. However, Cowles reviewed books in almost every issue of *BG*, 1896–1933. Warming's book had the same strong impact on Cowles that Drude's books had on Pound and Clements, and in the same year! Coulter had published a paper on flora of northern Indiana and took Cowles to glaciated dunes along Lake Michigan's south shore—largest dunes in the world—and apparently suggested Cowles study their vegetation for a dissertation.[255] In a paper from that dissertation, Cowles acknowledged:[256]

> [H]is great indebtedness to Dr. Eugene Warming, professor of botany at Copenhagen: his textbook on ecology and his treatises on the sand-dune flora of Denmark have helped greatly to make clear the true content of ecology.

Cowles' papers from his dissertation drew upon his geological studies to support his botanical research: "The influence of underlying Rocks on the Character of the Vegetation."[257] Clements explained in his summary history of succession studies that he and Pound had discussed succession of Nebraska plant formations, but not in depth, as Cowles did.[258] He followed succession from coastal dunes inward and saw successive stages representing different ages of vegetation. He introduced *climax* as an ecology term.[259]

In 1903, a plant physiologist called attention to progress in plant ecology,[260] and in 1904, Cowles did likewise and acknowledged that "the field of ecology is in chaos."[261] Impressive in his report are numerous citations, though without citing titles of journals where published. Plant ecology was obviously flourishing in both Europe and America. In 1909 Cowles thought it "essential that physiology and ecology break down the barrier erected for them in Madison in 1893, where it was declared that physiology is experimental and ecology observational. The ecologist must experiment."

As discussed earlier, plant ecology arose in Britain from influences of Warming, Schimper, and Flahault.[262] Vegetation maps were constructed for Edinburgh and Yorkshire, which inspired formation of a British Vegetation Committee in 1904.[263] Committee members constructed surveys and maps, which led to a massive *Types of British Vegetation* (1911), edited by Tansley), committee chair. He did not think that work was definitive; a larger *British Islands and Their Vegetation*, which he edited in 1939 (930 pages, 418 photographs, 162 plates), appeared before World War II began. He used the 1911 volume as a guide for his International Phytogeographic Excursion in August 1911.[264] Foreign ecologists who participated were Europeans Drude, Ostenfeld, Rübel, and Schröter, with Clements, Cowles, and wives from the United States. At places where vegetation was studied, local botanists who had contributed to the volume met and guided excursioners. Tansley built upon Excursion success to organize the British Ecological Society (BES) in 1913, the first such society on Earth.[265] American ecologists were influenced by Europeans, and Cowles and Clements organized a similar excursion in 1913.[266] America's excursion was also successful, but the initiative impulse to organize ESA came from two Midwestern animal ecologists, though they sought Cowles for organizing.[267] A planning meeting was held in 1914, and ESA was organized at a meeting in 1915. Animal ecologist Victor E. Shelford was elected its first president.[268]

German botanists Drude and Adolf Engler (1844–1930) contributed separately to phytogeography,[269] but even more importantly, they collaborated in organizing and editing encyclopedic *Die Vegetation der Erde*, in 15 volumes, 1902–23. Both of them also contributed a volume to the project.[270] Two English-language volumes were: John Harshberger (1869–1929), *Phytogeographic Survey of North America* (1911) and Leonard Cockayne (1855–1934), *The Vegetation of New Zealand* (1921). Although Merritt Fernald reviewed Harshberger's *Survey* unfavorably, its bibliography on American phytogeog-raphy was monumental.[271] Engler was primarily a systematic botanist, but as an early convert to Darwinism, he developed a strong interest in phytogeography.[272] In 1880, he founded *Botanische Jahrbücher für Systematik, Pflanzengeschichte und Pflamzengeographie*, which he edited for almost 50 years. It was first scientific journal to name phytogeography as a specialization. His contribution to his and Drude's encyclopedia was *Die Pflanzenwelt Afrikas* (1910–15).

Shelford organized and edited an ESA *Naturalist's Guide to the Americas* (1926, xv + 761 pages + 16 regional maps). Each state and Canadian province has a chapter by its own author, as do large national forests; in Central and South America, south to the Amazon River, one chapter per country, with Mexico's chapter, for example, much longer than Cost Rica's; the Caribbean and Pacific islands also have chapters. It was an impressive ecological and geographical achievement.

German zoologist Hans Gadow (1855–1928) earned his Ph.D. (1878) under Haeckel, University of Jena, and in 1880, he went to London to catalog British Museum's bird collection.[273] In 1882, he went to Cambridge University, where he wrote *The Wanderings of Animals* (1913), a small book of 155 pages and 16 maps. He commented in chapter 5, "Former Configurations of Land and Water," "present distribution of animals cannot be accounted for without the assumption of enormous changes in the configurations of land and waters."[274] However, he thought one could account for land tortoises reaching Galapagos Islands by movement of continents.

Scottish zoogeographer Marion Newbigin (1869–1934) earned her B.Sc. and Ph.D. (1893, 1898) at the University of London and published *Life at the Seashore* (1901) before returning to Edinburgh in 1902 to become editor of *Scottish Geographical Magazine*, until she died.[275] Her pioneer work was *Animal Geography* (1913), with her major work, *Plant and Animal Geography* (1936) appearing posthumously. Her books were popular in UK.

Swedish botanist O. Gunnar Erdtman (1897–1973) was a pioneer in palynology; he published his dissertation on it in German (Uppsala, 1921).[276] In Sweden for about 20 years, botany professor, Gustav Lagerheim (1860–1926), University of Stockholm, and geologist Lennart von Post (1884–1951) had used a Hiller peat auger to drill in peat to understand vegetation history from preserved pollen grains.[277] Erdtman synthesized his and their studies in *An Introduction to Pollen Analysis* (1943).

An important developer of continental movement theory was Berliner Alfred Wegener (1880–1930), who began his university education in astronomy (B.S. 1905).[278] On a Danish expedition to northeast Greenland, 1906–08, his interests moved to meteorology and geology. He later went on three further Greenland expeditions and froze to death on the last one. In 1910, he became impressed with similarities between geology of South America's east coast and Africa's west coast. In 1911, he learned of similarities of fossils of their coasts, and in 1912 he announced his theory of continental drift. He was wounded during World War I, and during sick leave wrote *Die Entstehung der Kontinengte und Ozeane* (1915), his most important work, appearing during World War I.[279]

He argued that continents are granite bodies afloat on a pliable medium. Amid post-war controversy, he added new evidence in later editions of 1920, 1923, and 1929. The 1922 edition was translated into English, French, Spanish, Swedish, and Russian. A theoretical weakness was lack of a mechanism for continental movement. English botanist Henry N. Ridley (1855–1956!) avoided controversy by sticking to the facts in *The Dispersal of Plants throughout the World* (1930, 744 pages).[280] Likewise did a dozen authors who wrote important treatises on regional vegetation, including New Zealand, South Africa, and Philippines.[281] Russian botanist Evgenii Wulff (1885–1941) published *An Introduction to Historical Plant Geography* (1932, English 1943) and reviewed different hypotheses to account for existing plant distributions and judged Wegener's most plausible. After 1960 Wegener's theory gained plausibility with rise of plate tectonics.[282] A 1990 symposium on Biological Relationships between Africa and South America provided late documentation of his argument and evidence, even if its published papers never mentioned him.[283]

German zoogeographer Richard Hesse (1868–1944) earned his Ph.D. at the University of Tübingen and taught there and at two other universities before becoming professor of zoology at the University of Berlin, 1926–35. His most important work was *Tiergeographie auf Oekologischer Grundlage* (1924), which influenced zoogeographers to consider ecological factors more than they had. Chicago ecologists W.C. Allee and K.P. Schmidt translated and revised it into English (1937). The English-language edition was a success, which motivated Allee and Schmidt to produce a second edition, which they called "an essentially new work."[284] Their second edition (1951) increased from 597 to 715 pages.

Contributions from Europe between world wars seems somewhat limited. Could the millions who died in World War I and 1918 flu epidemic be relevant?

In 1928, American Geographical Society published *Problems of Polar Research: a Series of Papers by Thirtyone Authors*.[285] To answer questions on what problems should be addressed in the future, authors had to explain what had been achieved—amounting to progress reports. Two papers considered Arctic plant and animal geography and two more papers Antarctic plant and animal geography.[286] Some moss species were known from both Arctic and Antarctic lands, which might have spread on feet or feathers of migrating birds.

Minnesotan Royal Chapman (1889–1939) earned B.A. and M.A. degrees (1914, 1915) at the University of Minnesota and Ph.D. at Cornell University.[287] He joined Entomology Department at Minnesota, until 1930, when he moved to the University of Hawaii as Dean of Graduate School of Tropical Agriculture. His *Animal Ecology* (1931) had a chapter on zoogeography which was well illustrated, mostly with maps and diagrams from other cited works. It was a good summary of existing knowledge.[288]

Chicago zoologist-paleontologist George Gaylord Simpson (1902–84) earned his degrees at Yale University during the 1920s. He became the leading paleontologist of the 1900s, and he wrote four autobiographical works.[289] He explored Patagonia and learned that after the Panama Isthmus emerged, North American mammals invaded South America.[290] Lomolino et al. reprinted Simpson's "Mammals and Land Bridges" (1940).[291]

A worldwide war might expand opportunities to collect evidence on plant geography. E.D. Merrill's *Plant Life of the Pacific World* (1945), however, was based upon pre-war publications, and its appearance was delayed by war. War also constrained civilian travel.[292] Simpson's 20-page lecture, "Mammals around the Pacific" (1966) cited no references and no authors except Darwin and Wallace; it mostly just summarized his own work.

Cincinnati botanist E. Lucy Braun (1889–1971) was first woman president of ESA, in 1950.[293] She earned three degrees (B.A. 1910, M.A. 1912, Ph.D. 1914) at the University of Cincinnati, then taught there during her career. She studied two nearby environments: eastern deciduous forests and Midwestern prairies. Her studies on former culminated in *Deciduous Forests of Eastern North America* (1950). Her biographer, Ronald Stucky, collected her papers on prairies in *E. Lucy Braun (1889–1971), Ohio's Foremost Woman Botanist: Her Studies of Prairies and Their Phytogeographical Relationships* (2001). Stucky cited one example in which she challenged Edward Deevey Jr.'s conclusions on pollen grains from a lake bottom for his dissertation under Evelyn Hutchinson at Yale. Deevey's 1949 projection of his conclusions beyond what his evidence warranted, implicitly challenged her 1950 account, and she wrote a paper (1951) to explain her evidence and its implications. Stucky stated that his nonresponse to her arguments indicated he realized she had the stronger evidence.

C. MARINE BIOGEOGRAPHY, 1697–1940S

Oceans cover two-thirds of the Earth, yet vast majority of biogeography studies concern terrestrial organisms, which were more accessible than marine ones. In 1957,

FIGURE 6.6 Mississippi Flyway used by migrating waterfowl. Hawkins et al. 1984.

Geological Society of America published its Memoir 57, a two-volume encyclopedia, *Treatise on Marine Ecology and Paleoecology*, begun in 1940, for National Academy of Sciences. Marine biologist Joel Hedgpeth edited volume 1: *Ecology* (viii + 1296 pages), and geologist Harry Ladd edited volume 2: *Paleoecology* (viii + 1077 pages). Hedgpeth wrote an introduction and chapter 13 on marine biogeography. Ladd wrote chapters 1–2 and coauthored chapter 3.

Marine biogeography includes island life, since oceanic birds, seals, sea turtles, a Galapagos iguana, and *Laticauda* snakes reproduce on land, but eat oceanic animals. Before humans became ocean travelers, coconuts and mangroves spread by their seeds floating from one place to another.

Until the late 1800s naturalists who studied marine biogeography were taken on government ships. Four books which survey such voyages are cited in the following notes, as are two similar books, one on New Zealand and one on Australia. There is a strong emphasis upon Pacific explorations since that ocean covers a third of Earth's surface and has 25,000 islands, many more than elsewhere. Further books survey just Pacific explorations.

English adventurer William Dampier (1651–1715) seems first to have included natural history observations and illustrations in his memoirs on three voyages around the world, his first being *A New Voyage around the World* (1697, ed. 5, 1703 + Dutch, French and German translations).[294] He had a life-long fascination for workings of nature. He was England's first collector of Australian plants (about 40 specimens survive), which English botanists described.[295] He described (and illustrated some) animals. Alex George discussed Dampier's natural history in Australia.[296] Dampier was apparently first to publish on *Discourse on Winds, Breezes, Storms, Tides and Currents* (1699).

Ladd and Gunther identified the earliest treatise relevant to marine paleoecology as by Italian scholar-diplomat-military engineer Luigi Ferdinando Marsili (or Marsigli 1658–1730), educated in Bologna, traveled in Ottoman Empire, and later published a perceptive study on currents in Bosphorus Channel at Constantinople (1681, 1978).[297] His most important work was the first treatise on oceanography, *Histoire physique de la mer* (1724), which included "the first bottom profiles, serial temperatures, and observations on wind, currents, and surface conditions to be made in a comprehensive manner . . . [and] dredging for red coral in the Mediterranean." He also tried to use his dredging insights to understand fossils in lands above sea level. He founded and led Academia delle Scienze dell' Istituto di Bologna.

An Italian successor, Vitaliano Donati (1717–62), reported on sea life in *Della Storia natural marina dell'Adriatico* (1745, French 1758). He saw that organisms on the sea floor were similar to those in deposits below them and also to fossils above sea level.[298] He argued that coral is not a plant, but a product of marine "insects."

Bering's two Russian expeditions to Pacific Ocean (1727–29, 1741), discussed earlier, part A, concerned Siberia, Kamchatka, and Alaska, but it was the sea voyages that were strategically important to Russia. Yet, the natural history publications of Gmelin and Steller were significant for both terrestrial and marine biogeography.

A so-called "French and Indian War" (between British and French American colonists), 1754–63, lead to Europe's Seven Years War. Britain provided more resources to its army in America than France did for its army, and so France ceded its colonies in America to Britain. France then decided to look elsewhere for colonies, and gave command to the army officer who had lost the decisive Battle for Quebec (1759), Louis-Antoine de Bougainville (1729–1811).[299] He had two ships to sail around the world, 1766–69) Their most sensational discovery was Tahiti, where sailors freely consorted with the women.

Philibert Commerson (1727–73) was doctor-naturalist on Bougainville's smaller ship and was his own natural history artist.[300] He had been a precocious child. A widower, he brought along an assistant, Jeanne Baret, dressed as a boy.[301] They disembarked on the return trip at Mauritius, and explored it and Madagascar for plants and animals. He died on Mauritius. He had provided for her in his will, and his collections and records reached Paris.[302] Bernard Lacépède used Commerson's fish specimens when writing his *Histoire naturelle des poisons* (5 vols., 1798–1803).

Amateur American scientist Benjamin Franklin (1706–90) heard in 1768 from a cousin, Timothy Folger, Nantucket ship captain, why ships going from England to America took longer than ships from America to England: because Gulf Stream flows from America toward England.[303] Franklin obtained a chart of the North Atlantic Ocean and had Folger indicate on it Gulf Stream's path. Franklin published the chart in 1769 or 1770 for benefit of sea captains. A photo of it is on Wikipedia (and reprinted in Egerton 2019a: 10). In 1775, returning from England on a six-week voyage, Franklin repeatedly checked temperature of surface waters, to distinguish Gulf Stream from other waters. He did so again on two voyages from England in summers of 1778 and 1785, which led him to publish progressively better maps.[304]

The best known natural history exploration of the 1700s was by naturalists on three voyages around the world (1768–79), led by British Navy's Captain James Cook (1728–79), who took along on his first voyage (1768–71) wealthy English botanist, Joseph Banks (1743–1820), with two assistants, who acquired a large collection of plants, and some animals.[305] A general expectation was that Banks would publish lavish volumes describing his collections and experiences. However, Banks' charismatic personality and wealth enabled him to dominate the natural history community in London, and he never published. He was knighted and became president of the Royal Society of London and was too busy for scholarly endeavors.

Banks' expectations for his retinue on Cook's second voyage (1772–75) were so grandiose that another cabin was built on top of the existing cabins, but then the ship was topheavy and it was removed. Banks then withdrew from that voyage.[306] Instead, a German botanist in London, Johann Reinhold Forester (1726–98), was recruited, and he brought along as his assistant his son Georg (1754–94).[307] Johann Forester was a capable and industrious naturalist, with a difficult personality. Both he and Georg published accounts of their global voyage. Cook did not take a naturalist on his third voyage (1776–80), as he wanted to focus on geography; so, he depended upon capable surgeon William

Anderson (1750–78), who had been on his second voyage, to make natural history observations and collections, which he did.[308] However, both Cook and Anderson died during that voyage. The French,[309] Spanish,[310] and Russian[311] governments were inspired by Cook's voyages to send out their own voyages of exploration during the late 1700s and the early 1800s.

After Bougainville's return, France waited about 15 years before sending out four more expeditions, 1785–1804. France decided to fill in some gaps in the geography of the Pacific Basin which Cook had not done. An expedition of two ships was organized with naturalists, under command of war veteran Captain Jean-Françoise de la Pérouse (1741–88?), in 1785.[312] La Pérouse sent his chief engineer to London for advice from Cook's officers. Sir Joseph Banks provided two inclining compasses Cook had used. The expedition used Cook's procedures to determine longitude and latitude. Personnel included 10 scientists, four being naturalists and three illustrators. After visiting Hawaii, Alaska, and Monterey, California, La Pérouse sent home naturalist Louis Dufresne (1752–1832) with collections in 1786. Expedition sailed to Macao, China, from which more observations and collections, including La Pérouse's journal until then, were sent to Paris. The expeditioners stayed six weeks at Botany Bay, Australia, and left there on 10 March 1788. Expedition was last seen leaving northern tip of Australia. Both ships struck a coral reef beside a Pacific island and sank.[313] No one on this expedition was ever seen again. France sent out eight search expeditions, 1791–2008.

In 1791, a search expedition of two ships left France commanded by Joseph-Antoine d'Entrecasteaux (1730–93).[314] Aboard were three physician-naturalists and a gardenerbotanist. One of the three was botanist Jacques-Julien Houton de Labillardière (1755–1834), who before the voyage, went to London and studied Banks' collections from Cook's first voyage.[315] The French Revolution had begun. Popular d'Entrecasteaux had peace on his ship, but after he died, an aristocrat replacement, Alexandre d'Auribeau, feared a republican crew might cause conflict, and on the return voyage, when they reached Java in October 1793, he had Dutch authorities arrest them, including Labillardière. Later, they were released in a prisoner swap.

Labillardière published his account of the expedition in 1799 (English 1800, German 1802), which Duyker thought was a classic of French travel literature. His "great botanical work, *Novae Hollandiae plantarum specimen*" appeared in parts, 1804–07. His "pioneering" *Sertum Astro-Caledonicum* appeared in 1824–25.[316] His biographer, Edward Duyker, considered him a founder of Australian botany, zoology, and ethnography.[317]

In 1800, both France and Britain sponsored expeditions to Australia in hopes of gaining some strategic advantages over the other country.[318] Both expeditions encountered difficulties, but also returned with important observations on geography and plants and with numerous specimens.

In 1825, Admiral Thomas Manby, RN, went to Paris and reported that a whaling vessel had visited an island and a chief visited his ship with a metal cross in one ear. He said when he was young, a ship had struck a coral reef during a storm and those aboard had died. In 1826, Manby obtained some artifacts and traced both ships to Vanikoro Island. France eventually determined that was where the ships had sunk.[319]

Almost simultaneously with Humboldt and Bonpland's Spanish-American explorations were French expeditions to Egypt under Napoleon, 1798–1801,[320] and to Australia, 1800–03, under Nicolas-Thomas Baudin (1754–1803),[321] and a British expedition to Australia under Matthew Flinders (1774–1814).[322] Both Baudin and Flinders became interested in natural history collections, and on 8 April 02, they accidentally met in a bay which Flinders named Encounter Bay. Although their countries were at war, both had passports from the other's navy, and they exchanged news of their recent discoveries.[323]

Baudin was from an Atlantic island and went to sea as an adolescent on a merchant ships and later joined the French Navy.[324] In 1780, he received command of a ship, only to have it later given to an aristocrat, which led to his resignation and return to merchant shipping. At Cape of Good Hope, he met a gardener in 1787 who wanted to carry a botanical collection to the Austrian emperor. Baudin's involvement in that endeavor led to further involvement in further botanical expeditions, one to the West Indies in 1796 being very successful. He rejoined the post-Revolutionary Navy and received support from botanist Antoine-Laurent de Jussieu for him to lead a scientific expedition around the world.[325] Baudin was a capable navigator but a poor manager of men on the expedition, all of who—sailors and scientists—he alienated. He received two ships, *Géographie* and *Naturaliste*, which left Le Havre on 19 October 1800. He had more naturalists than he wanted. They reached Ile de France (now Mauritius) on 16 March 01. It took a month for Baudin to obtain all the provisions needed. Before resuming the voyage, two botanists and two zoologists deserted. They claimed sickness, but Baudin visited the hospital and found none of them there. Two other zoologists died at sea, leaving only François Péron (1775–1810) and painter-naturalist Charles-Alexandre Lesueur (1778–1846), Pèron's assistant.[326] Pèron and Lesueur brought back over 100,000 specimens that included 2,500 species new to science. Lesueur made over 1500 drawings, but Pèron died before publication of their findings were completed, and few drawings were published until 1988.[327] Josephine Elliot and Jane Johansen's biography of Lesueur emphasized his later life at New Harmony, Indiana, but contains a few plates from his voyage.

Pèron campaigned in Paris to be official narrator of the expedition and was chosen.[328] He had access to Baudin's manuscript journal, which contained an unfavorable comment about himself.[329] Pèron published volume one of his *Voyage de découvertes* (1807) before dying of tuberculosis; vols. 2, 3, and atlas (1815–17) were completed by other members of the expedition.[330]

Matthew Flinders (1774–1814) explored Australia, 1795–1800, and Flinders Island and Flinders Ranges commemorate his achievements.[331] He recommended to Banks and the Royal society that an expedition be sent to Australia to continue researched Banks had done on Cook's first voyage.[332] Baudin's expedition added motivation, and the Navy agreed to Flinders leading it. He studied the Great Barrier Reef and realized it was the greatest reef known, and he developed an insightful hypothesis on its growth.[333] His geographic chart of reefs was still used in the 1900s.[334] Flinders obtained an excellent Scotch botanist, Robert Brown (1773–1858) who had been Bank's librarian, and Austrian biological artist, Ferdinand Bauer (1760–1826).[335] Yet, Flinders was unfortunate in having a ship that leaked even on its voyage from England. In 1803 he had to land at Ile de France for repairs, where he was imprisoned until October 1810. In captivity, he worked on his memoirs. He had left Brown and Bauer in Australia, and they returned to England in 1805. Brown published *Prodromus Florae Novae* Hollandiae (1810).

Russians joined imperial rivalry with four voyages around the world, 1803–29,[336] the first under Admiral Ivan Kruzenstern, veteran of British Navy, 1803–06.[337] It was undertaken at his suggestion, and it included two naturalists: Dr. Tilesius von Tilenau and Baron G.H. von Angasdorff. The second circumnavigation began after Napoleon's final defeat in 1815, under Otto von Kotzebue (1788–1846), who had served during the first expedition. His expedition also had two naturalists, botanist Adelbert von Chamisso (1781–1838)[338] and physician-zoologist Johan Eschscholtz (1793–1831).[339] Near the Bering Strait, a newly discovered island was named for Chamisso and a bay for Eschscholtz. On return voyage, they stopped at Hawaii and San Francisco and reached St. Petersburg 3 August 18. Chamisso returned to Berlin and became conservator of the Botanical Garden and Royal Horticulturist and only publishing his expedition journal in 1835. Kotzebue also commanded the third circumnavigation, 1823–26; Eschscholtz returned as zoologist, and a separate physician. Fadden Bellinsgauze (1779–1852), veteran of Russia's first circumnavigation, later commanded ships on Baltic and Black Seas.[340] In 1819–21, he led a two-ship expedition to explore as close to the South Pole as possible.[341] On 27 January 20, he reached latitude $69^{0}25'$—20 miles from Antarctica—and on 15 December 20 captured a previously unknown emperor penguin, and discovered two Antarctic islands.

A French explorer whose discoveries were quite influential because he published very frequently was naval officer Jules-Sébastien-César d'Urville (1790–1842), who died (with wife and son) in a train wreck, while publishing results of his final expedition. His Australian biographer, Edward Duyker, discovered such a wealth of records, manuscripts, and publications that he wrote a detailed biography (2014, 671 pages). Duyker was pleased with Auguste Chevalier's studies on d'Urville's contributions to phytogeography. His father died and at age 9 he joined the French navy for two years. Later, he went to Collège de Bayeux, spent another two years in the navy, then to Lycée Imperial de Caen, where he studied science and mathematics. He became friends with director of Caen Jardin des Plantes and with an entomologist who later became his family doctor in Paris. In 1807 he rejoined the navy, passed an exam for officer training and went to Toulon, where British ships blockaded the harbor. At age 22, he finally went to sea, until Napoleon surrendered in 1814. He spent his spare time studying Toulon's flora. In 1819 he became naturalist on a hydrographic voyage to Greek islands. Afterwards, he and wife Adélie went to Paris where he helped edit expedition results, botanizing in spare time. He published his own expedition findings in 1821. In Paris, he became friends with Humboldt, botanists, and entomologists. King Louis XVIII named him Chevalier de la Légion d'Honneur, his entry into higher society. He received a naval promotion and increased salary.

In 1811 he had become friends with Ensign Louis-Isidore Duperrey, and in 1821 they became interested in exploring Pacific islands, with Duperrey taking nautical and hydrographical responsibilities and d'Urville handling natural history and expedition narration. They proposed to the navy a circumnavigation of the Earth, emphasizing Pacific exploration. The proposal was approved. Duperrey was four years older than d'Urville and given command, with d'Urville second in command. Their ship, *Coquille*, sailed 11 August 22 to Canary and Azores Islands, Brazil, Strait of Magellan, Peru, then Tahiti.[342] They explored little-visited islands just above and below the Equator, with d'Urville collecting plants and insects when possible. Being second in command ended his friendship with Duperrey, and he became friends with Charles-Hector Jacquinot, six years younger, who became d'Urville's second in command on subsequent voyages. Returning to France, d'Urville published articles, but not a book. His plant collection of about 3,000 species was deeply appreciated at Museum d'Histoire naturelle, and his insects more so. Latreille reported that he had collected about 1,200 insects, of about 1,100 species.[343]

Since his achievements were warmly received, d'Urville's proposal, 23 May 25, of another exploration under his command was eventually accepted, and he was given the same ship, with name of *Astrolabe*. Not until 25 April 26 did they depart for Canaries, Azores, and Brazil, then east to Australia.[344] D'Urville collected insects and recorded associated plants. He and others were interested in ethnography and collected both information and artifacts. At Vanikoro Island, natives showed where each of La Pérouse's ships sank, and told stories of survivors and conflicts with natives who wanted to scavenge the wrecks. After a successful voyage, they returned to France on 25 March 29. D'Urville published five volumes and atlas on the voyage (1830–33), and there were specialist volumes by others which he edited on botany, zoology (four volumes), entomology (2 parts), hydrography and meteorology and atlas, and philology by d'Urville—all published in 1830–35 He then wrote a history of discoveries on circumnavigations by Spanish, Dutch, British, French, and Russian explorers (2 vols., 1834–35). He and his officers were widely celebrated and rewarded, and he proposed leading another voyage.

In February 1837, King Louis-Philippe suggested he sail toward the South Pole, with two ships. D'Urville appointed Charles Jacquinot commander of *Zelée*, which was identical to *Astrolobe*. They left on 7 September 37.[345] They surveyed Strait of Magellan, collecting marine animals, and a few insects, mosses, lichens, grasses, and algae. They sailed south to latitude 64⁰23' on 24 January 38, when blocked by ice. They attempted to re-provision the ships at Telcahuano, Chile, but other ships had recently done that and they had to sail on to Valparaiso. They departed 29 May and, after several stops, reached Tahiti on 9 September, Fiji on 16 October, where they sought vengeance for massacre (and eating) of a French crew in 1834 by Solomon Islanders. They arrived at Guam on 1 January 1839 and re-provisioned at Ternate, Dutch East Indies on 29 January. On 18 January 40 they headed south again and reached land at about 660, which d'Urville named Terre Adélié, for his wife.[346] With men in declining in health, they returned to Toulon on 7 November. Two months later he, Adélié, and their son died in a train wreck. His publication preparations were then underway, and Charles Jacquinot was able to take over and see 29 volumes published, 1841–54. Does not d'Urville deserve as much or more celebration as Cook?

German Philipp Franz von Siebold (1796–1866), from Würzburg, was son and grandson of physicians.[347] He earned his medical doctorate at the University of Würzburg (1820). He was inspired by Humboldt's writings, and after practicing medicine for two years, he went to the Netherlands and became a physician in the Dutch military. He sailed to Batavia (Jakarta), Dutch East Indies in 1823. He was sent to the Dutch trading post in Japan. He was credited with curing a local official, and he introduced vaccination against small pox. He founded a medical school at Nagasaki, and he collected plant and animal specimens. In 1829, Japan's government imagined he was a Russian spy and sent him back to Batavia. He returned to the Netherlands and settled with plant specimens in Leiden, where botanists and zoologists helped him publish *Fauna Japonica* (2 vols, 1833–50) and *Flora Japonica* (2 vols., 1835–70). He also published his memoirs (7 vols., 1832–52).

British army surgeon J. Vaughan Thompson (1799–1847) was first to make and use a plankton net,[348] and he published his findings in his *Zoological Researches* (1830), which Darwin took with him on the *Beagle*. Darwin became the second to use one.[349]

Darwin's contributions to biogeography began with publication of his *Journal of Researches* from his voyage on *Beagle* (1839). The ship sailed to Brazil, south along South America's east coast to Magellan Strait, north along west coast to Ecuador, and west 600 miles to Galapagos Islands. He had collected information on coastal ranges of species, which was useful, without stimulating theoretical speculations. He did notice that as he traveled along, a species he had studied might give way to a similar species having a different range from the one he had first studied. However, these islands had terrestrial plants and animals, such as large turtles and lizards, which was puzzling.

His *Journal of Researches* established his reputation as a naturalist. His discussion of animals at the Galapagos Islands is a notable illustration of his biogeography, but other examples of it are throughout the book. His *On the Structure and Distribution of Coral Reefs* (1842), with map of the oceans, is a notable biogeography monograph. His four monographs on Earth's fossil and living barnacles, 1850s, had broader coverage than *Coral Reefs*, since barnacles are not confined to tropics.[350] Darwin conducted first biogeographical experiments, subjecting different kinds of seed to immersion in sea water for 12–137 days, and testing for viability afterwards.[351]

English botanist Joseph Hooker (1817–1911) was son of Britain's leading botanist, William Hooker (discussed earlier). As son of a botanist, he had a head start, even in geographical botany:[352]

> [W]hen still a child, my father used to take me [on] excursions in the Highlands, where I . . . botanized, and well I remember on one occasion, that, after returning home, I built up by a heap of stones as representation of one of the mountains I had ascended, and stuck upon it specimens of the mosses I had collected on it, at heights relative to those at which I had gathered them. This was the dawn of my love for geographical botany.

The simplest way to become a professional botanist at the time was to earn an M.D. degree, which is what Joseph did, at the University of Glasgow (1839).[353] That enabled his father to obtain a position for him as an assistant surgeon and naturalist on a naval ship, H.M.S. *Erebus*, bound with H.M.S. *Terror*, to the South Seas, September 1839–September 1843, to map shorelines and discover the location of the magnetic South Pole. The expedition leader was Captain Clark Ross (1800–62), who had located the magnetic North Pole.[354] Hooker was following Darwin's example and took along the just-published *Journal of Researches* to read on the voyage.[355] Besides many small islands, the ships visited Antarctica, Tasmania, New Zealand, and Falkland Islands. Before returning to England, Joseph's extensive reports to his father appeared in *London J. of Botany* (1843), which his father edited. Hooker's main reports, two volumes each, were: *Flora Antarctica* (1844–47), *Flora Nouva-Zealand* (1853–55), and *Flora Tasmaniae* (1855–60). Accompanying lithographs were based upon his specimens and sketches. Thus began the career of the most productive phytogeographer since Humboldt.

In the mid-1830s, Americans decided to emulate European voyages of discovery. Congress had not exerted much influence on preparations for land expeditions, but an ocean voyage of discovery, 1838–42, was on a larger scale, requiring Congressional appropriations.[356] It was not a matter of squeezing a few naturalists on a ship with other objectives—the naturalists' goals were the ships' goals. The U.S. Navy's commanding officer, Lieutenant Charles Wilkes (1798–1877), took his responsibilities very seriously, but he had a rigid personality and had difficult relations with some scientists and officers.[357] He limited the number of scientists to 7, plus two artists. One, Peale, served as both naturalist and artist.[358] After the expedition, its specimens were deposited at the U.S. Patent Office. The Smithsonian Institution was established in 1846, but its first Secretary (director), Joseph Henry, a physical scientist, resisted accepting Expedition specimens, because he did not want the Smithsonian to become a warehouse of specimens, and he thought its maintenance would consume too much of his annual Congressional appropriation.[359] However, in 1857 Congress provided funds to construct display cases, moving expenses for collections, and annual maintenance, and transfer from Patent Office to Smithsonian occurred.

Charles M. Scammon (1825–1911) was from a middle-class family in Pittston, Maine, but instead of obtaining a conventional higher education and occupation, like older brother Johnathon, who became a Chicago lawyer, he went to sea in 1850, apparently responding to the California gold rush.[360] In California, he decided grubbing for gold was not for him. Instead, in 1852, he took command of a brig heading out to hunt seals, elephant seals, and whales.[361] His ship, *Mary Hellen*, had a successful hunt, and he had found his profession. During his first decade hunting pinnipeds and cetaceans, the idea of writing a book about his victims developed, which indicates he likely had some higher education. During the Civil War, he took command of a U.S. Revenue Cutter, headquartered in San Francisco, and remained there until he retired with disability in 1895.

He apparently began about 1862 collecting observations and measurements of marine mammals killed. He wrote articles for a magazine, *Overland Monthly*. When he wrote a note, "On a New Species of *Blaenoptera*" (1872), before publication, he consulted Philadelphia zoologist Edward Drinker Cope! When writing *The Marine Mammals of the Northwestern Coast of North America, together with an Account of the American Whale-Fishery* (1874), he also consulted Smithsonian zoologist William H. Dall. He dedicated his book to Louis Agassiz, who died a year before publication.

It is an impressive, very well-illustrated book (reprinted, 1968). There was no one else who had knowledge needed to write it. Part I: Cetacea had a chapter each for 8 whales and 16 chapters on dolphins and porpoises; Part II: Pinnipedia had seven chapters on seals, walruses, and fur otters; Part III: American Whale-fishery. An appendix by Dall, "Catalogue of the Cetacea of the North Pacific Ocean," was 27 pages.

D. 1950–80

I never wrote a longer version of this section for *ESA Bulletin*, so it is not condensed from a previous publication. Lomolino et al., *Foundations of Biogeography: Classic Papers with Commentaries* (2004, discussed earlier) reprints many papers, 1947 to 1982, which assisted my selection of papers

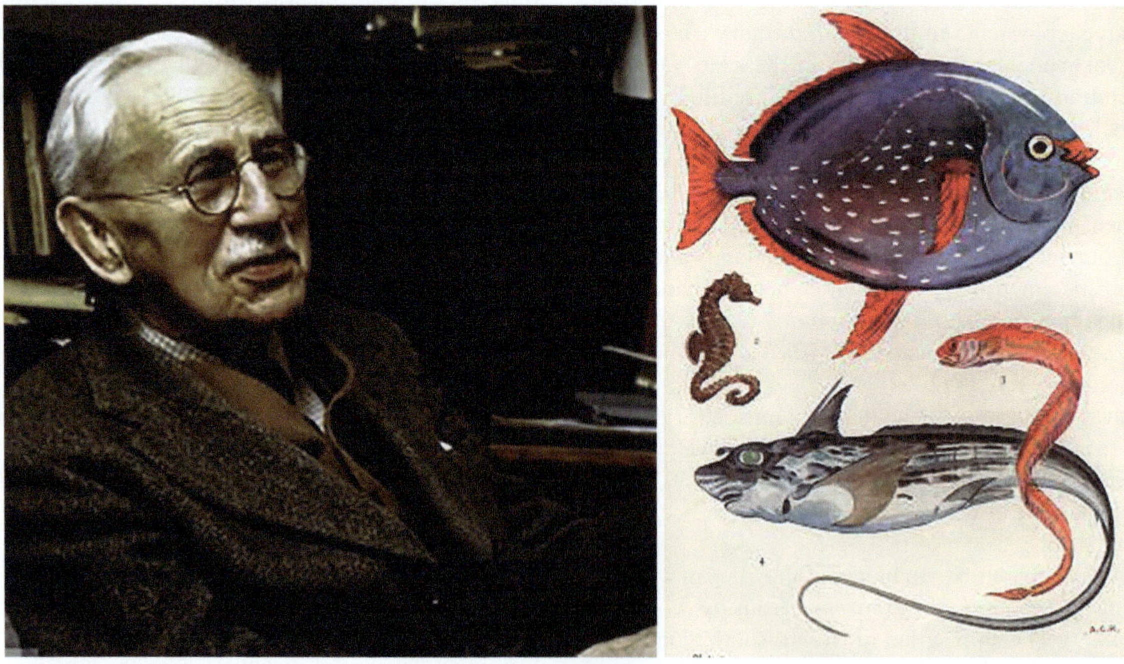

FIGURE 6.7 (a) Alister Hardy. Web. (b) Unusual oceanic fish around Britain. Hardy 1956–59: II, facing 19.

to fashion a coherent history. Their book also includes seven introductory essays for the period discussed here. Vicki Funk's essay on "Revolutions in Historical Biogeography," divided such revolutions into five categories: Descriptive (1820, A.P. de Candolle), Evolutionary (1859, C. Darwin), Phylogenic (1950, W. Hennig), Panbiogeography (1947–52, L. Croizat), and Vicariance (1974, G. Nelson). If German Hennig was so important, why did the editors omit him from their 1291-page anthology? Perhaps because he was primarily an insect taxonomist who developed phylogenetic systematics (cladistics). However, he did combine two subjects in *The Diptera Fauna of New Zealand as a Problem in Systematics and Zoogeography* (1966, 81 pages).[362] Even if this work was conceptually important, there was no stampede to try out Hennig's phylogenetic biogeography. Brown and Lomolino's textbook *Biogeography* (ed. 2, 1998) included a useful chapter, "The History of Biogeography."[363]

Uppsala native Sven Ekman (1876–1964) earned three degrees at Uppsala University (1899, 1903, 1904) and became a professor of zoology there in 1927. His research was in limnology and biogeography, and he published an article on zoogeography in German in 1940 before his *Zoogeography of the Sea* (1953) appeared in English. We may suspect his book's publication was delayed by World War II. It was probably the first comprehensive study of marine zoogeography, and the *Foundations* editors reprinted its conclusion.[364]

Charles Elton (1900–91, figure 4.8) was a fine animal ecologist whose higher education was at Oxford University, where he remained.[365] He gained wide recognition with his first book, *Animal Ecology* (1927), and he was an early member of British Ecological Society (BES).[366] Plant ecologists had founded BES, and it's *Journal of Ecology* was practically a plant ecology journal. In 1932, Elton founded at Oxford both a Bureau of Animal Population and *Journal of Animal Ecology*, with himself its first editor. Lomolino et al.'s reprinting (2004) of his "Competition and the Structure of Ecological Communities" (1947) seems odd, since it is not directly focused on biogeography. However, it can be viewed as a step toward Elton's *The Ecology of Invasions by Animals and Plants* (1958), which addressed two bio-geographical themes: invasions in general and dynamics of invasions on islands. These editors reprinted its chapter 4: "The Fate of Remote Islands," which initiated studies on this topic.[367]

Another English animal ecologist who began his research during the 1920s and achieved its final synthesis during the 1950s was Alister Hardy (1896–1985). After his service in World War I, he had studied under Julian Huxley at Oxford. He began studying whales and their plankton food aboard *Discovery*, 1925–27. His synthesis was in two volumes: *The Open Sea, Its Natural History: Part I, the World of Plankton* (1956) and *Part II, Fish and Fisheries* (1959).[368]

A third Oxford ecologist besides Elton and Hardy was Cambridge-educated David Lack (1910–73, figure 4.8), who went to Oxford in 1945 as new head of Edward Grey Institute of Field Ornithology.[369] Lack researched and published on bird ecology; he was primarily an ornithologist. If Lack and Elton had been Americans at an American university, one would presume they would have become pals. In England, heads of ecological institutions were concerned to protect their turf, and they never had a close relationship. Lack's best-known book was his second (of 12), *Darwin's Finches* (1947). Before moving to Oxford, Lack had been biology master, Dartington Hall School, in south Devon. He read P. R. Lowe's "The Finches of the Galapagos in Relation to Darwin's Conception of Species" (1936) and obtained a year's sabbatical to study them himself.[370] He and four assistants left England in fall 1936, and he collected data for a book, which would have been published before 1947 if an inconvenient war had not intervened. His research program, like Elton's, had also been established before Hennig's 1950 publication. Three more of Lack's books are notable here: *Natural Regulation of Animal Numbers* (1954), *Population Studies of Birds* (1966), and *Island Biology Illustrated by the Land Birds of Jamaica* (1976).

Another Englishman, Ronald Good (1896–1992), published an important book in 1947: *The Geography of Flowering Plants* (1947). In 1953, English botanist Turrill compiled an interesting list of works on world phytogeography, beginning with Grisebach's *Vegetation der Erde* (1872). Of the 17 books on his list, one was by an American and one by a Britain. The latter was Good's *Geography*—last on Turrill's list. My copy of Good's edition 3 (1964, + 518 pages) has 9 whole-page colored maps, 16 photographs, 86 drawings (including half-page and quarter-page black-and-white maps), and a bibliography of 838 titles. The preface of edition 1 states that his manuscript was completed in 1939 but could only be published after the war. Good was a professor of botany, University of Hull.

French-Canadian Pierre Dansereau (1911–2011!), from Montreal, earned B.A. and B.S. degrees (1931, 1936) at the University of Quebec, Montreal, then earned his D.Sc. at Université de Genève (1939) in plant systematics.[371] His early publications were on genus *Cistus*. In 1943, he established Service de Biogeographie, and in 1946–47 he published four papers on biogeography of Brazil. He synthesized his new field of research in a well-illustrated *Biogeography: an Ecological Perspective* (1957). It may have been the first modern attempt by one biogeographer to survey plant and animal biogeography. His book did discuss animals, more briefly than plants. In *Bio-Ecology* (1939) the problem of expertise in both fields was avoided because written by both plant ecologist Clements and animal ecologist Shelford.

Philadelphia-born Philip Jackson Darlington (1904–83) grew up in Hartford, Connecticut.[372] To become an entomologist, he earned three degrees at Harvard University (B.A. 1926, M.S. 1927, Ph.D. 1931), writing a dissertation on *Carabidae* beetles of New Hampshire. He worked for United Fruit Co., 1928–29, then returned to Harvard for the rest of his career. However, his studies went beyond entomology in "The Geographical Distribution of Cold-Blooded Vertebrates" (1948), which he further expanded into *The Geographic Distribution of Animals* (1957). He continued this line of research in "Area, Climate, and Evolution" (1959) and in *Biogeography of the Southern End of the World* (1965), with numerous maps. Lomolino et al. reprinted both his 1959 paper and part of his 1965 book.[373]

Another Harvard entomologist, Edward O. Wilson (b. 1929), from Birmingham, earned his B.S. and M.S. at the University of Alabama (1949, 1950), before going to Harvard to earn a Ph.D. (1955).[374] Ants have been one of Wilson's major research subjects, and our three editors reprinted his "Adaptative Shift and Dispersal in a Tropical Ant Fauna" (1959). Like Darlington, however, his interests were not limited to insects. Unlike Darlington, Wilson joined ESA and became as much ecologist as entomologist, and indeed, he became, and remains, one of America's most prominent ecologists.

American animal ecologist Robert MacArthur (1930–72, figure 4.) was born in Toronto, the son of genetics professor John MacArthur, who later moved his family to Marlboro, Vermont, where he taught at Marlboro College.[375] Robert developed an early interest in nature, yet earned two degrees in mathematics from Marlboro College (B.A. 1951) and Brown University (M.S. 1955), before earning a Ph.D. at Yale (1958) under Hutchinson. He spent 1957–58 studying with David Lack at Oxford. Among three important articles he published in 1955–58 was his dissertation on partitioning habitat by five species of genus *Dendroica*: "Population Ecology of Some Warblers of Northeastern Coniferous Forests" (1958) in *Ecology*. In 1961 at a AAAS meeting in New York, he and E.O. Wilson met, and Wilson was glad to meet an ecologist competent in mathematics. They first collaborated on "An Equilibrium Theory of Insula Zoogeography" (1961), later expanded into *The Theory of Island Biogeography* (1967), with much mathematical reasoning from their evidence.[376] Elton understood in general the dynamics of island biogeography, which MacArthur & Wilson explained mathematically. Despite living only 42 years, MacArthur became (as did Wilson) one of our greatest animal ecologists. When told by physicians he would live only one or two more years, he synthesized his knowledge in *Geographical Ecology* (1972).

Henry A. Gleason, discussed earlier, chapter 2C, in 1951 became Head Curator Emeritus at New York Botanical Garden. He began work on a plant geography book which, in 1961, he felt he could not complete and asked a Curator, Arthur J. Cronquist (1919–92), if he would. He did. The first 21 chapters in *The Natural Geography of Plants* (1964) are Gleason's and chapter 22 and modification of chapter 11 are by Cronquist. It contains numerous excellent photographs of plants and a map, but no bibliography or even mention of authors within the text. Since Gleason had an established reputation, the book was undoubtedly given serious attention, but it would have required supplements if used as a textbook.

Londoner Paul Colinvaux (1930–2016) studied ecology at Cambridge University (B.A. 1956, M.A. 1960), then went to New Brunswick, Canada and conducted a soil survey for the government. He met there marine botanist Llewellya Hillis, who was on a post-doc at the University of New Brunswick, and they married before emigrating to the United States. Paul enrolled at Duke University, Durham, North Carolina, to earn a Ph.D. in zoology (1962), under paleoecologist Daniel Livingstone. Colinvaux then went to Yale on a post-doc, working under Deevey and Hutchinson. In fall 1964 Colinvaux and Hillis joined the faculty of Ohio State University, he in Zoology and she in Botany Departments. He was a talented speaker, and his introductory ecology course soon drew over 1000 students a year. He also was a productive scholar, and published *Introduction to Ecology* (1973); 20 years later, it was just *Ecology* (1993). Colinvaux's major research began to take shape during his post-doc at Yale. He spent over 40 years taking cores from bottoms of lakes and ponds, to obtain fossil pollen grains of past vegetation at different periods, to determine the impact of glaciation in northern hemisphere upon climate in Amazon Basin. His fifth and last book told that story: *Amazon Expeditions: My quest for the Ice-Age Equator* (2007). In 1991, they moved to Smithsonian Tropical Research Institute, Balboa, Panama, and later to Woods Hole, Massachusetts.

Also getting underway in the mid-1960s was *vicariance biogeography*. It developed in reaction to general acceptance of plate tectonics. E.O. Wiley in 1988 stated that "the relationship between vicariance biogeography and ecological biogeography is still being defined."[377] Thirty years later, Ingi Agnarsson, Jason Ali, and David Barrington explained that vicariance biogeography "seeks to marry geology and biology in study of current and historical patterns of biodiversity,"[378] which is little different from Wiley's definition, but they at least provide guidance to relevant literature since Wiley's survey.

A noteworthy encyclopedia that has not attracted wide-spread attention (known to me) is E.J. Fittkau and four other editors, *Biogeography and Ecology in South America* (2 vols., 1968, xvi + 946 pp.). Two articles are in Spanish, and the rest are in English and German. It is unclear to me what magic led to completion of so many well-illustrated scientific accounts in time for publication. Its articles cover geology, geography, and different groups of plants and animals.

The only similar work for another continent known to me is Victor Shelford, editor, *Naturalist's Guide to the Americas* (1926, xv + 761 pages), sponsored by ESA; and in 1928, he began planning for a second edition written by him alone, *The Ecology of North America* (1963, xxii + 610 pages). There were some gaps in the first edition, and he had graduate students conduct doctoral research to fill them:[379]

> [H]emlock-cedar forest in Oregon, the cold desert in Utah, the Rocky Mountain forest and pinyon-juniper in Arizona, the mountain forest and oak bush in Utah, the mountain forest in Maine, the deciduous forest in Illinois, and the tall grass grassland in Illinois.

Graduate students who undertook those assignments for dissertations published their studies. Shelford's own new studies included tundra in northern Manitoba, Mississippi River floodplain in western Tennessee, and 14 years study of forest populations in central Illinois.

Augusta, Georgia native Wilfred T. Neill (1922–2001) earned a B.A. degree (1941) at the University of Georgia, joined the U.S. Army Airforce, and worked in South Pacific during World War II. Afterwards, he became a herpetologist (bitten by 41 venomous snakes!) and taught in Augusta colleges. In 1949, he became Research Director at E. Ross Allen's Reptile Institute, Silver Springs, Florida, where he published *The Story of Florida's Seminole Indians* (1956) and *Historical Biogeography of Present-Day Florida* (1957). His *The Geography of Life* (Columbia U.P., 1969, xv 480 pp.) is a well-illustrated biogeography textbook (with four pages of color photographs).

Three biologists at King's College London—Barry Cox, Ian Healey, and Peter Moore—published a paperback textbook, *Biogeography: an Ecological and Evolutionary Approach* (L and NY: 1973, ed. 2, 1976, ix + 194 pp.), with lots of maps, diagrams, drawings, and some photographs. Britons write more concise textbooks than do Americans.

English biogeographer David Watts (1987–2003) was founding editor of *Journal of Biogeography* (1974), in Canada.[380] He was editor until 1978, then co-editor until 1980. It appears monthly.

A remarkable marine biogeographer, Dutchman Geerat Vermeij (b. 1946), lost his sight at age 3, which did not deter him from becoming a very successful scientist.[381] He immigrated to America, earned a B.A. degree (1968) at Princeton University and M.Phil. (1970) and Ph.D. (1971) at Yale. Among his numerous publications is *Biogeography and Adaptation: Patterns of Marine Life* (1978, xv + 332 pp.), which could have been a textbook in marine biology. It was perhaps the first such work since Ekman's book of similar scope appeared in 1953. Vermeij emphasized mollusks, and his book has numerous photographs of sea shells plus two of fish. His final comment was: "I leave the reader with many unsolved problems and unanswered questions."[382]

Oregon State University's Thirty-seventh Annual Biology Colloquium (1976) was on *Historical Biogeography, Plate Tectonics, and the Changing Environment* (1979, xii + 500 pp.), which published 37 talks plus epilogue by editors Jane Gray and Arthur Boucot. Articles on geology and paleontology predominate, but many of them relate to modern distributions of species.

This brief survey-history of biogeography indicates some highlights of its developments, 1700–1980. Two new theories had an important impact: Darwin's theory of evolution (1859) and Wegener's theory of continental drift (1912). There are discussions in my *Roots of Ecology* (2012) that are relevant, but the subject is more organized and detailed in my parts 61A, B, and C (2018–2019) and more concise in this chapter. James H. Brown and Mark V. Lomolino's *Biogeography* (1998) is both an excellent summary of this discipline and a valuable history resource.

NOTES

1. Egerton 2018a: 192–194,
2. Egerton 2014c: 347–351, 2016c: 372, 2019a: 1–2.
3. Nicholls 2009: 253–254.
4. Von Hofsten 1916: 32, Hooykaas 1956, Browne 1983: 14, Ebach 2015: 27–28.
5. Raven 1942, Stressmann 1951: 41–45, 1975: 43–45. Webster 1975, Browne 1983: index, Egerton 2005, 2012a: 60–64, 2018a: 195.
6. English ed., 1975, Raven 1942: 81.
7. Ray 1717: 298–299.
8. Allen 1951: 461–463, Savage 1970: 27–36.
9. Stearn 1970: 306–315.
10. Lawson 1967: 96–120.
11. Feduccia 1985: 8.
12. Von Hofsten 1916: 231, 243, 245, De Virville 1954: index, Becker et al. 1957, Leroy 1976, Greene 1983: II, 938–964, Drouin 1991: 211, Magnin-Gonz 2004: 117–118.
13. Quoted in English in Linnaeus 1781: 91.
14. Savage 1970: 37–92, Frick 1971, Frick & Stearns 1961, Egerton 2006d: 345–349, 2012a: 78–79, Nelson & Elliott 2015.
15. Allen 1951, Brigham 1998: 95–96.
16. Feduccia 1985, Howard & Staples 1983.
17. Catesby 1747.
18. Golder 1922–25, Frost 2003, Kolchinsky 2004.
19. Von Hofsten 1916: 247–248, Kruta 1972, Egerton 2008b: 41–42, 2012a: 93–94, 2018a: 197, Jones 2014: index.
20. Hintzsche & Nickol 1996.
21. Gmelin 1911.
22. Stejneger 1936, Lindroth 1976, Mearns & Mearns 1988: 347–356, 1992: 409–414, Jones 2014: index.
23. Ford 1966, Steller 1974, 1988, Hintzsche & Nickol 1996, Littlepage 2006, Egerton 2008: 43–50, 2012a: 94–96.
24. Stejneger 1887, 1936: 353–357, Ford 1966: 152–165. Mattioli & Domning 2006.
25. Von Hofsten 1916: 239–240, 243–248, 1957, Stresemann 1951: 51–54, 1975: 50–52, Blunt 1971, Farber 1982: index, Browne 1983: index, Frängsmyr 1983, Lindroth 1973, 1983, Larson 1986: index, Drouin 1993: 40–44, Magnin-Gonze 2004: 120–128, Egerton 2007a, 2012a: index.
26. Linnaeus 1749.
27. Fries 1950: 18.
28. Caddy 1886–87.
29. Du Rietz 1957a: 161, slightly abridged.
30. Von Hofsten 1916: 237–242, Stresemann 1951: 60–61, 1975: 57–58, Piveteau 1954, Roger 1970, 1997a, b, Sloan 1976, 2002, Farber 1982: index, Mayr 1982: 440–442, Browne 1983: index, Egerton 2007, 2012a: index.
31. Buffon 1976: 96, 118, translated in Nelson 1978: 273–276, Larson 1986: 451–453, Ebach 2015: 23–24.
32. Aristotle, *Historia animalium* VIII: 28–29.
33. Granit 1973, Egerton 2006d: 349–353, 2012a: 79–80.
34. Hansen 1964: index, Eriksson 1972.
35. Fries 1950: 29.
36. Collander 1965: 14.
37. Holthuis & Sakai 1970: 16–20, Eriksson 1976a.
38. Von Hofsten 1916: 252–255, Mayr 1982: 442, Browne 1983: 25–26, Larson 1986: 454, 473–482, Feuerstein-Herz 2004, Jackson 2009: 252.
39. Bodenheimer 1955.

40. Camerini 1993b: 483–484.
41. Larson 1986: 453–454.
42. Browne 1983: 26.
43. Bylebyl 1976, Magnin-Gonze 2004: 155–156.
44. Von Hofsten 1916: 251–252, Browne 1983: 38–40, Larson 1986: 466–458.
45. Browne 1983, Lomolino et al. 2004.
46. Biermann 1972, Botting 1973, Stafleu & Cowan 1976–2009: II, 363–371, Browne 1983: 42–48, Rupke 2000, Helferich 2004, Egerton 2009a, b, 2012a: 121–125, Ebach 2015, Wulf 2015.
47. Sarton 1943, Stafleu & Cowan 1976–2009: I, 274–276, Hossard 2001, Cerrutti 2003, Lourteig 2003.
48. Pritzel 1871–77: 152–153, Botting 1973, Stafleu & Cowan 1976–2009: II, 363–371, Nicolson 1987: 174–186, 1996: 290–292, 2013: 96–97, Egerton 2009b, 2012a: 121–125.
49. Translated in Bowen 1981: 221.
50. James & Martin 1981: 113–1222, Helferich 2004, Muller-Schwarze 2004.
51. Stearn 1968: 116–128, Nelson 1978: 275–276, Browne 1983: 41–52, 58–62, Dajoz 1984: 10–13, Larson 1986: 469–471, Acot & Bourguet 2003, Drouin 2003, Drouin & Lalande 2003, Lalande 2003, Walls 2009: 1–147, Ebach 2015: 113.
52. Ebach 201: 38–40.
53. James & Martin 1981: 122–123, Egerton 2009a: 272–273,
54. Lowenberg 1873: 365–390, De Terra 1955: 283–303, Kellner 1963: 131–146, Botting 1973: 238–242.
55. Bowen 1981: 240–259, James & Martin 1981: 123–125.
56. Von Hofsten 1916: 256–257, Stressemann 1951: 67–72, 1975: 65–70, Esakov 1974, Mearns & Mearns 1988: 288–297, Wendland 1992, Egerton 2008: 50–55, 2012a: 96, Jones 2014: index.
57. Foster 1965: 13–14.
58. Urness 1974, Pallas 1812, 1967a, b.
59. Larson 1986: 461–462, 484, Sytin 1997, Sytin & Borkin 2007.
60. Ebach 2015: 165–173.
61. Ebach 2015: 49.
62. Ebach 2015: 61.
63. Pritzel 1871–77: 336–337, Fries 1950: 69–70, Eriksson 1976b.
64. Pritzel 1871–77: 52–55, Pilet 1971, Stafleu & Cowan 1976–2009: I, 438–452, Candolle 2004, Egerton 2010a: 26–29, 2012a: 132–136.
65. Candolle 1820.
66. Candolle 1820: 384, translated in Lyell 1830–33: II, 131, reprinted in Egerton 2015c: 434–435.
67. Egerton 2010a: 29.
68. Christensen 1924–26: 1, 253–276, Sanders 1975b.
69. Whittaker 1962: 9, Nicolson 1996: 293, 2013: 97.
70. Pritzel 1871–77: 287, Christensen 1924–26: II, 165–179.
71. Allan 1967, 1972, Stafleu & Cowan 1976–2009: II, 283–301.
72. Hooker 1834a.
73. Hooker 1834b.
74. Henriksen 1921–37: 84–97, 1932, Jespersen 1946, Tuxen 1967, 1973: 109–110, Landin 1971, D'Aguila 2006: 63–64, 199.
75. Fabricius 1778: 154–158, see Larson 1986: 480.
76. Latreille 1817, see von Hofsten 1916: 272, Wilson & Doner 1937: 15–17, Burkhardt 1973, Lindroth 1973: 121–123, Tuxen 1973: 112–113.
77. Lindroth 1973: 127.
78. Browne 1983: 69–72.
79. Smit 1976a.
80. Smit 1976b.
81. Larson 1986: 454, 469, 486.
82. Egerton 2006d, 2007d, 2009c, 2011a, b, 2012a, b, c, 2015b: 244–245.
83. Martin 1952, Bedini 1990, Dugatkin 2009.
84. Andrist 1967, Snyder 1970, Allen 1975, Ambrose 1996, Ambrose & Abell 1998, Stephens 1999, Cutright 2003, Tubbs & Jenkinson 2003, Dalton 2008.
85. Jackson 1962, Andrist 1967, Cutright 1969, Snyder 1970, Mearns & Mearns 1992: 143–146, 281–285, Botkin 1995, Ambrose 1996, Tubbs & Jenkinson 2003, Fresonke & Spence 2004.
86. Bartlette 1962, Goetzmann 1967, Meinig 1886–2004.
87. Jackson 1981: 246–248, also Hollon 1949, Billington 1967: index, Goetzmann 1986: index, Sterling 1997g, Nelson 1999, Egerton 2009c: 449–455, 2012a: 129–130.
88. Pike 1895: II, 431, 1966: I, 339.
89. Pike 1966: 353–354.
90. Pike 1895: II, 458, 1966: I, 351.
91. Flores 1984a, b, 1997, 2002.
92. Morton 1967, Flores 1984a: 213.
93. Flores 1984a: 208–210.
94. Cantwell 1961, Hunter 1983, Burtt & Peterson 1995, Atkin 1999, Egerton 2004, 2007c, 2009b: 454–459, 2012a: 127–128, Burtt & Davis 2013.
95. Burns 1909, Peattie 1936: 218–233, Allen 1951: 552–569, Welker 1955: 18–52, Savage 1970: 237–284, Stresemann 1976: 154–156, Elman 1977: 48–79, Hanley 1977: 47–54, Kastner 1977: 159–194, Porter 1986: 41–49, Gibbons & Strom 1988: 40–57, Blum 1993: 26–46, Peck 1999, Walters 2003: 98–100, Egerton 2009c : 434–487, 2012a: 127–128.
96. Graustein 1967, Beidleman 1960, 2006: 132–141, Coats 1969: 295–204, Thomas 1974, Hanley 1977: 103–108, Kastner 1977: 254–283, Stuckey 1997, 1999, Ewan & Ewan 2007: 521–540, Egerton 2009c: 459–461, 2012: 128–130.
97. Stafleu & Cowan 1976–2009: III, 781–787.
98. Nuttall 1818: I, 202–203.
99. Nuttall 1818: I: 23.
100. Nuttall 1951: 22–23.
101. Nuttall 1951: 53.
102. McKelvey 1955: 143–149.
103. McKelvey 1955: 1864–1887, Beidleman 1956, Graustein 1967: 141–145.
104. Nuttall 1821: 57–58.
105. Graustein 1967: 157, 161, MacPhail 1983: 17.
106. Graustein 1967: 134.
107. Weiss & Ziegler 1931, Jaffe 1958: 130–153, Mallis 1971: 19–25, Shor 1975, Hanley 1977: 112–118, Stroud 1992, Schwarz 1997.
108. Stroud 1992, 99, Schwarz 1997.
109. Weiss & Ziegler 1931, Shor 1975, Stroud 1992, 1999a.
110. Mallis 1971: 16–25.
111. Moore 1947, White 1973, Stroud 1992: 33–4, Doskey 1997, Baatz 1999.
112. Meisel 1924–29: II, 130–154, Gerstein 1967, Porter 1979, Bennett 1983.
113. For a map, see Weiss & Ziegler 1931: 189—reprinted in Egerton 2009c: 463.
114. Holthius 1969.
115. Jensen 1955: 42, Poesch 1961: 20–22, Stroud 1992: 59–66.

116. Dupree 1959: 35–36, Wood 1966, Billington 1967: index, Nicholas & Halley 1980, Goetzmann 1986: index, Sterling 1997e, Nichols 1999.
117. Weiss & Ziegler 1931, Shor 1975, Stroud 1992.
118. Poesch 1961, 1974.
119. Ewan & Ewan 1981: 10, Stroud 1992: 73 + portrait, Evans 1997: 20–21, 52–53.
120. Ewan 1950: 13–20, Evans 1993: 113–114, Thomas 1997.
121. Call 1895, Peattie 1936: 261–267, Jaffe 1958, Ewan 1975, Hanley 1977: 126–140, Kastner 1977: 240–253, Sterling 1978, Boewe 1982, 1997, 2003, Warren 2004.
122. Rafinesque 1836: 8, 1944: 299.
123. Rafinesque 1836: 18, 1944: 303, Warren 2004: 6–17.
124. Warren 2004: 159–160.
125. Warren 2004: 61.
126. Jordan 1877, Herrick 1917: I, 285–300, Markle 1997, Warren 2004: 72–75.
127. Stroud 2000: 214.
128. Jordan 1877.
129. Herrick 1917, Ford 1964, 1969, Mengel 1970, Elman 1977: 80–123, Durand & Harwood 1980, Gibbon & Strom 1988: 58–75, Blum 1993: 88–118, Fishman 2000: 136–156, Rhodes 2004, Egerton 2011a: 70–76. 2012a: index.
130. Lindsey 1985, Hart-Davis 2003, Souder 2004, Vedder 2006.
131. Audubon 1999, 2006.
132. Nelson 1975, Stresemann 1975: 153–179, Mearns & Mearns 1992: 94–104, 520–521, Stroud 2000.
133. Bonaparte 1824.
134. Stroud 2000: 56, 66–67.
135. Stroud 2000: 74.
136. Stroud 2000: 76.
137. Stroud 2000: 89.
138. Stroud 2000: 120, 328.
139. Ford 1964: 171, 348.
140. Stroud 2000: 138.
141. Ford 1964: 390.
142. Stroud 2000: index.
143. Ford 1951: 24–25, Fishman 2000: 144–146, Stephens 2000.
144. Audubon 1897: I, 447–532, II, 1–195, Harris 1951, McDermott 1965, Durant & Harwood 1980: 538–598.
145. Müller 1842, Glaubrecht 2002: 259–264, Egerton 2018b: 4.
146. Camerini 1993a, b.
147. Nicolson 2013: 102.
148. Von Hofsten 1916: 281, Smith 2005.
149. Kendeigh 1954: 154.
150. Mayr 1982: 562–64, Quammen 1996: 130–135, Smith 2005, Wagner et al. 2012.
151. Von Hofsten 1916: 301.
152. Schmidt 1955: 767–769, Ebach 2015: 99.
153. Ebach 2015: 97.
154. Austin 1975, Stresmann 19975: 203–204, 225, 260, Mayr 1982: 448, Mearns & Mearns 1992: 394–397, Innes 2004, Bircham 2007: 195–197.
155. Brown & Lomolino 1998: 25.
156. Bircham 2007: 219–220.
157. Thomson 1927, England 2004.
158. Brown & Lomolino 1998: 31.
159. Nelson 1978: 291–292, Ebach 2015: 131.
160. R.L. 1899.
161. Spruce 1908, Huxley 1918, Turrill 1963, Allan 1967: 97–250, Desmond 1977, 1999, Stafleu & Cowan 19976–2009: II, 267–283, Desmond 1977: 316, Endersby 2004, 2008, Egerton 2012a: 172–175, 2012b: 125–138, 2018b: 2, 10–11.
162. Reidy 2010: 28–29.
163. Hooker 1891: 30, Turrill 1953: 30 (quoted in Egerton 2018b: 132).
164. Turrill 1953.
165. Arnold 2005a: 185, 2005b.
166. Von Hagen 1945: 230–296, Desmond 1975, 1977: 378, Seaward & FitzGerald 1996.
167. Von Hagen 1948: 245–263.
168. Beddall 1969, McKinney 1972, 1976, Shermer 2002.
169. McKinney 1970.
170. Wallace 1905: I, 256–257.
171. Wallace 1890: 34.
172. Henderson 1996: 193.
173. Brown & Lomolino 1998: 28.
174. Brown & Lomolino 1998: 620.
175. Bates 1864: 11–18.
176. Belt 1985 [1888]: 71–84.
177. Bates 1864: 250–251.
178. Bates 1864: 415–425.
179. Bates 1862: 502–509, Kimler 1983, Ruxton et al. 2004: 139.
180. Bates 1862: 506–509, Blaisdell 1992: 63–139.
181. Bates 1862: 510–511, Darwin 1985–2019: vol. 10.
182. Darwin 1859: 84.
183. Darwin 1863.
184. Bates 1864: 113–115, illustrated.
185. Darwin 1985–2019: vol. 11.
186. Spruce 1908: II, 362–363.
187. Wallace 1905: I, 337–384, 1916: 297–298, Daws & Fujita 1999: 34.
188. Wallace 1905: I, 366.
189. Wallace 1855, Darwin 1839, Egerton 2012.
190. Schmidt 1955: 767, Darlington 1959: 309, Mayr 1982: 445–448, Browne 2003, 2004.
191. Darwin 1985–2019: vols. 6–7.
192. Dupree 1959: 237.
193. Wallace 1869: 68–69, 1890: 44.
194. Wallace 1869: 20–21 + map, 1890: 7–8 + map, Severin 1997.
195. Bastin 1986: vii.
196. Wallace 1876: I, vi.
197. Blyth 1871. On Blyth, McKinney 1970.
198. Mearns & Mearns 1992: 394–397.
199. Nelson 1978, 1983: 482.
200. Quammen 1996: index, Smith & Beccoloni 2008.
201. Schmidt 1955: 769.
202. Schmidt 1955: 773, Nelson 1975.
203. Wegmann 1976, Greene 1982: 191.
204. Schmidt 1955: 772–773, Wegman 1976: 148.
205. Fries 1950: 70, Collander 1965: 74–77, Stafleu & Cowan 1976–2009: I, 361, Egerton 2013b: 345.
206. Fries 1950: 70–71, Collander 1965: 77.
207. Söderqvist 1986: 50–51.
208. Mearns & Mearns 1998: 136–138.
209. Sharpe & Dresser 1871–72, McGhie 2017: 98.
210. McGhie 2017: 150.
211. Matagne 1998.
212. Christ 1893, Good 1955: 748–749, Pilet 1971a, Stafleu & Cowan 1976–2009: I, 433–437, Mayr 1982: 444–445, Browne 1983: 82–86, Drouin 1983: 60–62, Dajoz 1984: 39–46, Naef 1987: 352–355, Magnin-Gonz 2004: 165–167, Egerton 2010a: 26–29.
213. Nicolson 1996: 300, Matagne 1998: 433.

214. De Candolle 1862, Gray 1876: 197–198, Baehni 1955: 112–114, Drouin 1988: 159–161, 1993: 78–79, 2007: 265.
215. Kronfeld 1908, Stafleu & Cowan 1976–2009: II, 525–530, Cittadino 1990: 119, Nicolson 1996: 297–298.
216. Conard 1951: viii.
217. Goodland 1975, Billings 1985: 5, Nicolson 1996: 304–305.
218. Pritzel 1871–77: 129–130, Wagenitz 1972, Magnin-Gonze 2004: 202, Egerton 2013d: 343–344.
219. Acot 1988: II, 439–440.
220. Grisebach 1838: 160.
221. Grisebach 1846, 1849.
222. Emberger 1936, Jovet 1954: 256–257, Stafleu & Cowan 1976–2009: I, 84, Matagne 1998: 434–437, 1999: 49–50.
223. Matagne 1999: 169–171.
224. Stafleu & Cowan 1976–2009: 682–683, Tobey 1981: index, Matagne 2009: index.
225. Kendeigh 1954: 158.
226. Dupree 1959, 1972.
227. Watson & Goodale 1888.
228. Dupree 1959: 272–279.
229. Lurie 1960, 1970.
230. Allard 1970, 1997.
231. Bruce 1987: index, Rivinus & Youssef 1992: 56–80.
232. Stressmann 1975: 243–245, Rivinus & Youssef 1992: 81–97, Barrow 1998: index, Jackson 1999.
233. Grinnell 1930, Coan 1982, Brodhead 1999.
234. Osgood [& Grinnell] 1944, Sterling 1977, 1997, 1999, Barrow 1998: index, Hermann 2009, Egerton 2014a: 66–69.
235. Merriam 1884: 10–26.
236. Sterling 1977: 204–240.
237. Egerton 2009b: 268–269, 2012a: 123–124.
238. 1890 version, Kendeigh 1954: 161, Egerton 2018b: 34, 1893 version, Egerton 2014a: 67.
239. Brown & Lomolino 1998: 30.
240. Dice 1923, Shelford 1932, 1945, Daubenmire 1938, Peterson 1942, Peterson 1942, Schmidt 1955: 783–784.
241. Sterling 1977: 119–123, Goetzmann & Sloan 1982: 8–14.
242. Goetzmann & Sloan 1982: 16–30.
243. Goetzmann & Sloan 1982: 193–206.
244. Muir 1938, 1954, Wolfe 1945: 279–287. Also, Clarke 1979: 288–290, Worster 2008.
245. Burroughs 1951, Renehan 1992: 207–214, Warren 2006: 119–137.
246. Egerton 1976a: 340, 350, Cittadino 1980: 184–186, Burgess 1996: 70–71.
247. Kohler 2006: 100.
248. Schmidt 1955: 778, Simpson 1944, Colbert 1974, Shoshani 1997.
249. Granger 1933, De Querioz 2014: 43–45.
250. Darlington 1938, 1948, Schmidt 1955: 778–780.
251. Greg Mayer, personal communication.
252. Pool 1915, Ewan 1970.
253. Pound 1954, Clements 1960, Ewan 1971, Tobey 1981: 151–158, Burgess 1996: 26–27, Hagen 1999, Egerton 2009a, 2013b: 358–360.
254. Sears 1958, Ewan & Ewan 1981: 50–51, Tobey 1981: 104–108, Engel 1983: 137–150, Mitman 1992: 16–19, Cittadino 1993, Burgess 1996: 31–32, Cassidy 2007: 5–29.
255. Cassidy 2007: 33.
256. Cowles 1899: 97–98.
257. Cowles 1901, 2007.
258. Clements 1916: 26.
259. Drouin 2013: 55.
260. Spalding 1903.
261. Cowles 1904: 879.
262. Tansley 1947, Godwin 1977: 4, 9.
263. Good 1955: 53, Desmond 1977: 570, Ayres 2012: 72, 2018b: 19.
264. Sheail 1987: 30–35, Fischedick 2000, Ayres 2012: 83–86, Egerton 2013b: 355–357, 2018b: 19.
265. Sheail 1987.
266. Egerton 2015b: 246–248.
267. Egerton 2015e: 1–2.
268. Croker 1991, Egerton 2015e: 12–13.
269. Drude 1890, 1896, Engler 1899.
270. Drude 1902, Engler 1910–15.
271. Fernald 1911, Harshberger 1911.
272. Stafleu 1978, Stafleu & Cowan 1976–2009: II, 757–797, Lack 2000, Magnin-Gonz 2004: 197–199.
273. Smith 2005.
274. Gadow 1913: 76.
275. Creese 2004.
276. Fries 1950: 76.
277. Fries 1950: 60, 75, Söderqvist 1986: 42, 148.
278. Schmidt 1955: 775–777, Bullen 1976, Schwarzbach 1986, Greene 2015,
279. Hallam 1973: 1–21, 1983: 110–156, Colbert 1985, De Querioz 2014: 34–42.
280. Good 1955: 759, Desmond 1977: 520.
281. Good 1955.
282. Glen 1975: index, 1982: 4–5.
283. Goldblatt 1993.
284. Allee et al. 1951: vii.
285. Joerg 1928,
286. Harshberger 1928, Stejneger 1928, Brown 1928, Murphy 1928.
287. Howard 1930: 60, Graham 1941, Burgess 1996: 25.
288. Chapman 1931: 219–231.
289. Simpson 1934, 1964, 1978, 1987, see also Olson 1991, Laporte 2000.
290. Simpson 1934: 67.
291. Simpson 1940, Lomolino et al. 2004: 339–365.
292. Miyanishi 2014.
293. Stuckey 1997, 2001, Neumann 2009, Egerton 2015e: 62–63, 2015c: 455.
294. Lloyd 1966, Desmond 1977: 171, Baer 2004, Preston & Preston 2004, Williams 2011: 7–31.
295. Osborn & Gardner 1939.
296. George 1999.
297. Ladd & Gunther 1957: 67, Rodolico 1974, Stoye 1994, Mills 2001, Marsigli 1681, 1978.
298. Ladd & Gunther 1957: 67.
299. Bougainville 1964, Dunmore 1965–69: I, 57–113, Ross 1978, Brosse 1985: 24–32.
300. Oliver 1909, Laissus 1971, Brosse 1985: 25–30, Williams 2013: 54–72.
301. Dunmore 2002, Ridley 2010.
302. Oliver 1909: 85.
303. Richardson 1980.
304. Isaacson 2003: 290, 437, Chapin 2006: 196–200, 289–291, 319–322.
305. Villiers 1967, Foote 1970, Maclean 1972, Gray 1981, Carter 1988, Horwitz 2002, Williams 2013: 73–94, Egerton 2015b: 241–242.

306. Williams 2013: 73–94.
307. Williams 2013: 95–121.
308. Ledyard 2005, Williams 2013: 124–131.
309. Dunmore 1965–69, Williams 2013: 150–178.
310. Williams 2013: 179–200, 206–221.
311. Dodge 1971: 195–212, 1976: index.
312. Dunmore 1965–69: I, 250–282, Brosse 1985: 76–83, Beidleman 2006: 9–16.
313. Williams 2013: 150–178.
314. Brosse 1985: 84–92, Williams 2013: 164–178.
315. Horner 1987, Duyker 2003.
316. Pritzel 1871–77: 173, Duyker 2003, Williams 2013: 165–178.
317. Duyker 2003: 1, see Dayrat 2003: 196–201.
318. Williams 2013: 201–231.
319. Dunmore 1965–69: I, 339 with map.
320. Alleaume 1999, Drouin 1999, Fischer 1999.
321. Dunmore 1965–69: II, 9–40, Horner 1987, Bonnemains et al. 1988, Fornasiero et al. 2004, Hill 2012, Williams 2013: 205–221.
322. Rice 1999: 1989–229, Fornaskiero et al. 2004, Hill 2012, Williams 2013: 201–205, 221–231, Flinders 2015.
323. De Beer 1960: 12–120.
324. Fornasiero et al. 2004: 17–20.
325. Brosse 1985: 93–107, Horner 1987.
326. Wells 1973, Jovet & Mallet 1974, Fornasiero et al. 2004: index, Duyker 2006, Starbuck 2013: 106–109.
327. Bonnemains et al. 1988.
328. Horner 1987: 326–344, Duyker 2006: 209–220.
329. Baudin 1974: 509.
330. Duyker 2006: 311.
331. Laughton 1889, Keast 1966: 83, 85, Moorehead 1966: 153–156, Brosse 1985: 108–114, Estensen 2002.
332. Rosset al. 1933: 80–86, Kates 1934: 82–83.
333. Flinders 1814: II, 101–115, Sponsel 2009: 30.
334. McGregor 1975: 29.
335. Stearn 1970, Stafleu & Cowan 1976–2009: I, 364–370, Desmond 1977: 47–48. 96, Mabberley 1985, 1999, 2004a, b, Rice 1999: 198–22, Magnin-Gonze 2004: 169, Lack 2015.
336. Brosse 1985: 115–136, Jones 2014.
337. Florinsky 1961: 292, Brosse 1985: 115–124, Sponsel 2009: 32.
338. Rudnick 1978.
339. Blacher 1971, Sterling 1997b, Beidleman 2006: 48–60.
340. Fedoseyev 1970.
341. Brosse 1985: 132.
342. Dunmore 1965–69: II, 109–155, Brosse 1985: 146–151, Duyker 2014: 88–157, Egerton 2019: 16–20.
343. Duyker 2014: 163.
344. Duyker 2014: 177–183.
345. Dunmore 1965–69: II, 341–383, Brosse 1985: 183–194, Duyker 2014: 314–374.
346. Duyker 2014: 440 for map.
347. Holthuis & Sakai 1970: 23–78.
348. Hedgpeth 1957a: 1, Wheeler 1976, Stott 2003: 53–54.
349. Egerton 2010: 402.
350. Darwin 1851a, b, 1854a, b, Egerton 2011a, b, c: 353–355.
351. Darwin 1855, 1857.
352. Huxley 1918: I, 5.
353. Huxley 1918, Turrill 1953, 1963, Allan 1967: 112–250, Desmond 1972, 1999, Browne 1983: index, Endersby 2004, Griggs 2011, Egerton 2012a: 172–175, 2012c: 125–138.
354. Bravo 2000.
355. McCalman 2009: 85–148.
356. Tyler 1968, Stanton 1975, Viola & Margolis 1985, Philbrick 2003.
357. Leonhart 1985, Stann 1985.
358. Poesch 1961: 66–93, 119–203.
359. Fedunkiw, M.P. 2000.
360. Scheffer 1968.
361. Scammon 1874: 11.
362. Hennig 1966, Schmitt 2013.
363. Brown & Lomolino 1998: 13–35.
364. Ekman 1953, Lomolino et al. 2004: 245–248.
365. Burgess 1996: 39, Crowcroft 1991, and above, chapter 2B.
366. Sheail 1987.
367. Elton 1958, Lomolino et al. 2004: 575–593, Southwood & Clarke 1999, Egerton 2015d: 574–576.
368. Hardy 1956–59, Hay 2011.
369. Anderson 2013.
370. Lowe 1936.
371. Dansereau 2005, Monk 1998: 13–15, Egerton 2015e: 95–96.
372. Ball 1985, Wilson 1991.
373. Darlington 1948, 1957, 1965, Lomolino et al. 2004: 852–874, 407–418.
374. Wilson 1994, Burgess 1996: 114.
375. Fretwell 1975, Wilson 1973, 1990, Wilson & Hutchinson 1989, Burgess 1996: 69–70.
376. MacArthur & Wilson 1963, 1967.
377. Wiley 1988.
378. Agnarsson et al. 2019.
379. Shelford 1964: vii.
380. Flenley 2004.
381. Vermeij 1997.
382. Vermeij 1997: 265.

7 Biosphere Ecology

Biogeochemistry, Gaia, and Earth system science are akin to, or versions of, biosphere ecology. Global warming at a high temperature is a biosphere pathology.

PRE-BIOGEOCHEMISTRY

Photosynthesis is an important component of this history. During 400 AD, Rufinus of Aquileia translated from Greek into Latin pseudo-Clement's *Recognitions*, a conversation between a skeptical father and Christian sons. A question raised was "Does not the rebirth of seed from earth and water and its growth into plants for the use of man sufficiently demonstrate the workings of the province of God?"[1] A son proposed planting a seed in a known weight of dirt, with nothing added but water. After plants are grown, they can be weighed, and dirt weighed again. Dirt will have lost no weight, showing plants' substance came only from water.[2]

German scholar-diplomat Nikolaus von Cusa (c. 1401–(64 wrote, in 1450, *Idiota de staticis experimentis*, which included pseudo-Clement's experiment. Nikolaus added: "by the operation of the Sunne."[3] He also suggested weighing seeds or seedlings before the experiment and afterwards burning plants to determine their dry weight. Although both accounts were hypothetical, they seemed based on real trials. During the 1600s, such experiments were reported by Belgian physician Johann Baptisa von Helmont (1577–1644)[4] and English aristocratic physical scientist Robert Boyle (1627–91),[5] and the results were published. Van Helmont used a willow tree. Boyle wondered if his distilled water absorbed any substance from his glass container and his plant absorbed it with water.[6]

Their conclusion was undermined by studies on plant anatomy using microscopes by Italian medical professor Marcello Malpighi (1628–94)[7] and English physician Nehemiah Grew (1641–1712).[8] Malpighi conducted his experiments in 1675, published 1686; Grew conducted his in 1676, published 1682.[9]

Englishman John Evelyn (1620–1706), whose family wealth came from gunpower production, was a founder of the Royal Society of London,[10] as was Boyle, with Grew and Woodward joining later. Evelyn's reformist urges were expressed in two of his books. *Fumifugium, or, the Inconveniences of the Aer, and the Smoak of London Dissipated* (1661) proposed "removing certain trades and planting a green belt of fragrant trees and shrubs."[11] Londoners followed his advice in 1952, after the Great Smog killed 12,000.[12] In *Sylva, or, a Discourse of Forest Trees, and the Propagation of Timber in His Majesty's Dominions* (1664) he suggested that the "excessive humidity of Ireland and North America was due to excessive rain and mist attracted by the dense forests."[13] This was somewhat true, but in regard to his cure of cutting the forests, others may have thought it was worse than the situation.

English physician-naturalist John Woodward (1665–1728) in 1691–02 conducted controlled experiments.[14] He grew spearmint in three containers containing spring water, rainwater, and Thames River water. The last grew best, and he concluded impurities provided important nutrients. He postulated, without naming, the existence of transpiration: "The much greater part of the Fluid Mass that is drawn off and convey'd into the Plants, does not settle or abide there, but passes through the Pores of them, and exhales up into the Atmosphere."[15] However, he also concluded: "Water serves only for a vehicle to the terrestrial Matter which forms Vegetables; and does not in itself make any addition unto them."[16] He also reported that in America, early settlers were annoyed by humidity, but after they cut down forests, "Air mended and cleared up apace, changing into a Temper much more dry and serene than before."[17] Evelyn undoubtedly was pleased to read that.

Meteorology received a significant boost during the 1600s from the invention of meteorological instruments, led by Italian investigators.[18] Galileo led the way by developing a telescope and assisting the development of microscopes. Rain gauges and weather vanes predated the 1600s, but barometer and thermometer developed in Italy during the 1600s.[19]

Photosynthesis researchers contributed to and benefitted from 22 monographs published on photosynthesis and plant respiration by Springer, 1994–2006. Most relevant here is volume 20: *Discoveries in Photosynthesis* (2005), edited by J.T. Govindjee, H. Best, and J.F. Allen, especially "Discoveries in Photosynthesis (1727–2003): a perspective" by Govindjee and D. Krogman.

Their discussion began with English clergyman Stephen Hales (1677–1761), who conducted more sophisticated experiments than others, published in his *Vegetable Staticks*.[20] When he studied at Cambridge University, he was influenced by Isaac Newton's insistence on measurement in science. Hales' first experiments were on blood pressure in two horses and a deer. Van Helmont and Boyle had first conducted quantitative studies in plant physiology, but Hales went far beyond them, in studies on what we call transpiration, inventing a pneumatic trough for his experiments.[21] He showed that air and water are needed by plants, but he could not distinguish different kinds of "airs."

Parisian Antoine-Laurent Lavoisier (1743–95), from a wealthy family, received an excellent education.[22] He became Europe's leading chemist, until his life was cut short by a guillotine during France's Revolutionary War.

DOI: 10.1201/9780429342202-7

His interests were quite broad, including physiology, geology, economics, and social reform. Knowing relevant literature, he built upon Hales' achievements. His important early report was "Sur la nature de l'eau et sur les experiences par lesquelles on à pretend prouver la possibilité de changement en terre" (1773). In discrediting the ancient idea that water can be changed into Earth, he advanced chemical methods and reasoning. He defined an element as a substance that persists through chemical reactions and cannot be reduced to more fundamental substances. He effected a chemical revolution in his *Traité de Chemie* (1789, English 1952).

Englishman Joseph Priestley (1733–1804) was the son of a small-town cloth finisher and was a studious boy.[23] He became a Unitarian minister and opened a school. His intellectual interests were as broad as Lavoisier's, and he began scientific research on electricity and optics. When a fourth edition of Hales' *Vegetable Staticks* appeared in 1769, he read it and began reading chemical works. He collected air from an enclosed mouse and an enclosed plant.[24] The mouse's air became poisonous, killed the mouse, and extinguished a candle. When a plant, mouse, and candle were enclosed together, the mouse lived and the candle remained lit. A plant did well in air that killed a mouse.[25] He showed his experiments to an American, Benjamin Franklin, who commented: animals eat plants, and animal wastes provide fertilizer for plants. In 1774, he heated what we call mercuric oxide and released a gas he called "dephlogisticated air" (oxygen), which fit into an older theory of combustion.[26] In October, he visited French scientists, including Lavoisier, and told of his discovery. Lavoisier repeated Priestley's procedures and obtained the same results, but he offered a different interpretation.[27] Within his revolutionary chemistry, he called Priestley's gas "oxygen." Priestley was politically radical but scientifically conservative, and he never accepted Lavoisier's revolution.[28]

In 1778, Swedish apothecary Carl Scheele (1742–86) challenged some of Priestley's early discoveries.[29] Priestley repeated those experiments and did not obtain results previously published. Dutch physician Jan IngenHousz (1730–99) investigated the puzzle of plants and gases in summer 1779, conducting over 500 experiments.[30] He discovered plants only purify air in sunlight and that at night they render air as noxious as animals do. That was the first (1779) of his 19 publications on plants and atmosphere.[31] IngenHousz's biographer concluded he deserves credit for discovering photosynthesis.[32]

English scientists Henry Cavendish (1731–1810) and James Watt (1736–1819) conducted an experiment that they interpreted as indicating water is composed of two different substances.[33] Frenchman Lavoisier's chemical revolution of 1785 was built upon two English discoveries of oxygen and the composition of water.[34] Chemist-plant physiologist Nicolas-Theodore de Saussure (1767–1845) was the son of prominent Genevan scientist Horace Bénédict de Saussure (1740–99), who studied geology, meteorology, and botany.[35] Nicolas became his father's assistant. He began publishing on plant physiology in 1797, followed by his acclaimed *Recherches chimiques sur la Végétation* (1804, German 1805, English 2013), the first plant physiology textbook. His translator summarized his nine chapters and his contributions:[36]

> [W]ater is incorporated into the dry matter of plants, that plant carbon is derived from the carbon dioxide of the air, not from humus or soil, that the minerals of plants are absorbed from the soil, not created by a vital force, and that minerals are essential to plant growth.

GLOBAL QUESTIONS

Scotsman James Hutton (1726–97) of Edinburgh attended its university and liked studying physical sciences most but earned an M.D. degree in 1749, with a thesis *De sanguine et circulation microcosmi*.[37] When the Royal Society of Edinburgh was organized in 1783, he became active in its affairs. For some years, he thought about Earth history, and in 1785 he presented there a 28-page "Abstract of a Dissertation . . . concerning the System of the Earth, Its Duration, and Stability." It was about processes at work, with no data.[38] He then spent a decade expanding it into *Theory of the Earth* (2 vols., 1795). It introduced his uniformitarian principle, that geological processes at work now are key to understanding Earth history. The existence of land that once was under water indicates the Earth undergoes constant change. He finally asked, what forces caused these changes, to which he had no answer. Hutton's *Theory* was not an easy read, but a friend, John Playfair (1748–1819), wrote a readable *Illustrations of the Huttonian Theory of the Earth* (1802).

Parisian biologist Jean Baptiste de Lamarck (1744–1829) is remembered for his evolutionary theory, published in 1801 and 1809.[39] His interests were broad, however, one of which he discussed in *Hydrologie* (1802, English 1964). He also had a uniformitarian perspective, achieved independently of Hutton, whose treatise he never read. *Hydrologie* discussed four questions, three of which were geological. The fourth was: "What are the general effects of living organisms on the mineral substances which form the earth's crust and external surface?"[40] In answering this question, he stated: "The organic action of living organisms continuously creates combinations of substances which would never have existed otherwise." This claim is too broad; living organisms produce carbon dioxide, but so do fires. Neither Hutton's nor his book was easy to read. Lamarck also showed no mastery of Lavoisier's chemical revolution. His work was known but of limited influence.

FOOD CHAINS AND WEBS

A biosphere includes a global food web, and understanding earlier food chains and webs were steps toward that. Hyperparasitic food chains were discovered during the 1600s, and the concept was generalized, but not named, by English naturalist Richard Bradley (d. 1732), who published in 1718: "Insects which prey upon others are not without some others of lesser Rank to feed upon them likewise, and so to Infinity."[41] Jonathan Swift turned that into verse.[42]

Swedish professor Carl Linnaeus (1707–78) was the most important contributor to ecology during the 1700s.[43] He published essays that students defended for doctoral dissertations and published under their names. Most important here is "The Economy of Nature" (1775, Latin 1749), which included two food chains:[44]

> tree-louse lives upon plants. The fly called Musca aphidivora lives upon the tree-louse. The hornet and wasp fly upon the Musca aphidivora. The dragon fly upon the hornet and wasp fly. The spider upon the dragon fly. The small birds on the spider. And lastly, the hawk kind on the small birds.
>
> In like manner the monoclus delights in putrid water, the knat eats the monoclus, the frog eats the knat, the pike eats the frog, the sea calf eats the pike.

Linnaeus believed God designed nature and that food chains ensured no species became too numerous to be supported by its food species. The economy of nature is now called balance of nature.[45]

Very likely, there were discussions of food chains between Linnaeus and Darwin, as yet unknown. Darwin, in his *Journal of Researches into the Geology and Natural History of the various Countries Visited by* H.M.S. Beagle, may have first published a food web, at barren St. Paul's Rocks, in mid-Atlantic, where gannets, boobies, and noddy terns nested, the latter building a nest of seaweed:[46]

> By the side of many of these nests a small flying fish was placed: which, I suppose, had been brought by the male bird for its partner . . . quickly a large and active crab (*Graspus*), which inhabits the crevices of the rock, stole the fish from the side of the nest, as soon as we had disturbed the birds. Not a single plant, not even a lichen, grows on the island: yet it is inhabited by several insects and spiders. The following list completes, I believe, the terrestrial fauna: a species of Feronia and an acarus, which must have come here as parasites on the birds; a small brown moth, belonging to a genus that feeds on feathers; a staphlinus (*Quedius*) and a woodlouse from beneath the dung; and lastly, numerous spiders, which I suppose prey on the small attendants on, and scavengers of the waterfowl.

This was a specific example, much of which was observed by a knowledgeable observer, of particular species interacting, unlike the generalized examples described by Bradley and Linnaeus. Having a limited number of species involved was advantageous for Darwin's discovery. Rear-Admiral William Symonds read this and told Darwin he had seen St. Paul crabs drag young birds from nests and eat them. Darwin added his information to the second edition of his *Journal* (1845).[47]

In *The Origin*, however, Darwin cited a generalized food chain: humble bees obtain food while pollinating red clover, mice eat these bees, and domestic cats eat the mice.[48] Even this simple food chain was flawed since honey bees also pollinate clover.[49]

A remarkable German zoologist, Karl Semper (1832–93), delivered 12 lectures in Boston (1877), which he published: *Animal Life as Affected by the Conditions of Existence* (1881, German 1880). It was the first detailed synthesis of animal ecology, with a quantitative model of a food chain.[50] He had studied engineering and later physiology.[51] He was accustomed to thinking quantitatively when few other biologists did. He explained that when herbivores eat vegetation, due to oxidation there is loss of organic material; he estimated that 1,000 units of vegetation could produce 100 units of herbivore, and when predators eat herbivores, it produces 10 units of predator. His ratios were used by a more recent zoologist who may never have read Semper.[52]

Italian zoologist Lorenzo Camerano (1856–1917) grew up in Torino, attended its university, and later taught there.[53] The earliest known diagrams of food webs were two he published.[54] He was mainly an entomologist, and that paper was atypical. J.E. Cohen thought that paper showed the influence of Darwin's *Origin of Species* (1859).[55] Cohen had Camerano's paper translated into English.[56] His two food web diagrams are like none published later, indicating little influence.[57] They nevertheless indicate progress in studying food webs.

Illinoian Stephen A. Forbes (1844–1930) grew up on his family's farm along Des Plains River.[58] He served in the Union Army during the Civil War, and afterwards attended Rush Medical College, Chicago, and later Illinois State Normal University. He worked for Illinois in several positions, mainly state entomologist and professor. He investigated foods of birds and fishes and discussed the ecological implications of his findings.[59] His first general essay, "On Some Interactions of Organisms" (1880), emphasized interrelationships:

> The serious modification of any group, either in numbers, habits, or distribution, must modify considerably, various other groups; and each of these must transmit the change in turn, or initiate some other form of change, the disturbance thus propagating itself in a far extending circle.

His further researches and reasoning led him to compose his most memorable essay, "The Lake as a Microcosm" (1887, reprinted 1925, 1977, 1991). By *microcosm*, he meant *ecosystem*, a later term. This essay was the most concise overview of ecology to date.

The earliest American specific food web diagram was presumably by W. Dwight Pierce, Robert Asa Cushman, and C.E. Hood, in "The Boll Weevil Complex" (1912), a USDA *Bulletin*. The boll weevil *Anthonomus grandis* was a pest in cotton fields that migrated into the United States from Mexico during the late 1800s. Their bulletin and diagram of its parasites and predators were praised by Thomas Park.[60] I am unaware of any later web diagrammer following their easily understood design.

A year later, animal ecologist Victor Shelford (1877–1968) published *Animal Ecology in Temperate America, as illustrated in the Chicago Region* (1913), with diagrams

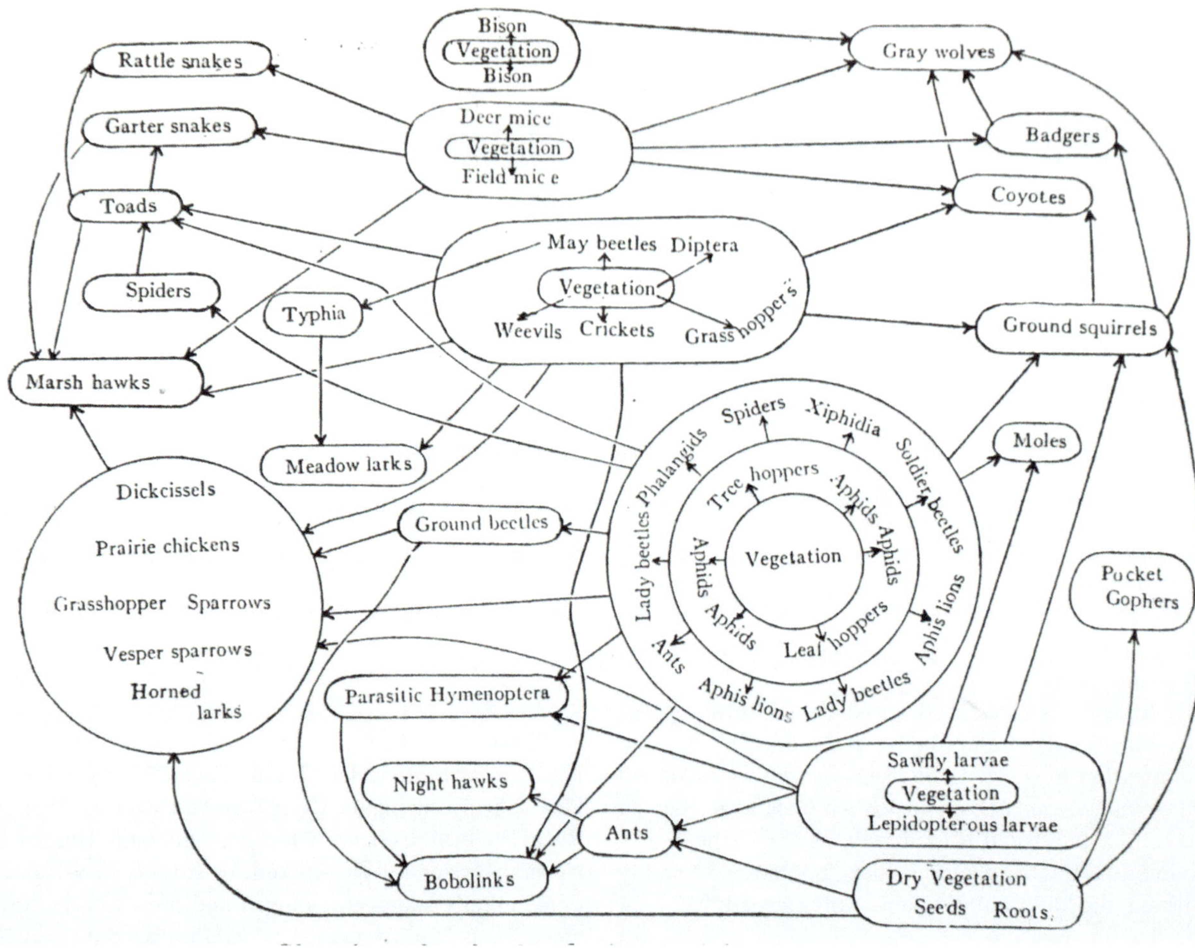

FIGURE 7.1 Victor Shelford's land food chain. Shelford 1913: 167.

of both land and aquatic food webs.[61] Later diagrammers did not follow his example either. Orlando Park reproduced four later food web diagrams, all quite different, along with a bibliography of food-web papers, 1929–38.[62] My later survey (2007) went from Richard Bradley (1718) to Clive Edwards (1970).[63] More recent surveys are by G.A. Polis, M.E. Power, and G.R. Huxel (2004), Mercedes Pascual and Jennifer Dunne (2006), and Carl Zimmer, "Ecosystems on the Brink" (2012). Craig Layman and 10 colleagues published "A Primer on the History of Food Web Ecology" (2015); Thomas Ings and Joseph Howes emphasized food networks in "The History of Ecological Networks" (2018). A journal, *Food Webs*, began in 2014.

The earliest known food web diagram for a marine community (Shelford's was fresh-water) was by Danish fishery biologist C.G. Johannes Peterson (1860–1928), "A Preliminary Result in the Investigations on the Valuation of the Sea" (1915).[64] He had studied under Steenstrup at the University of Copenhagen, and he used the mark-and-recapture procedure to estimate the size of a plaice population. He had co-founded Dansk Biologisk Station to study marine life. He studied Kattegat, a shallow bay between Denmark and Sweden about 150 miles long and about 90 miles wide, to learn its annual productivity.[65] His diagram was also original.[66] In 1916, his colleague, Harold Blegvad (1886–1951), conducted a similar study and drew a similar diagram.[67] Englishman Alistair Hardy showed similar productivity concerns in "The Herring in Relation to Its Animate Environment" (1924). As mentioned earlier, Charles Elton's studies on island food webs can be considered as land or marine webs.[68]

EARTH MOVING

This is a subject of convenience, not of logic, to include two very different phenomena discovered in the same time period: Alfred Wegener (1880–1930) on continental drift and Milutin Milankovich (1879–1958) on repeating irregularities in Earth's orbits. Berliner Wagener studied astronomy at Heidelberg and Innsbruck Universities before earning his Ph.D. at Berlin.[69] He became interested in meteorology and geology and in 1911 announced his theory of continental drift, published 1915: *Die Entstehung der Kontinente und Ozeane*. It was controversial and only gained acceptance during the 1960s, as plate tectonics was explained.[70] In 1930, he was in Greenland and on his birthday headed to the coast

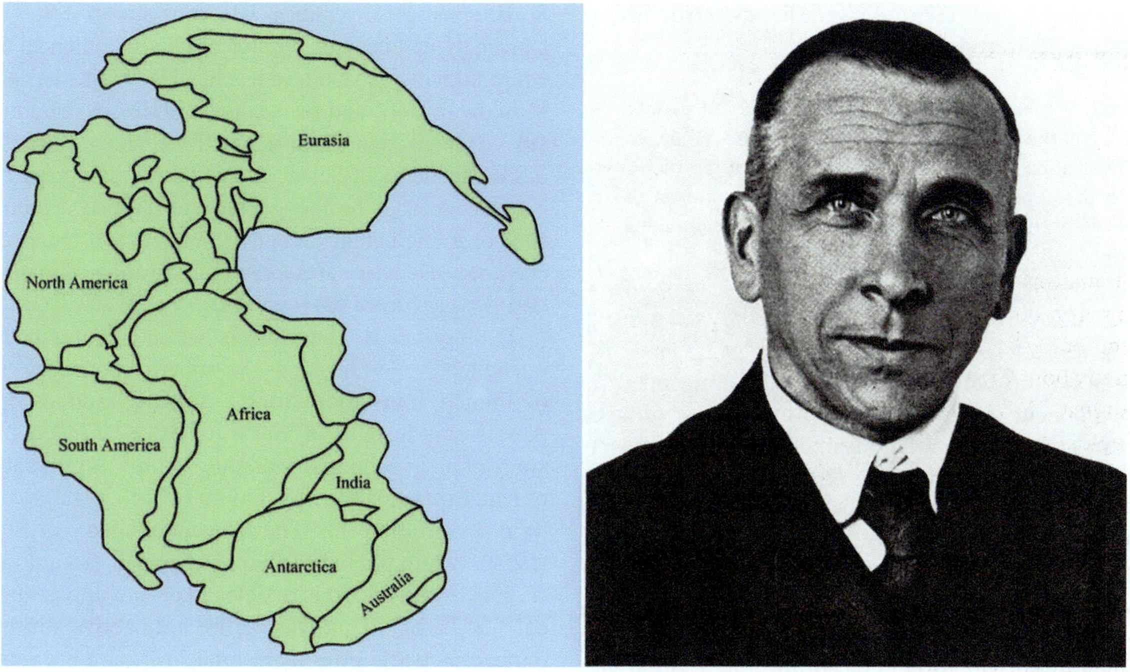

FIGURE 7.2 (a) Pangaea. 1915. (b) Alfred Wegener. Web.

but was never seen again. Milankovich was Serbian, within Austro-Hungary's Empire.[71] He attended Vienna University of Technology, earned an undergraduate degree and a Ph.D. (1904) in civil engineering. He became interested in causes of ice ages and concluded that three long-term cycles—within a cycle—of Earth around Sun, lasting in years—100,000 (elliptical), 42,000 (axial tilt), and 22,000 (axial precession)—influenced Earth climate, thought continental drift did also. He published in Serbian *Canon of Insolation of the Ice-Age Problem* (1941, English 1969).

ECOSYSTEMS

Ecosystems are discussed earlier, chapter 3. Here they are placed in a pre-biome context. An ecosystem is a distinct community that contributes to the whole, while also being influenced by the whole. A forest is an ecosystem that contributes to the whole by absorbing CO_2, except in a forest fire, when it expels CO_2. Since ecologists collectively are world travelers, few (small) ecosystems are still undescribed. Some ecosystems are isolated, such as an oasis in a desert, while others have significant interactions, such as a river flowing into an ocean.

Studied examples are in William Mc Shea and William Healy, eds., *Oak Forest Ecosystems: Ecology and Management for Wildlife* (2002), which has 20 chapters by 32 authors, plus editors' introduction and conclusion. Editors estimated about 50 *Quercus* species live in North America and over 500 species worldwide. Although acorns are important food for about 150 species of U.S. birds and mammals, only seven species of birds and 10 species of mammals are important for dispersing acorns/seeds.[72] Three authors discussed management in particular regions: Richard Standiford on California (2002), Peter Ffolliott on Arizona and New Mexico, and William Healy on eastern oak forests (2002).

BIOGEOCHEMISTRY

Biogeochemistry began, according to historian Eville Gorham, with "On the Chemical Statics of Organized Beings" (1841) by French chemist Jean-Baptiste-André Dumas (1800–84). In the 1830s Dumas was assistant professor at Sorbonne, in Paris, promoted to professor in 1841, and soon France's leading chemist.[73] He discussed the role of animals as well as plants:[74]

> [A]nimals constitute, in a chemical point of view, a real apparatus for combustion, by means of which burnt carbon incessantly returns to the atmosphere under the form of carbonic acid [CO_2], in which hydrogen burnt without ceasing, on its part continually engenders water; whence . . . free azote [nitrogen] is incessantly exhaled by respiration, and azote in the state of oxide of ammonium by the urine[.]

Yet, biogeochemistry did not flourish after Dumas published this in French and English.

Eduard Suess (1831–1914) was the son of a merchant, born in London; when he was 3, his family moved to Prague, and when 14, to Vienna.[75] He attended the Universities of Vienna and Prague. He never took courses in geology or paleontology, yet in 1856 he became a professor of paleontology at Vienna and in 1861 a professor of geology. His *Die*

Entstehung der Alpen (1875), on genesis and structure of Alps, stated in its introduction:[76]

> The plant, whose deep roots plunge into the soil to feed, and which at the same time rises into the air to breathe, is a good illustration of organic life in the region of interaction between the upper sphere and the lithosphere, and on the surface of continents it is possible to single out an independent biosphere.

Translator Vaclav Smil stated this was the only time Suess used *biosphere*, a term he coined.

The relationship between plants and the atmosphere is a vital component of global ecology, and photosynthesis is the main part of that relationship. Photosynthesis scientists Govindjee and Krogmann (2005) provided a chronology of photosynthesis advances from Hales (1727) to 2003.[77] An important step was by German physician Julius Robert Mayer (1814–78), who earned a medical degree at the University of Tübingen (1838).[78] In 1840–41, he was a physician on a Dutch ship to East Indies. At Djakarta, Java, physiological observations convinced him that motion and heat were manifestations of a force in nature, which was conserved in conversion from one form to the other. After returning, in 1845 he published his idea. Few scientists noticed his publication, so he published articles supporting his account.[79] He stated, "plants convert light energy into chemical energy" during photosynthesis. English physicist James P. Joule attacked Mayer's numerical value of the mechanical equivalent of heat, and Mayer attempted suicide. He was confined for a time in a mental institution. Another English physicist, John Tyndall, discovered the merit of Mayer's work and brought recognition to it. A German physicist, Rudolf Clausius, did not expect to take Mayer's scientific papers seriously, but after reading them, he wrote to Tyndall (17 May 1862): "I am amazed at the number of fine and correct ideas they contain."[80]

Eileen Crist argued that Darwin's last book, *The Formation of Vegetable Mould, through the Action of Worms* (1881), was a "scientific study about the reciprocal influence of life and environment."[81] "It was a perfect project for a man with his circle of landed connections. Perfect too for a slow moving country gentleman."[82] Darwin had published three earlier notes on this process.[83] In 1837, he had visited his uncle, Josiah Wedgewood, who showed him in a pasture where years before cinders and pieces of brick had been scattered but in 1837 they were a few inches below ground.[84] Westwood thought earthworms had by constant feeding in soil, buried objects. Darwin studied this and reported to the Geological Society of London, crediting his uncle's observations (1837). Amy Stewart discovered James Samuelson, *Humble Creatures: the Earthworm and the Common Housefly* (1858) "quoted liberally from Darwin's early paper on worms."[85] Darwin's worm book opened a new field of research, but only much later did others begin worm research.

Russian mineralogist-geochemist Vladimir Vernadsky (1863–1945) founded biogeochemistry and elaborated a biosphere concept.[86] His father taught at Kiev, Moscow, and St. Petersburg Universities. Vladimir attended St. Petersburg University and became curator of its mineralogical collection while studying to become a teacher. He obtained a doctorate in Moscow in 1897, and he became a professor there in 1898. In 1917–21 he lived in Ukraine, and in 1919 he organized Ukraine S.S.R. Academy of Sciences and was its first president. In his last two decades he recognized the important mutual influence of living beings and environments and founded biogeochemistry. He wrote articles, which culminated in his *Biosfera* (1929).[87] He coined *biogeochemistry* in 1926.

Ecologist G.E. Hutchinson, while a professor at Yale,[88] introduced Vernadsky's biosphere concept (available in French translation) to his students and others.[89] The American Museum of Natural History funded his 1940s study of biogeochemistry of aluminum.[90] A graduate student of Hutchinson, Howard T. Odum (1924–2002), wrote a doctoral dissertation on "The Biogeochemistry of Strontium" (1950). Odum published two brief papers from it in *Science* (1951a, b) and two long articles later in a university periodical (1957a, b).[91] That was before radioactive strontium as a dangerous waste product of nuclear fission became known. Howard took his Yale "gospel" to his older brother Eugene at the University of Georgia, who was writing an ecology textbook.[92] It was popular and had five editions.

Midwesterner Gene Likens (b. 1935) we met before as a successful limnology student under Hasler, whose career occurred in the Northeast, collaborated with forest ecologist Herman Bormann (1922–2012) at Hubbard Brook Experimental Forest, New Hampshire.[93] One collaboration was a multidiscipline treatise, with three other specialists, *Biogeochemistry of a Forested Ecosystem* (1977), followed two years later by Bormann and Likens, *Pattern and Process in a Forested Ecosystem*.[94] Their chapters discussed hydrology, chemistry, input-output budgets, weathering, nutrient cycles, and a northern hardwood ecosystem.

SOILS

Although soils had been discussed occasionally since antiquity, Darwin's *Formation of Vegetable Mould through the Actions of Worms* (1881) was early research into earthworm significance for improving soil fertility. He cited earlier, less detailed studies from the 1860s to 1870s. Later, zoologists and soil scientists studied other soil fauna. In 1958, the International Society of Soil Science appointed a Soil Zoology Committee to convene a Colloquium on Research Methods, which assembled 47 British and European authors of papers collected after the colloquium into a valuable reference work.[95] Also in 1962, German zoologist Friedrich Schaller wrote a brief textbook, *Soil Animals* (German 1962, English 1968), with numerous illustrations but (in English edition) no citations. J.B. Cragg, British Nature Conservancy, recruited two editors—Alan Burgess, New University of Ulster, Ireland, and F. Raw, Rothamsted Experimental Station, UK—to edit *Soil Biology*, 17 chapters.[96] Burgess edited microbiological and botanical chapters, and Raw edited zoological chapters.

More recently, soil biology became an aspect of biosphere ecology. Soil ecologists Daniel Richter Jr., Duke University, and his former student, Daniel Markewitz, University of Georgia, synthesized modern knowledge in *Understanding Soil Change: Soil Sustainability over Millennia, Centuries, and Decades* (2001). Their scope was broad:[97]

> One of the most extraordinary outcomes of biological evolution has been the co-evolution of soil and forest ecosystems . . . major components of the biosphere for nearly 400 million years . . . . While many biological species have long gone extinct, soil and forest systems have managed to sustain themselves.

Soils were unstable due to wind, ice, rain, and runoff until tree roots stabilized soils and dead trees were recycled by bacteria, fungi, and weathering. Richter's research team more recently has focused upon biogeochemistry.[98]

GAIA OR NOT?

Englishman James Lovelock, a centurian (b. 1919), is an independent scientist and independent thinker.[99] His autobiography did not mention Vernadsky but discussed development of his original theory, which he named Gaia.[100] He had not been a good student because he wanted to learn whatever interested him, rather than assignments. After graduating from high school in 1938, he joined consulting firm Murray, Bull and Spencer as a laboratory assistant. It was a good environment since he had opportunities for inventiveness when solving problems; he was a talented inventor. Later, he attended Manchester University and graduated in 1941.

Jon Turney, *Lovelock and Gaia* (2003), quoted an English science fiction author, Olaf Stapledon, whose *Star Maker* (1937) protagonist viewed Earth from space: "It displayed the delicacy and brilliance, the intricacy and harmony, of a living thing."[101] Turney stated that when Lovelock grew up, he read science fiction, and perhaps earlier Stapledon books, and so also *Star Maker*.

Lovelock's own retrospection did not extend back to his childhood:[102]

> Perhaps the most important event in my life as a scientist was the moment in 1957 when, as a staff member of the National Institute for Medical Research, I stumbled on the electron detector (ECD). This simple device that fits easily into the palm of my hand was without doubt the midwife to the infant environmental movement. Without it we would not have discovered that chlorinated pesticides like DDT and dieldrin had spread everywhere in the world.

He equated concern about pesticides with the environmental movement. That was an important component of that movement but was far from being all of it. Michael Lannoo explored different origins in *Leopold's Shack and Rickett's Lab: the Emergence of Environmentalism* (2010). Lovelock used his ECD to discover that chlorofluorocarbons (CFCs) were accumulating in the atmosphere. He did not take the next step, showing CFCs were destroying the protective ozone layer of our atmosphere. That step was taken by chemists Mario Molina and Sherwood Rowland (1974) and won a Nobel Prize in Chemistry in 1995.

Lovelock's next theoretical step was taken while he was at the Jet Propulsion Laboratory (JPL) in California in September 1965. News came from France that both Mars and Venus have atmospheres dominated by CO_2. He wondered what kept Earth's more complex atmosphere constant:[103]

> It came to me suddenly, just like a flash of enlightenment, that to persist and keep stable, something must be regulating the atmosphere and so keeping it at its constant composition . . . if most of the gases came from living organisms, then life at the surface of the earth must be doing the regulation.

Here was a broad-ranging scientific mind at work. Although he accumulated knowledge and expertise in atmospheric science, he did not think of himself as a specialist in any particular science.

He announced his Gaia hypothesis in 1.5 pages of text plus eight references (1972).[104] The references seem supportive of his reasoning, not sources from which he had developed his ideas. He argued that living organisms created Earth's atmosphere, and if all life died, the atmosphere would lose existing gases and devolve into an atmosphere similar to those of Mars and Venus. He began investigating how living organisms regulate Earth's atmosphere. He needed a collaborator. Chicago native Lynn Margulis (1938–2011) studied cell biology (B.A. 1957, University of Chicago; M.S. 1960, University of Wisconsin; Ph.D. 1963, University of California, Berkeley) and achieved recognition by convincing peers that cell organelles such as mitochondria and chloroplasts, were originally bacteria that imbedded into larger cells, through the process of endosymbiosis.[105] In 1970, she wanted to consult a scientist who was knowledgeable about oxygen in the atmosphere, and her husband, Carl Sagan, suggested a former colleague at JPL, James Lovelock.[106] They met in Boston in 1971 and found they were like-minded. Their collaboration yielded two Gaian papers: Lovelock and Margulis 1974, Margulis and Lovelock 1974.

In his first book, *Gaia: a new look at Life on Earth* (1979), Lovelock stated:[107] "early environment of the first photosynthesizers is likely to have been a reducing one, rich in hydrogen and hydrogen-bearing molecules . . . . Food for some species of primevals may have been oxidizing substances." One might expect him to cite reciprocal gas exchange between plants and animals, with each of their waste gases being essential for the life of the other, but I did not find he did so. In *Ages of Gaia* (1988), a novelty was a hypothetical planet, Daisyworld, inhabited only by black daisies and white daisies. He asked, what happened when temperature varied?[108] He left that question for Gaians and

FIGURE 7.3 (a) James Lovelock & Gaia statue. (b) Lynn Margulis & President Clinton, 24 November 2011. Paul Hosefos, Lynn Margulis, wearing her National Medal of Science Award. The New York Times. Available from: The New York Times, https://www.nytimes.com/2011/11/25/science/lynn-margulis-trailblazing-theorist-on-evolution-dies-at-73.html (accessed 2021).

Skeptics to debate. Gorham cited examples to illustrate Lovelock's claim about biogeochemistry,[109] including G.E. Hutchinson and V.T. Brown, *A Direct Determination of the Phosphorus Cycle in a Small Lake* (1947) and Alfred C. Redfield, *The Biological Control of Chemical Factors in the Environment* (1958). In a later book (2000), Lovelock asked what was the relation between Gaia theory and important predecessors Vernadsky and Hutchinson? He explained:[110]

> [B]iogeochemistry differs from geophysiology as platonic friendship differs from a happy marriage. In biogeochemistry, organisms and their material environment are recognized as coexisting and coevolving but still separately, like friends. In geophysiology, the organisms and their environment are so tightly coupled that they constitute a single system, Gaia.

Meanwhile, New Yorker climatologist Stephen Schneider (1945–2010) earned three degrees in mechanical engineering from Columbia University (B.S. 1966, M.S. 1967, Ph.D. 1971), and received a postdoc at NASA's Goddard Institute for Space Studies, where he studied greenhouse gases. He wrote numerous books; an early one, with Randi Londer, was *The Coevolution of Climate and Life* (1984). Most of his career was spent as a professor of environmental biology and global change at Stanford University. He was also a White House advisor during seven administrations. He suggested American Geophysical Union host a conference to explore Gaian theory. It convened at San Diego in March 1988. He co-edited 44 revised papers, by 173 participants, published by MIT Press.[111] No grand theory or conclusion emerged, but the conference led to important new investigations that advanced Earth system science. Lovelock answered some of his critics in *The Ages of Gaia* (1988).

Margulis disliked the notion of Earth being alive and offered an alternative in *The Symbiotic Planet* (1998): "Gaia is the series of interacting geosystems that compose a single huge ecosystem at the Earth's surface."[112] Lovelock responded by coining new terms for new aspects of Gaia—*geophysiology* and *planetary medicine*. These new aspects did not go as far from his original Gaia as Margolis suggested.

A second three-day Gaia conference occurred at Oxford University, March 1996. In 1998, New York University Professor of Biology Tyler Volk (b. 1950) published *Gaia's Body: toward a Physiology of Earth*, in which he explained: "I will not weave detailed arguments for or against the idea that Earth is alive, that Gaia self-regulates, and that it is a self-sustaining organism, or perhaps quasi-organism."[113] Volk considered Gaia simply "the interacting of systems of life, soil, atmosphere, and ocean." Its operating rules are not fully understood. How does Earth maintain its oxygen level at 21%?

American Geophysical Union cosponsored a third Gaian Conference in 2000 with the University of Valencia, held in Valencia, Spain. Lovelock and Margolis participated, and each wrote separate introductions to the resulting volume which Schneider, Boston, and two others edited, with 31 papers by 35 authors, also published by MIT Press. Still another conference met in 2001 at Amsterdam, sponsored by four international organizations, on global climate. The Amsterdam Declaration on Global Change stated: "The Earth System behaves as a single, self-regulating system comprised of physical, chemical, biological and human components."[114] That statement encouraged Lovelock to publish in *Nature* a two page-summary on progress in defending Gaia since 1980.[115]

English ecologist Toby Tyrrel had published a paper in *Nature* that interested Lovelock: "The Relative Influences of Nitrogen and Phosphorus on Oceanic Primary Production" (1999). That interest became mutual, leading to Tyrrel's *On Gaia: a Critical Investigation of the Relationship between Life and Earth* (2013). He was also influenced by the Amsterdam Declaration. He saw three competing hypotheses to explain relationships between life and Earth's environment: (1) Gaia, (2) geological, and (3) co-evolution. Gaia emphasized life-environment relationships; geological hypothesis depended on continental drift, volcanic activity, and "fluctuations in mid-ocean ridge spreading rates."[116] Coevolution emphasized evolutionary relationships between species, such as between flowers and pollinators or predators and prey. He judged coevolution the most compelling alternative. (Lovelock undoubtedly thought Gaia could absorb coevolution.)

In my opinion, Gaia is a recent version of the balance of nature, which differs from previous versions in providing far more details.[117] I think strong evidences against Gaia are past ice ages and present global warming.

EARTH SYSTEM SCIENCE

Eugene Odum published *Ecology and Our Endangered Life-Support Systems* (1989), partly an updating of his textbook, partly written for a broader audience concerned for our threatened environment. Its longest chapter, 5: "Material Cycles and Physical Conditions of Existence," updated biogeochemistry and devoted 3.5 pages to Lovelock's Gaia, which was about the same space he allotted for brother Howard's work. He complemented Lovelock without judging a limited version of Gaia, while rejecting the early broad version: "the earth is a superecosystem (but not a superorganism since its development is not genetically controlled)."[118] In a longer version of this chapter, I discussed eight more relevant worthy volumes from this period.[119]

GLOBAL WARMING

Phenology Focus

Global warming is an aspect of meteorology, a science emerging in the 1820s. Before that, phenology was well on its way. As noted earlier, John Evelyn thought a correlation existed between local climate and the extent of local forests.[120] Although no one responded to this possibility when published in his *Sylvia* (1664), his idea was not forgotten. An English family, Marsham, kept phenology records for over two centuries, 1736–1947.[121] Naturalist Robert Marsham (1798–97) lived north of Norwich, Norfolk. He became a member of the Royal Society of London in 1780, and in 1789 he published in its *Philosophical Transactions* "Indications of Spring" and listed 27 signs. He corresponded with parson-naturalist Gilbert White (discussed earlier), who also kept phenology records.[122]

Two "Calendars of Flora" appeared in the mid-1700s. Benjamin Stillingfleet (1702–71), also from Norfolk region, compiled his (1755) at the Marsham estate, without explaining the relation of his and the unpublished Marsham one, also being compiled there. He translated Linnaean dissertations (1759) and added his own after the Linnaean one, *Calendarium florae*, was defended as a doctoral dissertation by Alexander M. Berger (1737–1802). Two reputable Linnaean biographies discussed *Calendarium florae* without mentioning Berger.[123]

Thomas Jefferson owned a Virginia plantation, Monticello, and sought local climate data for when to plant crops and other needs. He recorded weather in his Garden Book, beginning with the opening of purple hyacinths on 30 March 1766, and continued doing so through 1816.[124] While France helped the United States fight a War of Independence, its ambassador asked each state for information on its features. Jefferson complied with *Notes on the State of Virginia* (1787), including state climate.[125] His instructions to Lewis and Clark in 1803 included climate observations in the West, in relation to plants and animals.[126]

Harvard University medical faculty member Jacob Bigelow (1787–1879) also taught botany.[127] In 1817, he corresponded with observers east of the Mississippi and published a table of dates of peach tree blooming in 11 places, from Alabama to Montreal and Brunswick, Maine, covering 14° of latitude and 17° of longitude.[128] The difference between progress of season in the north and south was about 2.5 months, with little difference between the longitudes east and west. This was America's first attempt to learn regional climate. Yale professor of chemistry Benjamin Silliman (1816–85) founded the *American Journal of Science*, and he summarized Bigelow's report in its first issue, to which he added blooming dates for London, Valencia, and Geneva, for comparison with American data.[129]

Use of such information was not limited to agriculture. During the first half of the 1800s, a major disease hypothesis was that some diseases were caused by unfavorable weather/climate.[130] Malaria is most common in warm and hot countries. In 1823 Meteorological Society of London arose, but only began publishing its *Transactions* in 1839, in which John Ruskin wrote that the society aspired to organize meteorological observations worldwide. It failed to do so, but in 1840 it published a rainfall map of England, using data from 52 stations.[131] In 1850 British Meteorological Society arose, to establish a uniform national system for regular reports. So much data accumulated that it later inspired historical discussions.[132]

France, in 1863 established a telegraph system to forecast weather.[133] Germany's Palatine-Bavaria state established Societas Meteorological Palatina in 1781, which lasted until 1795. Other German states established such organizations, until Prussia's Meteorological Institute, founded in 1847, expanded—after Prussia's unification of Germany in 1871—to all Germany. Englishman William Howitt (1817–79) at age 13, wrote "An Address to Spring," which appeared in *Monthly Magazine*. He became a professional author. His books included *The Book of the Seasons: or, the Calendar of Nature* (1831).

Massachusetts native Henry David Thoreau (1817— absorbed his mother's love of nature.[134] He attended Harvard University, where he took a natural history course from librarian-entomologist Thaddeus William Harris (1795–1856).[135] In 1836, Thoreau read Howitt's *Book of the Seasons*, which inspired Thoreau to write an essay, which quoted Howitt's phenological goals, followed by Thoreau's comments for every month of the year.[136] He also began keeping a journal, which he continued until 1858, commenting on plants, and soon also birds.[137] By 1852, he learned of others recording similar data which Smithsonian Institution coordinated and synthesized.[138] He continued making phenological observations but did not send them to Smithsonian. Some of them appeared in *Walden, or Life in the Woods* (1854), a "back to nature" book. The rest of them were in his *Journals*, published posthumously in four compiled volumes: *Early Spring in Massachusetts* (1881), Summer (1884), *Winter* (1888), and *Autumn* (1892). Thoreau collected phenological data because he thought some insight might emerge from it. He was right, but only in 1947, when Aldo Leopold used it, followed by seven others.[139]

Coincidentally, Franklin B. Hough (1822–85), later head of the U.S. Forest Service, USDA, also collected phenological data in 1851–59, unknown to Thoreau. Hough's data was published in a Congressional document and later used by Leopold.[140]

Iowan Aldo Leopold (1887–1948) studied forestry at Yale, though his main interest was in wildlife management—not yet taught anywhere. In 1928, he left U.S. Forestry Service to study game conditions for Sporting Arms and Ammunitions Manufactures' Institute.[141] His *Report on a Game Survey of the North Central States* (1931) could serve as a basis for his next project, a textbook, *Game Management* (1933)—still used in a game management course I took in 1959. That book helped him become the world's first professor of wildlife management at the University of Wisconsin.[142] He had become fond of Thoreau's writings since his teens, which would have predisposed him toward phenology. His mother had given him a first edition copy of Thoreau's *Journal* (14 volumes, 1906). He began keeping phenological records in 1935, and his records increased in 1938, while studying natural history of woodcocks.[143] In 1945, he organized his phenology data into an article, assisted by botany student Elizabeth Jones.[144] He used Thoreau's and Hough's data. Publishing their article in *Ecological Monographs* in 1947 helped his election as ESA's vice-president (1947) and president (1948).[145]

GLOBAL WARMING

French mathematician Jean-Baptiste Fourier (1768–1830) in 1798 was with Napoleon's Egyptian expedition, and in 1808 Napoleon made him a baron. Yet, he preferred to focus on heat diffusion, on which he published a long paper in 1807.[146] He was voted a member of Académie des Sciences in 1817. In 1822, he published *Théorie analytique de la chaleur* (English 1878), "one of the classics of mathematics."[147] He turned to the heating of Earth and concluded its heat originates from three sources:[148]

1. Solar radiation, which is unequally distributed over the year and which produces the diversity of climates
2. Temperature communicated by interplanetary space irradiated by the light of innumerable stars
3. Heat from the interior of the Earth remaining from its formation.

Solar radiation and interior heat are certainly important; heat from stars would be difficult to prove! He also explained that heat from sunlight penetrates the atmosphere more easily than heat reflected from Earth—seen now as the earliest reference to global warming.[149]

Northern Irishman John Tyndall (1820–93) began in 1839 as a surveyor and engineer, at the Irish Ordnance Survey, and then in 1842 was transferred to the English Survey, but in 1843 he complained of oppressive and incompetent management and was fired. In 1844–45 he worked on railroad surveys.[150] In 1847, he taught mathematics at Queenwood College. In 1848, a friend persuaded him to come along to study at Germany's University of Marburg, where Tyndall earned a doctorate in mathematics. He then worked for a time in a physics laboratory. He returned to England, and in 1852 Tyndall was elected to Royal Society. In 1853 he became a professor at the Royal Institute, where he was guided by Faraday. In 1860–70, he studied the effects of solar and heat radiation on atmospheric gases. He discovered that oxygen and nitrogen did not block heat from escaping beyond the atmosphere, but that CO_2 did.[151] He summarized his findings in a lecture, soon published.[152]

Wilmington, Connecticut native Elias Loomis (1811–89) graduated from Yale College (1830), and in 1846, he became a professor of mathematics and natural philosophy (physics) at Western Reserve College, Ohio.[153] He published an innovative weather map in 1846, which was generally adopted for weather predictions.[154] He also taught at several other colleges before returning to Yale in 1860. He compiled data for New Haven and found its weather quite stable.[155] His *A Treatise on Meteorology* (1868) was probably the first text-book on that written in the United States. There were then a few other contemporary American meteorologists.

Tyndall had discovered the potential for global warming, but concern over that possibility arose from Swedish physicist-chemist Svante Arrhenius (1859–1927).[156] His father was an official at Uppsala University, which is where Svante began his higher education, but he did not find there a professor under whom he wanted to study, so he continued his education in Stockholm. In a long career, he investigated numerous phenomena, explored by Elizabeth Crawford, *Arhenius: from Ionic Theory to the Greenhouse Effect* (1996). He began publishing on greenhouse effect with (English version) "On the Influence of Carbonic Acid in the Air upon the Temperature on the Ground" (1896).[157] He reviewed geological evidence of ice ages, followed by warming to temperatures comparable to recent times. He calculated CO_2 concentrations which might maintain ice ages versus contemporary climate. He also cited the amount of coal mined annually, 500 million metric tons, and estimated how much CO_2 that amount of coal might expel in combustion. His study was an excellent start for research on global warming. He won a Nobel Prize in 1903 for contributions to ionic theory.

Biosphere Ecology

Fellow Scandinavian Vilhelm Bjerknes (1862–1951), from Christiania (now Oslo), Norway, provided the next step.[158] While a boy, he helped his father, Carl Anton Bjerknes, with hydrodynamics research. Soon, Vilhelm began drawing his own conclusions. He began studying mathematics and physics at the University of Kristiana in 1880 and earned an M.S. degree in 1888. On a scholarship, he went to Paris and attended lectures on electro-dynamics by Henri Poincaré. Afterwards, he went to the University of Bonn, Germany, to become an assistant to Professor Henrich Hertz for almost two years. They collaborated on important studies. He then resumed studies in Norway and earned a Ph.D. (1892), with a dissertation on *Electtricitetsbevaegelsen i Hertz's primaere Leder* (Movement of Electricity in Hertz's primary Conductor). In 1893, he joined the faculty at Stockholm's School of Engineering as lecturer in applied mechanics. In 1895, he became a professor of applied mechanics and mathematical physics at the University of Stockholm. He associated with Arrhenius at Stockholm Physics Society. In 1898, Bjerknes lectured at the society on the circulation theorem of Kelvin and Helmholtz and suggested its application to movements in atmosphere and oceans. His goal was an ability to predict weather. He returned to the University of Kristiana, where he found four other professors with whom he collaborated. Yet, in 1912 the University of Leipzig lured him away with an offer of chairman of its new geophysical institute. In 1917, he received a still better offer from the University of Bergen and so went there, until he went to the University of Oslo, 1926–32. He and three colleagues published *Dynamic Meteorology and Hydrography* (2 vols., 1910–11).

English engineer Guy Callendar (1898–1964) worked for British Electrical and Allied Industries Research Association, collecting temperatures. He concluded that world temperature was increasing.[159] CO_2 levels seemed to be increasing by 10% in 100 years. He concluded the latter caused the former.[160] Although atmosphere contained more CO_2 than when Arrhenius published, most scientists thought the sea was a safety asset that absorbed excess CO_2.

Toronto native Gilbert Plass (1920–2004) earned a B.S. degree at Harvard University (1941) and a Ph.D. at Princeton University (1947).[161] His career developed at several U.S. institutions. He studied infrared spectrum of CO_2 and water vapor and concluded that if both CO_2 and atmospheric temperature continued to increase, it would indicate that the former caused the latter.[162] He explained that 50 years earlier, the most widely held theory of cause of climate change was CO_2, but by 1956 it was a minority view. Plass called for reexamination of CO_2 theory. He then gave a concise history of past studies.

Seattle native Roger Revelle (1909–91) earned a B.A. degree (1929) at Pomona College, Clairmont, California, and a Ph.D. (1936) at the University of California, Berkeley.[163] His scientific contributions and personality propelled him into academic prominence. By 1957, he was at Scripps Institute of Oceanography. He collaborated with chemist Hans Suess to study "Carbon Dioxide Exchange between Atmosphere and Ocean and the Question of an Increase of Atmospheric CO_2 during the Past Decades." Their literature search found no definite results, but they did find reasons to fear modern fuel consumption might cause an unfavorable increase in global temperatures since ocean's capacity to absorb CO_2 seemed limited. They suggested in 1957, "It therefore becomes of prime importance to determine the way in which carbon dioxide is partitioned between the atmosphere, the oceans, the biosphere and the lithosphere."[164]

To assist such research, Revelle initiated plans for an International Geophysical Year (IGY) for 1958. He had Pennsylvanian Charles David Keeling (1928–2005), who had studied chemistry at the University of Illinois (B.S., 1948) and Northwestern University (Ph.D. 1953), on Scripps Institute faculty, apply for IGY funding to build a research station on Mauna Loa Volcano, Hawaii, two miles above sea level.[165] Keeling invented in 1958 an instrument to measure CO_2 level in atmosphere and received funds for his station.[166] He began collecting data yearly, and did so for over five decades, putting CO_2 levels on a diagram now called Keeling Curve.[167] A colleague, in a Keeling eulogy, thought Keeling Curve provided the most important environmental data of the 1900s.[168] This conclusion could challenge Lovelock's claim that he started an environmental movement with his electron capture detector (ECD), which came into use when Keeling began collecting data, and using ECD produced useful data before Keeling's Curve did. Keeling and supporters might agree, but also point out that his curve data fulfilled expectations from the start, and continued doing so: "Keeling's data put the capstone on the structure built by Tyndall, Arrhenius, Callendar, Plass, and Revelle and Suess."[169]

Detroit native Reid Bryson (1940–2008) was a prominent climatologist who earned a Ph.D. in 1948 at the University of Chicago. He went to the University of Wisconsin in 1946 and founded its Department of Meteorology and Center for Climatic Research. He collaborated with author Thomas L. Murray to publish *Climates of Hunger: Mankind and the World's Changing Weather* (1977). He commented that greenhouse effect was a well-known danger, caused by increased CO_2—caused by industry and cars.[170] Bryson did not think data were sufficient to declare an immediate danger, since volcanic eruptions and particulate pollutants from human activities countered CO_2 effects. England's prominent climatologist H(ubert) H. Lamb wrote *Climate, History, and the Modern World* (1982) five years after Bryson's book appeared, yet did not yield more decisive data.

Iowan James Hansen (b. 1941) earned three degrees (B.S. 1963, M.S. 1965, Ph.D. 1967) at the University of Iowa in physics and astronomy.[171] He joined NASA's Goddard Institute for Space Studies and studied the atmosphere of Venus and then Earth. He was head of Goddard Institute in 1981–2013. He and six colleagues published "Climate

FIGURE 7.4 (a) Aldo Leopold. (b) James Hansen. (c) Al Gore.

Impact of Increasing Atmospheric Carbon Dioxide" (1981) in *Science*, announcing alarming discoveries on global warmth. On 23 June 88, Hansen testified at Senate Committee on Energy and Natural Resources: "the greenhouse effect is already large enough to begin to affect the probably of extreme events such as heat waves."[172] Soon afterwards, Hansen and his seven collaborators published on climate changes that Goddard Institute forecasted. Congress members who would never challenge a report from the National Institutes of Health were willing to repeat energy companies' skepticism of scientific conclusions unfavorable to company interests.[173] Most climatologists supported Hansen.

Also in 1988, the United Nations established its Intergovernmental Panel on Climate Change (IPCC) to provide the world's leading scientists an organization to discuss relevant issues and achieve impartial consensus on data from member countries. The United Nations had earlier experience coping with acid rain and ozone depletion in the atmosphere.[174] In 1995, IPCC published "Summary for Policymakers: the Science of Climate Change," which supported Hansen's alarm.[175]

Albert Gore Jr. (b. 1948) had parents from Tennessee, but his father was a Congressman, and he grew up in Washington, D.C., during school years and on his family farm in summers. He graduated (1969) from Harvard University, joined the U.S. Army, and served as a journalist during VietNam War. At Harvard he had taken a course under Revelle, which awakened his environmental concerns. In 1976, he ran for his father's former seat in Congress and won. After three terms in House, he ran for Senate and won that also. In 1992, he published *Earth in the Balance: Ecology and the Human Spirit*. He also ran for vice-president with Bill Clinton and also won that. In 1997 he attended and spoke at a global climate conference, in Kyoto, Japan. He signed the Kyoto Protocol, which Congress never ratified.[176] After Supreme Court declared Gore lost the presidential election in 2000 to Bush, he began giving talks on global warming around the world. He followed that with another book, *An Inconvenient Truth: the Planetary Emergency of Global Warming and What We Can Do about It* (2006). That book became the basis for a film that won an Academy Award in 2007. He also won a Nobel Prize in 2007 for environmental contributions.

In 1993–98, a Spanish encyclopedia of the biosphere appeared in Spain and soon was translated into English, with Raymond Folch project director (11 vols., 2000). It is well illustrated, and half of volume 11 is on the history of biosphere sciences. It may be in large public and university libraries.

Discussion of global warming after Hansen's 1988 testimony to Congress had seemed promising, yet, legislatively, little changed. In 2005, he was invited to give Keeling Memorial Lecture at the annual meeting of the American Geophysical Union. Hansen's data had become more precise in 17 years since his Congressional testimony.[177] He stated an increase of Earth's temperature by 10 C would make Earth hotter than it had been in one million years, and if Earth increased temperature by 2 or 30 C degrees, it would change Earth into a different planet. George W. Bush's administration had no response.

In 1988, the United Nations established its Intergovern mental Panel on Climate Change (IPCC) to provide leading climate scientists a forum to achieve impartial consensus on interpreting data from member countries. UN had earlier experiences combatting acid rain and ozone depletion.[178] In 1995, IPCC published a 12-page "Summary for Policymakers: the Science of Climate Change," which supported Hansen's dire conclusions on global warming.[179] He and his team never claimed infallibility, and in August 1998, they published "Global Climate Data and Models: a Reconciliation."[180] They explained: "The oversight was failure to account for decay of satellite altitude caused by atmospheric drag."[181]

A response to humans self-destructively using fuels that caused global warming was environmental education. A leader of this movement was William (Bill) McKibben (b. 1960), the son of an editor of *Boston Globe*. He graduated from Harvard in 1982 and became a staff writer at *The New Yorker*, 1982–87. He became a freelance author, and in 1988 he published a review of four works on global warming in *New York Review of Books* and included a brief sketch on investigators from Arrhenius to Hansen. It was a step toward McKibben's *The End of Nature* (1989), translated into 20 languages. His writing skills and previous books built his reputation further.

It is desirable in the hybrid field of ecology, economics, and political policy to have differing perspectives publicly discussed. However, the fruitfulness of that exercise depends upon both sides agreeing on facts, which seldom, if ever, happens. Danish author Bjorn Lomborg (b. 1965), who earned a Ph.D. at the University of Copenhagen (1994), taught at a university until he decided to be an independent author and head of Copenhagen Consensus Center, with funding doubtlessly coming from sources that agree with his outlook. Lomborg, according to his latest book, *False Alarm* (2020), accepts global warming but sees it as a world problem of about the same magnitude as AIDS, malaria, and malnutrition. All such problems can be solved by human inventiveness and free trade. Unfortunately, his previous numerous publications have been judged by The Climate Feedback, a network of scientists, as having frequent misrepresentations and cherry-picked evidence.

ESA sponsored almost two decades of planning and construction of 47 terrestrial and 14 aquatic field stations for the National Ecological Observatory Network (NEON). Sharon Collinge, Chief Scientist/Observatory Director, stated in 2018 that nearly 300 ecologists were in deserts, grasslands, forests, and streams measuring over 180 kinds of data. NEON's web site (neonscience.org) provides a map of locations, names, and other information. Stations add to the knowledge of global warming and other goals. The National Science Foundation sponsored a Global Lake Ecological Observatory Network, inspired by NEON's success.

Recently, news writers Justin Worland and N.M. Hobbs explained, "Oil Companies See Green."[182] Occidental Petroleum (Oxy) in Texas stated it pumps CO_2 underground to free oil deposits. Oxy estimated it annually stored underground as much CO_2 as Hawaii or Maine emit per year. Nice, but not enough to stop environmentalists from trying to shut down oil companies. So, companies formed Oil and Gas Climate Initiative (OGCI), which promised $1 billion to fund other ways to reduce greenhouse gas emissions. If true, "if you cannot beat them, join them."

What we see in this chapter is that naturalists and scientists gradually increased knowledge and understanding of biosphere ecology while society was increasing its degrading of Earth's environment. During most of the past three centuries, there was no obvious connection between increasing knowledge of biosphere and degrading of Earth's environment. During the 1900s, that connection became progressively more obvious. What was less obvious was the best way to apply that knowledge to saving human environments.

NOTES

1. Translated in Howe 1965: 409.
2. Howe 1965: 410, Egerton 2014a, b: 208.
3. Hoff 1964: 108, also Howe 1965, Hofmann 1971.
4. Partington 1961–64: II, index, Pagel 1972, Hill 2013: xv—xvi.
5. Partington 1961–64: II, index, Hall 1967, 1970, Hunter 2015.
6. Boyle 1661, Nash 1957: 328–335.
7. Adelmann 1966, Belloni 1974.
8. Metcalf 1972, White 2004.
9. Nash 1957: 335–336.
10. Avignon 1971.
11. Avignon 1971: 495.
12. Flannery 2005: 38.
13. Fleming 1998: 27.
14. Woodward 1669, Eyles 1971, 1976, Levine 1977.
15. Woodward 1669: 208.
16. Woodward 1669: 221.
17. Woodward 1669: 209.
18. Middleton 1969: 3–89.
19. Middleton 1964, 1966, 1969.
20. Partington 1962: 112–123, Guerlac 1972, Allan & Schofield 1980, Govindjee & Krogmann 2005: 64–66.
21. Nash 1957: 340–342, Parascandola & Ihde 1969: 353–357.
22. McKie 1962, Partington 1961–64: II, 363–494, Guerlac 1973: 67, 1975, Hill 2013: xvii—xviii.
23. Partington 1961–64: II, 237–296, Schofield 1963, 1975, 1997–2004.
24. Nash 1957: 350–358, Gorham 1991: 203.
25. Nash 1957: 355, Drouin 2010: 7–8.
26. Partington 1962: 256–263, Hill 2013: xvii—xviii.
27. Guerlac 1973: 75.
28. Holmes 2000.
29. Nash 1957: 358–369, Boklund 1975.
30. Ingen-Housz 1779.
31. Listed in Magiels 2010: 390–391.
32. Magiels 2010: 359.
33. Partington 1961–64: II, 344–348, McCormmach 1971, Dorn 1976.
34. Partington 1961–64: II, 440–452.
35. Nash 1957: 420–434, Partington 1962: 283–284, Carozzi 1975, Pilet 1975a, b, Drouin 2010: 8–10, Hill 2013: xxi—xxix.
36. Hill 2013: 155–159, xiii.
37. Eyles 1972.
38. Hutton 1987.
39. Landrieu 1909, Burlingame 1973: 589–590.
40. Lamarck 1964: 16, 78.
41. Bradley 1718, part 3: 60–61.
42. Swift 1733: lines 341–344.
43. Egerton 2007a, 2012: 80–84.
44. Linnaeus 1775: 114.
45. Egerton 1973: 335–337.
46. Darwin 1839: 10.
47. Edwards 1985: 34.
48. Darwin 1859: 73–74, Egerton 2007a, c: 51–52.
49. McAtee 1947.
50. Semper 1881: 51–52.
51. Mayr 1975.
52. Pequegnat 1958.
53. Cohen 1994.
54. Camerano 1880.
55. Cohen 1994: 353.
56. Camerano 1994.

57. Egerton 2007c, 2014a.
58. Croker 2001: 9–16.
59. Forbes 1977.
60. T. Park 1949: 52–53.
61. Both reproduced in Egerton 2007c: 54–55.
62. O. Park 1949: 508–514.
63. Egerton 2007c.
64. Spärck 1932.
65. Schlee 1973: 26–219.
66. Reproduced in Egerton 2007c: 56.
67. Blegvad 1916.
68. Hardy 1924, Summerhayes & Elton 1923, Elton 1927.
69. Bullen 1976, Schwarzbach 1986.
70. Glen 1975, 1982: 4–5.
71. Weart 2003: 17–18, 47–50.
72. Steele & Smallwood 2002: 184–185.
73. Partington 1964: 337–345, Kapor 1971.
74. Dumas 1841: 338.
75. Wegmann 1976.
76. Suess 1875: 3, translated in Smil 2002: 1.
77. Govindjee et al. 2005, Govindjee & Kingman 2005.
78. Turner 1974: 235.
79. Govindjee & Krogmann 2005: 67.
80. Quoted from Jackson 2018: 169.
81. Crist 2004: 162.
82. Browne 2003: 447.
83. Darwin 1840, 1884, 1869.
84. Stewart 2004: 4–10.
85. Stewart 2004: 144.
86. Fedoseyev 1976, Smil 2002: 2–13, Levit 2011.
87. Vernadsky 1923, 1924, 1926a, b, 1929, 1997.
88. Cevasco 2009, Slack 2010.
89. Slack 2010: 171–173, Egerton 2016b: 229.
90. Hutchinson 1945, Slack 2010: 175.
91. Limburg 2004.
92. Craig 2001: 39–42.
93. Egerton 2015a: 83–87, e.
94. Likens et al. 1977, Bormann & Likens 1979.
95. Murphy 1962.
96. Burgess & Raw 1967.
97. Richter & Markewitz 2001: 69.
98. Richter et al. 2018.
99. Lovelock 2000b.
100. Lovelock 2000a, b: 253.
101. Turney 2003: 46.
102. Lovelock 2000b: 191.
103. Lovelock 2000a, b: 252–253.
104. Lovelock 1972.
105. Sapp 1994: index, Wagensberg 2012: 12–13, Ruse 2013: 169–171, 191–198, Gray 2017.
106. Lovelock 2000b: 256.
107. Lovelock 1979: 29.
108. Lovelock 1988: index.
109. Gorham 1991: 231, 235.
110. Lovelock 2000a: 68.
111. Schneider & Boston 1991.
112. Margulis 1998, quoted from Turney 2003: 112.
113. Volk 1998: viii—ix.
114. Turney 2003: 139.
115. Lovelock 2003.
116. Tyrrel 2013: 8.
117. Egerton 1973.
118. Odum 1989: 61–62.
119. Egerton 2019c: 24–26.
120. Fleming 1998: 27.
121. Sparks & Carey 1995.
122. Egerton 2007e.
123. Hagberg 1952: 141–144, Blunt 1971: 213.
124. Stoller 1956: 175–176.
125. Jefferson 1787: 200–208.
126. Cutright 1969: 2.
127. Bordley & Harvey 1976: 36–37, Duffy 1976: 101.
128. Bigelow 1818.
129. Silliman 1818.
130. Fleming 1990: 13–16.
131. Fleming 1998: 35.
132. Listed in Fleming 1998: 37.
133. Fleming 1998: 37.
134. Walls 1995, 1999, Egerton 2011b, 2012a: 151–156.
135. Mallis 1971: 25–33.
136. Thoreau 1975: 26–36.
137. Stoller 1956: 172–174, Harding 1961: 81–85, Taylor 1963.
138. Fleming 1998: 41.
139. Leopold & Jones 1947, Whitford 1950, Stoller 1956, Nijhuis 2007, Miller-Rushing & Primack 2008, Willis et al. 2008, Primack & Miller-Rushing 2012, Ellwood et al. 2013.
140. Leopold & Jones 1947: 83, Curran 1997, 1999, Egerton 2019a: 10.
141. Meine 1988: 256.
142. Meine 1988: 307.
143. Meine 1988: 382.
144. Meine 1988: 471–472.
145. Egerton 2015a: 55, e.
146. Grattan-Guinness & Ravetz 1972, Ravetz & Grattan-Guinness 1972.
147. Kline 1972: 672, Ravetz & Grattan-Guinness 1972: 94–95.
148. Fleming 1998: 60.
149. Nuccitelli 2015: 1–2.
150. Tyndall 1900, MacLeod 1976, Jackson 2018.
151. Nuccitelli 2015: 3–4.
152. Tyndall 1861.
153. Newton 1895, Kutzbach 1973.
154. Whitnah 1965: 4–8, Fleming 1990, 1998.
155. Loomis & Newton 1866.
156. Snelders 1970, Fleming 1998: 74–82, McKibben 1988: 7–8, Weart 2003: 5–8, Flannery 2005: 27, 40, Nuccitelli 2015: 5–6.
157. Arrhenius 1896.
158. Pihl 1970, Friedman 1989, Fleming 2016: 16–17.
159. McKibben 1989: 8, Weart 2003: 11, 18–19, Fleming 2007, Nuccitelli 2015: 11–12.
160. Callendar 1938.
161. Fleming 1998: 121–125, Weart 2003: 223–25, Nuccitelli 2015: 13–14.
162. Plass 1958a, b, c.
163. Fleming 1998: 122–126, Weart 2003: 27–31.
164. Quoted in McKibben 2011: 42.
165. Fleming 1998: 126–127, Weart 2003: index.
166. Keeling 1960.
167. Flannery 2005: 24–26.
168. In McKibben 2011: 44.
169. Weart 2003: 38.
170. Bryson & Murray 1977: 144–152.
171. Bowen 2008: 71–78.

172. Hansen 1988.
173. Nuccitelli 2005: 38.
174. Howe 2014: 147–152.
175. McKibben 2011: 55–67.
176. Acot 2000, 2006: 118–120, McKibben 2011: 127–132, Howe 2017: 255–264.
177. Besel 2013.
178. Howe 2014: 147–152.
179. In McKibben 2011: 55–67.
180. Hansen et al. 1998.
181. Hansen et al. 1998: 930.
182. Worland & Hobbs 2018.

Abbreviations

Bull.	*Bulletin*
BDAC	*Biographical Dictionary of American and Canadian Naturalists and Environmentalists*
DAB	*Dictionary of American Biography*
DNB	*Dictionary of National Biography*
DNCB	*Dictionary of Nineteenth-Century British Scientists*
DSB	*Dictionary of Scientific Biography*
Ed.	Edition or Editor
ESAB	*Ecological Society of America Bulletin*
J.	*Journal*
Hist	*History*
L	London
NDSB	*New Dictionary of Scientific Biography*
NY	New York
ODNB	*Oxford Dictionary of National Biography*
P.	Press
Philos.	*Philosophical*
Plts.	*plates*
Proc.	*Proceedings*
Roy.	Royal
Sci.	*Science*
Soc.	*Society*
Trans.	*Transactions*
transl.	translator
U.	University
vol.	volume

References

LITERATURE CITED

Acosta, J. de. 2002. *The natural and moral history of the Indies*. F. Lopez-Morillas, transl. Durham NC: Duke U.P. Selections from 1604 ed. In von Hagen 1948: 56–75.

Acot, P., ed. 1998. *The European origins of scientific ecology*. B.P. Hamm, transl. 2 vols. Amsterdam: Gordon & Breach/ Éditions des Archives Contemporaines.

———. 2000. *The concept of biosphere*. Detroit: Gale Group.

———. 2006. *Catastrophes climatiques, désastres social*. Paris: P. U. de France.

———. 2008. Josias Braun-Blanquet (1884–1980), plant sociology, ecology, botany. *Complete Dictionary of Scientific Biography*, online.

——— & M.-N. Bourguet. 2003. Au Chimborazo: la géographie des plantes. *Musée des Arts et Métiers:La Revue* 39/40: 84–91.

Adams, C.C. 1913. *Guide to the study of animal ecology*. NY: Macmillan. 1977. NY: Arno P.

——— et al. 1920. Plants and animals of Mount Marcy, New York. *Ecology* 1: 71–94 + 1 plt., 204–233, 274–288.

Adelmann, H.B. 1966. *Marcello Malpighi and the evolution of embryology*. Vol. 1. (Biographical.) Ithaca: Cornell U.P.

Agassiz, A. 1903a. The coral reefs of the tropical Pacific. *Museum of Comparative Zoology Memoirs* 28: 1–400 + 238 plates.

———. 1903b. The coral reefs of the Maldives. *Museum of Comparative Zoology Memoirs* 29: 1–168 + 182 plates.

———. 1903c. On the formation of barrier reefs and of different types of atolls. *Royal Society of London Philos. Trans.* 71: 412–414.1.

Agassiz, G.R., ed. & an author. 1913. *Letters and recollections of Alexander Agassiz, with a sketch of his time and work*. Boston: Houghton Mifflin.

Agassiz, J.L.R. 1850. *Lake Superior: its physical character, vegetation, and animals, compared with those of other and similar regions*. Boston: Gould, Kendall & Lincoln. 1974. Hunting NY: Robert E. Krieger.

Agnarsson, I., Jason Ali & D.S. Barrington. 2019. Vicariance biogeography. *Oxford Bibliographies*, online.

Agrios, G.N. 2005. *Plant Pathology*. NY: Academic P. Ed. 1, 1969.

Ainsworth, G.C. 1969. History of plant pathology in Great Britain. *Ann. Rev. of Phytopathology* 7: 13–30.

———. 1976. *Introduction to the history of mycology*. L: Cambridge U.P.

———. 1981. *Introduction to the history of plant pathology*. L: Cambridge U.P.

Aleem, A.A. 2002. The Allan Hancock Pacific Explorations (1931–1962) and their contributions to marine biology. Pp. 316–319 in Benson & Rehbock 2002.

Alekseev, A.V. & I.D. Rostov. 2002. Russian oceanographic investigations of the Pacific Ocean: history and some results. Pp. 96–102 in Benson & Rehbock 2002.

Alexander, J. 2009. *Pandora's locks: the opening of the Great Lakes-St. Lawrence Seaway*. East Lansing: Michigan State U.P.

Allaby, M. 1994. *The concise Oxford dictionary of ecology*. NY: Oxford U.P.

Allan, D.G.C. 2004. Stephen Hales (1677–1761). *ODNB* 24: 553–556.

——— & R.E. Schofield. 1980. *Stephen Hales: scientist and philanthropist*. L: Scolar P.

Allan, M. 1967. *The Hookers of Kew, 1785–1911*. L: Michael Joseph.

———. 1972. William Jackson Hooker (1785–1865), botany. *DSB* 492–495.

Allard, D.C., Jr. 1967. *Spencer Fullerton Baird and the U.S. Fish Commission: a study in the history of American science*. Washington DC: Ph.D. dissertation, George Washington U. 1978. NY: Arno P.

———. 1970. Spencer Fullerton Baird (1823–87), zoology, scientific administration. *DSB* 1: 404–406.

———. 1997. Spencer Fullerton Baird (1823–87). *BDAC* 53–56.

Alleaume, G. 1999. Entre l'inventaire du territorire et la construction de la mémoire: l'oeuvre cartographique de l'expedition à Egypt. Pp. 279–294 in Bret 1999.

Allee, W.C. 1949. Ecological background and growth before 1900. Pp. 13–43 in Allee et al. 1949.

———, A.E. Emerson, O. Park, T. Park & K.P. Schmidt. 1949. *Principles of animal ecology*. Philadelphia: W. B. Saunders.

Allen, D.E. 1976. *The naturalist in Britain: a social history*. L: Allen Lane.

———. 1986. *The botanists: a history of the Botanical Society of the British Isles through a hundred and fifty years*. Winchester: St. Paul's Bibliographies.

Allen, D.L. 1979. *Wolves of Minong: their vital role in a wild community*. Boston: Houghton Mifflin.

Allen, E.G. 1951. The history of American ornithology before Audubon. *American Philos. Soc. Trans.* 41: 386–591.

Allen, J.A. 2004. Sir Charles Maurice Yonge (1899–1986), marine biologist. *ODNB* 60: online.

Allen, J.L. 1975. *Passage through the garden: Lewis and Clark and the image of the American Northwest*. Urbana: U. of Illinois P.

Almaça, C. 1997. Augusto Nobre and marine biology in Portugal. Pp. 125–134 in Saldanha & Ré 1997.

Amato, A.J., ed. 2017. *From conservation on the northern plains: new perspectives*. Sioux Falls SD: Augustana U.

Ambrose, S.E. 1996. *Undaunted courage: Meriwether Lewis, Thomas Jefferson, and the opening of the American West*. NY: Simon & Schuster.

——— & S. Abell. 1998. *Lewis and Clark: voyage of discovery*. Washington: National Geographic Society.

Amorim, I. 2009. Marine zoology: the second half of the 19th century in Portugal. *Studia Atlantica* 13: 42–59.

Anderson, J. 1794. *A practical treatise on peat-moss, considered as in its natural state fitted for affording fuel, or as susceptible of being converted into mold capable of yielding abundant crops . . .* Edinburgh/L: T. Chapman.

Anderson, T.R. 2013. *The life of David Lack, farther of evolutionary ecology*. Oxford: Oxford U.P.

——— & W.C. Gentleman. 2012. The legacy of Gordon Arthur Riley (1911–1985) and the development of mathematical models in biological oceanography. *J. of Marine Research* 70: 1–30.

Anderson, W. 2016. Postcolonial ecologies of parasite and host: making parasitism cosmopolitan. *J. of the Hist. of Biology 49: 241–259.*

Andrewartha, H.G. & L.C. Birch. 1973. The history of ecology. Pp. 229–266 in Smith, Mittler & Smith 1973.

Andrist, R.K. 1967. *To the Pacific with Lewis and Clark.* NY: American Heritage.

Anker, P. 2011. Plant community, Plantesfund. Pp. 325–331 in Schwarz & Jax 2011.

Anon. June 1961. A medal for Dr. Bigelow. *Oceanus* 7, no. 4: 16–21.

Appel, T.A. 1973. Felix-Joseph Henri de Lacaze-Duthiers (1821–1901), zoology. *DSB* 7: 545–546.

Applebaum, W., ed. 2000. *Encyclopedia of the scientific revolution from Copernicus to Newton.* NY: Garland.

Applegate, V.C. 1961. Downstream movement of lampreys and fishes in the Carp Lake River, Michigan. *U.S. Fish & Wildlife Service, Special Scientific Report—Fisheries* 387.

——— & C.L. Brynildson. 1951. Downstream movement of recently transformed sea lampreys *Petromyzson marinus*, in the Carp Lake River, Michigan. *American Fisheries Soc. Trans.* 81: 275–290.

———, J.H. Howell, A.E. Hall & M.A. Smith. 1957. Toxicity of 4,346 chemicals to larval lampreys and fishes. *U.S. Fish & Wildlife Service, Special Scientific Report: Fisheries* 207. 157 pp.

——— & B. Smith. 1950. Sea lamprey spawning runs in the Great Lakes in 1950. *U.S. Fish & Wildlife Service, Special Science Report: Fisheries no. 61.* 49 pp.

———. 1951. Movement and dispersion of a blocked spawning run of sea lampreys in the Great Lakes. *Sixteenth North American Wildlife Conference Trans.* 243–251.

——— & H.D. Van Meter. 1970. A brief history of commercial fishing in Lake Erie. *U.S. Fish & Wildlife Service, Fisheries Leaflet* 630. Iii + 28 pp.

Aristotle. 1965–91. *History of animals.* A.L. Peck, transl. vols. 1–2; D.M. Balme, transl. vol. 3. Cambridge MA: Harvard U.P.

Arnold, D. 2005a. *The tropics and the traveling gaze: India, landscape, and science, 1800–1856.* Seattle: U. of Washington P.

———. 2005b. Envisioning the tropics: Joseph Dalton Hooker in India and the Himalayas, 1848–1850. Pp. 137–155 in Driver & L. Martins 2005.

Aron, W.I. & S.H. Smith. 1971. Ship canals and aquatic ecosystems. *Science* 174: 13–20.

Arrhenius, S.A. 1896. On the influence of carbonic acid in the air upon the temperature of the ground. *Philos. Mag.* 41: 237–276.

Ashworth, J.H. & E.L. Mills. 2004. Sir John Murray (1841–1914), marine scientist and oceanographer. *ODNB* 39: 959–961.

Ashworth, W. 1986. *The late Great Lakes: an environmental history.* NY: Knopf. 1987. Detroit: Wayne State U.P.

Astro, R. 1973. *John Steinbeck and Edward Ricketts: the shaping of a novelist.* St. Paul: U. of Minnesota P.

Audubon, J.J. 1827–38. *Birds of America, from original drawings.* 4 vols. L: Author.

———. 1831–39. *Ornithological biography: an account of the habits of the birds of the United States of America.* 5 vols. Edinburgh: Adam & Charles Black.

———. 1839. *A synopsis of the birds of America.* Edinburgh: Adam & Charles Black.

———. 1840–44. *The birds of America.* 7 vols. NY: Author & J.B. Chevalier. 1967. NY: Dover. [text & plates]

———. 1999. *Writings and drawings.* C. Irmscher, ed. NY: Library of America.

———. 2006. *The Audubon reader.* R. Rhodes, ed. NY: Everyman's Library/Knopf.

——— & J. Bachman. 1845–46. *The viviparous quadrupeds of North America.* Illustrations by J.J. Audubon & J.W. Audubon. 2 vols, NY: Author. Ed. 2, 3 vols. 1846–54.

———. 1951. *Audubon's animals: the quadrupeds of North America.* A. Ford, ed. NY: Studio Publications/Thomas Y. Crowell.

———. 1989. *Audubon's Quadrupeds of North America: complete and unabridged.* Secaucus NJ: Wellfleet P.

Auer, N.A., ed. 2013. *Larval fishes of the Great Lakes Basin, with emphasis on the Lake Michigan Drainage.* Ann Arbor: Great Lakes Fishery Commission. Xxxi + 744 pp.

——— & D. Dempsey, eds. 2013. *The Great Lakes sturgeon.* East Lansing: Michigan State U.P.

Auerbach, S.I. 1995. Foreword—George M. Van Dyne—a reminiscence. Pp. xxvii—xxx in Patten & Jorgensen 1995.

Austin, O.L., Jr. 1975. Philip Lutley Sclater (1829–1913), ornithology. *DSB* 12: 257–258.

Avignon, P. 1971. John Evelyn (1620–1706), arboriculture, horticulture. *DSB* I4: 494–497.

Ayres, P. 2012. *Shaping ecology: the life of Arthur Tansley.* Oxford UK: John Wiley &Sons.

Baatz, S. 1999. William Maclure (1763–1840). *ANB* 14: 273–275.

Baehni, C. 1955. Correspondence de Charles Darwin et d'Alphonse de Candolle. *Gesnerus* 12: 109–156.

Baer, J.H. 2004. William Dampier (1651–1715), buccaneer and explorer. *ODNB* 15: 29–32.

Baigent, E. 2004. Jerard Boate [formerly Gerrit Boot] (1604–50), physician and writer and natural historian. *ODNB* 6: 384–385.

Bailes, K.E. 1981. Science, philosophy and politics in Soviet history: the case of Vladimir Vernadsky. *Russian Review* 40: 278–299.

Baker, E.A. & N.A. Auer. 2013. Habitat, foods, and feeding. Pp. 59–78 in Auer & Dempsey 2013.

Baker, J.R. 1952. *Abraham Trembley of Geneva: scientist and philosopher, 1710–1784.* L: Arnold, Edward.

———. 1976. Abraham Trembley (1710–84), zoology. *DSB* 13: 457–458.

Ball, G.E., ed. 1985. *Taxonomy, phylogeny, and zoogeography of beetles and ants.* Dordrecht Neth.: W. Junk.

Balme, D.M. 1970. Aristotle: natural history and zoology. *DSB* 1: 258–266.

Balon, E.K. 1994. The life and work of Eugenie Clark: devoted to diving and science. *Environmental Biology of Fishes* 41: 89–114.

Banks, E.M. 1985. Warder Clyde Allee and the Chicago school of animal behavior. *J. of Hist. of Behavioral Sciences* 21: 345–353.

Banks, J. 2009. *The scientific correspondence of Sir Joseph Banks, 1765–1820.* 6 vols. L: Pickering & Chatto.

Baranetsky, O.V. 1868. Independently living gonidia of lichens. [in Russian] *First Russian Congress of Naturalists and Physicians, Botanical Section Trans.* Pp. 45–59. St. Petersburg.

Barbour, M.G., J.H. Burk & W.D. Pitts. 1987. *Terrestrial plant ecology.* Menlo Park CA: Benjamin Cummings.

Barrow, M.V., Jr. 1998. *A passion for birds: American ornithology after Audubon.* Princeton: Princeton U.P.

———. 2009. *Nature's ghosts: confronting extinction from the age of Jefferson to the age of ecology*. Chicago: U. of Chicago P.

Barry, J.P. 1972. *The fate of the lakes: a portrait of the Great Lakes*.

Bartlette, R.A. 1962. *Great surveys of the American West*. Norman: U. of Oklahoma P.

Bartola, C. 1988. François-Alphonse Forel (1841–1912). Pp. 3–17 in *Découvir le Léman: 100 ans àpres François-Alphonse Forel. Actes du Colloque pluridisciplinaire, Nyon, 16–18 Setembere 1988*. Slakine & Nyons: Musée du Léman, Geneva.

Bary, A. de. 1866. *Morphologie und Physiologie der Pilze, Flechten und Myxomyceten*. Leipzig: Engelmann.

———. 1868. Beitrag zur Kenntnis des selfstënolohigen Lebens der Flechtingon, von J. Boranetzhy. *Botanische Zeitung* (Berlin) 26: 196–198.

———. 1879a. Die Ericheinung der Symbiose. Pp. 1–30 in *Vortrag auf der Versammlung der Naturforscher und Ärtze zu Cassel*. Strassburg: Trubner.

———. 1879b. De la symbiose. *Revue international des sciences* 3: 301–309.

Basalla, G. 1967. The spread of western science. *Science* 156: 611–622.

Bastin, J. 1986. Introduction, pp. vii—xvii in Wallace 1986.

Bates, H.W. 1862. Contribution to an insect fauna of the Amazon Valley, Lepidoptera: Heliconoidae. *Linnean Soc. of L. Trans.* 23: 495–566 + 4 plates. Reprinted in Bates 1978.

———. 1864. *The naturalist on the River Amazons*. L: John Murray.

———. 1978. *The principal contributions of Henry Walter Bates to a knowledge of the butterflies and longicorn beetles of the Amazon Valley*. E.G. Linsley, ed. NY: Arno P.

Bates, R.S. 1965. *Scientific societies in the United States*. Ed. 3. Cambridge: MIT P.

Baudin, N. 1974. *The journal of Post Captain Nicolas B audin, commander-in-chief of the corvettes Géographie and Naturaliste*. C. Cornell, transl. Adelaide: Libraries Board of S. Australia.

Baumann, P.C., J.F. Kitchell, J.J. Magnuson & T.B. Kayes. 1974. Lake Wingra, 1837–1973: a case history of human impact *Wisconsin Academy of Sciences, Arts, and Letters Trans.* 62: 57–94.

Baxter, S. 2003. *Ages in chaos: James Hutton and the discovery of time*. L: Weindonfeld & Nicolson.

Beach, E.L. 1969. Man beneath the sea. Pp. 232–261 in Idyll 1969.

Beale, N. & E. Beale. 2011. *Echoes of Ingen Housz: the long-lost story of the genius who rescued the Habsburgs from small-pox and became the father of photosynthesis*. Gloucester UK: Hobnob P.

Beck, H. 1959–61. *Alexander von Humboldt*. 2 vols. Wiessbaden: Steiner.

Beckel, A.L., author & ed. 1987. Breaking new waters: a century of limnology at the University of Wisconsin. *Wisconsin Academy of Sciences, Arts & Letters Trans.* Special issue.

Becker, G. et al. 1957. *Tournefort*. Paris: Museum National d'Histoire Naturelle.

Becking, R.W. 1957. The Zurich-Montpellier School of Phytosociology. *Botanical Review* 23: 411–488.

Beckinsale, R.P. 1971. Jean-André Deluc (1727–1817), geology, meteorology, physics, natural philosophy, theology. *DSB* 4: 27–29.

Beddall, B.G., ed. 1969. *Wallace and Bates in the tropics: an introduction to the theory of natural selection*. L: Collier-Macmillan.

Bedini, S.A. Thomas Jefferson (1743–1826), agriculture, botany, cartography, diplomacy, ethnology, meteorology, paleontology, surveying. *DSB* 7: 88–90.

———. 1990. *Thomas Jefferson: statesman of science*. NY: Macmillan.

Beebe, C.W. 1924. *Galapagos: world's end*. NY: Putnam.

———. 1926a. *The Arcturus adventure: an account of the New York Zoological Society's first oceanographic expedition*. NY: G.P. Putnam.

———. 1926b. The Arcturus oceanographic expedition. *Zoologica* 8: 1–45.

———. 1926c. A note on the Humboldt Current and the Sargasso Sea. *Science* 63: 91–92.

———. 1928. *Beneath tropical seas*. NY: Putnam.

———. 1931. A round trip to Davy Jones's Locker: peering into Mysteries a quarter mile down in the open sea, by means of the Bathysphere. *National Geographic* 59: 653–678.

———. 1932a. The depths of the sea: strange life forms a mile below the surface. *National Geographic* 61: 65–88. Paintings, E. Bostelmann.

———. 1932b. A wanderer under the sea. *National Geographic* 62: 741–758. Paintings, E. Bostelmann.

———. 1934. A half mile down: strange creatures, beautiful and grotesque as figments of fancy, reveal themselves at windows of the Bathysphere. *National Geographic* 66: 661–704.

———. 1951. *Half mile down*. NY: Duell, Sloane and Pearce. Ed. 1, 1934.

——— & J. Tee-Van. 1928. The fishes of Port-au-Prince, Haiti. *Zoologica* 10: 317–319.

———. 1933. *Field book of the shore fishes of Bermuda*. NY: G.P. Putnam's Sons.

Beeton, A.M. Limnological survey of Lake Erie, 1959 and 1960. *Great Lakes Fishery Commission*. Technical Report 6.

———. 1961. Environmental change in Lake Erie. *American Fisheries Soc. Trans.* 90: 153–159.

———. 1965. Eutrophication of the St. Lawrence Great Lakes. *Limnology and Oceanography* 10: 240–254.

——— & D.C. Chandler. 1963. The St. Lawrence Great Lakes. Pp. 535–558 in Frey 1963.

——— & W.T. Edmondson. 1972. The eutrophication problem. *Fisheries Research Board of Canada J.* 29: 673–682.

——— et al. 1981. *Report of Great Lakes Region Conference on Marine Pollution Problems*. Ann Arbor: Michigan Sea Grant & NOAA.

Beidleman, R.G. 1960. Some biographical sidelights on Thomas Nuttall, 1786–1859. *American Philos. Soc. Proc.* 104: 86–100.

———. 2006. *California's frontier naturalists*. Berkeley: U. of California P.

Bell, M.S. 2005. *Lavoisier in the year one: the birth of a new science in an age of revolution*. NY: Norton.

Bellevoye, A. & J. Laurent. 1897. Les plantations de pins dans la Marne et les parasites qui les attaquent. *Bull. de la Société d' Étude des sciences naturelles de Reims* 5: 70–126.

Bellon, R. 2000. Joseph Dalton Hooker, 1785–1911, British botanist and traveler. Pp. 348–349 in Hessenbruch 2000a.

Belloni, L. 1974. Marcello Malpighi (1628–94), medicine, microscopic and comparative anatomy. *DSB* 9: 62–66.

Belt, T. 1985. *The naturalist in Nicaragua*. Chicago: U. Chicago P. Ed. 1, 1874. Ed. 2, 1888.

Beneden, P.-J. van. 1875. *Les commensaux et les parasites dans la régne animal*. Paris: Baillière.

———. 1876. *Animal parasites and messmates*. L: Henry S. King. NY: Appleton.

Bengtsson, L.O. & C.U. Hammer, eds. 2001. *Geosphere-biosphere interactions and climate*. NY: Cambridge U.P.

Benito-Vessels, C. 1996. Francisco Hernández (ca. 1517–87), medical doctor and botanist. Pp. 492–493 in *Latin American Lives*. NY: Macmillan Compendium.

Benke, A.C. 2018. River food webs: an integrative approach to bottom up flow webs, top-down impact webs, and trophic position. *Ecology* 99: 1370–1381.

Bennett, G.W. 1958. Aquatic biology. *Illinois Natural History Survey Bull.* 27: 163–178.

Bennett, T.P. 1983. The history of the Academy of Natural Sciences of Philadelphia. Pp. 1–14 in Wheeler 1983.

Benson, K.R. 1988a. The Naples Zoologica and its impact on the emergence of American marine biology. *J of the Hist. of Biology* 21: 331–341.

———. 1988b. Laboratories on the New England shore: the somewhat different direction of American marine biology. *New England Quarterly* 61: 55–78.

———. 1988c. Why American marine stations? The teaching argument. *American Zoologist* 28: 7–14.

———. 1992. Experimental ecology on the Pacific Coast: Victor Shelford and his search for appropriate methods. *History & Philos. of the Life Sciences* 14: 73–91.

——— & P.F. Rehbock, eds. 2002. *Oceanographic Hist.: the Pacific and Beyond*. Seattle: U. of Washington P.

——— & H.M. Rozwadoski, eds. 2007. *Extremes: oceanography's adventures at the poles*. Sagamore Beach MA: Science History Publications.

Benson, N.G. 1970. *A century of fisheries in North America*. Washington: American Fisheries Society.

Berg, K. 1951. The content of limnology demonstrated by F.-A. Forel and August Thienemann on the shore of Lake Geneva. *International Association of Theoretical and Applied Limnology Proceedings* 11: 41–57. Science History Publications/Watson Publishing International.

Bergandi, D., ed. 2013. *The structural links between ecology, evolution and ethics: the virtuous epistemic circle*. NY: Springer.

Berger, A.M. 1756. *Calendarium florae. Uppsala: Höjer.* Reprinted in *Amoenitates Acqdemicae* 4: 387–414.

Berra, T.M. 1977. *William Beebe: an annotated bibliography*. Hamden CT: Archon Books.

Besel, R.D. 2013. Accommodating climate change science: James Hansen and the rhetorical/political emergence of global warming. *Science in Context* 26: 137–152.

Beverton, R.J. & S.J. Holt. 1957. *On the dynamics of exploited fish populations*. L: Ministry of Agriculture, Fisheries and Food.

Biermann, K.-R. 1972. Friedrich Wilhelm Heinrich Alexander von Humboldt (1769–1858), natural science. *DSB* 6: 549–565.

Bigelow, H.B. 1904. Medusae from the Maldive Islands. *Museum of Comparative Zoology Bull.* 39: 245–269 + 9 plts.

———. 1926. Plankton of the offshore waters of the Gulf of Maine. *U.S. Bureau of Fisheries Bull.* 40, part 2: 1–509 + 18 plts.

———. 1927. Physical oceanography of the Gulf of Maine. *U.S. Bureau of Fisheries Bull.* 40, no. 2: 511–1027.

———. 1928. Exploration of the waters of the Gulf of Maine. *Geographical Review* 18: 232–260.

———. 1964. *Memories of a long and active life*. Cambridge MA: Cosmos P.

——— & W.W. Welsh. 1925. Fishes of the Gulf of Maine. *U.S. Bureau of Fisheries Bull.* 40, part I: 1–567.

Bigelow, J. 1818. Facts serving to show the comparative forwardness of the spring in different parts of United States. *American Acad. of Arts & Sci. Mem.* 4: 77–85.

Billings, W.D. 1985. The historical development if physiological plant ecology. Pp. 1–15 in Mooney 1985.

Billington, R.A. 1967. *Westward expansion: a history of the American frontier*. Ed. 3. NY: Macmillan.

Bircham, P. 2007. *A history of ornithology*. L: Collins.

Birge, E.A. 1879. Notes on Cladocera. *Wisconsin Academy of Sciences, Arts, and Letters Trans.* 4: 77–112. Ed. 1. Madison: no publisher, 34 pp., 2 plates, 1878.

———. 1898. Plankton studies on Lake Mendota: II. The Crustacea of the plankton from July, 1894 to December, 1896. *Wisconsin Academy of Sciences, Arts and Letters Trans.* 11: 274–448.

———. 1904. The thermocline and its biological significance. *American Microscopical Soc. Trans.* 25: 5–33 + 2 plates.

———. 1910a. An unregarded factor in lake temperatures. *Wisconsin Academy of Sciences, Arts and Letters Trans.* 16: 989–1004.

———. 1910b. On the evidence for temperature seiches. *Wisconsin Academy of Sciences, Arts and Letters Trans.* 16: 1005–1016.

——— & C. Juday. 1911. The inland lakes of Wisconsin: the dissolved gases of the water and their biological significance. *Wisconsin Geological and Natural History Survey B* 22: xx + 1–259 pp. + 10 plates.

———. 1922. The inland lakes of Wisconsin: the plankton. I. Its quantity and chemical composition. *Wisconsin Geological and Natural History B* 64: 222 + ix pp.

———. 1934. Particulate and dissolved organic matter in inland lakes. *Ecological Monographs* 4: 440–474.

———. 1977. *Limnology in Wisconsin*. F.N. Egerton, compiler. NY: Arno P.

Birge, E.A, O.A. Olson & H.P. Harder. 1895. Plankton studies on Lake Mendota: I. The vertical distribution of the pelagic Crustacea during July, 1894. *Wisconsin Academy of Sciences, Arts & Letters Trans.* 10: 421–484 + 4 plates.

Birkhead, T., J. Wimpenny & B. Montgamerie. 2014. *Ten thousand birds: ornithology since Darwin*. Princeton: Princeton U.P.

Bjerknes, J. 1969. Atmospheric teleconnections from equatorial Pacific. *Monthly Weather Review* 97: 163–172.

Blacher, L.J. 1971. Johann Friedrich Eschscholtz (1793–1831), medicine, zoology. *DSB* 4: 406–407.

Blackman, F.E. & A.G. Tansley. 1903. Ecology in its physiological and topological aspects. *New Phytologist* 4: 232–253.

Blair, W.F. 1977. *Big biology: the US/IBP*. Stroudsburg PA: Dowden, Hutchinson & Ross.

Blaisdell, M.L. 1992. *Darwinism and its data: the adaptative coloration of animals*. NY: Garland.

Blanc, H. 1912. Le professeur Dr. François Alphonse Forel. *Société Helvétique des Sciences Naturelles Actes* 95: 110–148.

Blegvad, H.H. 1916. On the food of fish in the Danish waters within the Skaw. *Danish Biological Station Report* 24: 17–72 + 6 lists.

Blum, A.S. 1993. *Picturing nature: American Nineteenth-century zoological illustration*. Princeton: Princeton U.P.

Blunt, W. 1971. *The compleat naturalist:a life of Linnaeus*. L.: William Collins.

Blust, F.A. 1976. The U.S. Lake Survey, 1841–1974. *Inland Seas* 32: 91–104.

Blyth, E. 1871. A suggested new division of the Earth into zoological regions. *Nature* 3: 427–429.

Bocking, S. 1987. *The origins of aquatic ecological research in the Great Lakes Region*. Toronto: Thesis, U. of Toronto.

———. 1988. Alpheus Spring Packard and cave fauna in the evolution debate. *J. of the Hist. of Biology* 21: 425–456.

———. 1990a. Stephen Forbes, Jacob Reinhard, and the emergence of aquatic ecology in the Great Lakes region. *J. of the Hist. of Biology* 23: 461–498.

———. 1990b. Fisheries and fundamental science: Donald Rawson's Studies of Lake Productivity. *Scientia Canadensis* 14: 38–50.

———. 1997a. *Ecologists and environmental politics: a history of contemporary ecology*. New Haven: Yale U.P.

———. 1997b. Fishing the inland seas: Great Lakes research, fisheries management and environmental policy in Ontario. *Environ. Hist.* 2: 52–73.

——— & D. Heidt, eds. 2019. *Cold science: environmental knowledge in the North American Arctic during the Cold War*. NY: Routledge P.

Bodenheimer, F.S. 1955. Zimmermann's *Specimen zoologiae geographiae Quadrupedum*, a remarkable zoogeographical publication of the end of the 18th century. *Archives international d'Histoire des Sciences* 8: 351–357.

Boewe, C., ed. 1982. *Fitzpatrick's Rafinesque: a sketch of his life with bibliography*. Weston MA: M & S P.

———. 1997. Constantine Samuel Rafinesque (Schmaltz, 1783–1840). *BDAC* 654–655.

———. 1999. Constantine Samuel Rafinesque (1783–1840). *ANB* 18: 68–70.

———, ed. 2003. *Profiles of Rafinesque*. Knoxville: U. of Tennessee P.

Bogue, M.B. 2000. *Fishing the Great Lakes: an environmental history, 1783–1953*. Madison: U. of Wisconsin P.

Bohlin, I. 2000. Charles Darwin, 1809–1882, British naturalist. Pp. 163–165 in Hessenbruch 2000a.

Böhner, K. 1933–35. *Geschichte der Cecidologie: ein Beitrag zur Entwicklungsgeschichte naturwissenschaftlicher Forschung und ein Führer durch Cecidologie der Alten*. Mittenwald: Arthur Nemayer.

Boklund, U. 1975. Carl Wilhelm Scheele (1742–86), pharmacy, chemistry. *DSB* 12: 143–150.

Bolin, B. & R. Cook, eds. 1983. *The major biogeochemical cycles and their interactions*. NY: Wiley.

Bologa, A.S. 2004. Development of marine biological institutions around the Black Sea: an overview. Pp. 209–222 in Morcos et al. 2004.

——— & A. Marinescue. 2002. Romanian developmental contributions Emil Racovitza and Grigore Antipa to the scientific exploration of the Mediterranean. Pp. 275–279 in Benson & Rehbock 2002.

Bonaparte, C.-L. 1824. An account of four species of stormy petrels. *Academy of Natural Sciences of Philadelphia J.* 3: 227–233.

———. 1825–36. *American ornithology; or, the natural history of the birds inhabiting the United States not given by Wilson*. 4 vols. Philadelphia: Carey, Lea & Carey.

———. 1838. *A geographical and comparative list of the birds of Europe and North America*. L: John van Voorst.

Bonnemains, J., E. Forsyth & D. Smith, eds. 1988. *Baudin in Australian waters: the artwork of the French voyage of discovery in the southern lands, 1800–1804, with a descriptive catalogue of drawings and paintings of Australian subjects by C.A. Lesueure and N.M. Petit from the Lesueure Collection at the Museum d'Histoire Naturelle, Le Havre France.*

Bonneuil, C. & J.-B. Fressoz. 2016. *The shock of the Anthropocene: the Earth history and us*. L & NY: Verso Books.

Bonnier, G. & C.H.N. Flahault. 1878. Observations sur les modifications des végétaux suivant les conditions physiques du milieu. *Annales des science naturelles, Botanique* 7: 93–125. Partly reprinted in Acot 1998: 132–149.

Boran, E. 2004. Arnold Boate (1606–53), biblical scholar and writer on natural history. *ODNB* 6: 383–384.

Bordley, J. & A.M. Harvey. 1976. *Two centuries of American medicine, 1776–1976*. Philadelphia: W.B. Saunders.

Bormann, F.H. & G.E. Likens. 1966. Nutrient cycling. *Science* 155: 424–429.

———. 1979. *Pattern and Process in a forested ecosystem*. NY: Springer-Verlag.

Borrello, M.E. 2010. *Evolutionary restraints: the contentious history of group selection*. Chicago: U. of Chicago P.

Botting, D. 1973. *Humboldt and the Cosmos*. NY: Harper & Row.

Boucher, D.H., ed. 1985. *The biology of mutualism: ecology and evolution*. NY: Oxford U.P.

———, S. James & K.H. Keeler. 1982. The ecology of mutualism. *Annual Review of Ecology & Systematics* 13: 315–347.

Bougainville, L.-A. de. 1772. *A voyage round the World*. J.R. Forster, transl. London. 2011. Cambridge U.P.

———. 1964. *Adventure in the wilderness: the American journals*. E.P. Hamilton, transl. Norman: U. of Oklahoma P.

———. 2002. *The Pacific journals*. J. Dunmore, transl. L: Hakluyt Soc.

Bourne, G.C. 1892. Memoir of Henry Nottidge Moseley. Pp. v—xvi in Moseley 1892.

Bowen, M. 1981. *Empiricism and geographical thought: from Francis Bacon to Alexander von Humboldt*. NY & L: Cambridge U.P.

———. 2008. *Censoring science: inside the political attack on Dr. James Hansen and the truth of global warming*. NY: Dutton.

Bowen, S., ed. 1972. *Proceedings, 1972 tundra biome symposium at U. of Washington*. [Washington DC]: US/IBP Program.

Boyle, R. 1661. *The skeptical chemist*. Oxford UK. 1964. L: Dent & Sons.

Bradley, R.S. 1716. Observations and experiments relating to the sap in vegetables. *Roy. Soc. Philos. Trans.* 29: 486–460.

———. 1718. *New improvements of planting and gardening, both philosophical and practical*. Ed. 2., pt. 3. L: Mears.

———, M.K. Hughes, & H.F. Diaz. 2003. Climate change in Medieval times. *Science* 302: 404–405.

Braun, E.L. 1947. Development of the deciduous forests of eastern North America. *Ecological Monographs* 17: 211–220.

———. 1958. The development of association and climax concepts: their use in interpretation of the deciduous forest. Pp. 329–339 in Steere 1958.

———. 1994. Autobiographical letter and bibliography. Pp. 61–67 in Burgess 1994.

Braun-Blanquet, J. 1927. *Pflanzensoziologie*. Berlin: Bornträger.

———. 1928. *Pflanzensoziologie: Grunzilga der Vegetationkunde*. Berlin: Springer.

Breitwisch, R. March 1992. Tickling for ticks. *Natural History* 56–61.

Bremekampe, C.E.B. 1952. Linné's significance for the development of phytogeography. *Taxon* 2: 47–54.

Bremer, R.G. 1987. *Indian agent and wilderness scholar: the life of Henry Rowe Schoolcraft.* Mt. Pleasant: Clark Historical Library, Central Michigan U.

Bret, P., ed. 1999. *L'expedition d'Égypt, une enterprise des Lumières.* Paris: Technique et Documentation.

Bridges, W. 1974. *Gathering of animals: an unconventional history of the New York Zoological Society.* NY: Harper & Row.

Brinkley, D. 2011. *The quiet world: saving Alaska's wilderness kingdom, 1879–1960.* NY: Harper Collins.

Broberg, G. 2000. Carl Linnaeus, 1707–1778, Swedish naturalist and physician. Pp. 417–419 in Hessenbruch 2000a.

Brock, T.D., ed. 1961. *Milestones in microbiology.* NY: Prentice-Hall.

——— & E.S. Deevey, Jr. 1963. New England. Pp. 117–162 in Frey 1963a.

Bronstein, J.L. 1994. Our current understanding of mutualism. *Quart. Review of Biol.* 69: 1–31.

———, ed. 2015. *Mutualism.* NY: Oxford U.P.

———. Mutualism and symbiosis. *Oxford Bibliographies.* Ed. 2. Online. Ed. 1, 2015.

Brooks, J.L., III. 1998. *The eclectic legacy: academic philosophy and the human sciences in Nineteenth-Century France.* Newark: U. of Delaware P.

——— & E.S. Deevey, Jr. 1963. New England. Pp. 117–162 in Frey 1963a.

Brooks, P. 1972. *The house of life: Rachel Carson at work.* Boston: Houghton Mifflin.

Brooks, W.K. 1909. Alpheus Hyatt, 1838–1902. *National Academy of Sciences Biographical Memoirs* 6: 311–325. Online.

Brosco, J.P. 1989. Henry Bryant Bigelow, the U.S. Bureau of Fisheries, and intensive area study. *Social Studies of Science* 19: 349–264.

Brosse, J. 1985. *The great voyagers of discovery: circumnavigators and scientists, 1764–1843.* S. Hochman, transl. NY: Facts on File.

Brown, E.H., Jr., R.W. Rybichi & R.J. Poff. 1985. *Population dynamics and interagency management of the bloater* (Coregonus hoyi) *in Lake Michigan, 1967–1982.* Ann Arbor: Great Lakes Fishery Commission. Technical Report 44.

Brown, J.H. & M.V. Lomolino. 1998. *Biogeography.* Ed. 2. Sunderland MA: Sinauer Associates. Ed. 1, 1983.

———, P.C. Miller, L.L. Tieszen, F.L. Bunnell, eds. 1980. *An Arctic ecosystem: the coastal tnndra at Barrow, Alaska.* Stroudsburg PA: Dowden, Hutchinson & Ross.

Brown, M.T. & C.A.S. Hall, eds. 2004. Through the macroscope: the legacy of H.T. Odum. *Ecological Modelling* 178 (special issue): 1–297.

———, E.C. Odum & D.F. Campbell. 2007. Foreword. Pp. ix—x in Odum 2007.

Brown, R. 1868. On the nature of the discoloration of the Arctic Seas. *Quarterly J. of Microscopic Science* 8: 941–1010.

Brown, S.R. 2002. *Sightseers and scholars: scientific travelers in the golden age of natural history.* Toronto: Key Porter Books.

Browne, J. 1983. *The secular ark: studies in the history of biogeography.* New Haven: Yale U.P.

———. 1995. *Charles Darwin: voyaging.* L: Jonathan Cape. NY: Knopf.

———. 2003. *Charles Darwin: the power of place.* L: Jonathan Cape. NY: Knopf.

Bruce, R.V. 1987. *The launching of modern American science, 1846–1876.* NY: Knopf.

Bruhns, K., ed. 1873. *Life of Alexander von Humboldt.* 2 vols. J. Lassell & C. Lassell, transl. L: Longmans, Greene. German ed., 3 vols., 1872.

Bryson, R.A. & T.J. Murray. 1977. *Climates of hunger: mankind and the World's changing weather.* U. of Wisconsin P.

Buchner, P. 1965. *Endosymbiosis of animals with plant microorganisms.* B. Mueller & F.H. Foeckler, transls. Ed. 2. NY: Interscience Publishers.

Buchsbaum, R. 1940. Review: Ricketts & Calvin, Between Pacific tides. *Ecology* 21: 93–94.

Buell, L. 1995. Thoreau and the natural environment. Pp. 171–193 in Myerson 1995.

Buell, M.F. 1949. Primary succession in 1792. *Ecology* 30: 107–108.

Buffon, G.L.L. 1742. Mémoire sur la culture des forêts. *Académie des Sciences Mémoires* 233–246.

———. 1954. *Oeuvres philosophiques*: see Poveteau 1954.

Bullen, K.E. 1976. Alfred Wegener (1880–1930), meteorology, geophysics. *DSB* 14: 214–217.

Bulloch, W. 1938. The history of bacteriology. L: Oxford U.P.

Bulnheim, H.-P. 1990. A century of marine zoological and ecological research around Helogoland Island. Pp. 84–93 in Lenz & Deacon 1990.

Burgaleta, C.M. 1999. *José de Costa (1540–1600): his life and thought.* Chicago: U. of Chicago P.

Burgess, A. & F. Raw, eds. 1967. *Soil biology.* L: Academic P.

Burgess, R.L. 1977. The Ecological Society of America: historical data and some preliminary analysis. 24 pp. in Egerton 1977a.

———. 1993. John Thomas Curtis: botanist, ecologist, conservationist. Pp. 1–43 in Fralish et al. 1993.

———, ed. 1994. *The Murray Fife Buell seminar on world ecologists: a compilation of materials gathered by graduate students in ecology, 1953–1962.* Syracuse: SUNY College of Environmental Science.

———. 1996. American ecologists: a biographical catalog. *Huntia* 10: 5–116. Online.

Burkhardt, R.W., Jr. 1973. Pierre-André Latreille (1762–1833), entomology, zoology. *DSB* 8: 48–49.

———. 2005. *Patterns of behavior: Konrad Lorenz, Niko Tinbergen, and the founding of ethology.* Chicago: U. of Chicago P.

Burlingame, L.J. 1973. Jean Baptiste Pierre Antoine de Monet de Lamarck (1744–1829), botany, invertebrate zoology and paleontology, evolution. *DSB* 7: 584–594.

Burns, F.L. 1908a. Alexander Wilson I: the Audubon controversy. *Wilson Bull.* 20: 3–18.

———. 1908b. Alexander Wilson II: the mystery of the small-headed flycatcher. *Wilson Bull.* 20: 63–79.

———. 1909. The completion of the *American ornithology. Wilson Bull.* 21: 16–35.

Burroughs, J. 1951. *America: selections from the writings.* F.A. Wiley, ed. NY: Devin-Adair.

Burrows, M.J. 2014. Invading Dreissenid mussels transform the 100-year-old International Joint Commission. Pp. 187–194 in Nalepa & Schlosser 2014.

Burstyn, H.L. 1968. Science and government in the nineteenth-century: the *Challenger* Expedition and its report. *Bulletin de l'Institut Océanographie Monaco* 2, no. 2: 603–613.

———. 1971. Pioneering in large-scale scientific organization: the *Challenger* Expedition and its report. I. Launching the expedition. *Roy. Soc. of Edinburgh Proc.* B 72: 47–61.

———. 1974. John Murray (1841–1914), oceanography, marine geology. *DSB* 9: 588–590.

———. 1980. Reviving American oceanography: Frank Lillie, Wickliffe Rose and the founding of the Woods Hole Oceanographic Institution. Pp. 57–66 in Sears & Merriman 1980.

Burtt, H., Jr. & W.E. Davis, Jr. 2013. *Alexander Wilson: the Scot who founded American Ornithology.* Cambridge: Harvard U.P.

Butts, E.R. & J.R. Schwartz. 2000. *Eugenie Clark: adventures of a shark scientist.* North Haven CT: Linnet Books.

Bylebyl, J. 1976. Karl Ludwig Willdenow (1765–1812), botany, historical phytogeography. *DSB* 14: 386–388.

Bynum, H.J. 2004. Osbert Salvin (1835–98). *DNCB* 4: 1758–1759.

Caddy, F. 1886–87. *Through the fields with Linnaeus.* 2 vols. L: Longmans, Green.

Cairns, J., Jr., ed. 1980. *The recovery process in damaged ecosystems.* Ann Arbor MI: Ann Arbor Science.

Caldwell, L.K., ed. 1988. *Perspectives on ecosystem management: a reader.* Albany: State U. of New York P.

Calhoon, R.M. 2004. Arthur Dobbs (1689–1765), colonial governor and writer on trade. *ODNB* 16: 340–342.

Call, R.E. 1895. *The life and writings of Rafinesque.* Louisville KY: John P. Norton.

Callendar, G.S. 1938. The artificial production of carbon dioxide and its influence on temperature. *Roy. Meteorological Soc. Quarterly J.* 64: 223–240.

Calvin, J. July 1933. 'Nakwasina' goes north: a man, a woman, and a pup cruise from Tacoma to Juneau in a 17-foot canoe. *National Geographic* 64, no. 1: 1–16.

Camerano, L. 1880. Dell' equilibrio dei viventi mercè la reciproca destruzion. *Atti della Reale Accademia delle Scienze du Torino* 15: 393–414 + 4 plates.

———. 1994. On the equilibrium of living beings by means of reciprocal destruction. C.M. Jacobi, transl., J.E. Cohen, ed. Pp. 360–380 in Levin 1994.

Camerini, J.R. 1993a. Evolution, biogeography and maps: an early history of Wallace's Line. *Isis* 84: 700–727.

———. 1993b. The physical atlas of Heinrich Berghaus: distribution maps as scientific knowledge. Pp. 479–512 in Mazzolini 1993.

Cameron, J. 1929. *The Bureau of Biological Survey: its history, activities and organization.* Baltimore: Johns Hopkins U.P. 1974. NY: Arno P.

Cameron, L. & J. Matless. 2011. Translocal ecologies: the Norfolk Broads, the 'natural,' and the International phytogeographical excursion, 1914. *J. Hist. of Biology* 41: 15–41.

Campbell, C.L., P.D. Peterson & C.S. Griffith. 1999. *The formative years of plant pathology in the United States.* St. Paul MN: APS P.

Campbell, J. 1997. *In Darwin's wake: revisiting Beagle's South American anchorages.* Shrewsbury: Waterline Books.

Candolle, A.P. de. 1820. Géographie botanique. *Dictionnaire des sciences naturelles* 18: 359–422. Reprinted in Egerton 1977c.

———. 1855. *Géographie botanique raisonée, ou exposition des faits principaux et des lois concernant la distribution géographique des plantes de l'époque actualle.* 2 vols. Paris: Victor Masson.

———. 1862. Étude sur l'espèce à l'occasion d'une révision des cupulifères. *Archives des sciences physique et naturelles* 15: 211–237.

———. 2004. *Mémoires et souveniers (1778–1841).* J.-D. Candaux & J.-M. Drouin, eds. Geneva: Biblioithèque d'Histoire des Sciences.

Cangelosi, A. 2014. In recognition: John Glenn sparks a giant leap for environment protection. Pp. 5–6 in Nalepa & Schloesser 2014.

Canizeres-Esguerra, J. 2006. *Nature, empire, and nation: explorations of the history of science in the Iberian world.* Stanford U.P.

Cantwell, R. 1961. *Alexander Wilson: naturalist and pioneer.* Philadelphia/NY: J.B. Lippincott.

Carley, J.P. 2004. John Ireland (1503–52), poet and antiquary. *ODNB* 33: 297–301.

Carozzi, A.V. 1975. Horace Bénédict de Saussure (1740–99), geology, meteorology, botany, education. *DSB* 12: 114–123.

———. 1987. La geologie. Pp. 203–265 in J. Trembly, ed. *Les savants genevois dans l'Europe intellectuelle du XVIIe au milieu du XIXe siècle.* Geneva: Editions du *Journal de Genève.*

Carpine-Lancre, J. & L. Saldanha. 1992. *Don Carlos I, Roi de Portugal, Albert Ier, Prince de Monaco: Souverains océanographes.* Lisbon: Foundation Colouste Gulbenkian.

——— & A. McConnell. 2013. Prince Albert I and J.Y. Buchanan: Mediterranean investigations. Pp. 255–265 in Groeben 2013.

Carson, R.L. 1922. My favorite recreation. *St. Nicholas Magazine* 19: 999. Reprinted in Carson 1998: 12–13.

———. 1937. Undersea. *Atlantic Monthly* 160: 322–325. Reprinted in Carson 1998: 3–11.

———. 1941. *Under the sea wind: a naturalist's picture of ocean life.* NY: Simon & Schuster.

———. 1947. *Mattamuskeet: a National Wildlife Refuge.* Washington: U.S. Fish & Wildlife Service.

———. 1951. *The sea around us.* NY: Oxford U.P.

———. 1955. *The edge of the sea.* Boston: Houghton Mifflin.

———. 1962. *Silent spring.* Boston: Houghton Mifflin.

———. 1998. *Lost woods: the discovered writings.* L. Lear, ed. Boston: Beacon P.

Carter, H.B. 1988. *Sir Joseph Banks, 1743–1820.* L: British Museum (Natural History).

Casado, S. 2011. Early history of ecology in Spain, 1868–1936. Pp. 307–323 in Schwarz & Jax 2011.

Cassidy, V.M. 2007. *Henry Chandler Cowles: pioneer ecologist.* Chicago: Kedzie Siegel.

Catesby, M. 1727–49 (not 1731–43). *The natural history of Carolina, Florida and the Bahama Islands.* 2 vols. L: W. Innys & R. Manby.

———. 1747. Of Birds of Passage. *Roy. Soc of L. Philos. Trans.* 44: 435–444.

———. 1985. *Catesby's Birds of Colonial America.* A. Feduccia, ed. Chapel Hill: U. of North Carolina P.

Cattell, J.M. & D.R. Brimhall, eds. 1921. *American men of science: a biographical directory.* Ed. 3. Garrison: Science P.

——— & J. Cattell, eds. 1938. *American men of science: a biographical directory.* Ed. 6. NY: Science P.

Caullery, M. 1952. *Parasitism and symbiosis.* A.M. Lysaght, transl. L: Sidgwick & Jackson.

Cerrutti, C. 2003. Aimé Bonpland, botaniste et adventurier. *Musée des Arts et Métiers: la Revue* 39/40: 38–47.

Cevasco, G.A. 2009. Eugene Odum (1913–2002), biologist, naturalist, ornithologist, educator, and author. *Modern American environmentalists: a Biogr. Encyclopedia* 354–359.

Challinor, J. 1975. John Playfair (1748–1819), mathematics, physics, geology. *DSB* 11: 34–36.

Chandler, D.C. 1963. Michigan. Pp. 95–115 in Frey 1963a.

Chapin, J. 2006. *The first scientific American: Benjamin Franklin and the pursuit of genius.* NY: Basic Books.

Chapman, F.M. & T.S. Palmer, eds. 1933. Fifty years' progress of American Ornithology, 1883–1933. Lancaster PA: A.O.U.

Chapman, R.H. 1931. *Animal ecology: with especial reference to insects.* NY: McGraw-Hill. 1977. NY: Arno P.

Chardon, C.E. 1949. *Los naturalistas en la America Latina.* Ciudad Trujillo: Editora del Caribe.

Charlier, R.H. 2004. Thirteen decades of biological oceanography in Belgium: some highlights (1840s to 1970s). Pp. 369–383 in Morcos et al. 2004.

———. 2013. More Belgian marine contributions since 1830. Pp. 105–123 in Groeben 2013.

Cheke, A. & J. Hume. 2008. *Lost land of the dodo: an ecological history of Mauritius, Reunion and Rodrigues.* New Haven: Yale U.P.

Cheng, T.C. 1970. *Symbiosis: organisms living together.* NY: Pegasus.

———, ed. 1971. *Aspects of the biology of symbiosis.* Baltimore: U. Park P.

Chew, M.K. 2006. *Ending with Elton: preludes to invasion biology.* Tempe: Dissertation, U. of Arizona.

Chichester, H.M. 1921. Arthur Dobbs (1689–1765), of Castle Dobbs, County Antrim, governor of North Carolina, 1754–65, engineer-in-chief & surveyor-general in Ireland. *DNB* 5: 1035–1037.

Christ, H. 1893. Notice biographique sur Alphonse de Candolle. *Bull. De l'Herbier Boissier* 1: 224–234.

Christensen, C.F.A. 1924–26. *Den danske Botanik's Historie.* 2 vols. Copenhagen: Haderups.

———. 1927. Eug. Warming: en Levnedssskildring. *Botanisk Tidsskrift* 39: 7–30.

———. 1932a. Joakim Frederik Schouw (1789–1852). Pp. 100–103 in Meisen 1932.

———. 1932b. Eugenius Warming (1841–1924). Pp. 156–160 in Meisen 1932.

Christie, W.J. 1968. Possible influences of fishing in the decline of Great Lakes fishing stock. *Conference on Great Lakes Research Proceedings* 11: 31–38.

———. 1972. Lake Ontario: effects of exploitation, introductions, and eutrophication on the salmonid community. *Fisheries Research Board of Canada J.* 29: 913–929.

———. 1974. Changes in the fish species composition of the Great Lakes. *Fisheries Research Board of Canada J.* 31: 827–854.

———, G.R. Spangler et al. 1985. A perspective on Great Lakes fish community rehabilitation. *Canadian Bull. of Fisheries & Aquatic Sciences* 213: 187–199.

——— et al. 1986. Managing the Great Lakes as a home. *J. of Great Lakes Research* 12: 2–17.

Chroust, A.-H. 1973. *Aristotle: new light on his life and on some of his lost works.* Vol. 1. Notre Dame: U. Notre Dame P.

Chumley, J. 1910. Bibliography of limnological literature. Vol. 1, pp. 659–753 in Murray & Pullar 1910.

Cittadino, E. 1980. Ecology and the professionalization of botany in America, 1890–1905. *Studies in History of Biology* 4: 171–198.

———. 1990. *Nature as the laboratory: Darwinian plant ecology in the German Empire, 1880–1900.* NY: Cambridge U.P.

———. 1993. 'A marvelous cosmopolitan preserve': the dunes, Chicago and the dynamic ecology of Henry Chandler Cowles. *Perspectives on Science* 1: 520–559.

———. 1997a. Frederic Edward Clements (1874–1945), plant ecologist, botanist. Pp. 156–158 in Sterling et al. 1997.

———. 1997b. William Francis Ganong (1864–1941), botanist, naturalist, historian. Pp. 299–301 in Sterling et al. 1997.

———. 2015. Paul Sears: cautious 'subversive ecologist.' *ESAB* 96: 519–526.

Clark, E. 1953. *Lady with a spear.* NY: Henry & Brothers.

———. 1969. *The lady and the sharks.* NY: Harper & Row.

———. 1972. The Red Sea's Garden of Eels. *National Geographic* 142: 724–735.

———. 1974. The Red Sea's sharkproof fish. *National Geographic* 146: 719–727.

———. 1975. Into the lairs of sleeping sharks. *National Geographic* 147, no. 2: 570–584 + cover.

———. 1978. Flashlight fish of the Red Sea. *National Geographic* 154: 718–728.

———. August 1981. Sharks: magnificent and misunderstood. *National Geographic* 160, no. 2: 138–187.

———. July 1983. Hidden life of an undersea desert. *National Geographic* 164. No. 1: 129–144.

———. 1986. The world of sharks at 2,000 feet. *National Geographic* 170: 680–691.

———. November 1988. Down the Cayman wall. *National Geographic* 174: 712–730.

———. December 1992. Whale sharks: gentle monsters of the deep. *Nat. Geogr.* 182, no. 6: 123–138.

Clark, W.C., P.J. Crutzen & H.J. Schellnhuber. 2004. Sience for global sustainability: toward a new paradigm. Pp. 1–28 in Schellnhuber, Crutzen & Clark 2004.

Clarke, B., ed. 2015. *Earth, life, and system: evolution and ecology on a Gaian planet.* NY: Fordham U.P.

Clarke, G.L. 1954. *Elements of ecology.* NY: John Wiley & Sons.

Clarke, J.M. 1979. *The life and adventures of John Muir.* San Francisco: Sierra Club.

Clements, E.G.S. 1960. *Adventures in ecology: half a million miles . . . from mud to macadam.* NY: Pageant P.

Clements, F.E. 1897. Peculiar zonal formations of the Great Plains. *American Naturalist* 31: 968–970.

———. 1904. *The development and structure of vegetation.* Lincoln: Botanical Survey of Nebraska.

———. 1905. *Research methods in ecology.* Lincoln: U. Publishing Co. 1977. NY: Arno P.

———. 1907. *Plant physiology and ecology.* NY: Henry Holt.

———. 1913. The alpine laboratory. *Science* 37: 327–328.

———. 1916. *Plant succession: an analysis of the development of vegetation.* Washington: Carnegie Institution of Washington.

———. 1917. The development and structure of biotic communities. *J. Ecol.* 5: 120–121.

———. 1920. *Plant indicators: the relation of plant communities to process and practice.* Washington DC: Carnegie Institution.

———. 1928. *Plant succession and indicators.* NY. 1969. NY: Hafner.

———. 1949. *Dynamics of vegetation: selections from the writings.* B.W. Allred & E.S. Clements, eds. NY: H.W. Wilson.

——— & V.E. Shelford. 1939. *Bio-ecology.* NY: John Wiley & Sons.
Clendening, L. 1942. *Source book of medical history.* L: Constable. 1960. NY: Dover.
Coats, A.M. 1969. *The plant hunters.* L: Studio Vista.
Cockayne, L. 1898. On the freezing of New Zealand alpine plants: notes on an experiment conducted in the Freezing Chamber, Lyttelton. *New Zealand Institute Trans. & Proc.* 30: 435–442.
———. 1900. A sketch of the plant geography of the Wainakariri River Basin, considered chiefly from an oecological point of view. *New Zealand Institute Trans. & Proc.* 32: 243–325.
———. 1921. *The vegetation of New Zealand.* Leipzig: Wilhelm Engelmann. Ed. 2, 1928.
———. 1927. *New Zealand plants and their story.* Wellington: Government Printer.
Coe, W.R. 1932. Biographical memoir of Addison Emery Verrill, 1839–1926. *National Academy of Sciences Biographical Memoirs* 14: 19–66 + portrait. Online.
Cohen, J.E. 1978. *Food webs and niche space.* Princeton: Princeton U.P.
———. 1994. Lorenzo Camerano's contribution to early food web theory. Pp. 351–359 in S.A. Levin, ed. *Frontiers in mathematical biology.* NY: Springer.
———. E.F. Briand & C.M. Newman. 1990. *Community food webs: data and theory.* NY: Springer-Verlag.
Colbert, E.H. 1974. Henry Fairfield Osborne (1857–1935), vertebrate paleontology. *DSB* 10: 241.
Coleman, D.C. 2010. *Big ecology: the emergence of ecosystem science.* Berkeley: U. of California P.
Colin, P.L. 1980. A brief history of the Tortugas Marine Laboratory and the Department of Marine Biology, Carnegie Institution of Washington. Pp. 138–147 in Sears & Merriman 1980.
Colinvaux, P. 2007. *Amazon expeditions: my quest for the Ice-Age equator.* New Haven: Yale U.P.
Collander, R. 1965. *The history of botany in Finland, 1828–1918.* Helsinki: Societas Scientiarum Fennica.
Collinge, S.K. 2018. NEON is your observatory. *Frontiers* 16: 37.
Collins, S., comp. 1994. Charles Christopher Adams. Pp. 1–11 in Burgess 1994a.
Conant, J.B. 1957. The overthrow of the phlogiston theory: the Chemical Revolution of 1775–1789. Vol. 1, pp. 65–115 in Conant & Nash, eds. 1957.
——— & L.K. Nash, eds. 1957. *Harvard case studies in experimental science.* 2 vols. Cambridge MA: Harvard U.P.
Conard, H.S. 1951. Foreword. Pp. vii–viii in Kerner 1951.
Conklin, E.G. 1913. William Keith Brooks, 1848–1908. *National Academy of Sciences Biographical Memoirs* 7: 25–88 + portrait.
———. 1944. Charles Otis Whitman, 1843–1910. Pp. 157–161 in Lillie 1944.
Connolly, S.J. 2004. William King (1650–1729), Church of Ireland archbishop. *ODNB* 31: online.
Conzen, M.P. 1988. The historical and geographical development of the Illinois and Michigan Canal National Heritage Corridor. Pp. 3–25 in Conzen & Carr 1988.
——— & K.J. Carr, eds. 1988. *The Illinois and Michigan Canal Corridor—a guide to its history and sources.* DeKalb: Northern Illinois U.P.
Cook, N.D. 1998. Gonzalo Fernández Oviedo y Valdés (1478–1557), chronicler of the Indies. *Latin American Lives* 755–756.
Cooley, J. & R. Pacific. 1987. *Guidelines for fish habitat management and planning in the Great Lakes.* Ann Arbor: Great Lakes Fishery Commission.
Cooper, M. & M. Hunter, eds. 2006. *Robert Hooke: Tercentennial studies.* Aldershot UK: Ashgate.
Cooper, W.S. 1913. The climax forest of Isle Royale, Lake Superior, and its development. *Botanical Gazette* 55: 1–47, 115–140, 189–235.
———. 1926. The fundamentals of vegetation change. *Ecology* 7: 391–413.
Corfield, R.M. 2003. *The silent landscape: the scientific voyage of H.M.S.* Challenger. Washington DC: Joseph Henry P.
Coupland, R.T., ed. 1979. *Grasslnds of the worod: analysis of grasslands and their uses.* IBP publication 18. NY: Cambridge U.P.
Cousteau, J.-M. & D. Paisner. 2010. *My father, the captain: my life with Jacques Cousteau.* Washington: National Geographic.
Cousteau, J.-Y. October 1952. Fish men discover a new world undersea. *National Geographic* 102, no. 4: 431–472.
———. 1954. To the depths of the sea by Bathyscaphe. *National Geographic* 106: 67–79.
———. April 1960. Diving saucer takes to the deep. *National Geographic* 117, no. 4: 471–486.
———. 1979. *The ocean world.* NY: H.N. Abrams.
——— & P. Cousteau. 1987. *The shark: splendid savage of the sea.* NY: Arrowood P. 1970. NY: Doubleday.
——— & P. Diolé. 1971. *Life and death in a coral sea.* NY: A & W Visual Library.
———. 1973. *Three adventures: Galapagos, Titicaca, the Blue Holes.* NY: Doubleday.
———. 1987a. *The whale: mighty monarch of the sea.* J.F. Bernard, transl. NY: Arrowood P. French ed. 1972.
———. 1987b. *Dolphins.* J.F. Bernard, transl. NY: Arrowood P. French ed. 1974.
Cousteau, J.-Y. & J. Dugan. 1963. *The living sea.* NY: Harper & Row.
Cousteau, J.-Y. & with F. Dumas. 1953. *The silent world.* NY: Harper & Brothers.
Cousteau, J.-Y. & S. Schieffelbein. 2007. *The human, the orchid, and the octopus.* NY: Bloomsbury.
Cousteau, J.-Y. & A. Sivirine. 1983. *Jacque Cousteau's* Calypso. NY: Harry N. Abrams.
Cowles, H.C. 1898. Review: Pound & Clements 1897. *Botanical Gazette* 25: 370–372. Reprinted in Cassidy 2007: 271–273.
———. 1899. The ecological relations of vegetation on the sand dunes of Lake Michigan. *Botanical Gazette* 27: 95–117, 167–202, 281–308, 361–391. Reprinted in Egerton 1977c & in Real & Brown 1991: 28–58.
———. 1901a. The influence of underlying rocks on the character of the vegetation. *American Bureau of Geography Bulletin* 2: 163–176, 376–388. Reprinted in Cassidy 2007: 204–220.
———. 1901b. Physiographic ecology of Chicago and vicinity. *Botanical Gazette* 31: 73–108, 145–182. Reprinted in Egerton 1977c.
———. 1901c. The plant societies of Chicago and vicinity. *Geographical Soc. of Chicago Bull.* 2: 1–76.
———. 1904. The work of the year 1903 in ecology. *Science* 19: 879–885.
———. 1911. The causes of vegetative cycles. *Botanical Gazette* 51: 161–183.
———. 1919. Review: Clements, *Plant Succession. Botanical Gazette* 68: 477–478. Reprinted in Cassidy 2007: 276–277.

Cox, D.L. 1979. *Charles Elton and the emergence of modern ecology.* St. Louis: Ph.D. dissertation, Washington U.

Craige, B.J. 2001. *Eugene Odum: ecosystem ecologist and environmentalist.* Athens: U. of Georgia P.

Cramer, J.M. 1987. *Mission orientation ecology: the case of Dutch fresh-water biology.* Amsterdam: Rodopi.

Crawford, E.T. 1996. *Arrhenius: from ionic theory to the greenhouse effect.* Canton MA: Science History Publicatons.

———, T. Shinn & S. Sorlin, eds. 1993. *Denationalizing science.* Dordrecht & NY: Kluver Academic Publishers.

Crawford, S. 2001. *Salmonine introductions to the Laurentian Great Lakes: a historical review and evaluation of ecological effects.* Ottawa: National Research Council of Canada.

Creaser, C.W. 1932. The lamprey *Petromyzon marinus* in Michigan. *Copeia* 757.

Creese, R.A. 2004. Pioneer ecologist: the life and work of Marion Isabel Newbigin (1869–1934). *ODNB*, online.

Crist, E. 2004. Concern with trifles? A geophysical reading of Darwin's last book. Pp. 161–172 in Schneider et al. 2004.

Croce, N.D. 1990. Enrico Hillyer Giglioli and Italian Thalassography (13.6.1845 to 16.12.1909). Pp. 241–244 in Lenz & Deacon 1990.

Croker, R.A. 1991. *Pioneer ecologist: the life and work of Victor Ernest Shelford, 1877–1968.* Washington: Smithsonian Institution P.

———. 2001. *Stephen Forbes and the rise of American ecology.* Washington: Smithsonian Institution P.

Cronin, H. 1991. *The ant and the peacock: altruism and sexual selection from Darwin to today.* NY: Cambridge U.P.

Cronon, W. 1991. *Nature's metropolis: Chicago and the Great West.* NY: Norton.

Crosland, M. 2000. Antoine Laurent Lavoisier, 1743–1794, French chemist, financier, and administrator. Pp. 407–409 in Hessenbruch 2000a.

Crowe, M.J. 1974. Karl August Möbius. (1825–1908), zoology. *DSB* 9: 429–431.

Crowcroft, P. 1991, *Elton's ecologists: a history of the Bureau of Animal Populations.* Chicago: U. of Chicago P.

Crumly, C.R. 2009. Tortoises. Pp. 921–926 in Gillespie & Clague 2009.

Crutzen, P.J. 2002. The Anthropocene: geology of mankind. *Nature* 415: 23.

Cuddington, K. & B.E. Beisner, eds. 2005. *Ecological paradigms lost: routes of theory change.* Burlington MA: Elsevier Academic P.

Curtis, J.T. 1959. *The vegetation of Wisconsin: an ordination of plant communities.* Madison: U. of Wisconsin P.

——— & C. Juday. 1937. Photosynthesis of algae in Wisconsin lakes: observations of 1935. *Internationale Revue de Gesamten Hydrobiololgie und Hydrographie* 35: 122–133.

Cushing, D.H. 1988. *The provident sea.* NY: Cambridge U.P.

Cutright, P.R. *Lewis and Clark: pioneering naturalists.* Lincoln: U. of Nebraska P.

D'Aguilar, J. 2006. *Histoire de l'entomologie.* Paris: Delachaux & Niestlé.

Dajoz, R. 1984. Eléments pour une histoire de l'écologie: la naissance de l'écologie modern au XIXe siècle. *Histoire et Nature* 24–25: 5–111 + 6 plates.

Dalton, D.A. 2008. *The natural world of Lewis and Clark.* Columbia: U. of Missouri P.

Dana, J.D. 1853. On an isothermal oceanic chart, illustrating the geographical distribution of marine animals. *American J. of Science and Arts* 66: 153–168, 325–327, 391–392. Reprinted in Lomolino et al. 2004: 88–108.

———. 1885. Origins of coral reefs and islands. *American J. of Science* 30: 89–105, 158–159, 169–191.

Dammerman, K.W. 1945. A history of the Visitors' Laboratory ("Treub Laboratorium") of the Botanic Gardens, Beuitenzorog, 1884–1934. Pp. 59–75 in Honig & Verdorn 1945.

Daniel, R.J., ed. 1934. *James Johnstone Memorial Volume.* Livedrpool: U. of Liverpool P.

Dansereau, P. 1946. Os planos de biogeografia. *Revisits Brazielia de Geografia* 8: 189–210.

———. 1947. The vegetation of Brazilian sand dunes. *Roy. Soc. of Canada Trans.* 41: 200.

———. 1957. *Biogeography: an ecological perspective.* NY: Ronald P.

———. 2005. *Autobiographie. Vol. 1. La Lancée, 1911–1936.* Sainte-Foy Quebec: Éditions Mullimones.

Darley, G. 2006. *John Evelyn: living for ingenuity.* New Haven: Yale U.P.

Darlington, P.J., Jr. 1938. The origin of the fauna of the Greater Antilles, with discussion of animals over water and through the air. *Quart. Review of Biol.* 13: 274–300. Reprinted in Sterling 1974.

———. 1948. The geographical distribution of cold-blooded vertebrates. *Quart. Review of Biol.* 23: 1–26, 105–123.

———. 1957. *Zoogeography: the geographical distribution of animals.* NY: John Wiley & Sons.

———. 1959. Darwin and zoogeography. *American Philos. Soc. Proc.* 103: 307–319.

———. 1965. *Biogeography of the southern end of the World.* Cambridge MA: Harvard U.P. Selection reprinted by Lomollino et al. 2004: 407–418.

Darwin, C.R. 1839. *Journal of Researches into the geology and natural history of the countries visited during the voyage of H.M.S. Beagle round the world.* L: Henry Colburn. 1969. Brussels: Culture et Civilization.

———. 1840. On the formation of mould. *Geological Soc. of L. Trans.* 5: 505–509. 1977. I, 49–53.

———. 1842. *The structure and distribution of coral reefs.* L: Smith, Elder. Ed. 3, 1889.

———. 6 April 1844. On the origin of mould. *Gardeners Chronicle* no. 14: 218. 1977. I, 195–196.

———. 1851a. *A monograph of the sub-class Cirripedia, with figures of all the species: the Lapadidae, or pedunculated Cirripedes.* L: Ray Soc. Reprinted in Darwin. 1964. *A monograph on the sub-class Cirripedia.* Weinheim Ger.: J. Cramer.

———. 1851b. *A monograph of the fossil Lepadidae, or pedunculated Cirripedia species of Great Britain.* L: Paleontographical Society.

———. 1854a. *A monograph of the sub-class Cirripedia, with figures of all the species: the Balanidae (or sessile Cirripedes), the Verrucidae.* L: Ray Soc. Reprinted in Darwin. 1964. *A monograph on the sub-class Cirripedia.* Weinheim Ger.: J. Cramer.

———. 1854b. *A monograph on the fossil Balanidae and Verrucidae of Great Britain.* L: Paleontographical Society.

———. 1855a. Does sea-water kill seeds? *Gardenders' Chronicle* May 21, 356–357.

———. 1855b. Vitality of seeds. *Gardenders' Chronicle* November 17, 758.

———. 1857. On the action of sea-water on the germination of seeds. *Linnean Soc. J. (Botany)* 1: 130–140.

———. 1859. *On the origin of species by means of natural selection, or the preservation of favoured races in the struggle for life.* L: John Murray. 1964. Cambridge MA: Harvard U.P.

———. 9 June 1860/9 February 1861. Fertilization of British orchids by insect agency. *Gardeners' Chronicle & Agricultural Gazette* no. 29: 528/no. 6: 122. Reprinted in Darwin 1977: vol. 2: 32–35/38–39.

———. 1862. *The various contrivances by which British and foreign orchids are fertilized by insects.* L: John Murray. Ed. 2, 1877. 1984. Chicago: U. of Chicago P.

———. 1863. Review: H.W. Bates, "Contributions to an insect fauna of the Amazon Valley". *Natural History Review* 219–224.

———. 15 May1869. The formation of mould by worms. *Gardeners' Chronicle* no. 20: 530. 1977: II, 137–138.

———. 1972. *The expression of emotions in man and animals.* L: John Murray. 1965. U. of Chicago P.

———. 1875. *Insectivorous plants.* NY: Appleton.

———. 1985–2019. *The correspondence.* F. Burkhardt & S. Smith, eds. 27 vols. Cambridge: Cambridge U.P.

Darwin, F. 1922. Stephen Hales (1677–1761), physiologist and inventor. *DNB* 8: 916–920.

Dathan, W. 2012. *The reindeer botanist: Alf Erling Porsild, 1901–1977.* Calgary: U. of Calgary P.

Dáttilo, W. & V. Rico-Gray, eds. 2018. *Ecological networks in the tropics.* NY: Springer International.

Daubenmire, R. 1968. *Plant communities: a textbook of plant synecology.* NH: Harper & Row.

Davenport, C.B. 1926. Alfred Goldsborough Mayor, 1868–1922. *National Academy of Sciences Biographical Memoirs* 21: 1–10.

Davidson, M. 1983. *Uncommon sense: the life and thought of Ludwig von Bertalanffy (1901–1972), father of general systems.* Los Angles: J.P. Tarcher.

Daws, G. & M. Fujita. 1999. *Archipelago: the islands of Indonesia: from the Nineteenth-century discoveries of Alfred Russel Wallace to the fate of forests and reefs in the Twenty-first Century.* Berkeley: U. of California P.

Day, D. & E.L. Mills. 2013. Charles Atwood Kofoid and the biological stations of Europe. Pp. 233–254 in Groeben 2013.

Dayrat, B. 2003. *Les botanists et la flora de France: trois siècles de découvertes.* Paris: Museum National d'Histoire Naturelle.

Deacon, G.E.R. 1980. The Woods Hole Oceanographic Institution: an expanding influence. Pp. 25–31 in Sears & Merriman 1980.

Deacon, M.D. 1971. *Scientists and the sea, 1650–1900.* L: Academic P. Ed. 2, 1997. Aldershot: Ashgate.

———, ed. 1978. *Oceanography: concepts and history.* Stroudsburg: Dowden, Hutchinson & Ross.

———. 1993. Crisis and compromise: the foundation of marine stations in Britain during the late 19th century. *Earth Sciences History* 12: 19–47.

———. 1997a. *Scientists and the sea, 1650–1900: a study of marine science.* Ed. 2. Aldershot: Ashgate.

———. 1997b. British marine scientists in Portuguese seas. Pp. 65–110 in Saldanha & Ré 1997.

———. 2003. Oceanographic institutions. Pp. 220–223 in Heilbron 2003.

———. 2004a. Edgar Johnson Allen (1866–1942), marine zoologist and scientific administrator. *ODNB* 1.

———. 2004b. William Benjamin Carpenter (1813–85). *DNCB* 1: 372–378.

———. 2004c. Charles Wyville Thomson (1830–82). *DNCB* 4: 1989–1990.

Dean, D.R. 1992. *James Hutton and the history of geology.* Ithaca: Cornell U.P.

———. 2004a. Charles Lyell (1797–1875). *DNCB* 3: 1281–1287.

———. 2004b. John Playfair (1748–1819). *DNCB* 3: 1604–1606.

DeBach, P. 1974. *Biological control by natural enemies.* NY: Cambridge U.P.

De Bary: see Bary.

Debaz, J. 2005. *Les stations françaises de biologie marine et leurs périodiques entre 1872 et 1914.* Paris: Ph.D. dissertation, Ecole des Hautes Etude en Sciences Sociales.

De Beer, G. 1953. *Sir Hans Sloane and the British Museum.* L: Oxford U.P.

———. 1973. Edwin Ray Lankester (1847–1929), zoology, natural history. *DSB* 8: 26–27.

———. 1975. Sir Hans Sloane (1660–1753), medicine, natural history. *DSB* 12: 456–459.

De Candolle: see Candolle.

Deevey, E.S., Jr. 1942. A re-examination of Thoreau's 'Walden.' *Quarterly Review of Biology* 17: 1–11.

Delaunay, P. 1926a. L'adventureuse existence de Pierre Belon du Mons. Paris: Edouard Champion.

———. 1926b. *Pierre Belon, naturaliste.* Le Mans: Monnoyer.

———. 1962. *La zoologie au seizième siècle.* Paris: Hermann.

Dempsey, D. 2001. *Ruin and recovery: Michigan's rise as a conservation leader.* Ann Arbor: U. of Michigan P.

———. 2004. *On the brink: the Great Lakes in the 21*st *century.* East Lansing: Michigan State U.P.

Dennis, J. 2003. *The living Great Lakes: searching for the heart of the inland seas.* NY: St. Martin's P.

De Querioz, A. 2014. *The monkey's voyage: how improbable journeys shaped he history of life.* NY: Basic Books.

De Renzi, S. 2000a. Accademia del Cimento. Pp. 3–5 in Hessenbruch 2000a.

———. 2000b. Accademia dei Lincei. Pp. 2–3 in Hessebbruch 2000.

De Saussure: see Saussure.

Desmond, R.G.C. 1972. Sir Joseph Dalton Hooker (1817–1911), botany. *DSB* 6: 488–492.

———. 1975. Richard Spruce (1817–93), botany. *DSB* 12: 594.

———. 1994. *Dictionary of British and Irish botanists and horticulturists, including plant collectors, flower painters and garden designers.* Ed. 2. L: Taylor & Francis. Ed. 1, 1977.

———. 1999. *Sir Joseph Dalton Hooker, traveler and plant collector.* Woodbridge: Royal Botanic Garden.

De Terra, H. 1955. *Humboldt: the life and times of Alexander von Humboldt, 1769–1859.* NY: Knopf.

De Waal, F. September 2014. One for all. *Scientific American* 311, no. 3: 69–71.

Dexter, R.W. 1952. The Annisquam Sea-side Laboratory of Alpheus Hyatt. *Scientific Monthly* 112–116.

———. 1956. Views of Alpheus Hyatt's Sea-side Laboratory and excerpts from his expeditionary correspondence. *The Biologist* 39: 5–11.

———. 1988. History of American marine biology and marine biology institutions: introduction. *American Zoologist* 28: 3–6.

———. 1997. Addison Emery Verrill, 1839–1926. *BDAC*, 796–798.

Dhombres, J., ed. 1995. *Adventures scientifiques: savants en Poitou-Charentes du XVIe au XXe siècle.* Poitou-Charentes: Les Éditions de l'Actualité.

Dice, L.R. 1923. Life zones and mammalian distribution. *J. of Mammalogy* 4: 39–47. Reprinted in Sterling 1974.

Dickenson, J. 2004. Henry Walter Bates (1825–92), naturalist. *ODNB* 4: 316–319.

Dickmann, D.I. & L.A. Leefers. 2003. *The forests of Michigan.* Ann Arbor: U. of Michigan P.

Dobbs, A. 1750. Concerning bees, and their method of gathering wax and honey. *Roy. Soc. of London Philos. Trans.* 46: 536–549.

Dobbs, D. 2005. *Reef madness: Charles Darwin, Alexander Agassiz, and the meaning of coral.* NY: Pantheon Books.

Dodge, E.S. 1971. *Beyond the capes: Pacific exploration from Captain Cook to the* Challenger, *1776–1877.* Boston: Little Brown.

———. 1976. *Islands and empires: western impact on the Pacific and East Asia.* Minneapolis: U. of Minnesota P.

Dohrn, A. 1872. The foundation of zoological stations. *Nature* 5: 277–298.

Dolan, J.R. 2019a. Unmasking "The eldest son of the father of protozoology": Charles King. *Protist* 170: 374–384.

———. 2019b. Ernst Haeckel's Radiolarians and medusa: the influence of his visit to Villefranche on his science and his art. *OpenScience* 1–14.

Dolling, L.M., ed. *George Perkins Marsh: an American for all seasons.* Hoboken: Stevens Institute of Technology.

Dorn, H. 1976. James Watt (1739–1819), engineering, chemistry. *DSB* 14: 196–199.

Doskey, J.S. 1997. William (James) Maclure (1763–1840). *BDAC*, 493–497.

Dressler, R.L. 1982. Biology of the orchid bees (Euglossini). *Ann. Rev. of Ecology and Systematics* 13: 373–394.

Dritschilo, W. 2006. Rachel Carson and Mid-Twentieth Century ecology. *ESAB* 87: 357–366.

Driver, F. & L. Martins, eds. 2005. *Tropical visions in the age of empire.* Chicago: U. Chicago P.

Drobnicki, J.A. 1997. Karl Patterson Schmidt, 1890–1957, herpetologist, zoogeographer, taxonomist. *BDAC* 706–707.

Drouin, J.-M. 1991. Quelques figures de l'insularité: reflections sur la biogéoraphie. Pp. 197–216 in Roger & Guéry 1991.

———. 1993. *L'écologie et son histoire.* Paris: Champ Flammarion. Ed. 1.

———. 1994. Histoire et écologie végétale: les origins du concept du succession. *Ecologie* 25: 147–155.

———. 1995. Réne-Antoine Ferchault de Réaumur, les curiosités d'un physiciene. Pp. 197–209 in Dhombres 1995.

———. 1997. Paul Jovet: les concepts de l'écologie végétale à l'épeuve de la Ville. *J. de Agriculture traditionnelle et de Botanique appliquée* 39: 75–89.

———. 1998. Botanical geography. Pp. 9–18 in Acot 1998.

———. 1999. Récolter, décrire et raconteur: Delile et Rozière. Pp. 261–277 in Bret 1999.

———. 2003. Humboldt et la popularization des sciences. *Musée des Arts et Métiers: La Revue* 39/40: 54–63.

———. 2008. *L'herbier des philosophies.* Paris: Editions du Seuil.

———. 2013. Some conceptions of time in ecology. Pp. 49–62 in Bergandi 2013.

——— & T. Lalande. 2003. Le parcours d'une exposition. *Musée des Arts et Métiers* 39/40: 6–11.

Drude, C.G.O. 1890. *Handbuch der Pflanzengeographie.* Stuttgart: J. Engelhorn. NY: Arno P. 1997.

———. 1896. *Deutschlands der Pflanzenengeographie: ein geographes charakterbid der Flora von Deutschland und angrenzender Alpen-Sowie Karphandern.* Stuttgart: J. Engelhorn. 1977. NY: Arno P.

———. 1913. *Die Oekologie der Pflanzen.* Braunschweig: Vieweg & Sohn.

Duff, A.G. & P.D. Lowe. 1981. Great Britain. Pp. 141–156 in Kormondy & McCormick 1981.

Duffy, J. 1976. *The healers: a history of American medicine.* NY: McGraw-Hill. 1979. Urbana: U. of Illinois P.

Dugatkin, L.A. 2006. *The altruism equation: seven scientists search for the origin of goodness.* Princeton: Princeton U.P.

———. 2009a. *Principles of animal behavior.* Ed. 2. Norton.

———. 2009b. *Mr. Jefferson and the giant moose: natural history in early America.* Chicago: U. of Chicago P.

Dumas, J.-B.-A. 1841. On the chemical statics of organized beings. Transl. anon. *Philos. Magazine* 19: 337–347, 456–469.

Dungan, M.T., T.E. Miller & D.A. Thomson. 1982. Catastrophic decline of a top carnivore in the Gulf of California's rocky intertidal zone. *Science* 216: 989–991.

Dunmore, J. 1965–69. *French explorations I the Pacific:* vol 1. 18th century; vol. 2. 19th century. Oxford: Clarendon P.

Dunn, G.A. 1996. *Insects of the Great Lakes region.* Ann Arbor: U. of Michigan P.

Dunne, J.A. 2006. The network structure of food webs. Pp. 27–86 in Pascual & Dunne 2006.

Dupree, A.H. 1959. *Asa Gray: American botanist, friend of Darwin.* Cambridge MA: Harvard U.P. 1988. Baltimore: Johns Hopkins U.P.

———. 1970. Alexander Agassiz (1835–1910), zoology, oceanography, engineering. *DSB* 1: 71–72.

———. 1972. Asa Gray (1810–88), botany. *DSB* 5: 511–514.

Du Rietz, G.E. 1921. *Zur methodologischen Grundlage der modernen Pflanzensoziologie.* Vienna: Adolph Holzhausen.

———. 1957a. Linnaeus as a phytogeographer. *Vegetatio* 7: 161–168.

———. 1957b. Linné som Myrforskare. *Uppsala Universitats Årsskrift* 5: 1–80.

Duyker, E. 2003. *Citizen Labillardiere: a naturalist's life in revolution and exploration (1755–1834).* Melbourne: Melbourne U.P.

———. 2006. *François Péron, an impetuous life: naturalist and voyager.* Melbourne: Melbourne U.P.

———. 2014. *Dumont d'Urville: explorer and polymath.* Otago NZ: Rachel Scott.

Ebach, M.C. 2015. *Origins of biogeography: the role of biological classification in early plant and animal geography.* NY: Springer.

——— & D.F. Goujet. 2006. The first biogeographical map. *J. of Biogeography* 33: 761–769.

Ebert, J.D. 1985. Carnegie Institution of Washington and marine biology: Naples, Woods Hole and Tortugas. *Biological Bulletin* 168, supplement: 172–182.

Eck, G.W. & L.R. Wells. 1983. Biology, population structure, and estimated forage requirements of lake trout in Lake Michigan. *U.S. Fish & Wildlife Service.* Technical Paper 111. 18 pages.

Edds, M.V., Jr. 1970. William Keith Brooks (1848–1908), zoology, embryology. *DSB* 2: 501–502.

Edgerton, H.E. 1955. Photographing the sea's dark underworld. *National Geographic* 107: 523–537.

Edwards, A.J. 1985. St. Paul's Rocks: a bibliographical review of the natural history of a mid-Atlantic island. *Archives of Nat. Hist.* 12: 31–49.

Edwards, C.J. & H.A. Regier, eds. 1990. *An ecosystem approach to the integrity of the Great Lakes in turbulent times.* Ann Arbor: Great Lakes Fishery Commission.

Edwards, W.H. 1847. *A voyage up the River Amazon, including a residence at Pará.* NY: D. Appleton.

Egan, D. 2017. *The death and life of the Great Lakes.* NY: Norton.

Egerton, F.N. 1962a. The scientific contributions of François Alphonse Forel, the founder of limnology. *Schweizerische Zeitschrift für Hydrologie* 24: 181–199.

———. 1962b. Review: Cantwell 1961. *Wilson Bull.* 74: 104–107.

———. 1963. The Birge-Juday limnological collection. *Wisconsin Academy Review* 10: 105–107.

———. 1967a. *Observations and studies of animal populations: a survey concluding with Darwin's Origin of Species.* Madison: Ph.D. dissertation, U. of Wisconsin-Madison.

———. 1967b. Leeuwenhoek as a founder of animal demography. *J. of the Hist. of Biology* 1: 1–22.

———. 1968. Ancient sources for animal demography. *Isis* 59: 175–189.

———. 1973. Changing concepts of the balance of nature. *Quarterly Review of Biology* 48: 322–350.

———. 1975. Aristotle' population biology. *Arethusa* 8: 307–330.

———. 1976a. Ecological studies and observations before 1900. Pp. 311–351 in B.J. Taylor & T.J. White, eds. *Issues and ideas in America.* Norman: U. of Oklahoma P. Reprinted in Egerton 1977a.

———. 1976b. Hewett Cottrell Watson (1804–81), evolution, phytogeography. *DSB* 14: 189–191.

———, ed. 1977a. *History of American ecology.* NY: Arno P.

———, ed. 1977b. *Ecological studies on insect parasitism.* NY: Arno P.

———, ed. 1977c. *Ecological phytography in the Nineteenth Century.* NY: Arno P.

———, ed. 1977d. *Early marine ecology.* NY: Arno P.

———. 1978. François Alphonse Forel (1841–1912), limnology, earth sciences. *DSB* 15: 158–159.

———. 1979. Hewett C. Watson, Great Britain's first phytogeographer. *Huntia* 3: 87–102.

———. 1983. Appendix B. Medieval knowledge of plants. Vol. 1, pp. 426–452 in Greene 1983.

———. 1985. *Overfishing or pollution? Case history of a controversy on the Great Lakes.* Ann Arbor: Great Lakes Fishery Commission.

———. 1987a. The Wisconsin limnological community. Pp. 85–122 in Beckel 1987.

———. 1987b. Pollution and aquatic life in Lake Erie: some scientific studies. *Environmental Review* 11: 189–205.

———. 1999a. Edward Asahel Birge (1851–1950), limnologist and president of University of Wisconsin. *ANB* 2: 809–810.

———. 1999b. Chancey Juday (1841–1944), limnologist. *ANB* 12: 294–295.

———. 1999c. Victor Ernest Shelford (1877–1968), ecologist. *ANB* 19: 785–786.

———. 2001a. History of ecological sciences, part 1: early Greek origins. *ESAB* 82: 93–97.

———. 2001b. History of ecological sciences, part 2: Aristotle and Theophrastos. *ESAB* 82: 149–152.

———. 2003a. History of ecological sciences, part 9: Albertus Magnus: a scholastic naturalist. *ESAB* 84: 87–91.

———. 2003b. History of ecological sciences, part 10: Botany during the Italian Renaissance and beginnings of the Scientific Revolution. *ESAB* 84: 130–137.

———. 2003c. History of ecological sciences, part 11: emergence of vertebrate zoology during the 1500s. *ESAB* 84: 206–212.

———. 2003d. *Hewett Cottrell Watson, Victorian plant ecologist and evolutionist.* Aldershot UK/Burlington VT: Ashgate.

———. 2004a. Richard Bradley (1688?—1732), botanist and writer. *ODNB* 7: 221.

———. 2004b. History of ecological sciences, part 12: invertebrate zoology and parasitology during the 1500s. *ESAB* 85: 27–31.

———. 2005. History of ecological sciences, part 17: invertebrate zoology and parasitology during the 1600s. *ESAB* 86: 133–149.

———. 2006a. History of ecological sciences, part 19: Leeuwenhoek's microscopic natural history. *ESAB* 87: 47–58.

———. 2006b. History of ecological sciences, part 20: Richard Bradley, entrepreneurial naturalist. *ESAB* 87: 117–127.

———. 2006c. History of ecological sciences, part 21: Réaumur and his history of insects. *ESAB* 87: 212–224.

———. 2006d. History of ecological sciences, part 22: early European naturalists in eastern North America. *ESAB* 87: 341–356.

———. 2007a. History of ecological sciences, part 23: Linnaeus and the economy of nature. *ESAB* 88: 72–88.

———. 2007b. History of ecological sciences, part 24: Buffon and environmental influences on animals. *ESAB* 88: 146–159.

———. 2007c. Understanding food chains and food webs, 1700–1970. *ESAB* 88: 50–69.

———. 2007d. History of ecological sciences, part 25: American naturalists explore eastern North America: John and William Bartram. *ESAB* 88: 253–268.

———. 2007e. History of ecological sciences, part 26: Gilbert White, naturalist extraordinaire. *ESAB* 88: 385–398.

———. 2008a. August Friedrich Thienemann (1882–1960), limnology. *NDSB* 7: 35–38.

———. 2008b. History of ecological sciences, part 27: naturalists explore Russia and the north Pacific during the 1700s. *ESAB* 89: 39–60.

———. 2008c. History of ecological sciences, part 28: plant growth studies during the 1700s. *ESAB* 89: 159–173.

———. 2008d. History of ecological sciences, part 29: plant disease studies during the 1700s. *ESAB* 89: 231–244.

———. 2008e. History of ecological sciences, part 30: invertebrate zoology and parasitology during the 1700s. *ESAB* 89: 407–433.

———. 2009a. Homage to Frederic E. Clements, historian of plant succession studies. *ESAB* 90: 43–79.

———. 2009b. History of ecological sciences, part 32: Humboldt, nature's geographer. *ESAB* 90: 253–282.

———. 2009c. History of ecological sciences, part 33: naturalists explore North America, mid-1780s-mid-1920s. *ESAB* 90: 434–487.

———. 2010a. History of ecological sciences, part 34: a changing economy of nature. *ESAB* 91: 21–41.

———. 2010b. History of ecological sciences, part 35: beginnings of British marine biology: Edward Forbes and Philip Gosse. *ESAB* 91: 176–201.

———. 2010c. Hewett Watson: plant geographer and evolutionist. *ESAB* 91: 294–312.
———. 2010d. History of ecological sciences, part 37: Charles Darwin's voyage on the *Beagle*. *ESAB* 91: 398–431.
———. 2011a. History of ecological sciences, part 38B: naturalists explore North America, 1838–1850s. *ESAB* 92: 153–184.
———. 2011b. History of ecological sciences, part 39: Henry David Thoreau, ecologist. *ESAB* 92: 251–275.
———. 2011c. History of ecological sciences, part 40: Darwin's evolutionary ecology. *ESAB* 92: 351–374 + cover.
———. 2012a. *Roots of ecology: from antiquity to Haeckel*. Berkeley: U. California P.
———. 2012b. History of ecological sciences, part 41: Victorian naturalists in Amazonia—Wallace, Bates, Spruce. *ESAB* 93: 35–59.
———. 2012c. History of ecological sciences, part 42: Victorian naturalists abroad—Hooker, Huxley, Wallace. *ESAB* 93: 125–159.
———. 2012d. History of ecological sciences, part 43: plant physiology, 1800s. *ESAB* 93: 197–219.
———. 2012e. History of ecological sciences, part 44: phytopathology during the 1800s. *ESAB* 93: 303–339.
———. 2013a. History of ecological sciences, part 45: ecological aspects of entomology during the 1800s. *ESAB* 94: 36–88.
———. 2013b. History of ecological sciences, part 46: from parasitology to germ theory. *ESAB* 94: 136–164.
———. 2013c. History of ecological sciences, part 47: Ernst Haeckel's ecology. *ESAB* 94: 222–244.
———. 2013d. History of ecological sciences, part 48: formalizing plant ecology, about 1870 to mid-1920s. *ESAB* 94: 341–378.
———. 2014a. History of ecological sciences, part 49: formalizing animal ecology, 1870s to 1920s. *ESAB* 95: 59–81.
———. 2014b. History of ecological sciences, part 50: formalizing limnology, 1870s to 1920s. *ESAB* 95: 131–153.
———. 2014c. History of ecological sciences, part 51: formalizing marine ecology, 1870s to 1920s. *ESAB* 95: 347–430.
———. 2015a. History of ecological sciences, part 52: symbiosis. *ESAB* 96: 80–139.
———. 2015b. History of ecological sciences, part 53: Organizing ecologists before 1946. *ESAB* 96: 239–311.
———. 2015c. History of ecological sciences, part 54: succession, community, and continuum. *ESAB* 96: 426–474.
———. 2015d. History of ecological sciences, part 55: animal population ecology. *ESAB* 96: 560–626.
———. 2015e. *A centennial history of the Ecological Society of America*. Boca Raton FL: CRC P./Taylor & Francis.
———. 2016a. History of ecological sciences, part 56: ethology until 1973. *ESAB* 97: 31–88.
———. 2016b. History of ecological sciences. Part 57: aspects of limnology in America, 1930s to about 1990, led by Hutchinson and Hasler. *ESAB* 97: 228–282.
———. 2016c. History of ecological sciences, part 58A: marine ecology, mid-1920s to about 1990, featuring Beebe, Bigelow, Ricketts. *ESAB* 97: 372–402.
———. 2017a. History of ecological sciences, part 58B: marine ecology, mid 1920s to about 1990: Carson, Riley, Cousteau, and Clark. *ESAB* 98: 113–149.
———. 2017b. History of ecological sciences, part 59: niches, biomes, ecosystems, and systems. *ESAB* 98: 298–337.
———. 2018a. History of ecological sciences, part 61A: terrestrial biogeography and paleobiogeography, 1700–1830s. *ESAB* 99: 192–241.
———. 2018b. History of ecological sciences, part 61B: terrestrial biogeography and paleobiogeography, 1840s-1940s. *ESAB* 99: separate, 1–63.
———. 2019a. History of ecological sciences, part 61C: marine biogeography, 1690s-1940s. *ESAB* 100: separate, 1–55.
———. 2019b. History of ecological sciences, part 62: saving habitats and managing wildlife in North America before 2000. *ESAB* 100: separate 1–63.
———. 2019c. History of ecological sciences, part 63: biosphere ecology. *ESAB* 100: separate 1–56.
———. 2019d. History of ecological sciences, part 64: history of physiological ecology of animals. *ESAB* 100: separate 1–28.
——— & G. Mayer. 2018. Animal population ecology. *Oxford Bibliographies—Ecology*. Online.
——— & L.D. Walls. 1997. Rethinking Thoreau and the history of American ecology. *The Concord Saunterer* 5: 64–91.
Ehrenberg, C.G. 1834. *Ueber die Natur und Bildung des Corallen-Bänke des Rothen des Rothen Meere*. Berlin.
Ehrlich, M.D. & R.P. Sánchez. 1990. Lights and shadows in biological oceanography research in Argentina: a historical review. Pp. 481–498 in Lenz & Deacon 1990.
Eilperin, J. 26 February 2015. Eugenie Clark, 'shark lady' who explored ocean depths, dies at 92. *Washington Post*.
Ekman, S. 1935. *Tiergeographie des Meeres*. Leipzig: Aka. Verlag.
———. 1953. *Zoogeography of the sea*. E. Palmer, transl. L: Sidgwick & Jackson.
Ellis, R. 2005. *Singing whales and flying squid: the discovery of marine life*. Guilford CT: Lyons P.
Ellison, A.M. & N.J. Gotelli. 2009. Energetics and the evolution of carnivorous plants—Darwin's "most wonderful plants in the world". *J. of Experimental Botany* 60: 19–42. Online.
Ellwood, E.R. et al. 2013. Record-breaking early flowering in the eastern United States. *PLOS ONE* 8: 1–9.
Elman, R. 1977. *First in the field: America's pioneering naturalists*. NY: Van Nostrand Reinhold.
Elster, H.-J. 1974. History of limnology. T.T. Macan, transl. *Internationale Vereingung für Mitterlungen theoretischen und angewandte Limnologie* 20: 7–30.
Elton, C.S. 1927. *Animal ecology*. L: Sidgwich & Jackson. Ed. 2, 1935. Ed. 3, 1947.
———. 1958. *The ecology of invasions by animals*. L: Methuen.
Emberger, L. 1936. Charles Flahault (1852–1935). *Revue general de botanique* 48: 1–48.
Emerson, A.E. & T. Park. 1955. Warder Clyde Allee: ecologist and ethologist. *Science* 121: 686–687.
Emery, L. 1985. *Review of fish species introduced into the Great Lakes, 1819–1974*. Ann Arbor: Great Lakes Fishery Commission. Technical Report 45. 31 pp.
Endersby, J. 2004a. Joseph Dalton Hooker (1817–1911). *DNBS* 2: 994–1001.
———. 2004b. Joseph Dalton Hooker (1817–1911). *ODNB* 27: 963–968.
———. 2008. *Imperial nature: Joseph Hooker and the practices of Victorian Science*. Chicago: U. of Chicago P.
Engel, J.R. 1983. *Sacred sands: the struggle for community in the Indiana dunes*. Middleton CT: Wesleyan U.P. 1983.

Engler, A. 1899. *Entwicklung der Pflanzengeographie in den letzen hundert Jahren. Wisenschaftliche Beitrtäge zum Gedächtniss der hunderjährigen Wiederkehr des Antritts von Alexander von Humboldt's Reisen ach Amerika.* Berlin: W.H. Kühl.

———. 1910–15. *Die Pflanzenwwelt Afrikes, insbesondere seiner tropischen Gebiete.* Leipzig: W. Engelman.

Erdtman, O.G.E. 1931. Pollen statistics: a new research method in paleoecology. *Science* 73: 399–401.

———. 1943. *An introduction to pollen analysis.* Waltham MA: Chronica Botanica.

Erickson, R.F. 1999. Henry Chandler Cowles (1869–1939), botanist. *ANB* 5: 617–618.

———, M.A. Font & B. Schwartz, eds. 2004. *Alexander von Humboldt: from Americas to the cosmos.* NY: Bildner Center for Western Hemisphere Studies, City U. of New York.

Eriksson, G. 1972. Peter Forsskål (also Forsskål or Forskål, 1732–63), botany. *DSB* 5: 74.

———. 1976a. Carl Peter Thunberg (1743–1828), botany. *DSB* 13: 391–393.

———. 1976b. Göran (Georg) Wahlenberg (1780–1851), botany. *DSB* 14: 116–117.

Esakov, V.A. 1974. Pytor Simon Pallas (1741–1811), natual science, geography. *DSB* 10: 283–285.

Eshenroder, R.L., T.P. Poe & C.H. Oliver. 1984a. *Working papers developed at the August 1983 conference on lake trout research.* Ann Arbor: Great Lakes Fishery Commission, Special Publication 84.

———, eds. 1984b. *Strategies for rehabilitation of lake trout in the Great Lakes.* Ann Arbor: Great Lakes Fishery Commission, Technical Report 40. 63 pages.

Espinas, A.V. 1878. *Des sociétés animal.* Paris: Germer Baillière.

Estensen, M. 2002. *The life of Matthew Flinders.* Sydney: Allen & Unwin.

Evans, F.C. 1978. Obituary: Lee Raymond Dice (1887–1977). *J. of Mammalogy* 59: 635–644.

Evans, H.E. 1997. *The natural history of the Long Expedition to the Rocky Mountains, 1819–1820.* NY: Oxford U.P.

Evans, M.A. & H.E. Evans. 1970. *William Morton Wheeler, biologist.* Cambridge MA: Harvard U.P.

Ewan, J. 1950. *Rocky Mountain naturalists.* Denver: U. of Denver P.

———. 1970. Charles Edwin Bessey (1845–1915), botany, education. *DSB* 2: 102–104.

———. 1971. Frederic Edward Clements (1874–945), botany. *DSB* 3: 317–318.

———. 1975. Constantine Samuel Rafinesque (1783–1830), natural history, archaeology. *DSB* 11: 262–264.

Ewan, 1978. Albert Spear Hitchcock (1865–1935), botany. *DSB* 15: 224–225.

——— & N.D. Ewan. 1981. *Biographical dictionary of Rocky Mountain naturalists.* Utrecht & The Hague: Bohn Scheltema & Holkema/Junk.

———. 2007. *Benjamin Smith Barton: naturalist and physician in Jeffersonian America.* St. Louis: Missouri Botanical Garden.

Ewing, S.A. 2006. *Organized veterinary medicine in the 46*th *State: The Oklahoma Veterinary Medical Association, 1907–2006.* Oklahoma City: Oklahoma Veterinary Medical Association.

Eyferth, B. & W. Schoenichen. 1909. *Einfachste Lebensformen das Tierund Pflanzenreiches: Naturgeschichte mikroscopischen.* Braunschweig: Benno Goeritz & Alfred Kalberlak.

Eyles, V.A. 1971. John Woodward, F.R.S., F.R.C.P., M.D. (1665–1728): a biobibliographical account of his life and work. *Soc. for the Bibliog. of Nat. Hist.* 5: 399–437.

———. 1972. James Hutton (1726–97), geology, agriculture, physical sciences, philosophy. *DSB* 6: 577–589.

———. 1976. John Woodward (1665–1728), geology, mineralogy, botany. *DSB* 14: 500–503.

Fabricius, J.C. 1778. *Philosophie entomologia.* Hamburg-Kiel.

Fagan, B. 2008. *The great warming: climate change and the rise and fall of civilizations.* NY: Bloomsbury P.

Fairley, R.E. 2004. Francis Wall Oliver (1864–1951). *DNCB* 3: 1506–1507.

Falkowski, P.G. & D. Tchernow. 2004. Human footprints in the ecological landscape. Pp. 211–226 in Schellnhuber et al. 2004.

Famintzin, A.S. & O.V. Baranetsky. 1867a. Sur le changement des gonidies en zoospores. *Annales des Sciences naturelles Botanique* 8: 137–144.

———. 1867b. Beitrag zur Entwickelungsgeschichte der Gonidien und zoosporenbildung bei *Physica parietina* D.N. Botanische Zeitung (Berlin) 25: 181–190.

Farber, P.L. 1982. *Discovering birds: the emergence of ornithology as a scientific discipline, 1760–1850.* Dordrecht Netherlands: D. Reidel. 1997. Baltimore: Johns Hopkins U.P.

Fautin, D.G. 2002. Beyond Darwin: coral reef research in the Twentieth Century. Pp. 446–449 in Benson & Rehbock 2002.

Feder, H.M. 1966. Cleaning symbiosis in the marine environment. Vol. 1, pp. 327–380 in Henry 1966.

Fedoseyev, I.A. 1970. Faddei Bellingsgazen (1779–1857), navigation, oceanography. *DSB* 1: 594–595.

———. 1976. Vladimir Ivanovich Vernadsky (1863–1945), mineralogy, geochemistry, biogeochemistry. *DSB* 13: 616–620.

Feduccia, A. 1985. Introduction. Pp. 3–13 in Catesby 1985.

Fedunkiw, M.P. 2000. Smithsonian Institution. Pp. 690–691 in Hessenbruch 2000.

Fernald, M.L. 1911. Review: Harshberger 1911. *Rhodora* 18: 213–224.

Feuerstein-Herz, P. 2004. *Eberhard August Wilhelm von Zimmermann (1743–1815) und die Tiergeographie.* Dissertation. Braunschweig.

Ffolliott, P.F. 2002. Ecology and management of evergreen oak woodlands in Arizona and New Mexico. Pp. 304–316 in Mc Shea & Healy 2002.

Fielder, D.G. 2008. Examination of factors contributing to the yellow perch population. *J. of Great Lakes Research* 34: 506–523.

———. 2010a. Response to Diana commentary. *J. of Great Lakes Research* 36: 195–198.

———. 2010b. Response of Yellow perch in Les Cheneaux Islands. *J. of Great Lakes Research* 36: 207–214.

——— et al. 2010. Status of walleye in Lake Huron. Pp. 71–90 in Roseman et al. 2010.

Findlen, P. 1993. Controlling the experiment: rhetoric, court patronage and the experimental method of Francisco Redi. *Hist. of Sci.* 31: 35–64.

———. 2000a. Ulisse Aldrovandi (1522–1605). Pp. 21–22 in Applebaum 2000.
———. 2000b. Francisco Redi (1626–97). Pp. 554–556 in Applebaum 2000.
Fink, B. 1896. Contributions to a knowledge of the lichens of Minnesota—I. Lichens of the Lake of the Woods. *Minnesota Botanical Studies* 1: 693–701.
———. 1910. The lichens of Minnesota. *U.S. National Herbarium Contributions* 14: viii + 269 + xvii.
Fischedick, K.S. 2000. From survey to ecology: the role of the British Vegetation Committee. *J. of the History of Biology* 33: 291–314.
——— & T. Shinn. 1993. The International Phytographical Excursions, 1911–1923; intellectual convergence in vegetation science. Pp. 107–136 in Crawford et al. 1993.
Fischer, J.-L. 1999. Les manuscrits Égyptiens d'Étienne Geoffroy Saint-Hilaire. Pp. 243–259 in Bret 1999.
Fishman, G. 2000. *Journeys through paradise: pioneering naturalists in the Southeast.* Gainesville: U.P. of Florida.
Flader, S.L., ed. 1983. *The Great Lakes forest: an environmental and social history.* Minneapolis: U. of Minnesota P.
Flahault, C.-H.-M. 1900. Projet de nomenclature phytogeographie. Pp. 429–446 in *1er Congrèes International de Botanique Actes.* Lons-de Saunier, France: E. Perrot. Reprinted in Acot 1998: II, 502–519.
Flannery, T. 2005. *The weather makers: how man is changing the climate and what it means for life on Earth.* NY: Atlantic Monthly P.
Fleming, J.R. 1990. *Meteorology in America, 1800–1870.* Baltimore: Johns Hopkins U.P.
———. 1996. *Historical essays on meteorology, 1919–1995.* Boston: Amer. Meteorological Soc.
———. 1998. *Historical perspectives on climate change.* NY: Oxford U.P.
Flenley, J. 2004. David Watts, 1935–2003. *J. of Biogeography* 31: 679–682. Online.
Flinders, M. 1814. *A voyage to Terra Australis, with an accompanying atlas.* 2 vols. L: G & W Nicol.
———. 2015. *Australia circumnavigated: the journal of HMS* Investigator, *1800–1803.* K. Morgan, ed. 2 vols. L: Hakluyt Soc.
Florinsky, M.T., ed. 1961. *McGraw-Hill Encyclopedia of Russia and the Soviet Union.* NY: McGraw-Hill.
Florkin, M. 1970. Pierre-Joseph van Beneden (1809–94), zoology. *DSB* 1: 602–603.
Folch, R., project director. 2000. *Encyclopedia of the biosphere.* T. Foskett, transl. 11 vols. Detroit: Gale Group.
Fontaine, T.D. & D.J. Stewart. 1990. Trophic dynamics and ecosystem integrity in the Great Lakes: past, present, and possibilities. Pp. 153–167 in Edwards & Regier 1990.
Foote, G.A. 1970. Joseph Banks (1743–1820), botany. *DSB* 1: 433–437.
Forbes, S.A. 1877–79. The food of birds. *Illinois Horticultural Soc. Trans.* 10: 37–44, 12: 140–145.
———. 1878. The food of Illinois fishes. *Illinois Laboratory of Natural History Bull.* 1: 71–89. Reprinted in Forbes 1977.
———. 1880a. On some interactions of organisms. *Illinois State Laboratory of Natural History Bull.* 1: 3–17. Reprinted in Forbes 1977.
———. 1880b. Notes on insectivorous Coleoptera. *Illinois Laboratory of Natural History Bull.* 1: 153–160. Reprinted in Forbes 1977.
———. 1887. The lake as a microcosm. *Peoria Scientific Association B.* 77–87. Reprinted in 1925, 1977, 1991.
———. 1913. The *Simulium*-pellagra problem in Illinois. *Science* 37: 86–91.
———. 1925. The lake as a microcosm. *Illinois Natural History Survey B.* 15: 537–550. Reprinted in Real & Brown 1991: 14–27.
———. 1977. *Ecological investigations.* F.N. Egerton, compiler. NY: Arno P.
Ford, A. 1951. Audubon and the *Quadrupeds.* Pp. 15–60 in Audubon [& Bachman] 1951.
———. 1963. *John James Audubon.* Norman: U. of Oklahoma P.
———, ed. 1969. *Audubon by himself: a profile of John James Audubon from writings selected.* NY: Natural History P.
Ford, C. 1966. *Where the sea breaks its back: the epic story of early naturalist Georg Steller and the Russian exploration of Alaska.* Boston: Little, Brown.
Forel, F.A. 1869. Introduction à l'étude de la faune profounde du lac Léman. *Société Vaudoise des Sciences naturelles* 10: 217–223.
———. 1892–1904. *Le Léman monographie limnologique.* 3 vols. Lausanne: F. Ronge.
———. 1901. *Handbuch des Seenkunde: Allgemaine Limnologie.* Stuttgart: J. Engelhorn. 1977. NY: Arno P.
Fornasiero, J., P. Monteath, & J. West-Sooby. 2004. *Encountering* Terra Australis: *the Australian voyages of Nicolas Baudin and Matthew Flinders.* Kent Town Aus.: Wakefield P.
Fortenbaugh, W.W. & D. Gutas, eds. 1992. *Theophrastus: his psychological, doxographical, and scientific writings.* New Brunswick: Transactions Publishers.
———, P.M. Huby & A.A. Long, eds. 1985. *Theophrastus of Eresus: his life and work.* New Brunswick NJ: Transactions Publishers.
——— & A.W. Sharples., eds. 1988. *Theophrastus of Eresus: sources for his life, writings, thought, and influence.* New Brunswick NJ: Transactions Publishers.
Foster, W.D. 1965. *A history of parasitology.* Edinburgh: E. & S. Livingstone.
Fourier, J.-B.J. 1824. Remarques générales sur les températurs du globe Terrestre et des spaces planétaires. *Ann. Chimie physique* 27: 136–167.
Fralish, J.S., R.P. McIntosh & O.L. Loucks, eds. 1993. *John T. Curtis: fifty years of Wisconsin plant ecology.* Madison: Wisconsin Academy of Sciences, Arts & Letters.
Francis, G.R., J.J. Magnuson, H.A. Regier & D.R. Talhelm. 1979. *Rehabilitating Great Lakes ecosystems.* Ann Arbor: Great Lakes Fishery Commission.
Franco, J.L. & J.A. Drummond. 2008. Wilderness and the Brazilian mind (1) nation and nature in Brazil from the 1920s to the 1940s. *Environmental Hist.* 13: 724–750.
Frängsmyr, T., ed. 1983. *Linnaeus: the man and his work.* Berkeley: U. of California P.
Fretwell, S.D. 1975. The impact of Robert MacArthur on ecology. *Ann. Rev. of Syst. & Ecol.* 6: 1–13.
Frey, D.G., ed. 1963a. *Limnology in North America.* Madison: U. Wisconsin P.
———. 1963b. Wisconsin: the Birge-Juday era. Pp. 3–54 in Frey 1963a.
Frick, G.F. 1971. Mark Catesby (1683–1749), natural history. *DSB* 3: 127–130.
——— & R.P. Stearns. 1961. *Mark Catesby: the colonial Audubon.* Athens: U. of Georgia P.

Friedman, M.R. 1989. *Appropriating the weather: Vilhelm Bjerknes and the construction of a modern meteorology.* Ithaca: Cornell U.P.

Fries, R.E. 1950. *A short history of botany in Sweden.* Uppsala: Almqvist & Wiksells.

Frost, O.W. 2003. *Bering: the Russian discovery of America.* New Haven: Yale U.P.

Fry, F.E.J. & V. Legendre. 1963. Ontario and Quebec. Pp. 487–519 in Frey 1963a.

Fuller, G.D. 1918. Units of vegetation and their classification. *Botanical Gazette* 66: 385–388.

———. 1935. Raunkiaer's ecological papers. *Ecology* 16: 111–113.

Furley, P.A. & W.W. Newey. 1989. *Geography of the biosphere: an introduction to the nature, distribution and evolution of the world's life zones.* L: Butterworths.

Fye, P.M. 1980. The Woods Hole Oceanographic Institution: a commentary. Pp. 1–9 in Sears & Merriman 1980.

Gadeceau, E. 1909. *Le Lac de Grand-Lieu: monographie.* Nantes: A. Degas.

Gadow, H. 1913. *The wanderings of animals.* Cambridge UK: Cambridge U.P.

Gage, A.T. 1938. *A history of the Linnean Society of London.* L: Taylor & Francis.

——— & W.T. Stearn. 1988. *A bicentenary history of the Linnean Society.* L: Academic P.

Galløe, O. 1908. Dansk likeners økologie. *Botanisk Tidsskrift* 28: 285–372.

Galtsoff, P.S. 1962. *The story of the Bureau of Commercial Fisheries Biographical Laboratory, Woods Hole, Massachusetts.* Washington: U.S. Bureau of Commercial Fisheries.

Gärtner, K.F. 1844–49. *Beiträge zur Kenntniss der Befruchtung.* 2 vols. Stuttgart: Schweizerbart.

Gates, D.M., C.H.O. Clarke & J.T. Harris. 1983. Wildlife in a changing environment. Pp. 52–80 in Flader 1983.

Gause, G.F. 1936. The principles of biocoenology. *Quarterly Review of Biology* 11: 320–336.

Geiger, R.L. 1999. David Starr Jordan (1851–1931), naturalist and educator. *ANB* 12: 266–268.

Geiser, S.W. 1937. *Naturalists of the frontier.* Dallas: Southern Methodist U.P. Edition 2, 1948.

George, A.S. 1999. *William Dampier in New Holland: Australia's first natural historian.* Hawthorn, Vic.: Bloomings Books.

Gerking, S.D. 1963. Central states. Pp. 239–268 in Frey 1963.

Gerlach, J. 2009. Seychelles. Pp. 829–833 in Gillespie & Clague 2009.

Ghiselin, M.T. 2008. British use and support of the Naples Zoological Station prior to the First World War. *California Academy of Sciences Proc.* 59: S155—S157.

Gibbons, F. & D. Strom. 1988. *Neighbors to the Birds: a history of birdwatching.* NY: Norton.

Gibbs, F.W. 1965. *Joseph Priestley: adventurer in science and champion of truth.* L.

Gibson-Reinemer, D.K. 2015. A vacant niche: how a central ecological concept emerged in the 19th century. *ESAB* 96: 324–334.

Gigon, A., F. Klötzli & U. Rahm. 1981. Switzerland. Pp. 313–329 in Kormondy & McCormick 1981.

Gillespie, R.G. & D.A. Clague, eds. 2009. *Encyclopedia of islands.* Berkeley: U. of California P.

Glaubrecht, M. 2002. From Solomon Müller to Ernst Mayr, or the insights of travelling naturalists toward a zoological geography and evolutionary biology. *Verhandlung zur Geschichte und Theorie der Biology* 9: 245–282.

Gleason, H.A. 1910. The vegetation of the inland sand deposits of Illinois. *Illinois State Laboratory of Natural History Bull.* 9: 20–173.

———. 1917. The structure and development of the plant association. *Torrey Botanical Club Bull.* 44: 463–481.

———. 1929. The significance of Raunkiaer's law of frequency. *Ecology* 10: 406–408.

———. 1939. The Individualistic concept of the plant association. *American Midland Naturalist* 21: 92–110.

———. 1975. Delving into the history of American ecology, 18 Feb. 52. *ESAB* 56: 7–10.

——— & A. Cronquist. 1964. *The natural geography of plants.* NY: Columbia U.P.

Glen, W. 1975. *Continental drift and plate tectonics.* Columbus OH: Charles E. Merrill.

———. 1982. *The road to Jaramillo: critical years of the revolution in earth science.* Stanford: Stanford U.P.

Gmelin, J.G. 1911. *Der Erforscher Sibeeriens: ein Gedenbuch.* R. Grandmann, transl. Munich: O. Gmelin.

Godwin, H. 1977. Sir Arthur Tansley: the man and the subject. *J. of Ecology* 65: 1–26.

Goerke, H. *Linnaeus.* D. Lindley, trans. NY: Charles Scribner's Sons.

Goetzmann, W.H. 1967. *Exploration and empire: the explorer and the scientist in the winning of the American West.* NY: Alfred A. Knopf.

———. 1986. *New lands, new men: America and the second great age of discovery.* NY: Viking.

——— & K. Sloan. 1982. *Looking far north: the Harriman Expedition to Alaska.* NY: Viking.

Goldblatt, P., ed. 1993. *Biological relationships between Africa and South America.* New Haven: Yale U.P.

Golder, F.A., ed. 1922–25. *Bering's voyages: an account of the efforts of the Russians to determine the relation of Asia and America.* L. Stejneger, trans. 2 vols. NY: American Geographical Soc. 1968. NY: Octagon Books.

Goldschmidt, R.B. 1951. Charles Atwood Kofoid, 1865–1947. *National Academy Sciences Biographical Memoirs* 26: 121–151 + photo. Online.

Goldstein, J. 2019. *Georg Forster.* Chicago: U. of Chicago P. German edition, 2015.

Golley, F.B., ed. 1977. *Ecological succession.* Stroudsburg PA: Dowden, Hutchinson & Ross.

———. 1993. *A history of the ecosystem concept in ecology: more than the sum of the parts.* New Haven: Yale U.P.

Good, R. 1955. Plant geography. Pp. 747–765 in Kessel 1955.

———. 1964. *The geography of flowering plants.* L: Longmans. Ed. 1, 1947.

Goodale, G.L. 1912. Alexander Agassiz, 1835–1910. *National Academy of Sciences Biographical Memoirs* 7: 291–305 + portrait. Online.

Goodland, R.J. 1975. The tropical origin of ecology: Eugene Warming's jubilee. *Oikos* 26: 240–245.

Goodman, E.J. 1972. *The explorers of South America.* NY: Macmillan.

Gore, A.A. 1992. *Earth in the balance: ecology and the human spirit.* Boston: Houghton Mifflin.

———. 2006. *An inconvenient truth: the planetary emergency of global warming and what we can do about it.* Emmaus PA: Rodale Books.

———. 2013. *The future: six drivers of global warming.* NY: Random House.

Gore, R. 1990. Between Monterey tides. *National Geographic* 177: 2–43.

Goreau, T.F. 1963. Calcium carbonate deposition by coralline algae and corals in relation to their roles as reef builders. *New York Academy of Sciences Annals* 109: 127–167.

Gorham, E. 1953. Some early ideas concerning the nature, origin and development of peat lands. *J. of Ecology* 41: 257–274.

———. 1991. Biogeochemistry: its origins and development. *Biogeochemistry* 13: 199–239.

Goryashko, N.A. & S.A. Fokin. 2013. Konstantin Karlovitch Saint-Hilaire (1866–1941) and his biological station at the White Sea (1911–1940). Pp. 299–309 in Groeben.

Gough, J.B. 1975. René-Antoine Ferchault de Réaumur (1683–1757), mathematics, technology, natural history, biology, experimental physics. *DSB* 11: 327–335.

Gould, C.G. 2004. *The remarkable life of William Beebe: explorer and naturalist.* Washington DC: Island P./Shearwater Books.

Gould, S.J. 1972. Alpheus Hyatt (1838–1902), invertebrate paleontology, zoology. *DSB* 6: 613–614.

Govindjee, J.T., H. Beatty & J.A. Allen, eds. 2005. *Discoveries in photosynthesis.* Dordrecht: Springer.

——— & D. Krogmann. 2005. Discoveries in oxygenic photosynthesis (172–2003): a perspective. Pp. 64–105 in Govindjee et al. 2005.

Grady, W. 2007. *The Great Lakes: the natural history of a changing region.* Vancouver: Greystone Books.

Grafen, A. 2004. William Donald Hamilton, 1 Aug 36–7 Mar 2000. *Roy. Soc. L. Biographical Memoirs of Fellows* 50: 109–132.

Grafmeyer, Y. & I. Joseph, eds. 1979. *L'école de Chicago: naissance de l'écologie urbaine.* Paris: Ed. Du Champ Urbain.

Graham, S.A. 1941. Royal Norton Graham, 1889–1939. *Entomological Soc. of America Ann.* 34: 521–524 + portrait.

Granger, W. 1933. William Diller Matthew (1871–1930), vertebrate paleontology. *DAB* 12: 411–412.

Granit, R. 1973. Pehr Kalm (1716–79), natural history. *DSB* 7: 210–211.

Grant, R. September 2019. From the tide pool to the stars. *Smithsonian* 50, no. 5: 26–37, 82–86.

Grant, V. 1949. Arthur Dobbs (1750) and the discovery of the pollination of flowers by insects. *Torrey Botanical Club Bull.* 76: 217–219.

Grasse, P.-P., ed. 1962. *La vie et l'ueuvre de Réaumur (1683–1757).* Paris: P.U. de France.

Grattan-Guinness, I., ed. 2005. *Landmark writings in western mathematics, 1640–1940.* Amsterdam: Elsevier.

——— & J. Ravetz. 1972. *Joseph Fourier, 1768–1830: a survey of his life and work.* Cambridge MA: MIT P.

Graustein, J.E. 1967. *Thomas Nuttall, naturalist: explorations in America, 1808–1841.* Cambridge: Harvard U.P.

Gray, A. 1848. *A manual of the botany of the northern United States . . . from New England to Wisconsin and south to Ohio and Pennsylvania.* Boston: James Munroe.

———. 1859. On the botany of Japan, and its relation to North America. *American Acad. of Arts & Sciences Mem.* 6: 377–452. An 1872 version of this paper was reprinted in Gray 1876: 169–194.

———. 1876. *Darwiniana: essays and reviews pertaining to Darwinism.* NY: D. Appleton.

Gray, M. & K.G. Kozsninski. 2017. Lynn Margulis and the endosymliont hypothesis: 50 years later. *Molecular Biology of the Cell* 28: 1285–1287.

Gray, W.R. 1981. *Voyages to paradise: exploring in the wake of Captain Cook.* Washington: National Geographic Society.

Greene, M.T. 1982. *Geology in the Nineteenth Century.* Ithaca NY: Cornell U.P.

———. 2015. *Alfred Wegener: science, exploration, and the theory of continental drift.* Baltimore: Johns Hopkins U.P.

Gribbin, J. & M. Gribbin. 2009. *James Lovelock: in search of Gaia.* Princeton: Princeton U.P.

Gridgeman, N.T. 1973. Alfred James Lotka (1880–1949), demography, statistics. *DSB* 8: 512.

Grinevald, J. 1987. On a holistic concept for deep and global ecology: the biosphere. *Fundamenta Scientiae* 8: 197–226.

Grinnell, G.B. 1943. *Philosophy of nature: selected writings of a western naturalist.* A.H. Miller, ed. 1968. NY: Books for Libraries P.

Grinnell, H.W. 1990. Joseph Grinnell (1877–1939). *The Condor* 42: 1–34 + photo.

Grinnell, J. 1917. The niche-relationship of the California thrasher. *Auk* 34: 427–433.

———. 1930. James Graham Cooper (1830–1902), naturalist. *DAB* 4: 406–407.

——— & A.H. Miller. 1944. *The distribution of the birds of California.* Berkeley: Cooper Ornithological Club.

——— & H.S. Swarth. 1913. *Birds and mammals in the San Jacinto area of southern California with remarks upon the behavior of geographic races on the margins of their habitats.* Berkeley: U. of California Publications in Zoology, 10: no. 10.

Grisebach, A.H.R. 1838. Uber den Einfluss der Climas auf die Begränzung der natural. *Linnaea* 12: 159–200. Reprinted in Egerton 1977c.

———. 1846. Report on the contributions to botanical geography during the years 1842 and 1843. W.B. MacDonald & George Busk, transls. *Ray Society Reports and Papers on Botany* 1: 55–212. Reprinted in Egerton 1977c.

———. 1849. Report on the progress of geographical botany during the years 1844 and 1845. W.B. MacDonald, transl. *Ray Society Reports and Papers on Botany* 2: 315–493. Reprinted in Egerton 1977c.

———. 1864. *Flora of the British West Indian Islands.* L: Lovell Reve.

———. 1872. *Die Vegetation der Erde nach klimatischen Anordnung.* 2 vols. Leipzig: Englemann. Ed. 2, 1884–85. French ed., 1877–78.

Groeben, C. 1984. The Naples Zoological Station and Woods Hole. *Oceanus* 27: 60–69.

———. 1985. Anton Dohrn—the statesman of Darwinism. *Biological Bulletin* 168, supplement 4: 4–25.

———, ed. 2013. *Places, people, tools: oceanography in the Mediterranean and beyond.* Naples: Giovanni Editore.

——— & S.I. Fokin. 2013. From Russia with love: Russian scientists at the Naples Zoological Station (1874–1934). Pp. 141–172 in Groeben 2013.

——— & I.O. Müller. 1975. *The Naples Zoological Station at the time of Anton Dohrn.* Naples: Stazione Zoologica di Napoli.

Grove, G.I. 1990. *A history of human helminthology.* Wallingford UK: C.A.B. International.

Grove, N. 1984. *Wild lands for wildlife: America's national refuges.* Washington DC: National Geographic.
Guenther, K. 1931. *A naturalist in Brazil: the record of a year's observations of her flora, her fauna and her people.* B. Miall, transl. L: Allen & Unwin. German edition 1927.
Guerlac, H. 1972. Stephen Hales (1677–1761), physiology, public health. *DSB* 6: 35–48.
———. 1973. Antoine-Laurent Lavoisier (1743–94), chemistry, physiology, geology. *DSB* 8: 66–91. Reprinted as *Antoine-Laurent Lavoisier (1743–74), chemistry, physiology, geology, economics, social reform.* New York: Charles Scribner's Sons, 1975.
Guiral, P. 1957. Tournefort et son voyage au Levant. Pp. 77–96 in Becker et al. 1957.
Gunning, G.E. 1963. Illinois. Pp. 163–189 in Frey 1963.
Güttler, N.R. 2011. Scaling the period eye: Oscar Drude and the cartographical practice of plant geography. *Science in Context* 24: 1–41.
Hachey, H.B. 1965. *History of the Fisheries Research Board of Canada.* Ottawa: Fisheries Research Board of Canada.
Haeckel, E. 1890. *Plankton-Studien: Vergleichende Untersuchungen über die Bedentung und Zussammeksetzing der pelagischen Fauna und Flora.* Jena: Gustav Fischer.
Haedrich, R.L. & K.O. Emery. 1980. Growth of an oceanographic institution. Pp. 67–82 in Sears & Merriman 1980.
Hafertepe, K. 1984. *America's castle: the evolution of the Smithsonian building and its institution, 1840–1878.* Washington: Smithsonian Institution P.
Hagberg, K. 1952. *Carl Linnaeus.* A. Blair, transl. L: Jonathan Cape.
Hagen, J.B. 1986. Ecologists and taxonomists: divergent traditions in Twentieth-Century plant geography. *J. of the History of Biology* 19: 197–214.
———. 1988. Organism and environment: Frederic Clement's vision of a unified physiological ecology. Pp. 257–280 in Rainger et al. 1988.
———. 1992. *An entangled bank: the origins of ecosystem ecology.* New Brunswick: Rutgers U.P.
———. 1993. Clementsian ecologists: the internal dynamics of a research school. *Osiris* 8: 178–195.
———. 1999. Frederic Edward Clements (1874–1945), botanist and ecologist. *ANB* 5: 53–54.
Hahn, R. 1971. *The anatomy of a scientific institution: the Paris Academy of Sciences, 1666–1803.* Berkeley: U. California P.
Haigh, E.V. 2000. Russia: Pp. 664–665 in Hessenbruch 2000.
Hale, P.J. 2014. Political descent: Malthus, mutualism, and the politics of evolution in Victorian England. Chicago: U. of Chicago P.
Hales, S. 1727. *Vegetable Staticks: or, an account of some statistical experiments on the sap of vegetables.* L: W. & J. Innys. Ed. 2, 1731, ed. 3, 1733, 1961 (reprint of ed. 1., M.A. Hoskin, introd.)
Hall, C.A.S., ed. 1995. *Maximum power: the ideas and applications of H.T. Odum.* Niwot: U.P. of Colorado.
Hall, M.B. August 1967. Robert Boyle. *Scientific America* 217, no. 2: 97–102.
———. 1971. Robert Boyle (1627–91), natural philosophy, physics, chemistry. *DSB* 2: 377–382.
Hallam, A. 1973. *A revolution in the earth sciences: from continental drift to plate tectonics.* Oxford: Clarendon P.
———. 1983. Great geological controversies. Oxford: Oxford U.P.
Hamilton, W.D. 1963. The evolution of altruistic behavior. *American Naturalist* 97: 354–356.
———. 1964. The genetical evolution of social behavior. *J. of Theoretical Biology* 7: 1–52.
Hamilton, W.J., Jr. 1955. Mammalogy in North America. Pp. 662–688 in Kessel 1955.
Hammer, U.T. 2005. Donald Strathearn Rawson (1905–61). *The Encyclopedia of Saskatchewan.* Regina: U. of Regina. Online.
Hanley, W. 1977. *Natural history in America: from Mark Catesby to Rachel Carson.* NY: Quadrangle/New York Times Books.
Hannah, G. 2013. *The Silwood Circle: a history of ecology and the making of scientific careers in late Twentieth-Century Britain.* L: Imperial College P.
Hansen, J. 1988. *Testimony before the Senate Committee on Energy and Natural Resources, June 23, 1988.* Hearings on Greenhouse Effect and Global Climate Change. Reprinted in Howe 2017: 207–208.
———. et al. 1981. Climate impact of increasing atmospheric carbon dioxide. *Science* 213: 957–966.
———. et al. 1998. Global climate data and models: a reconciliation. *Science* 281: 930–932.
Hansen, T. 1964. *Arabia Felix: the Danish expedition of 1761–1767.* NY: Harper & Row.
Harding, W. 1961. *A Thoreau handbook.* NY: New York U.P.
Hardy, A.C. 1924. The herring in relation to its animate environment. Part 1. The food and habits of the herring with special reference to the east coast of England. *Fisheries Investigations*, series 2, 7: 1–53.
———. 1951. Obituary: Walter Garstang. *J. of the Marine Biol. Assoc. of the UK* 29: 561–566.
———. 1956–59. *The open sea: its natural history.* 2 vols. L: Collins.
———. 1967. *Great waters: a voyage of natural history to study whales, plankton and the waters of the Southern Sea.* NY: Harper & Row.
Harmond, R. 1997. Charles Christopher Adams, 1873–1955. *BDAC* 8–9.
Harris, E. 1951. *Up the Missouri with Audubon.* Norman: U. of Oklahoma P.
Harris, H. et al. 1982. *Green Bay in the future—a rehabilitative prospectus.* Ann Arbor: Great Lakes Fishery Commission.
Harshberger, J.AW.C. 1911. *Phytogeographic survey of North America: a consideration of the North American continent, including Mexico, Central America, and the West Indies, together with the evolution of North American plant distribution.* Leipzig: W. Englemann/NY: Stechert.
———. 1928. Unsolved problems in Arctic plant geography. Pp. 143–153 in Joerg 1928.
Hart, H. 1930. Nicolas Théodore de Saussure. *Plant Physiology* 5: 424–429.
Hart-Davis, D. 2003. *Audubon's elephant.* L: Weindenfield & S. Nicolson.
Hartman, W.L. 1972. Lake Erie: effects of exploitation, environmental changes and new species in the fishery resources. *J. of Fisheries Research Board of Canada* 29: 899–912.
Hartridge, M.L. 1997. *The anandrous journey: revealing letters to a mentor.* Amherst WI: Amherst P.

Hasler, A.D. 1945. Observations on the winter perch population of Lake Mendota. *Ecology* 26: 90–94.
———. 1946. A war time view of European biological stations. *Biologist* 28: 81–93.
———. 1954. Odor perception and orientation in fishes. *Fisheries Research Board of Canada J.* 11: 107–129.
———. 1963a. Wisconsin, 1940–1961. Pp. 55–93 in Frey 1963.
———. 1963b. The Laboratory of Limnology and associated field units of the University of Wisconsin. *American Zoologist* 3: 337–340.
———. 1965. Perception of pathways of fishes in migration. *Quarterly Review of Biology* 31: 200–209.
———. 1966. *Underwater guideposts: homing in salmon.* Madison: U. of Wisconsin P.
———. May 1969. Cultural eutrophication is reversible. *BioScience* 425–431.
———, ed. 1975a. *Coupling of land and water systems.* NY: Springer-Verlag.
———. 1975b. How the salmon comes home. Pp. 184–191 in *Science year: the World Book science annual.* Chicago: Field Enterprises Educational Corp.
——— & A.T. Scholz. 1983. *Olfactory imprinting and homing in salmon: investigation into the mechanism of the imprinting process.* Berlin: Springer-Verlag.
Hass, H. 1953. *Manta: under the Red Sea with spear and camera.* J. Cleugh, transl. Chicago: Rand McNally.
Hatch, M.H. 1972. Leland Ossian Howard (1857–1950), applied entomology. *DSB* 6: 524–525.
Hawkins, A.S. et al. 1984. *Flyways: Pioneering waterfowl management of North America.* Washington: U.S. Fish & Wildllife Service.
Hazlett, M. 2008. Science & spirit: struggles of the early Rachel Carson. Pp. 149–167 in Sideris & Moore 2008.
Healy, W.M. 2002. Managing eastern oak forests for wildlife. Pp. 317–332 in Mc Shea & Healy 2002.
Hedgpeth, J.W., ed. 1957a. *Treatise on marine ecology and paleontology.* Vol. 1: Ecology. NY: Geological Society of America.
———. 1957b. Introduction. Pp. 1–16 in Hedgpeth 1957a.
———. 1978. *The outer shores.* 2 parts. Eureka CA: Mad River P.
Hedwig, J. 1784. *Theoria generationis et fructificationis plantarum cryptogamicarum Linnaei.* St. Petersburg: Academia Imperialis Scientiarum.
Heilbron, J.L., editor. 2003. *The Oxford companion to the history of modern science.* NY: Oxford U.P.
Helferich, G. 2004. *Humboldt's cosmos: Alexander von Humboldt and the Latin American journey that changed the way we see the world.* NY: Gotham Books.
Hellman, G/ 1968. *Bankers, bones, and beetles: the first century of the American Museum of Natural History.* Garden City NY: Natural History P.
Hellwarth, B. 2012. *Sealab: America's forgotten quest to live and work on the ocean floor.* NY: Simon & Schuster.
Hemming, J. 2015. *Naturalists in paradise: Wallace, Bates, and Spruce in the Amazon.* L: Thames & Hudson.
Henderson, A. 1996. Richard Spruce and the Palms of the Amazon and Andes. Pp. 188–196 in Seaward & FitzGerald 1996.
Henriksen, M.S. 1932. Johann Christensen Fabricius (1745–1808). Pp. 76–80 in Meisen 1932.
Henry, S.M., ed. 1966–67. *Symbiosis: I. Associations of microorganisms, plants, and marine organisms. II. Associations of invertebrates, birds, ruminants and other biota.* NY: Academic P.
Hepp, P. 1860. *Abbildungen und beschreibung der sporen zum IX. X. XI. und XII. Band der flechten Europas in getrokeneten mikroscopisch untersuchten.* Zurich: Author.
Heppell, D. 1973. John Gwyn Jeffreys (1809–85), marine zoology. *DSB* 7: 91–92.
Herbert, W. 1837. *Amaryllidacea, with a treatise on cross-bred vegetables.* L: Ridgeway.
Herdman, W.A. 1923. *Founders of oceanography and their work: an introduction to the science of the sea.* L: Edward Arnold.
Herodotos. 1926–38. *History.* 4 vols. A.D. Godley, transl. Cambridge MA: Harvard U.P.
Herrera, C.M. 2002. Seed dispersal by vertebrates: an evolutionary approach. Pp. 185–208 in Herrera & Pellmyr 2002.
——— & O. Pellmyr, eds. 2002. *Plant-animal interactions: an evolutionary approach.* Oxford: Blackwell Science.
Herrick, F.H. 1917. *Audubon the naturalist: a history of his life and time.* NY: D. Appleton. Ed. 2. 1938, 1968.
Hershey, D.R. 2004. *The widespread misconception that the tambalacoque* or calvaria tree absolutely required the dodo bird for its seeds to germinate. *Plant Sci. Bull.* 50, no. 4: 105–109.
Hesse, R., W.C. Allee & K. Schmidt. 1937. *Ecological animal geography.* NY: John Wiley.
Hessenbruch, A. ed., 2000a. *Reader's guide to the history of science.* L. & Chicago: Fitzroy Dearborn.
———. 2000b. Joseph Priestley, 1733–1804, British chemist and clergyman. Pp. 594–595 in Hessenbruch 2000a.
Heuss, T. 1991., ed. *Anton Dorhn: a life for science.* L. Gieckmann, transl. Berlin: Springer-Verlag.
Hiatt, R.W. 1954. *Directory of hydrobiological laboratories and personnel in North America.* Honolulu: U. of Hawaii P.
———. 1963. *World Directory of hydrobiological and fisheries institutions.* Washington: American Institute of Biological Sciences.
Hill, A.W. 1935. Leonard Cockayne (1855–1934). *Royal Society of London Obituary Notices of Fellows* 1: 443–457.
Hill, J. 2013. Translator's introduction. Pp. xiii—xlvii in Saussure 2013.
Hill, M.N., ed. 2012. *The seas: ideas, and observations.* NY: John Wiley & Sons.
Hinde, R.A. 1970. *Animal behaviour: a synthesis of ethology and comparative psychology.* Ed. 2. NY: McGraw Hill. Ed. 1, 1966.
Hintzsche, W. & T. Nickol, eds. 1996. *Die Grosse Nordische Expedition: Georg Wilhelm Steller (1709–1746): ein Lutheraner Erforscht Siberian und Alaska.* Halle: Lutz Grumbach.
Hoare, M.E. 1972a. (Johann) Georg Adam Forster (1754–94), natural philosophy, geography. *DSB* 5: 75–76.
———. 1972b. Johann Reinhold Forster (1729–98), natural philosophy, geography. *DSB* 5: 76–77.
Hobart, W.H., ed. 1995. *Baird's legacy: the history and accomplishments of NOAA's National Marine Fisheries Service, 1871–1996.* Silver Springs MD: U.S. Department of Commerce, NOAA.
Hoeppli, R. 1959. *Parasites and parasitic infections in early medicine and science.* Singapore: U. of Malaya P.
Hoff, H.E. 1964. Nicholas of Cusa, van Helmont, and Boyle: the first experiment of the Renaissance in quantitative biology and medicine. *J of the History of Medicine* 19: 99–117.
Hoffman, W. 2017. The tallgrass prairie: Raymond Lindeman, a Minnesota bog lake, and the birth of ecosystems ecology. Pp. 1–31 in Amato 2017.
Hofmann, J.E. 1971. Nicolas Cusa or Nikolaus von Cusa, Nicolaus Cusanus (c. 1401–64), philosophy, mathematics. *DSB* 3: 512–516.

Hofsten, N. von. 1916. Zur alteren Geschichte der Diskontinuitätsproblems in der Biogeographie. *Zoologishe Annalen* 7: 197–353.

———. 1957. Linnaeus' conception of nature. *Kungliga Vetenskaps-Societentens Årsbok* 65–105.

Holland, N.D. 2011. Walter Garstang: a retrospective. *Theory in Biosciences* 247–258.

Holman, J.A. 1999. *Ancient life of the Great Lakes Basin: Precambrian to Pleistocene*. Ann Arbor: U. of Michigan P.

Holmes, F.L. 2000. The "Revolution in chemistry and physics": overthrow of a reigning paradigm or competition between contemporary research programs? *Isis* 91: 735–753.

Holthuis, L.B. 1969. Thomas Say as a carcinologist. Pp. v—xi in Say 1969.

——— & T. Sakai. 1970. *Ph. F. von Siebold and Fauna Japonica: a history of early Japanese zoology*. Tokyo: Academic P. of Japan.

Honig, P. & F. Verdorn, eds. 1945. *Science and scientists in the Netherlands Indies*. NY: Board of the Netherlands Indies, Surinam & Curaçao.

Honohane, J. 1697. Part of a letter dated June 7, 1697 giving an account of a moving bog in Ireland. *Roy. Soc. of London Philos. Trans.* 19: 714–716.

Hooker, J.D. 1844. Diatoms. Vol. 1, pp. 503–519 in Hooker, *The botany of the Antarctic voyage of H.M. Discovery ships* Erebus *and* Terror *in the years 1839–1843 under the command of Captain Sir James Clark Ross*. Vol. 1: *Flora Antarctica*. L: Reeve Brothers.

———. 1891. *Himalayan journals*. L: Minerva Library.

Hooker, W.J. 1834a. Geography considered in relation to the distribution of plants. Pp. 227–246 in *An encyclopaedia of geography*. L: Longman. Reprinted in Egerton 1977c.

———. 1834b. England. Pp. 323–336 in *An encyclopaedia of geography*. L: Longman. Reprinted in Egerton 1977c.

Hooykaas, R. 1956. The zoogeography of Abraham van Mijle. *Archives internationales d'Histoire des Sciences* 9: 125–132.

Horner, F. 1987. *The French reconnaissance: Baudin in Australia, 1801–1803*. Melbourne: Melbourne U.P.

Horwitz, T. 2002. *Blue latitudes: boldly going where Captain Cook has gone before*. NY: Henry Holt.

Hossard, N. 2001. *Aimé Bonpland, 1773–1858, medicine, naturaliste, explorateur en Amérique du Sud*. Paris: L'Harmettan.

Houck, L.D. & L.C. Drickamer, eds. 1996. *Foundations of animal behavior: classic papers with commentaries*. Chicago: U. of Chicago P.

Houot, G.S. 1954. Two and a half miles down. *National Geographic* 106: 80–86.

Houston, C.S. 1997. Joseph Grinnell (1877–1939), zoologist, university professor. *BDAC*, 329–333.

Houvenaghel, G.T. 1980. Belgium and the early development of modern oceanography, including note on A.F. Renard. Pp. 667–681 in Sears & Merriman 1980.

Howard, L.O. 1897. *A study in insect parasitism: a consideration of the parasites of the white-tussock moth, with an account of their habits and interrelations*. USDA Bureau of Entomology Technology Series 5, 57 pp. Reprint in Egerton 1977b.

———. 1930. *A history of applied entomology (somewhat anecdotal)*. Washington: Smithsonian Institution.

———. 1932. Biographical memoir of Stephen Alfred Forbes, 1844–1930. *National Academy of Science Biographical Memoirs* 15: 1–54. Reprinted in Forbes 1977.

———. 1933. *Fighting the insects: the story of an entomologist, telling of the life and experiences of the writer*. NY: Macmillan.

———. 1934. Alpheus Spring Packard (1839–1905), entomologist, teacher. *DAB* 14: 126–127.

——— & W.F. Fiske. 1911. *The importation into the United States of the Parasites of the Gypsy Moth and the Browntailed moth*. USDA Bureau of Entomology Bulletin 91, 345 pp. 1977. NY: Arno P.

Howe, H.F. & J. Smallwood. 1982. Ecology of seed dispersal. *Annual Review of Ecology and Systematics* 13: 201–228.

Howe, H.M. 1965. A root of van Helmont's tree. *Isis* 56: 408–419.

Howe, J.P. 2014. *Behind the curve: science and the politics of global warming*. Seattle: U. of Washington P.

———, ed. 2017. *Making climate change history: documents from global warming's past*. Seattle: U. of Washington P.

Hubbard, J.M. 1997. Archibald Gowanlock Huntsman, 1883–1973, marine ecologist, zoologist, oceanographer, educator. *BDAC* 386–388. [1 vol.]

Hudson, J.C. & S. Ziegler. 2014. Environment, culture, and the Great Lakes fisheries. *Geographical Review* 104: 391–413.

Humboldt, F.W.H.A. July 1820–21. On isothermal lines and the distribution of heat over the Globe. *Edinburgh Philosophical J.* 3: 1–20, 256–274, 4: 23–37, 262–281, 5: 28–39. Reprinted in Egerton 1977c.

———. 2019. *Selected writings*. A. Wulf, ed. NY: Everyman's Library. 792 pp.

——— & A. Bonpland. 2009. *Essay on the geography of plants*. S.T. Jackson, ed., S. Romanowski, transl. Chicago: U. of Chicago P.

Humphrey, H.B. 1961. *Makers of North American botany*. NY: Ronald P.

Hunter, C. 1983. *The life and letters of Alexander Wilson*. Philadelphia: American Philos. Soc.

Hunter, M. 2015. *Boyle studies: aspects of the life and thought of Robert Boyle (162 7–91)*. Farnham UK: Ashgate.

Hutchinson, G.E. 1918. A swimming grasshopper. *Entomological Record & J. of Variation* 30: 138. Reprinted in Hutchinson 2010: 57.

———. 1933. Limnological studies at high altitudes in Ladakh. *Nature* 132: 136–138. Reprinted in Hutchinson 2010: 86.

———. 1943. The biogeochemistry of aluminum and of certain related elements. *Quart. Rev. of Biology* 18: 1–29, 128–153,242–262, 331–363.

———. 1945. Aluminum in soils, plants and animals. *Soil Science* 60: 29–40.

———. 1954. The biochemistry of the terrestrial atmosphere. Pp. 371–433 in Kuiper 1954.

———. 1957–93. *A treatise on limnology*. 4 vols. NY: Wiley.

———. 1969. Eutrophication, past and present. Pp. 17–24 in Rohlich 1969.

———. Sept. 1970. The biosphere. *Scientific American* 223, no. 3: 44–53.

———. 1978. *An introduction to population ecology*. New Haven: Yale U.P.

———. 1979. *The kindly fruits of the earth: reflections of an embryo-ecologist*. New Haven: Yale U.P.

———. 1982. Reminiscences and notes on some otherwise undiscussed papers. Pp. vii—ix in Riley 1982.

———. 2010. *The art of ecology: writings*. D.K. Skelly, D.M. Post & M.D. Smith, eds. New Haven: Yale U.P.

———, G.E. Pickford & J.F.M. Schuurman. 1932. A contribution to the hydrobiology of pans and other inland waters of South Africa. *Archive für Hydrobiologie* 24: 1–154.

Hutton, J. 1987. *The 1789 abstract of James Hutton's theory of the Earth.* Edinburgh: Scottish Academic P.

Huxley, L., ed. 1918. *Life and letters of Sir Joseph Dalton Hooker.* 2 vols. L: John Murray.

Idyll, C.P., ed. 1969a. *Exploring the ocean world: a history of oceanography.* NY: Thomas Y. Crowell.

———. 1969b. The science of the sea. Pp. 2–21 in Idyll 1969a.

Ihde, A.J. 1997. Rachel (Louise) Carson (1907–64), marine biologist, ecologist, writer. *BDAC* 137–140.

Ilerbaig, J. 1999. Allied sciences and fundamental problems: C.C. Adams and the search for method in early American ecology. *J. of the History of Biology* 32: 439–463.

Ingen-Housz, J. 1779. Account of a new kind of inflammable air or gas, which can be made in a moment without apparatus, and it is as fit for explosions as other inflammable gases in use for that purpose: with a new theory of gun powder. *Roy. Soc. L Philos. Trans.* 69: 376–418.

Ings, T.C. & J.E. Howes. 2018. The history of ecological networks. Pp. 15–28 in Dáttilo & Rico-Gray 2018.

Innes, S. 2004. Philip Lutley Sclater (1829–1913). *DNCB* 4: 1784–1785.

Isaacson, W. 2003. *Benjamin Franklin: an American life.* NY: Simoon & Schuster.

Israel, G. 1988. On the contribution of Volterra and Lotka to the development of modern biomathematics. *History and Philosophy of the Life Sciences* 10: 37–49.

———. 1993. The emergence of biomathematics and the case of population dynamics: a revival of mechanical reductionism and Darwinism. *Science in Context* 6: 469–509.

——— & A.M. Gesco, eds. 2002. *The biology of numbers: the correspondence of Vito Volterra on mathematical biology.* Basel: Birkhäuser.

———. 2005. Vito Volterra, *Book on mathematical biology.* Pp. 936–944 in Grattan-Guinness 2005.

Jack, H.A. 1945. Biological field stations of the world. *Chronica Botanica* 9: 1–73.

Jackson, D. 1981. *Thomas Jefferson and the Stony Mountains.* Urbana: U. of Illinois P.

Jackson, J. 2005. *The world on fire: a heretic, an aristocrat, and the race to discover oxygen.* NY: Viking.

Jackson, R. 2018. *The ascent of John Tyndall: Victorian scientist, mountaineer and Public intellectual.* NY: Oxford U.P.

Jackson, S.T. 2009. Biographical sketches. Pp. 227–252 in Humboldt & Bonpland 2009.

Jaffe, B. 1958. *Men of science in America.* NY: Simon &Schuster.

Jahn, I. 1971. Christian Gottfried Ehrenberg (1795–1876), biology, micropaleontology. *DSB* 4: 288–292.

———. 1989. Das Ornithologicon des Arztes Daniel Gottlieb Messerschmidt: Univeröffentlichte zoologische Ergebnisse seiner Forschungsreise. *Leopoldina* 33: 103–135.

James, P.E. 1972. *All possible worlds: a history of geographical ideas.* Indianapolis: Odyssey P.

——— & G.J. Martin. 1981. *All possible worlds: a history of geographical ideas.* NY: John Wiley & Sons.

Jamison, A. 1993. National political cultures and the exchange of knowledge: the case of systems ecology. Pp. 187–208 in Crawford et al. 1993.

Jansen, A.J. 1972. An analysis of 'balance of nature' as an ecological concept. *Acta Biotheoretica* 21: 86–114.

Jax, K. 2011. Formation of scientific societies. Pp. 171–173 in Schwaez & Jax 2911.

Jefferson, T. 1787. *Notes on the State of Virginia.* L: John Stockdale. In Jefferson 1984: 123–325.

———. 1984. *Writings.* D.P. Merrill, ed. NY: Library of America.

Jenny, H. 1961. *E.W. Hilgard and the birth of soil science.* Pisa, Italy: Agrochimica.

Jensen, A.S. 1932. Peter Wilhelm Lund, 1801–1880. Pp. 110–114 in Meisen 1932.

Jensen, O. 1955. The Peales. *American Heritage* 6: 40–51, 97–101.

Jespersen, P.H. 1946. J.C. Fabricius as an evolutionist. *Svenska Linné-Sällskapets Årsskrift* 29: 35–56.

Joerg, W.L.G., ed. 1928. *Problems of polar research: a series of papers by thirty-one authors.* NY: American Geographical Soc.

Johnston, D.W. & E.P. Odum. 1956. Breeding bird populations in relation to plant succession on the Piedmont of Georgia. *Ecology* 37: 50–62.

Johnson, R.I. 1969. Semper's Reisen im Archipel der Philippenen, Wissenschaftliche Resultate, 1867–1916. *Soc. for the Bibliography of Natural History J.* 5: 144–147.

Johnson, S. 2008. *The invention of air: a story of science, faith, revolution, and the birth of America.* NY: Riverhead Books/Penguin.

Johnson, W.C. & N.R. French. 1981. Soviet Union. Pp. 343–383 in Kormondy & McCormick 1981.

Johnstone, K. 1977. *The Aquatic explorers: a history of the Fisheries Research Board of Canada.* Toronto: U. of Toronto P.

Jones, R.T. 2014. *Empire of extinction: Russians and the north Pacific's strange beasts of the sea, 1741–1867.* NY: Oxford U.P.

Jordan, D.S. 1877. Review of Rafinesque's memoirs on North American fishes. *Contributions to North American ichthyology: United States Museum Bull.* 9.

———. 1922. *The days of a man: being memoires of a naturalist, teacher, and minor prophet of democracy.* 2 vols. Yonkers NY: World Book Co.

——— & B.W. Evermann. 1923. *American food and game fishes.* NY: Doubleday, Page. 1969. NY: Dover.

Joseph, L.E. 1990. *Gaia: the growth of an idea.* NY: St. Martin's P.

Jost, L. 1930. Zum hundertsten Geburtstag Anton de Bary's. *Zeitschrift für Botanik* 24: 1–74 + portrait.

Jovet, P. 1954. Flore et phytogéographie de la France. Pp. 242–268 in Virville 1954.

——— & J. Mallet. 1974. François Péron (1775–1810), zoology, natural history. *DSB* 517–518.

Joyce, K.A. 1993. Underwater Walden. Pp. 196–198 in Schofield & Baron 1993.

Juan, J. & A. de Ulloa. 1964. *A voyage to South America.* J. Adams, transl. NY: Knopf.

Juday, C. 1910. Some European biological stations. *Wisconsin Academy of Sciences, Arts and Letters Trans.* 16: 1257–77 + 4 plates.

——— & Hasler, A.D. 1944. List of publications dealing with Wisconsin limnology, 1871–1945. *Wisconsin Academy of Sciences, Arts and Letters Trans.* 36: 469–490. Published 1946.

———, C.L. Schhloemer & C. Livingston. Effect of fertilizers on plankton production and growth of fish in a Wisconsin lake. *Progressive Fish Culture* 40: 24–27.

Kajak, Z. & E. Pieczyska. 1981. Poland. Pp. 287–302 in Kormandy & McCormick 1981.

Kalm, P. 1770–71. *Travels into North America; containing its natural history, and a circumstantial account of its plantations and agriculture.* J.R. Forster, transl. 3 vols. L: J.T. Loendes.

———. 1937. *Travels in North America.* J.R., transl., A.B. Benson, revisor. NY: Wilson-Erickson.

Kapor, S.C. 1971. Jean-Baptiste-André (1800–84), chemistry. *DSB* 4: 242–248.

Kärnefelt, I., P. Scholz, M.R.D. Seaward & A. Thell. 2012. Lichenology in Germany: past, present and future. *Schleahtendalia* 23: 1–94.

Kastner, J. 1969, 1977. *A species of eternity.* NY: Dutton. Knopf.

Kates, J.S. 1934. *Minute stories of famous explorers.* NY: Grosset & Dunlap.

———. 2001. *Planning a wilderness: regenerating the Great Lakes cutover region.* Minneapolis: U. of Minnesota P.

Katz, R.W. 2002. Sir Gilbert Walker and a connection between El Nino and statistics. *Statistical Science* 17: 97–112.

Kaups, M. 1978. North Shore commercial fishing, 1848–70. *Minnesota History* 46: 43–58.

Kean, B.H., K.E. Mott & A.J. Russell, eds. 1978. *Tropical medicine and parasitology: classic investigations.* 2 vols. Ithaca: Cornell U.P.

Keast, A. 1966. *Australia and the Pacific Islands: a natural history.* NY: Random House.

Keddy, P.A. 1987. Beyond reductionism and scholasticism in plant community ecology. *Vegetatio* 69: 209–211.

———. 2001. *Competition.* Ed. 2. Dordrecht: Kluwer. Ed. 1, 1989.

———. 2005. Milestones in ecological thought—a canon for plant ecology. *J. of Vegetation Science* 16: 145–150.

———. 2007. *Plants and vegetation: origins, processes, consequences.* NY: Cambridge U. P.

Keeling, C.D. 1960. The concentration and isotopic abundance of carbon dioxide in the atmosphere. *Tellus* 12: 200–203.

———. 1997. Climate change and carbon dioxide: an introduction. *National Acad. of Sci. Proc.* 94: 8273–8274.

Kehoe, T. 1997. *Cleaning up the Great Lakes: from cooperation to confrontation.* De Kalb: Northern Illinois U.P.

Keller, A.G. 1975. Guillaume Rondelet (1507–66), medicine, anatomy. *DSB* 11: 527–528.

Kellner, L. 1963. *Alexander von Humboldt.* L: Oxford U.P.

Kemna, A. 1897. *P.J. Beneden, la vie et l'Oeuvre d'un zoologist.* Anvers: J.E. Buschmann.

Kemp, S. & A.V. Hill. 1943. Edgar Johnson Allen, 1866–1942. *Roy. Soc. of L. Obituary Notices of Fellows* 4: 357–367.

Kendeigh, S.C. 1954. History and evaluation of various concepts of plant and animal communities in North America. *Ecology* 35: 152–171. Reprinted in Sterling 1974.

———. 1961. *Animal ecology.* Englewood Cliffs NJ: Prentice-Hall.

Kenoyer, L.A. 1927. A study of Raunkiaer's law of frequence. *Ecology* 8: 341–349.

Kerner von Marilain, A.J. 1863. *Das pflanzenleben der Donaulaender.* Innsbrück: Wagner. Extracts in Acot 1998: vol. 2, 465–479.

———. 1896. *The natural history of plants: their forms, growth, reproduction, and distribution.* F.W. Oliver, transl. L: Blackie.

———. 1951. *The background of plant ecology: a translation of the plant life of the Danube Basin.* H.S. Conrad, transl. Ames: Iowa State U.P.

Kessel, E.L., ed. 1955. *A century of progress in the natural sciences.* San Francisco: California Academy of Sciences.

Khakhina, L.N. 1992. *Concepts of symbiogenesis: a historical and critical study of the research of Russian botanists.* L. Margulis & M. McMenamin, transl. New Haven: Yale U.P.

Kiger, J.C. 1962. The American Society of Parasitologists: a short history. *J. of Parasitology* 48: 641–650.

Kimler, W.C. 1988. Warder Clyde Allee (1885–1955), ecology. *DSB* 17: 16–17.

Kinch, M.P. 1980.Geographical distribution and origin of life: the development of early nineteenth-century British explanations. *JHB* 13: 91–119.0

King, C.S., ed. 1906. *A great archbishop of Dublin, William King, D.D., 1650–1729: his autobiography, family, and a selection from his correspondence.* Longman Green.

King, L.J. 1975. Christian Konrad Sprengel (1750–1816), botany. *DSB* 12: 587–591.

King, W. 1685. Of the bogs and loughs of Ireland. *Roy. Soc. L. Philos. Trans.* 15: 948–960.

Kingsland, S. 1982. The refractory model: the logistic curve and the history of population ecology. *Quarterly Review of Biology* 57: 29–52.

———. 1985. *Modeling nature: episodes in the history of population ecology.* Chicago: U. of Chicago P.

———. 2005. *The evolution of American ecology: 1890–2000.* Baltimore: Johns Hopkins P.

Kingsolver, J.G. & R.T. Paine. 1991. Conversational biology and ecological debate. Pp. 309–317 in Real & Brown 1991.

Kish, G. 1970. José de Acosta (1539–1600), geography. *DSB* 1: 48. [or born 1540.]

Kitchell, J.F. & G.G. Sass. 2000. Great Lakes ecosystems: invasions, food web dynamics, and the challenge of ecological restoration. Pp. 157–170 in Waller & Rooney 2008.

Kitching, R.L. 1983. *Systems ecology: an introduction to ecological modelling.* St. Lucia: U. of Queensland P.

Kline, M. 1972. *Mathematical thought from ancient to modern times.* NY: Oxford U.P.

Klopfer, P.H. & J.P. Hailman. 1967. *An introduction to animal behavior: ethology's first century.* Englewood Cliffs NJ: Prentice-Hall.

Knight, T.A. 1799. An account of some experiments on the fecundation of vegetables. *Roy. Soc. L. Philos. Trans.* 889: 195–204.

Knowles, M.C. 1913. The maritime and marine lichens of Howth. *Royal Dublin Soc. Scientific Proc.* 14: 79–143.

Koelz, W.F. 1925. Fishing industry of the Great Lakes. Pp. 553–617 in *U.S. Commission of Fisheries Report for 1925.* Washington DC.

———. 1929 (for 1927). Coregonid fishes of the Great Lakes. Pp. 297–643 in *U.S. Bureau of Fisheries Bulletin* 43.

Koerner, L. 1999. *Linnaeus: nature and nation.* Cambridge MA: Harvard U.P.

Kofoid, C.A. 1898. The fresh-water biological stations of America. *American Naturalist* 32: 391–406.

———. 1903. The plankton of the Illinois River, 1894–1899, with introductory notes upon the hydrography of the Illinois River and its basin. *Illinois State Laboratory of Natural History Bull.* 6 & 8: 95–631. 1977. NY: Arno P.

———. 1910. *The biological stations of Europe*. Washington: U.S. Bureau of Education, B 4.

———. 1911. Contributions of Alexander Agassiz to marine biology. *Hydrobiologie und Hydrographie* 4: 39–43.

Kohler, R.E. 2002. *Landscapes and labscapes: exploring the lab-field border in biology*. Chicago: U. of Chicago P.

———. 2006. *All creatures: naturalists, collectors, and biodiversity, 1850–1959*. Princeton: Princeton U.P.

Kolchinsky, E.I. 2004. The role of Eighteenth Century Russian expeditions in the development of natural history. *California Acad. of Sci. Proc.* 55, supplement 2: 106–116.

Kölmel, R. 1988. Zwischen Universalismus und Empire—die Begrundung der nodernen Ökologieund Biozöse-Konzeption durch Möbius. *Mitteilungen aus dem zoologischen Museum der Universität Kiel* 1, no. 7: 17–34.

———. 1990. The Prussian 'Kommission zur wissenschaftlichen Unterclhung der deutschen Meere in Kiel' and the origin of modern concepts in marine biology in Germany. Pp. 399–407 in *Ocean sciences: their history*.

König, R. et al. 1981. *Karl Möbius: Beiträge zu Leben und Werk*. Kiel: Goecke & Evers.

Konishi, M. & Y. Itô. 1973. *Early Entomology in East Asia*. Pp. 1–18 in R.F. Smith, T.E. Mittler & C.N. Smith, eds. *History of entomology*. Palo Alto CA: Annual Reviews.

Kormondy, E.J. & J.F. McCormick, eds. 1981. *Handbook of contemporary developments in world ecology*. Westport CT: Greenwood P.

Kors, A.C., ed. 2002. *Encyclopedia of the enlightenment*. NY Oxford U.P.

Kouba, T.F. 1973. *Wisconsin's amazing woods: then and now*. Madison: Wisconsin House.

Kozlowski, S. & A. Sevedrzushi. 2009. Proffessor Józel Paszoski-tworca fitosajologii (creator of phytosbiology). *Lakarstwo w Police* (grassland science in Poland) 121: 83–95.

Krause, D.J. 1989. Testing a tradition: Douglass Houghton and the native copper of copper of Lake Superior. Isis 80: 622–639.

Kroll, G. 2008. *America's ocean wilderness: a cultural history of Twentieth Century exploration*. Lawrence: U.P. of Kansas.

Kronfeld, E.M. 1908. *Anton Kerner von Marilaun: Leben und Arbeit eines deutschen Naturforschers*. Leipzig: Chr. H. Tauchnitz.

Kropotkin, P. 1902. *Mutual aid: a factor in evolution*. L: Heinemann. Ed. 2, 1904.

Krumbein, W.E., ed. 1978. *Environmental biogeochemistry and geomicrobiology*. 3 vols. Ann Arbor: Ann Arbor Science.

Kruta, V. 1972. Johann Georg Gmelin (1709–55), botany, natural history, geography. *DSB* 5: 427–429.

Kuchenberg, T. 1978. *Reflections in a tarnished mirror: use and abuse of the Great Lakes*. Sturgeon Bay WI: Golden Glow Publishing.

Kuhn, T.S. 1970. *The structure of scientific revolutions*. Ed. 2. Chicago: U. of Chicago P.

Kuiper, G.P. 1954. *The Earth as a planet*. Chicago: U. Chicago P.

Kutzbach, G. 1970. John Young Buchanan (1844–1925), oceanography, chemistry. *DSB* 2: 557–558.

———. 1973. Elias Loomis (1811–89), meteorology, mathematics, astronomy. *DSB* 8: 487.

Kwa, C. 1987. Representation of nature mediating between ecology and science policy: the case of the International Biological Programme. *Social Studies of Science* 17: 413–442.

———. 1989. *Mimicking nature: the development of systems ecology in the United States, 1950–1975*. Ph.D. dissertation, U. of Amsterdam.

———. 1993a. Radiation ecology, systems ecology and the management of the environment. Pp. 213–249 in Shortland 1993. Revision of chapter 2 of Kwa 1989.

———. 1993b. Modeling the grasslands. *Hist. Studies in the Physical and Biological Sciences* 24: 125–155.

———. 2005. Local ecologies, global science: discourse and strategies of the International Geosphere-Biosphere Programme. *Social Studies of Science* 35: 923–950.

La Bastille, A. 1980. *Women and Wilderness*. San Francisco: Sierra Club Books.

Lack, D. 1946. Clutch and brood size in the robin. *British Birds* 39: 98–109, 130–135.

———. 1947. *Darwin's finches: an essay on the general biological theory of evolution*. Cambridge UK: Cambridge U.P. 1961. NY: Harper & Row.

———. 1954. *The natural regulation of animal numbers*. Oxford: Clarendon P.

———. 1966. *Population studies in birds*. Oxford: Oxford U.P.

———. 1973. My life as an amateur ornithologist. *Isis* 115: 421–431.

La Condamine, C.M. de. 1745. *Relation abrégée d'un voyage fait dans l'interieur de l'Amerique méridionale*. Paris.

———. 1751. *Journal du voyage fait par ordre du roi à l'équateur*. Paris.

Lacordaire, T. 1834–38. *Introduction à entomologie*. 3 vols. Paris: Librairie encyclopedique de Roret.

———. 1838–39. On the geographical distribution of insects. *Edinburgh New Philos. J.* 27: 94–111, 333–351, 28: 170–183, 224–238.

Ladd, H.S., ed. 1957. *Treatise on marine ecology and paleoecology, vol. 2: Paleoecology. Geol. Soc. Amer. Mem. 67.*

——— & G. Gunther. 1957. Development of marine paleoecology. Pp. 67–73 in Ladd 1957.

Laissus, Y. 1971. Philibert Commerson (1727–73), natural history. *DSB* 3: 365–366.

———. 1978. Charles-Marie de La Condamine (1701–74), mathematics, natural history. *DSB* 15: 269–273.

Lalande, T. 2003. Parcors géographique d'une adventure savant. *Musée des Arts et Métiers* 39/40: 12–20.

Lamarck, J.E.A. de. 1964. *Hydrogeology*. A.V. Carozzi, transl. Urbana: U. of Illinois P.

Lamb, H.H. 1982. *Climate, history and the modern world*. L: Methuen.

Lampert, K. 1925. *Das Leben der Bennenwässer*. Ed. 3. Leipzig: Tauchnitz.

Landin, B.-O. 1971. Johann Christian Fabricius (1745–1808), entomology. *DSB* 4: 512–513.

Landrieu, M. 1909. *Lamarck, la fondateur du transformism: savie, son oeuvre*. Paris: La Société zoologique de France.

Lane, C.E. 1969. Biology of the sea. Pp. 42–61 in Idyll 1969.

Langlois, T.H. 1949. The biological station of the Ohio University. *Franz Theodore Stone Laboratory Contribution* 11: 1–55.

Lanner, R. 1996. *Made for each other: a symbiosis of birds and pines*. NY: Oxford U.P.

Lannoo, M.J. 2010. *Leopold's shack and Ricketts lab: the emergence of environmentalism*. Berkeley: U. of California P.

———. 2018. *This land is your land: the story of field biology in America*. Chicago: U. Chicago P.

Laporte, L. 2000. *George Gaylord Simpson: paleontologist and evolutionist*. New Haven: Yale U.P.

Large, E.C. 1940 *Advance of the Fungi*. L: Jonathan Cape.

Larson, J.L. 1980. Not without a plan: geography and natural history in the late Eighteenth-Century. *J. of the Hist. of Biology* 19: 447–488.

Latreille, P.A. 1817. Introduction à la géographie générale des arachides et des insectes. *Mémoires de Museum d'Histoire Naturelle* 3: 37–67.

———. 1819. Introduction à la géographie générale des arachnids et des insects, ou des climats propres à ces animaux. *Mémoires sur divers sujets de l'Historie naturelle*. Paris: Deterville.

Lauff, G.H. 1963. A history of the American Society of Limnology and Oceanography. Pp. 667–682 in Frey 1963a.

Laughton, J.K. [1889 [reprinted 1921]. Matthew Flinders (1774–1814), captain in navy, hydrographer and discoverer. *DNB* 7: 325–329.

———. 1921–22 [reprint of 1889]. Sir James Clark Ross (1800–62), rear admiral and Arctic and Antarctic navigator. *DNB* 17: 265–266.

Lavoisier, A.-L. 1773. Sur la nature de l'eau et sur les experiences par lesquelles on a pretend pouver la possibilité de son changement en terre. *Académie des sciences mémoires. Oeuvres* 2 (1862) 1–28.

Lawrence, G.H.M. 1974. William Nylander (1822–99), botany. *DSB* 10: 165–166.

Lawson, J. 1967. *A new voyage to Carolina*. H.T. Lefler, ed. Chapel Hill: U. of North Carolina P. Ed. 1, 1709.

Lawton, J. 1996. Patterns in ecology. *Oikos* 75: 145–147.

———. 1999. Are there general laws in ecology? *Oikos* 84: 177–192.

———. 2001. Earth system science. *Science* 292: 1965.

Laycock, G. 1965. *The sign of the Flying Goose: a guide to the National Wildlife Refuges*. Garden City NY: Natural History P.

Layman, C. et al. 2015. A primer on the history of food web ecology: fundamental contributions of fourteen researchers. *Food Webs* 4: 14–24.

Lear, L.J. 1997. *Rachel Carson: witness for nature*. NY: Henry Holt.

———. 1999. Rachel Louise Carson (1907–64), writer and scientist. *ANB* 4: 474–476.

Lecoq, H. 1854–58. *Études sur la géographie botanique de l'Europe et, en particulier, sur la végétation du plateau central de la France*. 9 vols. Paris: Baillière.

Ledyard, J. 2005. *The last voyage of Captain Cook: the collected writings*. J. Zug, ed. Washington: National Geographic.

Lee, S. 1921. John Leland or Leyland (1506?-52), antiquary. *DNB* 11: 892–896.

Leeuwenhoek, A. van. 1939–99. *Collected letters*. Committee of Dutch scientists, eds. 17 vols. Amsterdam: Swets & Zellinger.

Lenz, J. 2004. Victor Hensen (1835–1924): founder of quantitative plankton research. Pp. 29–34 in Morcos, Zhu et al. 2004.

Lenz, W. 1990. German marine research in the Atlantic Ocean between World War I and II. *Deutsche Hydrographische Zeitschrift* 22: 114–121.

——— & M. Deacon, eds. 1990. *Ocean sciences: their history and relation to man*. Hamburg: *Deutsche Hydrographische Zeitschrift*, number 22.

Leonard, I.A. 1964. Introduction. Pp. 3–19 in Juan & Ulloa 1964.

Leonhart, J. 1985. Charles Wilkes: a biography. Pp. 189–203 in Viola & Margolis 1985.

Leopold, A. 1933. *Game management*. NY: Charles Scribner's Sons.

——— & S.E. Jones. 1947. A phenological record for Sauk and Dane Counties, Wisconsin, 1935–1945. *Ecol. Monogr.* 17: 81–122.

Leroy, J.F. 1976. Joseph Pitton de Tournefort (1650–1708), botany, medicine. *DSB* 13: 442–444.

Lévêque, C. 2003. *Ecology: from ecosystem to biosphere*. Anonymous translation. Enfield NH/Plymouth UK: Science Publishers.

Levin, S.A., ed. 1994. *Frontiers in mathematical biology*. NY: Springer.

Levine, J.M. 1977. *Dr. Woodward's shield: history, science, and satire in Augustan England*. Berkeley: U. of California P.

Levit, G.S. 2011. Looking at Russian ecology through the biosphere theory. Pp. 333–347 in Schwarz & Jax 2011.

Lewandowski, S.A. 1999. Henry Bryant Bigelow (1879–1967), zoologist and oceanographer. *ANB* 2: 750–751.

Leydig, F. 1860. *Naturgeschichte der Daphniden (Crustacea Cladocera)*. Tübingen: Laupp und Siebeck. Online.

Likens, G.E. 2001. Eugene Odum, the ecosystem approach, and the future. Pp. 309–328 in Barrett & Barrett 2001.

———. 2002. Arthur Davis Hasler, 1908–2001. *National Academy of Sciences Biographical Memoirs* 82: 1–14.

———. 2013a. Resolution of respect: F. Herman Bormann, 1922–2012. *ESAB* 94: 304–308.

———. 2013b. The Hubbard Brook Ecosystem study: celebrating fifty years. *ESAB* 94: 336–337.

——— & F.H. Bormann. 1974. Acid rain: a serious regional environmental problem. *Science* 184: 1176–1179.

Likens, G.E. & F.H. Bormann. 1975. An experimental approach in New England landscapes. Pp. 7–29 in Hasler 1975.

———. 1977. *Biogeochemistry of a forested ecosystem*. NY: Springer-Verlag.

Likens, G.E. & N.M. Johnson. 1972. Acid rain. *Environment* 14: 33–40.

Lillie, F.R. 1944. *The Woods Hole Marine Biological Laboratory*. Chicago: U. of Chicago P.

Limburg, K.E. 2004. The biogeochemistry of strontium: a review of H.T. Odum's contributions. *Ecological Modelling* 178: 31–33.

Limoges, C. 1970. *La sélection naturelle: étude sur la première constitution d'un concept (1837–1859)*. Paris: Presses Universitaires de France.

Lindeman, R.L. 1942. The trophic dynamic aspect of ecology. *Ecology* 23: 399–418. Reprinted in Real & Brown (1991), 157–176.

Lindroth, C.H. 1973. Systematics specializes between Fabricius and Darwin: 1800–1859. Pp. 119–154 in Smith et al. 1973.

Lindroth, S. 1973. Carl Linnaeus (or von Linné, 1707–78), botany, zoology, geology, medicine. *DSB* 8: 574–581.

———. 1976. Georg Wilhelm Steller (1709–46), geography, biology. *DSB* 13: 28–29.

———. 1983. The two faces of Linnaeus. Pp. 1–62 in Frängsmyr 1983.

Lindsey, A.A. et al. 1985. *The bicentennial of John James Audubon*. Bloomington: Indiana U.P.

Linklater, E. 1972. *The voyage of the Challenger*. NY: Doubleday.

Linnaeus, C. 1749. *Specimen academicum de oeconomia naturae*. Uppsala. English, 1759.

———. 1754. *Dissertatio botanica, sistens stations plantarum.* Uppsala: L.M. Höjer.

———. 1760. *De polita naturae.* Uppsala. English, 1781.

———. 1775. *Miscellaneous tracts relating to natural history, husbandry, and physic.* B. Stillingfleet, ed. & transl. London: J. Dodsley.

———. 1781. *Select dissertations from the Amoenitates Academica.* F.W. Brand, transl. L: G. Robinson & J. Robson. 1977: NY: Arno P.

Lippincott, J.S. 1863. Geography of plants. Pp. 464–525 in *Report of the Commissioner of Agriculture for 1863.* Washington: Government Printing Office. Reprinted in Egerton 1977c.

Littlepage, D. 2006. *Steller's Island: adventures of a pioneer naturalist in Alaska.* Seattle: Mountaineers Books.

Lloyd, C. 1966. *William Dampier.* L: Faber & Faber.

Lomborg, B. 2020. *False alarm! How climate change panic costs us trillions, hurts the poor, and fails to fix the planet.* NY: Basic Books.

Lomolino, M.V., D.F. Sax & J.H. Brown, eds. 2004. *Foundations of biogeography: classic papers with commentaries.* Chicago: U. Chicago P.

Longley, W.H. January 1927. Life on a coral reef: the fertility and mystery of the sea, studied beneath the waters surround Dry Tortugas. *National Geographic* 51, no. i: 61–83.

——— & C. Martin. January 1927. The first autochromes from the ocean bottom: marine life in its natural habitat along the Florida Keys is successfully photographed in colors. *National Geographic* 51, no. 1: 56–60.

Longstreet, S. 1973. *Chicago, 1860–1919.* NY: David McKay.

Loomis, E. 1868. *A treatise on meteorology: with a collection of meteorological tables.* NY.

——— & H.A. Newton. 1866. On the mean temperature and on the fluctuations of temperature at New Haven, Conn., Lat 41⁰18' N, Long. 72⁰55' of Greenwich. *Conn. Acad. of Arts & Sci. Trans.* 1: 194–246.

López de Azcona, J.M. 1973. Jorge Juan Santacilla (1713–73), geodesy. *DSB* 7: 183–184.

Lotka, A.J. 1912. Quantitative studies in epidemiology. *Nature* 88: 497–498.

———. 1925. *Elements of physical biology.* Baltimore: Williams & Wilkins. Retitled: *Elements of mathematical biology.*

Lourteig, A. 2003. À la mémoire d'Aimé Bonpland. *Musée des Arts et Métiers: la Revue* 39/40: 48–53.

Lovelock, J. 1972. Gaia as seen through the atmosphere. *Atmospheric Environment* 6: 579–580.

———. 2000a. *Gaia: the practical science of planetary medicine.* Ed. 2. L & NY: Oxford U.P. Ed. 1, 1991. L: Gaia Books.

———. 2000b. *Homage to Gaia: the life of an independent scientist.* Oxford: Oxford U.P.

———. 2003. The living Earth. *Nature* 425: 769–770.

——— & L. Margolis. 1974. Atmospheric homeostasis by and for the biosphere: the Gaia hypothesis. *Tellus* 26: 2–9.

Lovely, R.A. 1995. *Mastering nature's harmony: Stephen Forbes and the roots of American ecology.* Madison: Ph.D. dissertation, U. of Wisconsin.

Lowe, P.R. 1936. The finches of the Galapagos in relation to Darwin's conception of species. *Ibis* 310–321.

Lowenberg, J. 1873. Alexander von Humboldt: youth and early manhood, travels in America and Asia. Vol. 1, pp. 3–390 in Bruhns 1873.

Lowenthal, D. 2000. *George Perkins Marsh: prophet of conservation.* Seattle: U. of Washington P.

Lucas, C.E. 1980. On the environmental unity in marine research. Pp. 496–508 in Sears & Merriman 1980.

Lurie, E. 1960. *Louis Agassiz: a life in science.* Chicago: U. Chicago P.

———. 1999. Alexander Agassiz (1835–1910), marine biologist, oceanographer, and industrial engineer. *ANB* 1: 172–173.

Lyell, C. 1830–33. *Principles of geology.* 3 vols. L: John Murray.

Lytle, M.H. 2007. *The gentle subversive: new narratives in American History.* NY: Oxford U.P.

Maarel, E. van der. 1975. The Braun-Blanquet approach in perspective. *Vegetatio* 30: 213–219.

Mabberley, D.J. 1985. *Jupiter Botanicus: Robert Brown of the British Museum.* Braunschweig Ger.: J. Cramer.

———. 1999. *Ferdinand Bauer: the nature of discovery.* L: Merrell Holberton/Natural History Museum.

———. 2004a. Ferdinand Lukas Bauer (1760–1826). *DNCB* 1: 144–148.

———. 2004b. Robert Brown (1773–1858). *DNCB* 1: 309–315.

MacDougall, D. 2019. *Endless novelties of extraordinary interest: the voyage of* H.M.S. Challenger *and the birth of oceanography.* New Haven: Yale U.P.

Macfarlane, D. 2014. *Negotiating a river: Canada, the U.S. and the creation of the St. Lawrence Seaway.* Vancouver: British Columbia U.P.

Maclean, A. 1972. *Captain Cook.* Garden City NY: Doubleday.

MacLeod, R. 1976. John Tyndall (1820–93), natural philosophy, microbiology, popularization of science. *DSB* 1 3: 521–524.

———., ed. 2001. Nature and empire. *Osiris* 16: 1–323.

MacMillan, M.C. 1892. *The Metaspermae of the Minnesota Valley: a list of the higher seed-producing plants indigenous to the drainage of the Minnesota River.* Reports of the Geological and Natural History Survey of Minnesota. Botanical series, no. 1. 826 pp.

———. 1897. Observations on the distribution of plants along shore oat Lake of the Woods. *Minnesota Botanical Studies* 1: 949–1023.

MacPhail, I. 1983. *Sterling Morton Library Bibliographies in Botany and Horticulture. II. Thomas Nuttall.* Lisle IL: Morton Arboretum.

Madrell, A. 2009. Marion Isabel Newbigin (1869–1968!). *Geographers: Bibliographical Studies* 28: 119–130.

Madsen, A. 1986. *Cousteau: an unauthorized biography.* NY: Beaufort Books.

Madsen, B. 2009. *Jacques Cousteau: the sea king.* NY: Pantheon Books.

Magiels, G. 2010. *From sunlight to Insight: Jan Ingen Housz, the discovery of photosynthesis and science in the light of ecology.* Brussels: VUB P.

Magnin-Gonze, J. 2004. *Histoire de la botanique.* Paris: Delachaux & Niestlé.

Magnuson, J.J. 1990. Long-term ecological research and the invisible present. *BioScience* 40: 495–501.

———. 1992. A sense of beginning: the Laurentian Great Lakes in the process of rehabilitation. *Restoration & Management Notes* 10: 29–32.

——— & T.P. Quinn. 2005. Arthur D. Hasler: he showed us the way. *Environmental Biology of Fishes* 74: 667–677.

———, H.A. Regier et al. 1980. To Rehabilitate and restore Great Lakes ecosystems. Pp. 95–112 in Cairns 1980.

Maienschein, J. 1985a. Agassiz, Hyatt, Whitman, and the birth of the Marine Biological Laboratory. *Biological Bulletin* 168, supplement: 26–34.

———. 1958b. First impressions: American biologists at Naples. *Biological Bulletin* 168, supplement: 187–191.

———. 1958c. Early struggles at the Marine Biological Laboratory over mission and money. *Biological Bulletin* 168, supplement: 192–196.

———. 1988. History of American marine laboratories: why do research at the seashore? *American Zoologist* 28: 15–25.

———. 1989. *100 years exploring life, 1888–1988: the Marine Biological Laboratory at Woods Hole*. Boston: Jones & Bartlett.

Major, J. 1962. Autecological evaluation of Russian plants. *Ecology* 43: 177–179.

Malkus, A. 1960. *Blue-water boundary: epic highway of the Great Lakes and the Saint Lawrence*. NY: Hastings House.

Mallis, A. 1971. *American entomologists*. New Brunswick NJ: Rutgers U.P.

Mancall, P.C. 2019. *The early explorers*. Oxford Bibliographies. Online. Ed. 1, 2017.

Mann, C. 1991. Lynn Margulis: science's unruly Earth mother. *Science* 252: 378–381.

Marchal, P. 1897. L'equilibre numérique des espèces et ses relations avec les parasites chez les insects. *Société Biologique Comptes Rendus* 49: 129–130.

Margulis, L. 1970. *Origin of eukaryotic cells*. New Haven: Yale U.P.

———. 1981. *Symbiosis in cell evolution: life and its environment on the early Earth*. San Francisco: W.H. Freeman.

———. 1998. *The symbiotic planet: a new look at evolution*. L: Weidenfeld & Nicolson. 1999. NY: Phoenix.

——— & R. Fester, eds. 1991. *Symbiosis as a Source of evolutionary innovation: speciation and morphogenesis*. Cambridge MA: MIT P.

——— & J. Lovelock. 1974. Biological modulation of the Earth's atmosphere. *Icarus* 21: 471–489.

Markle, D.F. 1997. Audubon's hoax: Ohio River's fishes described by Rafinesque. *Archives of Natural Hist.* 24: 439–447.

Marsigli (Marsili), L.F. 1725. *Histoire physique de la mer.* Leclerc, transl. Amsterdam.

Marsigli, L.F. 1681. *Osservazioni intorno al Bosforo yracio overo Canale di Constantinopoli.* Rome: Angelo Tinassi.

———. 1978. Internal observations on the Thracian Bosporus, or true channel of Constantinople. E. Hudson, transl. Pp. 33–44 in Deacon 1978.

———. 1999. *Histoire physique de la mer.* G. Dragoni, ed. A. McConnell, commentary. Bologna: Museo di Fisica dell'Universita di Bologna. Ed. 1, Amsterdam, 1725.

Martins, A.M.F. 1997. Afonso Chaves and the marine sciences in the Azores. Pp. 135–148 in Salanha & Ré 1997.

Martin, E.T. 1952. *Thomas Jefferson: scientist.* NY: Henry Schuman.

Maslow, J. 1996. *Footsteps in the jungle: adventures in the scientific exploration of the American tropics*. Chicago: Ivan R. Dee.

Matagne, P. 1998. The taxonomy and nomenclature of plant groups. Vol. 2, 427–437 in Acot 1998.

———. 1999. *Aux origins de l'écologie: les naturalists en France de 1800 à 1914*. Paris: Comité des travaux historiques et scientifiques—histoire des sciences et des techniques.

———. 2009. *La naissance de l'écologie*. Paris: Ellipse.

Matsen, B. 2003. *The incredible record—setting deep-sea dive of the Bathysphere*. Berkeley Heights NJ: Evslow.

———. 2005. *Descent: the heroic discovery of the abyss*. NY: Pantheon.

———. 2009. *Jacques Cousteau: the sea king*. NY: Pantheon.

Mattioli, S. & D.P. Domning. 2006. An annotated list of extant skeleton material of Steller's sea cow (*Hydradahalis gigas*) (Sirena: Dugongidamalis) from the Commander Islands. *Aquatic Mammals* 33: 273–288.

Maycock, P.F. 1967. Joseph Paczosk: founder of the science of phytosociology. *Ecology* 48: 1031–1034.

Mayer [Mayor], A.G. 1903a. The Tortugas, Florida, as a station for research in biology. *Science* 17: 190–192.

———. 1903b. A tropical marine laboratory for research? *Science* 17: 655–660.

———. 1911. Alpheus Hyatt, 1838–1902. *Popular Science Monthly* 78: 126–146.

———. 1918. The growth rate of Samoan coral reefs. *National Academy of Sciences Proc.* 4: 390–393.

———. 1924. Structure and ecology of Samoan reefs. *Tortugas Laboratory Papers* 19: 1–25.

Mayer, J.R. 1845. *Die organische Bewegung in ihrem zur zussamenhag mit den Stoffwechsee: ein Beitrag zur Naturkunde.* Heilbronn: C. Drechsler'schen Buchhandlung.

Mayr, E. 1975. Carl Gottfried Semper (1832–93), zoology, ecology. *DSB* 12: 299.

———. 1982. *The growth of biological thought: diversity, evolution, and inheritance*. Cambridge MA: Harvard U.P.

———. 1986. Joseph Gottlieb Kolreuter's contributions to biology. *Osiris* 2: 135–176.

Mazzolini, R.G., ed. 1993. *Non-verbal communication in science prior to 1900*. Florence: Leo S. Olschki.

McAtee, W.L. 1947. The cats-to-clover chain. *Scientific Monthly* 65: 241–242.

McCalman, I. 2009. *Darwin's armada: four voyagers and the battle for the theory of evolution*. NY: Norton.

———. May 2014. The great coral grief. *Scientific American* 310, no. 5: 66–69.

McClellan, J.E., III. 1985. *Science reorganized: scientific societies in the Eighteenth Century*. NY: Columbia U.P.

———. 2000a. Académie des Sciences. Pp. 1–2 in Hessenbruch 2000.

———. 2000b. Colonialism and science. Pp. 143–144 in Hessenbruch 2000.

McCormmach, R. 1971. Henry Cavendish (1731–1810), natural philosophy. *DSB* 3: 155–159.

———. 2004. *Speculative truth: Henry Cavendish, natural philosophy, and the rise of modern theoretical science*. NY: Oxford U.P.

McDiarmid, J.B. 1976. Theophrastus (c. 371-c. 287 BC), botany, mineralogy, philosophy. *DSB* 13: 328–334.

McFadden, R.D. 26 February 2015. Eugenie Clark, scholar of the life aquatic, dies at 92. *New York Times*.

McGhie, H.A. 2017. *Henry Dresser and Victorian ornithology: birds, books and business*. Manchester UK: Manchester U.P.

McGill, T.E., ed. 1965. *Readings in animal behavior.* NY: Holt, Rinehart & Winston.

———. 1972. *Alfred Wallace and natural selection*. New Haven: Yale U.P.

———. 1976. Alfred Russel Wallace (1823–1913), natural history. *DSB* 14: 133–140.

McGovern, A. 1978. *Shark lady: true adventures of Eugenie Clark*. NY: Scholastic.

———. 1998. *Adventures of the shark lady: Eugenie Clark around the world.* NY: Scholastic.

———& E. Clark. 1991. *The desert beneath the sea.* NY: Scholastic.

McIntosh, R.P. 1975. H.A. Gleason, "individualist ecologist," 1882–1975. *Torrey Botanical Club Bull.* 102: 253–275.

———. 1976. Ecology since 1900. Pp. 353–372 in Taylor & White 1976. Reprinted in Egerton 1977a.

———. 1983. Excerpts from the work of L.G. Ramensky. *ESAB* 64: 7–12.

———. 1985. *The background of ecology: concept and theory.* NY: Cambridge U.P.

McKenna, H.L. 1970. Henry Walter Bates (1825–92), natural history. *DSB* 1: 500–504.

McKibben, W.E. 8 December 1988. Is the world getting hotter? *NY Review of Books*, online.

———. 1989. *The end of nature.* NY: Anchor.

———, ed. 2011. *The global warming reader: a century of writing about climate change.* NY: Penguin Books.

———. 22 Nov. 2018a. A very grim forecast. *NY Review of Books* 65 no. 18: 4–8.

———. 26 Nov. 2018b. Life on a shrinking planet. *New Yorker* 46–55.

McKinney, H.L. 1970. Edward Blyth (1810–73), natural history. *DSB* 2: 205–207.

McKee, R. 1966. *Great Lakes country.* NY: Thomas Y. Crowell.

McKelvey, S.D. 1955. *Botanical exploration of the Trans-Mississippi West, 1790–1850.* Jamaica Plains MA: Arnold Arboretum. 1991. Corvallis: Oregon State U.P.

McKie, D. 1962. *Antoine Lavoisier: scientist, economist, social reformer.* NY: Collier Books. Ed. 1, 1952.

McKinney, H.L. 1970. Henry Walter Bates (1825–92), natural history. *DSB* 1: 500–504.

———. 1972. *Alfred Russell Wallace and natural selection.* New Haven: Yale U.P.

———. 1974. Fritz (Johann Friedrich Theodor) Müller (1822–97), natural history. *DSB* 9: 559–561.

———. 1976. Alfred Russel Wallace (1823–1913), natural history. *DSB* 14: 133–140.

McLoone, M. 2000. *Women explorers of the oceans: Ann Davidson, Eugenie Clark, Sylvia Earle, Naomi James, Tania Albi.* Mankato MN: Capstone Books.

McShea, W.J. & W.M. Healy, eds. 2002. *Oak forest ecosystems: ecology and management for wildlife.* Baltimore: Johns Hopkins U.P.

Mearns, B. & R. Mearns. 1988. *Biographies for birdwatchers: the lives of those commemorated in western Palearctic bird names.* NY & L: Academic P.

———. 1992. *Audubon to Xantus: the lives of those commemorated in North American bird names.* NY & L: Academic P.

———. 1998. *The bird collectors*, NY & L: Academic P.

Meine, C. 1988. *Aldo Leopold: his life and work.* Madison: U. of Wisconsin P.

Meinig, D.W. 1986–2004. *The shaping of America: a geographical perspective on 500 years of history.* 4 vols. New Haven: Yale U.P.

Meisel, M. 1967. *A bibliography of American natural history: the pioneer century, 1769–1865.* 3 vols. NY: Premier.

Meisen, V., ed. 1932. *Prominent Danish scientists through the ages.* Copenhagen: Levin & Munksgaard.

Mengel, R.M. May 1967. How good are Audubon's bird pictures in light of modern ornithology? *Scientific American* 216, no. 5: 155–159.

———. 1970. John James Audubon (1785–1851), ornithology. *DSB* 1: 329–332.

Merriam, C.H. 1873. Report on the mammals and birds of the expedition. Pp. 661–715 in *U.S. Geological survey of the territories, Sixth Annual Report.* Washington: Government Printing Office. Reprinted in Merriman 1974.

———. 1890. *Results of a biological survey of the San Francisco Mountain region and Desert of the Little Colorado. Arizona. North American Fauna, no. 3.* Washington: Government Printing Office. Reprinted in Merriman 1974.

———. 1894. Report of the ornithologist and mammologist. Pp. 227–232 + map, in *Report of the Secretary of Agriculture for 1893.* Washington: Government Printing Office.

———. 1974. *Selected works.* K.B. Sterling, compiler. NY: Arno P.

Merriman, D. 1972a. Johan Hjort (1869–1948), marine biology. *DSB* 6: 441–442.

———. 1972b. Challengers of Neptune: the "Philosophers." *Roy. Soc. of Edinburgh Proc.* B 72: 15–454.

Metcalf, C.R. 1972. Nehemiah Grew (1641–1712), plant morphology, plant anatomy. *DSB* 5: 534–536.

Meyen, F.J.F. 1846. *Outlines of the geography of plants, with particular inquiries concerning the native country, the culture, and the uses of the principle cultivated plants on which the prosperity of nations is based.* M. Johnston, transl. L: Ray Soc. 1977. NY: Arno P.

Meyer, G.F.W. 1825. *Die entwickelung metamorphose und fortpflanzung der flechten in anwendung auf ihre systematische anordung und zur nachweisung des allgemeinen ganges der formbildung in den unteren ordnungen cryptogamische gewächse.* Göttingen: Vanderhoeck und Ruprecht.

Meyer, H.A. & K. Möbius. 1865–72. *Fauna der Kieler Bucht.* 2 vols. Leipzig: Engelmann.

Micheli, P. 1729. *Nova plantarum genera iuxta Tourneforti methodum disposita.* Florence: Paperini.

Middleton, W.E. 1964. *The history of the barometer.* Baltimore MD: Johns Hopkins U.P.

———. 1966. *A history of the thermometer and its use in meteorology.* Baltimore: Johns Hopkins U.P.

———. 1969. *Invention of the Meteorological Instruments.* Baltimore: Johns Hopkins U.P.

Miller, M.A. 1976. *Kropotkin.* Chicago: U. of Chicago P.

Miller-Primack, A.J. & R.B. Primack. 2008. Global warming and flowering times in Thoreau's Concord: a commentary. *Ecology* 89: 332–341.

Mills, E.L. 1958. A century of biological research: from 1858–1958. *Illinois Natural History Survey Bull.* 27: 84–103.

———. 1964. Stephen Alfred Forbes. *Systematic Zoology* 13: 208–214.

———, ed. 1975a. *One hundred years of oceanography: essays commemorating the visit of H.M.S. Challenger to Halifax, May 9–14, 1873.* Halifax: Dalhousie U.

———. 1975b. The *Challenger* expedition: how it started and what it did. Pp. 1–23 in Mills 1975a.

———. 1978. Edward Forbes, John Gwyn Jeffreys, and British dredging before the *Challenger* Expedition. *Soc. for Bibliography of Natural Hist. J.* 8: 507–536.

———. 1980a. Alexander Agassiz, Carl Chun and the problem of the intermediate fauna. Pp. 360–372 in Sears & Merriman 1980.

———. 1980b. One 'different kind of gentleman'; Alfred Merle Nariman (1831–1918), invertebrate zoologist. *Linnean Soc. of London Zoology J.* 68: 69–98.

———. 1982. Saint-Simon and the oceanographers: patterns of change in the study of plankton dynamics, 1897–1946. Pp. 3–18 in Riley 1982.

———. 1983. Problems of deep-sea biology: an historical perspective. Pp. 1–79 in G.T. Rowe, ed. *The sea*. Vol. 8: *Deep sea biology*. NY: John Wiley & Sons.

———. 1984. A view of Edward Forbes. naturalist. *Archives of Natural History* 11: 365–393.

———. 1989. *Biological oceanography: an early history, 1870–1960*. Ithaca: Cornell U.P.

———. 2000. Oceanography. Pp. 531–532 in Hessenbruch 2000a.

———. 2001. Enlightened natural history or the beginnings of oceanic science? *Annals of Science* 58: 403–408.

———. 2014. Canadian marine science from before *Titanic* to the establishment of the Bedford Institute of Oceanography in 1962. Pp. 13–21 in Nettleship et al. 2014.

——— & J. Carpine-Lancre. 1992. *Ocean frontiers: explorations on five continents*. E.M. Borgese, ed. NY: Harry N. Abrams.

Miloy, J. 1983. *Creating the College of the Sea: the origin of the Sea Grant Program*. College Station: Texas A & M U. Sea Grant Program.

Minns, C.K. et al. 1987. Productive biomass size-spectra in the Bay of Quinte, a eutrophic ecosystem. *Canadian J. of Fisheries & Aquatic Sciences* 44: 148–155.

Mitchell, A. 1990. *The fragile South Pacific: an ecological odyssey*. Austin: U. of Texas P.

Mitchell, M.E. 2002. "Such a strange theory": Anglophone attitudes to the discovery that lichens are composite organisms, 1871–1890. *Huntia* 11: 193–207.

———. 2005. The inside story: a commentary on the study of lichen structure in the 18th and 19th centuries. *Huntia* 12: 13–29.

———. 2006. "Function is smother'd in surmise": a survey of observations on the role of lichen conidia, 1850–2000. *Huntia* 12: 149–167.

———. 2007. Signposts to symbiosis: a review of early attempts to establish the constitution of lichens. *Huntia* 13: 101–120.

———. 2011. A somewhat esoteric pursuit: the course of lichenological inquiry from 1700 to 1950. *Huntia* 14: 15–145.

———. 2015. De Bary's legacy: the emergence of different perspectives on lichen symbiosis. *Huntia* 15: 5–33.

———. 2018. Idëe fixe: a commentary on the opposition in France to the theory of lichen duality, 1870–1900. *Huntia* 16: 165–182.

Mitchell, R., R.A. Mayer & J. Downhower. 1976. An evaluation of Three biome programs. *Science* 192: 859–865.

Mitchison, R. 2004. James Anderson (1739–1808), agriculturist and political economist. *ODNB* 2: 40–42.

Mitman, G. 1988. From the population to society: the cooperative metaphors of W.C. Allee and A.E. Emerson. *J. of the History of Biology* 21: 173–194.

———. 1992. *The state of nature: ecology, community, and American social thought*. Chicago: U. of Chicago P.

———. 1997. (Ernest) Victor Shelford (1877–1968), animal and community ecologist. *BDAC*, 728–729.

———. 1999. Warder Clyde Allee (1885–1955), biologist. *ANB* 1: 298–299.

Mitsch, W.T., ed. 1994. Energy flow in a pulsing system. *Ecological Engineering* 3: 77–105.

——— & J.W. Day, Jr. 2004. Thinking big with whole-ecosystem studies and ecosystem restoration—a legacy of H.T. Odum. *Ecological Modelling* 178: 133–155.

Miyanishi, K. 2014. ESA's third decade: the war years. *ESA Bull.* 95: 103–105.

Möbius, K.A. 1877. *Die Auster und die Austernwirthschaft*. Berlin: Wiegandt, Hemple & Parey.

———. 1880. The oyster and oyster culture. *The U.S. Fish Comm. Report*, 683–751.

Monk, C.D., comp. 1994. Pierre Dansereau. Pp. 161–171 in Burgess 1994.

Monnier, A.M. 1975. Paul Portier (1866–1962), biology, physiology. *DSB* 11: 101–102.

Moore, B.L. 1938. The beginnings of ecology. *Ecology* 19: 592.

———. 2009. A. Starker Leopold (1913–83), wildlife biologist, ecologist, educator, and author. Pp. 250–253 in Cevasco & Harmond 2009.

Moore, J.P. 1947. William Maclure—scientist and humanitarian. *American Philos. Soc. Proc.* 91, no. 3.

Moore, L.B. 1967. The Cockayne Memorial Lecture, 1965: Leonard Cockyne, botanist. *Royal Soc. Of New Zealand Trans.* 2: 1–18.

Moorehead, A. 1966. *The fatal impact: an account of the Invasion of the South Pacific, 1767–1840*. NY: Harper & Row.

Morcos, S.A.M., X.A. Fraga & G. Parrilla-Barrera. 2013. The dawn of Spanish oceanography: Odón de Buen and his letters to Henri de Lacaze-Duthiers (Banyuls-sur-Mer), Prince Albert I and Jules Richard (Monaco), 1895–1945. Pp. 267–298 in Groeben 2013.

———, J. Smed &A. Svansson. 2004. The eel and Mediterranean oceanography: a Danish contribution to ocean science. Pp. 197–207 in Morcos et al. 2004.

———, Zhu and four others, eds. 2004. *Ocean sciences bridging the millennia: a spectrum of historical accounts*. Paris: UNESCO.

Morge, G. 1973. Entomology in the Western World in Antiquity and in Medieval Times. Pp. 37–80 in Smith, Mittler & Smith 1973.

Moring, J. 2002. *Early American naturalists: exploring the American West, 1804–1900*. NY: Cooper Square P.

Morris, C. 2014. *Milestones in ecology*. Princeton: Princeton U.P.

Morse, E.S. 1912. Charles Otis Whitman, 1842–1910. *National Academy of Sciences Biog. Mem.* 7: 269–288 + portrait.

Mortimer, C.H. 1956. An explorer of lakes. Pp. 165–211 in Sellery 1956.

———. 1981. *The Lake Michigan pollution case: a review and commentary on the limnology and other issues*. Madison: U. of Wisconsin-Madison Sea Grant Institute/ Center for Great Lakes Studies, U.W.-Madison & U.W.-Milwaukee.

Morton, A.G. 1981. *History of botanical science: an account of the development of botany from ancient times to the present day*. L: Academic P.

Moseley, H.N. 1892. *Notes of a naturalist on H.M.S.* Challenger. Ed. 2. L: John Murray. Ed. 1, 1879.

Moyal, A.M. 1976. *Scientists in Nineteenth-Century Australia: a documentary history*. Melbourne: Cassell Australia.

Muir, J. 1915. *Travels in Alaska*. Boston: Houghton Mifflin.

———. 1938. *John of the mountains: the unpublished journals of John Muir*. L.M. Wolfe, ed. Boston: Houghton Mifflin.

———. 1954. *The wilderness world*. E.W. Teale, ed. Boston: Houghton Mifflin.

Mullen, P.C. 1973. Charles Atwood Kofoid (1865–1947), zoology. *DSB* 7: 447.

Müller, D. 1976. Johannes Eugenius Bülow Warming (1841–1924), botany. *DSB* 14: 181–182.

Müller, G.H. 1990. Friedrich Alfred Krupp (1854–1902), promoter of natural sciences, and his deep-sea researches on board the *Maja* and *Puritan* in the Gulf of Naples. Pp. 235–240 in Lenz & Deacon 1990.

Müller, H. 1883. *The fertilization of flowers.* L: Macmillan. German edition, 1873. Reprint. 1977. NY: Arno P.

Müller, S. 1842. Geographie der Thiere: Ueber die geographische Vebreitung der Saugethiere im Indischen Archipelagus. *Annalen der Erd-, Völker-, und Steatenkunde* 1: 251–333.

Muller-Schwarze, D. 2004. Trailblazer for ecology. Pp. 453–469 in Erickson et al. 2004.

Mullett, C.F. 1968. A village Aristotle and harmony of interests: James Anderson (1739–1808) of Monks Hill. *J. of British Studies* 8: 94–118.

Munson, R. 1989. *Cousteau: the captain and his world.* NY: William Morrow.

Murphy, P.W., ed. 1962. *Progress in soil zoology: papers from a colloquium on research methods.* L: Butterworths.

Murphy, R.C. 1926. Oceanic and climatic phenomena along the west coast of South America during 1925. *Geographic Review* 16: 26–54.

———. 1936. *Oceanic birds of South America.* 2 vols. NY: Macmillan/American Museum of Natural History.

Murphy, S.L. 1968. *Wild sanctuaries: our National Wildlife Refuges—a heritage restored.* NY: Dutton.

Murray, A.D. 1866. *The geographical distribution of mammals.* L: Day & Son.

Murray, J. 1880a. On the structure and origin of coral reefs and islands. *Nature* 25: 351–355.

———. 1880b. On the structure and origin of coral reefs and islands. *Royal Soc. of Edinburgh Proc.* 10: 505–518.

———. 1889. On the structure and origin of coral reefs and islands. *Nature* 39: 873–887.

———. 1895. *Report on the scientific results of the voyage of H.M.S.* Challenger *during the years 1872–76.* L: Her Majesty's Stationary Office. 1977. NY: Arno P.

———. 1911. Alexander Agassiz: his life and scientific work. *Science* 33: 873–887.

——— & J. Hjort. 1912. *The depths of the ocean: a general account of the modern science of oceanography based largely on the scientific researches of the Norwegian steamer Michael Sars.* L: Macmillan.

——— & L. Pular, eds. 1910. *Bathymetrical survey of the Scottish fresh-water lochs, during the years 1897 to 1909.* 6 vols. Edinburgh: Challenger Office.

Muscatine, L. & J.W. Porter. 1977. Reef corals: mutualistic symbiosis adapted to nutrient-poor environments. *BioScience* 27: 454–460.

Myers, N., ed. 1984. *Gaia: an atlas of planet management.* NY: Doubleday.

Myers, R.A. & B. Worm. 2003. Rapid worldwide depletion of predator fish communities. *Nature* 4223: 280–283.

Myerson, J., ed. 1995. *The Cambridge companion to Henry David Thoreau.* Cambridge UK: Cambridge U.P.

Nalepa, T.F. & D.W. Schloesser, eds. 2014. *Quagga and Zebra Mussels: biology, impacts, and control.* Boca Raton: CRC P./Taylor & Francis.

Nash, L.K. 1957. Plants and the atmosphere. Vol. 2, pp. 323–436 in J.B. Conant & L.K. Nash, eds. *Harvard case histories in experimental science.* Cambridge MA: Harvard U.P.

Naumann, E. 1919. Några synpunker angående limnoplanktons ökologie med särskild sn till phytoplankton. *Svensk Botanisk Tidshrift* 13: 129–163.

———. 1929. The scope and chief problems of regional limnology. *Internationale Revue der gesamten Hydrobiologie* 22: 423–444.

Naumov, G.V. 1973. Patr Alekseevich Kropotkin (1842–1921), geography. *DSB* 7: 510–512.

Neave, S.A. & F.J. Griffin. 1933. *The history of the Entomological Society of London, 1833–1933.* L: Entomological Society of London.

Needham, J.G. 1956. *Science and civilization in China*, vol. 2: *History of scientific thought.* Cambridge UK: Cambridge U.P.

——— & J.T. Lloyd. 1916. *The life of inland waters.* Ithaca NY: Comstock P.

Needham, J.T. 1749. A summary of late observations upon the generation, composition and decomposition of animal and vegetable substances. *Roy. Soc. of London Philos. Trans.* 45: 615–666. Selection in Brock 1961: 11–12.

———. 1750. *Novelles observations microscopiques, avec des découvertes intéressantes sur la composition et la decomposition des corps organizes.* Paris: Louis-Étienne Ganeau. 324 pp.

Needler, A.W.H. 1958. Fisheries Research Board of Canada Biological Station, Nanaimo, B.C., 1908–1958. *Fisheries Research Board of Canada J.* 15: 759–777.

Nelson, C.M. 1975. Karl Ludwig Rütimeyer (1825–95), vertebrate paleontology, geography. *DSB* 12: 37–39.

Nelson, E.C. & D.J. Elliott, eds. 2015. *The curious Mister Catesby: a "truly ingenious" naturalist explores new worlds.* Athens: U. of Georgia P.

Nelson, G. 1978. From Candolle to Croizat: comments on the history of biogeography. *J. of the Hist. of Biology* 11: 269–305.

———. 1983. Vicariance and cladistics: historical perspectives with implications for the future. Pp. 469–492 in Sims et al. 1983.

Nettleship, D.N. et. al., eds. 2014. *Fifty years of marine research at Canada's Bedford Institute of Oceanography, 1962–2012.* Dartmouth, Nova Scotia: Bedford Institute of Oceanography.

Neumann, C.E. 2009. E. Lucy Braun (1889–1971), pioneer ecologist of the 1900s. *Modern American Environmentalists* 1: 48–51.

Newton, H.A. 1895. Biographical memoir of Elias Loomis. *Natiolnal Acad. of Sciences Biog. Mem.* 3: 213–252. Online.

Nicholls, S. 2009. *Paradise found: nature in America at the time of discovery.* Chicago: U. of Chicago P.

Nichols, R.L. & P.L. Halley. *Stephen Long and American frontier exploration.* Newark DE: U. of Newark P.

Nicolson, M. 1984. *The development of plant ecology, 1790–1960.* Edinburgh: Ph.D. dissertation, U. of Edinburgh.

———. 1987. Alexander von Humboldt, Humboldtian science, and the origins of the study of vegetation. *History of Science* 25: 167–194.

———. 1989. National styles, divergent classifications: a comparative case study from the history of French and American plant ecology. *Knowledge and Society: Studies in the Sociology of Science Past and Present* 8: 139–186.

———. 1990. Henry Allan Gleason and the individualistic hypothesis: the structure of a botanist's career. *Botanical Review* 56: 91–161.

———. 1996. Humboldtian plant geography after Humboldt: the link to ecology. *British J. for History of Science* 29: 289–310.

———. 2001. "Towards establishing ecology as a science instead of an art": the work of John T. Curtis on the plant community continuum. *Web Ecology* 2: 1–6.

———. 2013. Community concepts in plant ecology: from Humboldtian plant geography to the superorganism and beyond. *Web Ecology* 13: 95–102.

——— & R.P. McIntosh. 2002. H.A. Gleason and the individualistic hypothesis revisited. *ESAB* 83: 133–142.

Nikolaus (Nicholas) von Cusa. *Complete philosophical and theological treatises.* 2 vols. J. Hopkins, ed. Minneapolis: A.J. Banning P.

Noland, L.E. 1970. Edward Asahel Birge (1851–1950), limnology. *DSB* 2: 141–142.

———. 1973. Chancey Juday (1871–1944), limnology. *DSB* 7: 184.

Norland, C.E. 1974. Alpheus Spring Packard, Jr. (1839–1905), entomology. *DSB* 10; 272–274.

Northcote, T.G. & P.A. Larkin. 1963. Western Canada. Pp. 451–485 in Frey 1963a.

Novlyanskaya, M.G. 1970. *Danill Gotlib Messerschmidt I ego rabony po issledoviyu Sibiri.* Leningrad: Nauka.

Nuccitelli, D. 2015. *Climatology versus pseudoscience: exposing the failed predictions of global warming skeptics.* Santa Barbara: ABC-Clio/Praeger.

Nute, G.L. 1926. The American Fur Company's fishing enterprises on Lake Superior. *Mississippi Valley Historical Review* 12: 483–503.

Nuttall, T. 1818. *The genera of North American plants, and a catalogue of species to the year 1817.* 2 vols. Philadelphia: Author. 1971. NY: Hafner.

———. 1821. *A Journal of travels into the Arkansas Territory, during the year 1819, with occasional observations on the manners of the aborigines.* Philadelphia: Palmer.

———. 1832–34. *A manual of the ornithology of the United States and Canada.* Vol. 1: Land birds. Vol. 2: Water birds. Cambridge MA: Hillard & Brown.

———. 1951. *Travels in the old Northwest: an unpublished 1810 diary.* J.E. Graustein, ed. Waltham MA: Chronica Botanica.

Nyhart, L.K. 2009. *Modern nature: the rise of the biological perspective in Germany.* Chicago: U. Chicago P.

Odum, E.P. 1959. *Fundamentals of ecology.* Philadelphia: W.B. Saunders. Ed. 2 (with H. Odum), 1959. Ed. 3, 1971.

———. 1987. Early University of Georgia research, 1952–1962. Pp. 43–83 in *The Savannah River and its environs: Proceedings of a symposium.* Aiken SC: Savannah River Laboratory. Cited from Craig 2001: 214.

———. 1989. *Ecology and our endangered life-support system.* Sunderland MA: Sinauer Associates.

Odum, H.T. 1951a. Notes on the strontium content of sea water, celesite Radiolaria, and strontianite snail shells. *Science* 114: 211–213.

———. 1951b. The stability of the world strontium cycle. *Science* 114: 407–411.

———. 1956. Primary production in flowing waters. *Limnology & Oceanography* 1: 102–117.

———. 1957a. Strontium in natural waters. *Institute of Marine Sci. Publ., U. of Texas* 4, no. 2: 22–37.

———. 1957b. Biogeochemical deposition of strontium. *Institute of Marine Sci. Publ., U. of Texas* 4, no. 2: 38–114.

———. 1983. *Systems ecology: an introduction.* NY: Wiley-Interscience/John Wiley & Sons.

———. 2007. *Environment, power and society, for the Twenty-first Century: the hierarchy of energy.* NY: Columbia U.P.

——— & E.P. Odum. 1955. Trophic structure and productivity of a windward coral reef community on Eniwetok Atoll. *Ecological Monographs* 25: 291–320.

Oehser, P.H. 1949. *Sons of science: the story of the Smithsonian Institution and its leaders.* NY: Henry Schuman.

Ohle, W. 1961. August Thienemann—sein Werk und Vermächtnis. *Archive für Hydrobiologie* 57: 1–12.

Olby, R. 1973. Joseph Gottlieb Kolreuter (1733–1806), botany. *DSB* 7: 440–442.

Oldroyd, D. 2000. Charles Lyell, 1797–1875, British geologist. Pp. 423–425 in Hessenbruch 2000a.

Oliver, F.W. 1891. The effects of urban fog upon cultivated plants. *J. Royal Horticultural Soc.* 13: 139–151.

Oliver, S.P. 1909. *The life of Philibert Commerson, D.M., naturalist du Roi.* G.F.S. Elliot, ed. L: John Murray.

Olson, E.C. 1991. George Gaylord Nelson, June 16, 1902—October 6, 1984. *National Academy of Sciences Biographical Memoirs* 60: 331–353.

Ophuijsen, J.M. van & M. von Realte, eds. 1998. *Theophrastus: reappraising the sources.* New Brunswick: Transaction Publishers.

Oppenheimer, J. 1978. Felix Anton Dohrn (1840–1909), zoology. *DSB* 15: 122–125.

———. 1980. Some historical backgrounds for the establishment of the Stazione Zoologica. Pp. 179–187 in Sears & Merriman 1980.

Oosting, H.J. 1948. *The study of plant communities: an introduction to plant ecology.* San Francisco: W.H. Freeman.

Oreskes, N. 2004. The scientific consensus on climate change. *Science* 306: 1686.

Orlob, G.B. 1964. The concepts of etiology in the history of plant pathology. *Pflanenschutz Nachrichten "Bayer"* 17: 185–268.

———. 1973. Ancient and medieval plant pathology. *Pflangenschutz Nachrichten "Bayer"* 26: 63–294.

———. History of plant pathology in the Middle Ages. *Ann. Rev. of Phytopathology* 9: 7–20.

Ornstein, M. 1938. *The role of scientific societies in the Seventeenth Century.* Chicago: U. of Chicago P.

Osgood, W.H. & H.W. Grinnell. 1947. Biographical Memoir of Clinton Hart Merriman, 1855–1942. *National Academy of Sciences Biographical Memoirs* 24: 1–26.

Overfield, R.A. 1993. *Science with practice: Charles E. Bessey and the maturing of American botany.* Ames: Iowa State U.P.

Oviedo, G.F. 1535. *La historia natural y general de las Indias.* Seville. Selections in von Hagen 1948: 17–30, transl. R. Eden.

Owen, G.E.L. et al. 1970. Aristotle (384–322 BC), science and philosophy. *DSB* 1: 250–281.

Packard, A.S. 1888. The cave fauna of North America, with remarks on the anatomy of the brain and origin of the blind species. *National Academy of Sciences Memoirs* 4, part 1: 3–156 + 27 plates.

Pagel, W. 1972. Johannes (Joan) Baptista von Helmont (1579–1644), chemistry, natural philosophy, medicine, mysticism. *DSB* 6: 253–259.

Pallas, P.S. 1812. *Travels through the southern provinces of the Russian Empire in the years 1783 and 1794.* Anon. transl. L: Longman & Rees. Ed. 1, 1802.

———. 1967a. *A naturalist in Russia: letters from Peter Simon Pallas to Thomas Pennant*. C. Urness, ed. Minneapolis: U. of Minnesota P.

———. 1967b. *Reise durch verschildene Provinzen der Russischen Reichs*. 3 vols. Graz: Akademische Druck-u. Verlagsanstalt.

Palmer, R.S. 1956. Resolution of respect: Dr. Charles C. Adams (1873–1955). *ESAB* 17: 103–105.

Palumbi, S.R. & C. Sotka. 2011. *The death and life of Monterey Bay: a story of revival*. Washington: Island P.

Paracer, S. & V. Ahmadjian. 2000. *Symbiosis: an introduction to biological associations*. Ed. 2. NY: Oxford U.P.

Park, O. 1949. Community organization: metabolism. Pp. 495–528 in Allee et al. 1949.

Park, T. 1949. The history of ecology. Pp. 13–72 in Allee et al. 1949.

———. 1967. Alfred Edwards Emerson, eminent ecologist—1967. *ESAB* 48: 104–107.

Parsons, J.W. 1973. History of salmon in the Great Lakes, 1850–1970. *U.S. Fish & Wildlife Service Technical Paper* 68.

Partington, J.R. 1961–64. *A history of chemistry*. Vols. 2–4. L: Macmillan.

Pascual, M. & J.A. Dunne, eds. 2006. *Ecological networks: linking structure to dynamics in food webs*. NY: Oxford U.P.

Pasteur, G. 1982. A classificatory review of mimicry systems. *Annual Rev. of Ecology & Systematics* 13: 169–199.

Patten, B.C. 1959. An introduction to the cybernetics of the ecosystem: the trophic-dynamic aspect. *Ecology* 40: 221–231.

———. 1966. Systems ecology: a course sequence in mathematical ecology. *BioScience* 16: 593–599.

———., ed. 1971–75. *Systems analysis and simulation in ecology*. 3 vols. NY: Academic P. [vols. 4–5 later].

——— & E.P. Odum. 1981. The cybernetic nature of ecosystems. *American Naturalist* 118: 886–895.

———. 1995. Dedication and remembrance—complex person, complex ecologist—father of systems ecology? Pp. xxi—xxiii + 2 pp. of photos in Patten & Jørgensen 1995.

Patten, B.C & S.E. Jørgensen, eds. 1995. *Complex ecology: the whole-part relation in ecosystems*. Englewood Cliffs NJ: Prentice Hall.

Pauly, P.J. 1988. Summer resort and scientific discipline: Woods Hole and the structure of American biology, 1882–1924. Pp. 121–150 in Rainger et al. 1988.

———. 2000. *Biologists and the promise of American life: from Meriwether Louis to Alfred Kinsey*. Princeton: Princeton U.P.

Pearsall, W.H. 1964. The development of ecology un Britain. *J. of Ecology* 52: supplement, 1–12. *J. of Animal Ecology* 23: 12–44.

Pearse, A.S. 1926. *Animal ecology*. NY: McGraw-Hill. Ed. 2, 1939.

———. 1927. The loves of plants. *Classical Philology* 22: 94–98.

———. 1944. Chancey Juday. *Science* 99: 506.

———. 1952a. *Adventure . . . trying to be a zoologist*. Durham: North Carolina.

———. 1952b. A sketch of the development of ancient botany. *Phoenix* 44: 44–51.

Peattie, D.C. 1936. *Green Laurels: the lives and achievements of the great naturalists*. NY: Simon & Schuster.

Peck, R.M. 1999. George Ord (1781–1866), naturalist, writer, and lexicographer. *ANB* 16: 755–756.

Penso, G. 1981. *La conquête du monde invisible: parasites et microles è travers les siècles*. M.M. Rampa, transl. Paris: Les Éditions Roger Dacosta.

Pequegnat, W.E. January 1958. Whales, plankton, and man. *Scientific American* 198, no 1: 84–90.

Perrein, C. 1994. Emile Gadeceau (1845–1928) et la naissance de l'écologie. Pp. 179–192 in *La Bretagne des Savants et des ingénieurs, 1825–1900*. Rennes: Editions Ouest-France.

Petersen, C.G.I. 1913. Valuation of the sea. II. The animal communities of the sea body and their importance for marine zoogeography. *Danish Biological Station Reports* 21: 1–44.

Peterson, G.D. 2005. Ecological management: control, uncertainty, and understanding. Pp. 371–395 in Cuddington et al. 2005.

Petit, G. 1965. Conrad Gesner, zoologist. *Gesnerus* 22: 195–204.

———. 1970a. Albert I of Monaco (Honoré Charles Grimaldi) (1848–1922), oceanography. *DSB* 1: 92–93.

———. 1970b. Lucien Jules Laurent (called Charles Lucien) Bonaparte (1803–57), zoology. *DSB* 2: 281–283.

——— & J. Théodoridès. 1962. *Histoire de la zoologie des origins à Linné*. Paris: Hermann.

Philbrick, N. 2003. *Sea of glory: America's voyage of discovery: the U.S. Exploring Expedition, 1838–1842*. NY: Viking.

Philip, C.B. & L.E. Rozeboom. 1973. Medico-Veterinary entomology: a generation of progress. Pp. 333–360 in Smith, Mittler & Smith 1973.

Phillips, B. 1880. In memory of the late professor James Wood Milner. *American Fisheries Soc. Trans*. 9: 4–8.

Piavis, G.W. 1961. Embryological stages in the sea lamprey and effects of temperature on development. *U.S. Dept. of Interior Fishery Bull*. 182: 111–143.

Pierce, W.D., R.R. Cushman & C.E. Hood. 1912. The insect enemies of the cotton boll weaver. *Washington: USDA Bureau of Entomology Bull*. 100: 1–99.

Pihl, M. 1970. Vilhelm Frimann Koren Bjerknes (1862–1951), physics, geophysics. *DSB* 2: 167–169.

Pike, Z.M., ed. 1895. *The expeditions of . . . to the sources of the Mississippi River, through Louisiana Territory and in New Spain, during the years 1805–6–7*. E. Coues, ed. 3 vols. NY. Ed. 1, 1810.

———. 1966. *The Journals of Jeb. Mont. Pile, with letters and related documents*. D. Jackson, ed. Norman: U. Oklahoma P.

Pilet, P.E. 1971a. Alphonse de Candolle (1806–93), botany, phytogeography. *DSB* 3: 42–43.

———. 1971b. Augustin-Pyramus de Candolle (1778–1841), botany, agronomy. *DSB* 3: 43–45.

———. 1972. Konrad Gesner (1516–65), natural sciences, medicine, philology. *DSB* 5: 378–379.

———. 1975a. Nicolas-Théodore de Saussure (1767–1845), chemistry, plant physiology. *DSB* 12: 123–124.

———. 1975b. Jean Senebier (1742–1809), physiology. *DSB* 12: 308–309.

Pimm, S.L. 1980. Bounds on food web connectance. *Nature* 284: 591.

———. 1982. *Food webs*. L: Chapman & Hall.

———. et al. 1991. Food web patterns and their consequences. *Nature* 350: 669–674.

Piveteau, J., ed. 1954. *Oeuvres philosophiques de Buffon*. Paris: Universitaries de France.

Plass, G.N. 1956a. The carbon dioxide theory on climate change. *Tellus* 8: 140–154.

———. 1956b. Infrared radiation in the atmosphere. *American J. of Physics* 24: 303–321.

———. 1956c. Effect of carbon dioxide variations on climate. *American J. of Physics* 24: 376–3878.

Playfair, J. 1802. *Illustrations of the Huttonian theory of the Earth.* Edinburgh: Cadell & Davies, London: S. William Creech. 1956. NY: Dover.

Plitt, C.C. 1934. Albert Schneider. *Bryologist* 37: 5–8.

Poesch, J. 1961. *Titian Ramsay Peale and his journals of the Wilkes Expedition.* Philadelphia: American Philos. Soc.

———. 1974. Titian Ramsay Peale (1799–1885), natural history. *DSB* 10: 439–440.

Polis, G.A., M.E. Power, & G.E. Huxel. 2004. *Food webs at the landscape level.* Chicago: U. of Chicago P.

Pool, R.J. 1915. A brief sketch of the life and work of Charles Edwin Bessey. *American J. of Botany* 2: 505–518.

———. 1954. Frederic Edward Clements. *Ecology* 35: 108–112.

Poole, A. & F. Gill, eds. 1996. *The birds of North America.* Philadelphia: Academy of Natural Sciences/Washington: American Ornithologists Union.

Popper, K.R. 1962. *Conjectures and Refutations: the growth of scientific knowledge.* NY: Basic Books.

Porep, R. 1970. Der physiologe und planktonforscher Victor Hensen (1835–1924): sein leben und sein werk. *Keiler Beiträg zur Geschichte der Medizin und Pharmazie* 9: 1–147. Also: Neumünster: Karl Wachholtz.

Porter, C.M. 1979. The concussion of revolution: publications and reform at the early Academy of Natural Sciences, Philadelphia, 1812–1842. *J. Hist. Biol.* 12: 273–292.

———. 1986. *The eagle nest: natural history and American ideas, 1812–1842.* Tuscaloosa: U. of Alabama P.

Pound, R. 1893. Symbiosis and mutualism. *American Naturalist* 27: 509–520.

———. 1954. Frederic Clements as I knew him. *Ecology* 35: 112–113.

———. & F.E. Clements. 1898. A method of determining the abundance of secondary species. *Minnesota Botanical Studies* 2: 19–24.

Pound & F.E. Clements. 1900. *The Phytogeography of Nebraska: general survey.* Lincoln NB: Botanical Seminar. Ed. 2.

Praeger, R.L. 1949. *Some Irish naturalists: a biographical notebook.* Dundalk: W. Tempest/Dundalgan P.

Prantner, E.R. Oleszko, M. Vesley & R. Sweeny. 1974. *Annotated bibliography of limnological and related studies concerning Lake Erie and influent tributaries.* 5 vols. Buffalo: U.S. Army District.

Prentice, I.C. 2001. Interactions of climate change and the terrestrial biosphere. Pp. 176–195 in Bengtsson & Hammer 2001.

Preston, D. & M. Preston. 2004. *A pirate of exquisite mind: explorer, naturalist, and buccaneer: the life of William Dampier.* NY: Walker.

Preston, F.W. 1948. The commonness and rarity of species. *Ecology* 29: 254–283.

Price, N.B. 2004. John Murray (1841–1914). *DNCB* 3: 1454–1458.

Priestley, J. 1772. Observations on different kinds of air. *Roy. Soc. L. Philos. Trans.* 62: 147–264.

———. 1775. An account of further discoveries in air. *Roy. Soc. L. Philos. Trans.* 65: 384–394.

———. 1779. *Experiments and observations relating to various branches of natural philosophy.* L: T. Johnson.

Primack, R.B. & A.J. Miller-Rushing. 2012. Uncovering, collecting, and analyzing records to investigate the ecological impacts of climate change: a template from Thoreau's Concord. *BioScience* 62: 170–181.

Pritzel, G.A. 1871–77. *Thesaurus literaturae botanicae.* Leipzig: Brockhaus.

Pursell, C. 1973. William Emerson Ritter (1856–1944), naturalist and science administer. *DAB* supplement 3: 635–636.

Purver, M. 1967. *The Royal Society: concept and creation.* Cambridge: MIT P.

Quammen, D. 1996. *The song of the dodo: island biogeography in an age of extinctions.* NY: Scribner.

Querner, H. 1974. Karl August Möbius (1825–1908), zoology. *DSB* 9: 431–432.

Quick, D.G. 1972. Thoreau as limnologist. *Thoreau Quarterly J.* 4: 13–20.

Rafinesque, C.S. 1820. *Ichthologia Ohiensis: history of the fishes inhabiting the River Ohio and its tributary streams, preceded by a physical description of the Ohio and its tributaries.* Lexington KY: Author. 1970. NY: Arno P.

———. 1836. *A life of travels and researches in North America and South Europe.* Philadelphia: Author. 1944. E.D. Merrill, Ed. Chronica Botanica 8: 292–353.

Rainger, R., K.R. Benson & J. Maienschein, eds. 1988. *The American development of biology.* Philadelphia: U. of Pennsylvania P.

Ramensky, L.G. 1983. Basic regularities of vegetation cover and their study. J.L. Brooks, transl. *ESAB* 64: 8–12.

Ramos, R. 2006. *D. Carlos: 1863–1908.* Lisbon: Ciruculo de de Leitores.

Raunkiaer, C.C. 1904 [for 1903]. Om biologiske typer . . . *Botanisk Tidsskrift* 26.

———. 1907. *Planterigrts livaformer og deres detyning for geografien.* Copenhagen.

———. 1934. *The life forms of plants and statistical plant geography.* H. Gilbert-Carter, A.G. Tansley & A. Fausbøll, transl. Oxford: Clarendon P.

Raven, C.E. 1942. *John Ray, naturalist: his life and works.* Cambridge: Cambridge U.P.

Ravetz, J.R. & I. Grattan-Guinness. 1972. Jean Baptiste Joseph Fourier (1768–1830), mathematics, mathematical physics. *DSB* 5: 93–99.

Rawson, D.S. 1928. Preliminary studies of the bottom fauna of Lake Simcoe, Ontario. *U. of Toronto Studies, Biological Series*, no. 31/*Ontario Fisheries Research Laboratories Publication* 36: 75–102.

———. 1930. The bottom fauna of Lake Simcoe and its role in the ecology of the lake. *U. of Toronto Studies, Biological Series*, no. 34, *Ontario Fisheries Research Laboratories Publication* no. 40: 1–183.

Ray, J. 1717. *The wisdom of God manifested in the works of Creation.* Ed. 7. London: Innys. 1977. NY: Arno P. Ed. 1. 1691.

———. 1975. *Flora of Cambridgeshire.* A.H. Ewen & C.T. Prime, transl. Hitchin UK: Wheldon & Wesley.

Read, C.P. 1970. *Parasitism and symbiology: a introductory text.* NY: Ronald P.

Real, L.A. & J.H. Brown, eds. 1991. *Foundations of ecology: classic papers with commentaries.* Chicago: U. of Chicago P.

Réaumur, R.-A.F. de. 1800. *Short history of Bees in two parts: the natural history of bees, with directions for management of them.* L: Verenar & Hood.

Redfield, A.C. 1934. On the proportions of organic derivatives in sea water and their relation to the composition of the plankton. Pp. 176–192 in Daniel 1934.

———. 1958. The biological control of chemical factors in the environment. *American Scientist* 46: 204–221.

———. 1976. Henry Bryant Bigelow, 1879–1967. *National Academy of Sciences Biographical Memoirs* 98: 50–80. Online.

Redi, F. 1684. *Observazioni intorno agli animali viventi*. Firenze: Pier Matini.

———. 1909. *Experiments on the generation of insects*. Chicago: Open Court.

Reess, M. 1872. Über die Eutstehung der Flechte *Collema glaucescens* Hoff. durch Ausoat der Sporen derselben auf *Nostoc lichenoides* Monatsblatt der Königlich preussischen. *Akademie der Wissenschaften zu Berlin* 523–533.

Regal, P.J. 1982. Pollination by wind and animals: ecology of geographic patterns. *Annual Review of Ecology and Systematics* 13: 497–524.

Regier, H.A. & V.C. Applegate. 1972. Historical review of the management approach to exploitation and introduction in SCOL lakes. *Fisheries Research Board of Canada J.* 29: 683–692.

———, R.A. Ryder et al. 1969. *The ecology and management of the walleye in western Lake Erie*. Ann Arbor: Great Lakes Fishery Commission Technical Report 15. vii + 101 Pp.

——— & J.A. Goodier. 1992. Irruption of sea lamprey in the upper Great Lakes: analogous events to those that may follow climate warming. Pp. 185–212 in Glanz 1992.

———, J.J. Magnuson et al. 1989. Likely effects of global climate change on fish associations of the Great Lakes. Pp. 234–255 in Topping 1989.

Reichle, D.E., ed. 1970. *Analysis of temperate forest ecosystems*. NY: Springler-Verlag.

———. 1975. Advances in ecosystem analysis. *BioScience* 25: 257–264.

———, ed. 1981. *Dynamic properties of forest ecosystems*. NY: Cambridge U.P.

———, B.E. Dinger, N.T. Edwards, W.F. Harris & P. Sollins. 1973. Carbon flow and storage in a forest ecosystem. Pp. 345–365 in Woodwell & Pecan 1973.

——— & W.F. Harris. 2004. Resolution of respect: Stanley I. Auerbach, 1922–2004. *ESAB* 85: 91–95.

Reid, D.F. & D.W. Wilkinson. 2014. Catalyst for change: the little Dreissenid that did (change national policy on aquatic invasive species. Pp. 177–185 in Nalepa & Schloesser 2014.

Reidl, R. 1980. Marine ecology—a century of changes. *Marine ecology: Publicazioni della Stazione Zoologica di Napoli* 1: 3–46.

Reidy, M.S. 2010. From the oceans to the mountains. Pp. 17–38 in Vetter 2010.

Reighard, J.E. 1894. A biological examination of Lake St. Clair. *Michigan Fish Commission Bull.* No. 4, 68 pp.

Reingold, N. & M. Rothenberg. 1985. The Exploring Expedition and the Smithsonian Institution. Pp. 242–253 in Viola & Margolis 1985.

Reise, K. 1981. Hundert Jahre Biozönose: die Evolution eines ökologische Begriffes. *Naturwissenschaftliche Rundschau* 33: 328–335.

———. 1990. Karl Möbius: dredging the first community concept from the bottom of the sea. Pp. 149–152 in Lenz & Deacon 1990.

Reise, R.A. 2005. *Eugenie Clark: marine biologist*. NY: Ferguson/Facts on File.

Reissek, S. 1856. Vortrag über die Bildungeschichte der Donauinseln im mittleren Lauf dieses Stromes. *Flora* 39: 622–623.

Reiter, H. 1885. *Die Consolidation der Physiognomik als Versuch einer Oekologie der Gewaechse; mit einem Angang: das System der Erdkunde*. Graz: Luchner & Lubensky.

Renehan, E., Jr. 1992. *John Burroughs: an American naturalist*. Post Mills VT: Chelsea Green Publishing.

Repcheck, J. 2003. *The man who found time: James Hutton and the discovery of the Earth's antiquity*. NY: Simon & Schuster.

Revelle, R. 1980. The Oceanographic and how it grew. Pp. 10–24 in Sears & Merriman 1980.

——— & H.E. Suess. 1957. Carbon dioxide exchange between atmosphere and ocean and the question of an increase of atmospheric CO_2 during the past decades. *Tellus* 9: 18–27.

Rezneck, S. 1972. Douglass Houghton (1809–45), medicine, geology. *DSB* 6: 523–524.

Rhodes, R. 2004. *John James Audubon: the making of an American*. NY: Knopf.

———, ed. 2006. *The Audubon reader*. NY: Everyman's Library.

Rice, A.L. 2004. John Gwyn Jeffreys (1809–85). *DNBS* 2: 1067–1068.

——— & J.B. Wilson. 1980. The British Association Dredging Committee: a brief history. Pp. 373–385 in Sears & Merriman 1980.

Rice, T. 1999. *Voyages of discovery: three centuries of natural history exploration*. L: Natural History Museum.

Richard, G. 1973. The historical development of nineteenth and twentieth century studies on the behavior of insects. Pp. 477–502 in Smith, Mittler & Smith 1973.

Richards, P.W. 1972. Johann Hedwig (1730–99), botany. *DSB* 6: 218–220.

Richardson, P.L. 1980. The Benjamin Franklin and Timothy Folger charts of the Gulf Stream. Pp. 703–717 in Sears & Merriman 1980.

Richter Jr, D. de B. & D. Markewitz. 2001. *Understanding soil change: soil sustainability over millennia, centuries, and decades*. NY: Cambridge U.P.

——— & M.L. Mobley. 2009. Monitoring Earth's critical zone. *Science* 326: 1067–1068.

——— et al. 2018. Ideas and perspectives: strengthening the biogeosciences in environmental research networks. *Biogeosciences* 15: 4815–4832.

Ricketts, E.F. 7 March 1947. Science studies the sardine. *Monterey Peninsula Herald*, 3 pp.

———. 1978. *The outer shores*. J.W. Hedgpeth, ed. Stanford: Stanford U.P.

———. 2006. *Breakthrough: essays, journals, and travelogues*. K.A. Rodgers, ed. Berkeley: U. of California P.

——— & J. Calvin. 1939. *Between Pacific tides: an account of the habits and habitats of some five hundred of the common, conspicuous sea-shore invertebrates of the Pacific Coast between Sitka, Alaska, and northern Mexico*. Stanford: Stanford U.P. Ed. 2, 1948.

———, J.W. Hedgpeth & D.W. Phillips. 1985. *Between Pacific tides*. Ed. 5. Stanford: Stanford U.P.

Ridley, G. 2010. *The discovery of Jeanne Baret: a story of science, the high seas, and the first woman to circumnavigate the globe*. NY: Crown/Random House.

Rikkinen, K. 1988. Ragnar Hult and the emergence of geography in Finland, 1880–1900. *Fennia* 166: 3–192.

Riley, G.A. 1941. Plankton studies: V. regional summary. *J. of Marine Research* 4: 162–171. Reprinted in Riley 1982.

———. 1946. Factors controlling phytoplankton populations on Georges Bank. *J. of Marine Research* 6: 54–73. Reprinted in Riley 1982.

———. 1947. A theoretical analysis of the zooplankton population of Georges Bank. *J. of Marine Research* 6: 104–113. Reprinted in Riley 1982.

———. 1956a. Introduction: oceanography of Long Island Sound, 1952–1954. *Bingham Oceanographic Collection Bull.* 15: 9–14.

———. 1956b. II. Physical oceanography of Long Island Sound, 1952–1954. *Bingham Oceanographic Collection Bull.* 15: 47–61.

———. 1956c. IX. Production and utilization of organic matter: oceanography of Long Island Sound, 1952–1954. *Bingham Oceanographic Collection Bull.* 15: 324–344.

———, ed. 1963a. *Marine Biology I: Proceedings of the First International Interdisciplinary Conference on Marine Biology.* Baltimore: Port City P.

———. 1963b. Theory of food chain relations in the ocean. Pp. 438–463 in Hill 1963.

———. 1979. Summation. Pp. 74–78 in *Second informal workshop on the oceanography of the Gulf of Maine and adjacent seas.* Halifax NS: Dalhousie U.

———. 1982. *Selected works.* J.S. Wroblewski, ed. Halifax: Dalhousie U.P.

———. 1982–83. *Reminiscences of an oceanographer*, online.

——— & D.F. Bumpus. 1946. Phytoplankton-zooplankton relationships on Georges Bank. *J. Marine Research* 6: 33–47.

———, H. Stommel & D.F. Bumpus. 1949. Quantitative ecology of the plankton of the Western North Atlantic. *Bingham Oceanographic Collection Bull.* 12, no. 3: 1–169.

Riley, J.L. 2013. *The once and future Great Lakes country.* Montreal: McGill-Queen's U.P.

Riley, J.L. & W. Riley. 1979. *Guide to the National Wildlife Refuges.* NY: Anchor P./Doubleday.

Rivinus, E.E. & E.M. Youssef. 1992. *Spencer Baird of the Smithsonian.* Washington: Smithsonian Institution.

Rizhinashvili, A.L. 2019. Ernst Haeckel's 'ecology' in Russia of the First Half of the Twentieth Century. *Theory in Biosciences* 138: 89–103.

———. 2020. Production hydrobiology in the USSR under the pressure of Lysenkoism: Vladimir I. Zhadin's forgotten theory of biological productivity (1940). *JHB* 53: 105–139.

R.L. 1899. Review: Sclater & Sclater 1899. *Nature* 60: 217–219.

Robinson, G. 1971. (Heinrich) Anton de Bary (1831–88), botany. *DSB* 3: 611–614.

Rodger, K.A., ed. 2002. *Renaissance man of Cannery Row: the life and letters of Edward F. Ricketts.* Tuscaloosa: U. of Alabama P.

———., ed. 2006. *Breaking through: essays, journals, and travelogues of Edward F. Ricketts.* Tuscaloosa: U. of Alabama P.

Rodgers, A.D., III. 1944. *John Merle Coulter: missionary in science.* Princeton: Princeton U.P.

Rodhe, H. & R. Carlson, eds. 1998. *The legacy of Svante Arrhenius: understanding the greenhouse effect.* Stockholm: Royal Swedish Acad. of Sciences.

Rodhe, W. 1969. Crystallization of eutrophication concepts in northern Europe. Pp. 50–64 in Rohlich 1969.

———. 1974a. Limnology turns to warm lakes. *Archiv für Hydrobiologie* 73: 537–546.

———. 1974b. The International Association of Limnology: creation and functions. *Mitteilungen International Vereinigung für theoretische und angewandte Limnologie* 20 [sic]: 44–70.

———. 1975. President's lecture: the SIL founders and our fundament. *Verhandlungen Internationale Vereinigung Limnologie* 19 [sic]: 16–25.

Rodolico, F. 1974. Luigi Ferdinando Marsili (Marsigli, 1658–1730), natural history, *DSB* 9: 134–136.

Roger, A. & F. Guéry, eds. 1991. *Maires et protecteurs de la nature.* Champs Vällon Fr.: Collection Milleux.

Roger, J. 1963. *Les sciences de la vie dans la pensée français du XVIIIe: la generation des animaux de Descartes à l'Encyclopédie.* Paris: Armand Colin. Ed. 2, 1971.

———. 1970. Georges-Louis Leclerc, Comte de Buffon (1707–88), natural history. *DSB* 2: 576–582.

———. 1989. *Buffon: un philosophe au Jardin du Roi.* Paris: Fayard.

———. 1997a. *Buffon: a life in natural history.* S.L. Bonnefoi, transl. K.R. Benson, ed. Ithaca: Cornell U.P.

———. 1997b. *Life sciences in Eighteenth-century French thought.* K.R. Benson, ed., R. Ellrich, trans. Stanford: Stanford U.P.

Rohlich, G.A., ed. 1969. *Eutrophication: causes, consequences, corrections.* Washington: National Academy of Sciences.

Roll, H.U. 1990. On the roots of oceanography in Germany. Pp. 3–19 in Lenz & Deacon 1990.

Romanes, G. 1882. *Animal intelligence.* L: Kegan, Paul. Trench. Extracts in Houck & Drickamer 1996: 38–44.

Roseman, E., P. Kocovsky & C.S. Vandergoot, eds. 2010. *Status of walleye in the Great Lakes: Proceedings of the 2006 Symposium.* Ann Arbor: Great Lakes Fishery Commission Technical Report 69.

Rosenvinge, L.K. et al. 1927. Eugenius Warming, 3 November 1841–2 April 1924. *Botanisk Tidsskrift* 39: 1–56.

Ross, C.C., ed. 1976. *Paleogeography.* Stroudsburg: Dowden, Hutchinson & Ross.

Ross, K. 1996. José de Acosta (1540–1600), Spanish Jesuit historian. Pp. 4–5 in *Latin American Lives.* NY: Macmillan Compendium.

Ross, M.E. 1978. *Bougainville.* L: Gordon & Cremonesi.

———. 2000. *Fish watching with Eugenie Clark.* Minneapolis: Carol Rhoda Books.

Rothschuh, K.E. 1972. (Christian Andreas) Victor Hensen (1835–1924), physiology, marine biology. *DSB* 6: 187–188.

Rozwadowski, H.M. 2002. *The sea knows no boundaries: a century of marine science under ICES.* Seattle: U. Washington P.

———. 2005. *Fathoming the ocean: the discovery and exploration of the deep sea.* Cambridge MA: Harvard U.P.

———. 2011. Playing by end on and under the sea: the importance of play for knowing the ocean. Pp. 162–188 in Vetter 2011.

Rudnick, D. 1978. Adelbert von Chamisso (1781–1838), natural history, botany. *DSB* 15: 81–83.

Rudwick, M.J.S. 2004. Sir Charles Lyell, first baronet (1792–1875), geologist. *ODNB* 34: 852–857.

Rupke, N.A. 2000. Alexander von Humboldt, 1769–1859, German naturalist and traveler. Pp. 356–359 in Hessenbruch 2000a.

Ruse, M. 1979. *Sociobiology: sense or nonsense?* Boston: D. Reidel.

———. 1980. Charles Darwin and group selection. *Annals of Science* 37: 615–630.
———. 2013. *The Gaia hypothesis: science on a pagan planet.* Chicago: U. Chicago P.
Rush, S.A. et al. 2015. Changes in heron and egret populations on the Laurentian Great Lakes and connecting channels. *Avian Conservation and Ecology* 10: 18 pp.
Russell-Hunter, W.D. 1985. From Woods Hole to the World: *The Biological Bulletin. Biological Bull.* 168, supplement: 200–202.
Sagan, D. 1990. *Biospheres: metamorphosis of planet Earth.* NY: St. Martin's P.
———. 2012. *Lynn Margulis: life and legacy of a scientific rebel.* White River Junction VT: Chelsea Green.
——— & L. Margulis. 1983. The Gaian perspective of ecology. *The Ecologist* 13: 160–167.
Sagarin, R.D., W.F. Gilly et al. 2008. Remembering the Gulf: changes to the marine communities of the Sea of Cortez since the Steinbeck and Ricketts Expedition of 1940. *Frontiers in Ecology & the Environment* 6: 372–379.
Saldanha, L. 1980. King Carlos of Portugal, a pioneer in European oceanography. Pp. 606–613 in Merriman & Deacon 1980.
———. 1997. King Carlos de Braganca, the father of Portuguese oceanography. Pp. 19–38 in Salanha & Ré, eds. 1997.
——— & P. Ré, eds. 1997. *One hundred years of Portuguese oceanography.* Lisbon: Publicacões Avulas Portugal.
Salisbury, E.J. 1952. Francis Wall Oliver, 1864–1951. *Roy. Soc. L. Obituary Notices of Fellows* 8: 229–240.
Samuelson, J. 1859. *Humble creatures: the earthworms and the common housefly, in eight letters.* L: John van Voorst.
Sanders, A.P.M. 1974. Karl Friedrich Philipp von Martius (1794–1868), botany, ethnology. *DSB* 9: 148–149.
———. 1975a. Andreas Franz Wilhelm Schimper (1856–1901), botany. *DSB* 12: 165–167.
———. 1975b. Joakim Frederik Schouw (1789–1852), plant geography, climatology. *DSB* 12: 214–215.
———. 1975c. Simon Schwendener (1829–1919), plant anatomy. *DSB* 12: 255–256.
———. 1975d. Johann Baptiste von Spix (1781–1826), zoology. *DSB* 12: 578–579.
Santavy, S.G. 2014. My story on finding the first zebra mussel (*Dreissena polymorpha*) in North America. Pp. 3–4 in Nalepa & Schloesser 2014.
Sapp, J. 1991. Living together: symbiosis and cytoplasmic inheritance. Pp. 15–25 in Margulis & Fester 1991.
———. 1994. *Evolution by association: a history of symbiosis.* NY: Oxford U.P.
Sarton, G. 1943. Aimé Bonpland. *Isis* 34: 385–399.
Sarukhán, J. 1981. Mexico. Pp. 35–51 in Kormondy & McCormick 1981.
Satchell, J.E. 1985. *Earthworm ecology: from Darwin to vermiculture.* L: Chapman & Hall.
Saussure, N.-T. de. 1804. *Recherches chimique sur la végétation.* Paris: Chez la V. Nyon.
———. 2013. *Chemical research on plant growth.* NY: Springer.
Savage, H., Jr. 1970. *Lost heritage.* NY: Morrow.
———. 1978. *Discovering America, 1700–1875.* NY: Harper & Row.
Say, T. 1859. *The complete writings . . . . on the entomology of North America.* J.L. Le Conte, ed. NY. 1978. NY: Arno P. 1978.
———. 1969. *An account of the Crustacea of the United States.* NY: J. Cramer. Ed. 1, 1818.
Scammon, C.M. 1968. *The marine mammals of the northwestern coast of Northwestern America: together with an account of the American whale fishery, 1825–18??* NY: Dover. Ed. 1, 1874.
Scaps, P. 2005. *Histoire de la biologie marine.* Paris: Ellipses.
Scarboro, J. 1977. Some beetles in Pliny's *Natural history. Coleopterist Bulletin* 31: 293–296.
———. 1979. On the history of early entomology, chiefly Greek and Roman, with a preliminary bibliography. *Melsheimer Entomological Series* 26: 17–27.
Scharz, A. & K. Jax, eds. 2011. *Ecology revisted: reflecting on concepts, advancing science.* L: Springer.
Scheele, I. & C. Hünemörder. 1990. On the contribution of zoologists from Hamburg to marine biology. Pp. 102–107 in Lenz & Deacon 1990.
Scheffer, V. 1968. Introduction. Pp. v—x in Scammon 1968.
Schellnhuber, H.J., P.J. Crutzen, W.C. Clark, M. Claussen & H. Held, eds. 2004. *Earth system analysis for sustainability.* Cambridge MA: MIT P.
Schlee, S. 1973. *The edge of an unfamiliar world: a history of oceanography.* NY: Dutton. Ed. 1., 1972.
———. 1978. *On almost any wind: the saga of the oceanographic research vessel Atlantis.* Ithaca: Cornell U.P.
———. 1980. The R/V *Atlantis* and her first oceanographic institution. Pp. 49–56 in Sears & Merriman 1980.
Schmidt, E.J. 1912. Danish researches in the Atlantic and Mediterranean on the life history of the fresh-water eel (*Anguilla vulgaris*). *Internationale Revue du gusamten Hydrobiologie und Hydrographie* 5: 317–342.
———. 1922. The breeding places of the eel. *Roy. Soc. of London Philos. Trans. Series B, Biological Sciences* 205: 179–208.
Schmidt, K.P. 1942. Review: Steinbeck & Ricketts, *Sea of Cortez* (1941). *Ecology* 23: 242–243.
———. Animal geography. Pp. 767–794 in Kessel 1955.
———. 1957. Warder Clyde Allee, 1885–1955. *National Academy of Sciences Biographical Memoirs* 30: 2–40.
Schneider, A. 1897. The phenomena of symbiosis. *Minnesota Botanical Studies* 1: 923–948.
Schneider, D.W. 2000. Local knowledge, environmental politics, and the founding of ecology in the United States: Stephen Forbes and the lake as a microcosm. *Isis* 91: 681–705.
Schneider, J.C. & J.H. Leach. 1979. *Walleye stocks in the Great Lakes, 1800–1975: fluctuations and possible causes.* Ann Arbor: Great Lakes Fishery Commission. Technical Report 31. Iii + 51 pp.
Schneider, S.H. & P.J. Boston, eds. 1991. *Scientists on Gaia.* Cambridge MA: MIT P.
——— & R. Londer. 1984. *The coevolution of climate and life.* San Francisco: Sierra Club Books.
———,J.R. Miller, E. Crist & P.J. Boston. 2004. *Scientists debate Gaia: the next century.* Cambridge MA: MIT P.
Schnitten, C., A. Masseran & P. Chavot, 1981. Muséum National d'Histoire Naturelle. Transl. C. Rodine. Pp. 489–490 in Hessenbruch 2000a.
Schoener 1989. The ecological niche. Pp. 79–113 in *Ecological concepts: the contribution ecology to understanding of the natural world.* J.M. Cherrett, ed. Oxford: Blackwell.
Schofield, R.E., ed. 1963. *A scientific autobiography of Joseph Priestley, 1733–1804.* Cambridge MA: MIT P.
———. 1975. Joseph Priestley (1733–1804), chemistry, electricity, natural philosophy, theology. *DSB* 11: 139–147.

———. 1997–2004. *The enlightenment of Joseph Priestley*. 2 vols. University Park: Penn State U.P.
——— & Baron, eds. 1993. *Thoreau's world and ours: a natural legacy*. Golden CO: North American P.
Schoolcraft, H.R. 1821. *Narrative journal of travels through the northwestern regions of the United States extending from Detroit through the great chain of American lakes, to the sources of the Mississippi*. Albany: E. & E. Hosford.
Schouw, J.F. 1822. *Grundtraek tile en almindelig Plantgeographie*. Copenhagen: Gyldendal. German edition, Berlin 1823.
Schröter, C. 1908. *Pflanzenleben der Alpen: eine Schildering der Hochebirgsflora*. Zurich: Albert Raustein.
——— & C.-H.-M. Flahault. 1910. Rapports et propositions: nomenclature phytogéographique. *Actes du IIIe Congrès International de Botanique*. Brussels.
——— & O. Kirchner. 1896–1902. Die Vegetation des Bodensees. *Schriften: Vereins für Geschichte des Bodensees und seiner Umgebung* 25: 1–119, 31: 1–86.
Schwach, V. 2002. Internationalist and Norwegian at the same time: Johan Hjort and ICES. *ICES Marine Science Symposium* 215: 39–44.
———. 2007. Faded glory. Pp. 31–70 in Benson & Rozwadoski 2007.
———. 2012. The sea around Norway: science, resource management and environmental concerns, 1860–1970. *Environmental Hist.* 18: 101–110.
———. 2014. A sea change: Johan Hjort and the natural fluctuations in the fish stocks. *ICES J. of Marine Science* 71: 1993–1999.
——— & J. Hubbard. Johan Hjort and the birth of fisheries biology: the construction and transfer of Knowledge, approaches, and attitudes, Canada and Norway, 1890–1920. *Studia Atlantica* 13: 22–41.
——— & K. Jax, eds. 2011. *Ecology revisited: reflecting on concepts. Advancing science*. Dordrecht: Springer.
Schwarz, A.E. 2003. *Wasserwüste-Mikrokosmos-Ökosystem: eine Geschichte der "Eroberung" des Wasserraums*. Freiburg: Rombach-Verlag.
——— & K. Jax, eds. 2011. *Ecology revisited: reflecting on concept, advancing science*. Dordrecht: Springer.
Schwarz, R.G. 1997. Thomas Say, 1787–1834, naturalist. *BDAC* 703–706.
Schwarzbach, M. 1986. *Alfred Wegener: the father of continental drift*. Madison WI: Science Tech.
Schwendener, S. 1867. Ueber der Bau des Flechtenthallus. *Verhandlungen der Schweizerischen Naturforscheaden Gessellschaft Aarau* 51: 88–89.
Schwerdteger, F. 1973. Forest entomology. Pp. 361–386 in Smith et al. 1936. *Sustainability*. Cambridge MA: MIT P.
Sclater, P.L. 1858. On the geographical distribution of the members of the Class Aves. *Linnean Soc. J. of Zoology* 2: 130–148.
———. 1897. On the distribution of marine mammals. *Zool. Soc. of London Proc.* 41: 347–359.
Sclater, W.L. & P.L. Sclater. 1899. *The geography of mammals*. L: Kegan Paul, Trench & Trübner. 1978. NY: Arno P.
Scott, W.B. & E.J. Grossman. 1973. *Freshwater fishes of Canada*. Ottawa: Fisheries Research Board of Canada.
Scudo, F.M. & J.R. Ziegler, eds. 1978. *The golden Age of theoretical ecology: Volterra, V.A. Kostitzin, A.J. Lotka & A.N. Kolmlgoroff*. NY: Springer-Verlag.
Sears, M., ed. 1961. *Oceanography: invited lectures presented at the International Congress held in New York, 31 August-12 September 1959*. Washington: AAAS.
——— & D. Merriman, eds. 1980. *Oceanography: the past*. NY: Springer-Verlag.
Sears, P.B. 1958. Henry Chandler Cowles (1869–1939), botanist, geographer, a founder of *Ecology*. *DAB* suppl. 2: 127–128.
———. 1973. Frederic Edward Clements (1874–1945), botanist, pioneer ecologist. *DAB* suppl. 3: 168–170.
Seaward, M.R.D. & S.M.D. FitzGerald. 1996. *Richard Spruce (1817–1893), botanist and explorer*. Kew UK: Royal Botanical Garden.
Seckbach, J., ed. 2002. *Symbiosis: mechanisms and model systems*. Boston: Kluwer Academic Publishers.
Segerstrâle, U.C.O. 2000. *Defenders of the truth: the battle for science in the sociobiology debate and Beyond*. NY: Oxford U.P.
Sellery, G.C. 1956. *E.A. Birge: a memoir*. Madison: U. of Wisconsin P.
Semper, C. [or K] G. 1881. *Animal life as affected by the natural conditions of existence*. NY: D. Appleton. 1977. NY: Arno P. German ed., 1880.
Serpoinau, G. & V. Malciu. 2002. The pioneers of oceanographic research in Romania. Pp. 271–274 in Benson & Rehbock 2002.
Severin, T. 1997. *The Spice Islands voyage: in search of Wallace*. L: Little, Brown.
Shaw, R.E. 1966. *Erie water west: a history of the Erie Canal, 1792–1854*. Lexington: U. of Kentucky P.
Sheail, J. 1987. *Seventy-five years in ecology: the British Ecological Society*. Oxford: Blackwell Scientific Publications.
Shelford, V.E. 1913. *Animal communities in temperate America, as illustrated in the Chicago region*. Chicago: *Geog. Soc. Chicago Bull.* 1937. Chicago: U. Chicago P. 1977. NY: Arno P.
———. 1938. The organization of the Ecological Society of America. *Ecology* 19: 164–166.
Sheriff, C. 1997. *The artificial river: the Erie Canal and the paradox of progress, 1817–1862*. NY: Hill & Wang.
Shermer, M. 2002. *In Darwin's shadow: the life and science of Alfred Russel Wallace: a biographical study on the psychology of history*. L & NY: Oxford U.P.
Shetler, S.G. 1967. *The Komarov Botanical Institute: 250 years of Russian research*. Washington: Smithsonian Institution P.
Shimwell, D.W. 1991. *The description and classification of vegetation*. Seattle: U. of Washington P.
Shor, E.N. 1974a. Charles Atwood Kofoid (1865–1947), zoologist. *DAB* suppl. 4: 461–462.
———. 1974b. Clinton Hart Merriam (1855–1942), biology. *DSB* 9: 313–314.
———. 1975. Thomas Say (1787–1834), entomology. *DSB* 12: 132–134.
———. 1976. Addison Emery Verrill (1839–1926), zoology. *DSB* 14: 1–2.
———. 1988. Henry Bryant Bigelow (1879–1967), oceanographer. *DAB* suppl. 8: 37–38.
Short, P. 2004. *The pursuit of plants: experiences of Nineteenth Century plant collectors*. Portland ME: Timber P.
Shoshani, E.N. 1997. Henry Fairfield Osborn (1859–1935), vertebrate paleontologist, museum administrator. *BDAC* 592–596.

Shugart, H.H. & V. O'Neill, eds. 1979. *Systems ecology.* Stroudsburg PA: Dowden, Hutchinson & Ross.

Shuler, J. 1995. *Had I the wings: the friendship of Bachman and Audubon.* Athens: U. of Georgia P.

Sideris, L.H. & K.D. Moore, eds. 2008. *Rachel Carson: legacy and challenge.* Albany: State U. of New York P.

Silliman, B. 1818. On the comparative forwardness of the spring, in different parts of the United States, in 1817. *American J. of Sci.* 1: 76–77.

Simpkins, D.M. 1975. James Edward Smith (1759–1828), botany. *DSB* 12: 571–572.

Simpson, G.G. 1934. *Attending marvels: a Patagonian journal.* NY: Macmillan.

———. 1940. Mammals and land bridges. *Washington Academy of Sciences J.* 30: 137–163. Reprinted in Sterling 1975 & Lomolino et al. 2004: 339–365.

———. 1944. *Tempo and mode in evolution.* NY: Columbia U.P.

———. 1964. *This view of life: the world of an evolutionist.* NY: Harcourt, Brace & World.

———. 1978. *Concessions to the improbable: an unconventional autobiography.* New Haven: Yale U.P.

———. 1987. *Simple curiosity: letters from George Gaylord Simpson to his family, 1921–1970.* Berkeley: U. of California P.

Sims, R.W., J.H. Price & P.E.S. Whalley, eds. 1983. *Evolution, time and space: the emergence of the biosphere.* NY & L: Academic P.

Sjörs, H. 1981. Sweden. Pp. 303–312 in Kormondy & McCormick 1981.

Skutch, A.F. 1999. *Helpers at bird's nests: a worldwide survey of cooperative breeding and related behavior.* Ed. 2. Iowa City: U. of Iowa P. Ed. 1, 1987.

Slack, N.G. 2010. *G. Evelyn Hutchinson and the invention of modern ecology.* New Haven CT: Yale U.P.

Sloan, P.R. 1976. The Buffon-Linnaeus controversy. *Isis* 67: 356–375.

———. 2002. Georges-Louis Leclerc de Buffon (1707–88). Pp. 176–180 in Kors 2002.

Smed, J. & J. Ramster. 2002. Overfishing, science, and politics: the background in the 1890s to the formation of the International Council for the Exploration of the Sea. *ICES Symposia* 215: 13–21.

Smil, V. 2001. *Cycles of life: civilization and the biosphere.* NY: Scientific American Linrary.

———. 2002. *The Earth's biosphere.* Cambridge MA: MIT P.

Smiley, C.W. 1882. Changes in the fisheries of the Great Lakes during the decade 1870–1880. *American Fisheries Soc. Trans.* 11: 28–37.

Smit, P. 1976a. Gottfried Reinhold Treviranus (1776–1837), zootomy, physiology. *DSB* 13: 460–462.

———. 1976b. Ludolph Christian Treviranus (1779–1864), plant anatomy, plant physiology. *DSB* 13: 462–463.

Smith, C.H., J. Woleben & C. Rodgers. 2005. *Some biogeographers, evolutionists, and ecologists: chrono biographical sketches.* Bowling Green: Western Kentucky U. Online.

Smith, R.F., T.E. Mittler & C.N. Smith, eds. 1973. *History of entomology.* Palo Alto: Annual Reviews.

Smith, S.H. [published as by J.W. Moffett]. 1957a. Recent changes in the deep-water fish populations of Lake Michigan. *American Fisheries Soc. Trans.* 86: 393–408.

———. 1957b. Limnological surveys of the Great Lakes—early and recent. *American Fisheries Soc. Trans.* 86: 409–418.

———. 1970. Trends in fishery management of the Great Lakes. Pp. 107–114 in Benson 1970.

Smith, T.D. 1994. *Scaling fisheries: a science driven by economics and politics, 1885–1955.* L: Cambridge U.P.

Smith, W.G. 1913. Raunkiaer's "life-forms" and statistical methods. *J. of Ecology* 1: 16–26.

Snapp, J.R. 1999. Arthur Dobbs (1689–1765), colonial governor of North Carolina. *ANB* 6: 657–659.

Snelders, H.A.M. 1970. Svante August Arrhenius (1859–1927), chemistry, physics. *DSB* 1: 296–302.

Snyder, G.S. 1970. *In the footsteps of Lewis and Clark.* Washington: National Geographic Society.

Sobolev, L.N. & V.D. Utekhin. 1973. Russian (Remensky) approaches to community systemization. Pp. 75–103 in Whittaker 1973.

Söderqvist, T. 1986. *The ecologists: from merry naturalists to saviours of the nation. A sociologically informed narrative survey of ecologization of Sweden, 1895–1975.* Stockholm: Almqvist & Wiksell International.

Sømme, L. 2003 [2001]. Den biologiske stasjon I Drøbak perspektiv. *Det Norske Videnskaps-Akademi Årbok* 2001: 375–392.

Sorrenson, R. 2000. Royal Society of London. Pp. 662–664 in Hessenbruch 2000a.

Souder, L. 2000. Aristotle, 384–322 BC. Pp. 40–41 in Hessenbruch 2000a.

Souder, W. 2004. *Under a wild sky: John James Audubon and the making of the Birds of America.* NY: North Point P./Farrer, Straus & Giroux.

———. 2012. *On a farther shore: the life and legacy of Rachel Carson.* NY: Crown.

Southwick, C.H., ed. 1985. *Global ecology.* Sunderland MA: Sinauer Associates.

Southwood, R. & J.R. Clarke. 1999. Charles Sutherland Elton, 29 March 1900–1 May 1991. *Royal Society Biographical Memoirs of Fellows* 45: 131–146.

Southwood, T.R.E. 1985. The state of ecology. *Science* 228: 871–873.

Spalding, V.M. 1903. The rise and progress of ecology. *Science* 17: 201–210.

Spallanzani, L. 1765. *Spaggio di osserazioni microscopiche relative al Sistema della generazione dei signori Needham e Buffon.* Modena, Italy.

———. 1777. *Opuscules de physique animal et végétale.* 2 vols. Geneva: Barthelemi Chirol.

———. 1799. *Tracts on the nature of animals and vegetables; observations and experiments upon the animalcula of infusions.* Anon. transl. Edinburgh. Selection in Brock 1961: 13–16.

Spärck, R. 1932a. Japetus Steenstrup, 1813–1897. Pp. 115–119 in Meisen 1932.

———. 1932b. C.G. Johannes Petersen, 1860–1928. Pp. 186–189 in Meisen 1932.

Sparks, T.H. & P.D. Carey. 1995. The responses of species to climate over the centuries: an analysis of the Marsham phenological record, 1736–1947. *J. of Ecology* 83: 321–329.

Spencer, L.T. 2002. Four men and an *Albatross:* the growth of American oceanography, 1882–1921. Pp. 288–297 in Benson & Rehbock 2002.

Sponsel, A.W. 2009. *Coral reef formation and the sciences of earth, life and sea, c. 1770–1952.* Princeton: Ph.D. dissertation, Princeton U.

Spruce, R. 1908. *Notes of a botanist on the Amazon and Andes.* A.R. Wallace, ed. 2 vols. L: Macmillan.

Spurr, S.H. 1952. Origin of the concept of forest succession. *Ecology* 33: 426–427.

Stafleu, F.A. & R.S. Cowan. 1976–2009. *Taxonomic literature: a selective guide to botanical publications and collections, with dates.* Edition 2. 8 vols. Utrecht: Bohn, Scheltema, & Holkema.

Stagg, R. 2010. *The golden dream: a history of the St. Lawrence Seaway.* Toronto: Dundurn P.

Stahl, E. 1874. Beiträge zue Entwickelungsgeschichte der Flechten. *Botanische Zeitung (Berlin)* 32: 177–180.

Standiford, R. 2002. *California's oak woodlands.* Pp. 280–303 in Shea & Healy 2002.

Stanley, G.D. 2006. Photosymbiosis and the evolution of modern coral reefs. *Science* 312: 857–858.

Stann, E.J. 1985. Charles Wilkes as diplomat. Pp. 204–224 in Viola & Margolis 1985.

Stanton, W. 1975. *The great United States Exploring Expedition of 1838–1842.* Berkeley: U. of California P.

Stearn, W.T. 1970. Robert Brown (1773–1858), botany. *DSB* 2: 516–522.

Steedman, R.J. & H.A. Regier. 1990. Ecological bases for an understanding of ecosystem integrity. Pp. 239–255 in Edwards & Regier 1990.

Steele, M.A. & P.D. Smallwood. 2002. Acorn dispersal by birds and mammals. Pp. 182–195 in Mc Shea & Healy 2002.

Steenstrup, J.J.S. 1842. Geognostick-geologisk untersgelse af Skovmiserne vidnesdam og lillemose I det nordlige Sjaelland. *Kongelige Dansk Videnskabernes Selskab Skrifter* 9: 17–120.

Steere, E.C., ed. 1958. *Fifty years of botany: golden jubilee volume of the Botanical Society of America.* NY: McGraw-Hill.

Steffen, J.O. 1999. William Clark (1770–1838), explorer, Indian agent, and governor of Missouri Territory. *ANB* 4: 953–955.

Steinbeck, J. 1995a. *The log from the Sea of Cortez: the narrative portion of the book,* Sea of Cortez, *by John Steinbeck and E.F. Ricketts, 1941.* NY: Viking P.

———. 1995b. About Ed Ricketts. Pp. vii—lxvii in Steinbeck 1995a.

Steinbeck, J. & E.F. Ricketts. 2009. *Sea of Cortez: a leisurely journal of travel and research, with a scientific appendix comprisimg materials for a source book om the marine animals of the Panamic Faunal Province.* NY: Penguin Books. 1941. Ed 1, NY: Viking Books.

Stejneger, L.H. 1887. How the great northern sea-cow (*Rytina*) became exterminated. *American Naturalist* 21: 1047–1054.

———. 1890. Annotated list of reptiles and batrachians collected by Dr. C. Hart Merriam and Vernon Bailey on the San Francisco Mountain Plateau and Desert of the Little Colorado, Arizona. Pp. 103–118 in *North American Fauna,* number 3. Washington: Government Printing Office. Reprinted in Merriman 1974.

———. 1936. *Georg Wilhelm Steller: the pioneer of Alaskan natural history.* Cambridge MA: Harvard U.P.

Steleanu, A. 1989. *Geschichte der Limnologie und ihrer Grundlagen.* Frankfurt am Main: Haag & Herche.

Steller, G.W. 1974. *Beschreibung von dem Lande Kamtchatka, Reise von Kamtchatka nach Amerika, Ausführliche Beschreibung von sonderbaren Merthiereb.* H. Beck, transl. Stuttgart: Brockhaus.

———. 1988. *Journal of a voyage with Bering, 1741–1742.* M.A. Engel & O.W. Frost, transl. Stanford: Stanford U.P.

Stephens, L.D. 2000. *Science, race, and religion: John Bachman and the Charleston circle of naturalists, 1815–1895.* Chapel Hill: U. of North Carolina P.

———. & D.R. Calder. 2006. *Seafaring scientist: Alfred Goldsborough Mayor, pioneer in marine biology.* Columbia: U. of South Carolina P.

Sterling, K.B., compiler & intro. 1974. *Selections from the literature of American biogeography.* NY: Arno P.

———. 1977. *Last of the naturalists: the career of C. Hart Merriam.* Ed. 2, NY: Arno P.

———., comp. 1978. *Rafinesque and lives.* NY: Arno P.

———. 1981. (Charles) William Beebe, 1877–1962, naturalist, oceanographer, zoological society executive. *DAB* 7 (supplement) 45–46.

———. 1997a. Warder Clyde Allee, 1885–1955, ecologist. *BDAC* 17–18. [1 vol.]

———. 1997b. (Charles) William Beebe, 1877–1962, naturalist, oceanographer, zoological society executive. *BDAC* 71–73.

———. 1997c. Johann Friedrich Eschseholtz (1793–1831), physician, naturalist. *BDAC* 256–258.

———. 1997d. Leland Ossian Howard, 1857–1950, entomologist. *BDAC* 383–384.

———. 1997e. Stephen Harrison Long (1784–1864), explorer, engineer, army officer. *BDAC* 476–478.

———. 1997f. Clinton Hart Merriam, 1855–1942, mammologist, zoologist, federal government administrator. *BDAC* 526–529.

———. 1997g. Zebulon Montgomery Pike (1779–1813), army officer, explorer. *BDAC* 629–630.

———. 1997h. Merriweather Lewis (1774–1809), explorer, engineer, army officer, territorial governor. *BDAC* 464–467.

———. 1999a. William Beebe (1877–1962), naturalist, oceanographer, and zoological society executive. *ANB* 2: 461–462.

———. 1999b. Clinton Hart Merriam (1855–1942), zoologist, government official. *ANB* 15: 345–347.

Sterner, R.W. 2012. Raymond Laurel Lindeman and the trophic dynamic viewpoint. *Limnol. & Oceanogr. Bull.* 21: 38–51.

Stewart, A. 2004. *The Earth moved: on the remarkable achievements of earthworms.* Chapel Hill: Algonquin Books.

Stillingfleet, B. 1755. *Calendar flora.* Reprinted in Linnaeus 1759.

Stimson, D. 1948. *Scientists and amateurs: a history of the Royal Society.* NY: Henry Schuman.

Stoller, L. 1956. A note on Thoreau's place in the history of phenology. *Isis* 47: 172–181.

Stott, R. 1984. History of biogeography. Pp. 1–24 in Taylor 1984.

———. 2003. *Darwin and the barnacle: the story of one tiny creature and history's most spectacular breakthrough.* L: Faber & Faber.

Stoye, J. 1994. *Marsigli's Europe: the life and times of Luigi Ferdinando Marsigli, soldier and virtuoso.* New Haven: Yale UP.

Streshinsky, S. 1998. *Audubon: life and art in the American wilderness.* Athens: U. of Georgia P.

Stroud, P.T. 1992. *Thomas Say: New World naturalist.* Philadelphia: U. of Pennsylvania P.

———. 1995. The founding of the Academy of Natural Sciences of Philadelphia in 1812 and its *Journal* in 1817. *Archives of Natural History* 22: 221–233.

———. 1999. Thomas Say (1787–1834), naturalist and explorer. *ANB* 19: 331–332.

———. 2000. *The emperor of nature: Charles Lucien Bonaparte and his world.* Philadelphia: U. of Pennsylvania P.

Stuckey, R.L. 1973. E. Lucy Braun (1889–1971), outstanding botanist and conservationist: a biographical sketch, with bibliography. *Michigan Botanist* 12: 93–106. Reprinted in Stuckey 2001.

———, compiler. 1994. *E. Lucy Braun: Ohio's foremost woman botanist, a collection of biographical accounts, maps, and photos.* Knoxville: Botanical Society of America.

———. 1997. (Emma) Lucy Braun, 1889–1971, botanist, ecologist, conservationist, educator. *BDAC* 103–106.

———. 2001. *E. Lucy Braun (1889–1971): Ohio's foremost woman botanist: her studies of prairies and phytogeographical relationships, an anthology of papers.* Columbus: Ohio State U.P.

Styran, R.M. & R.R. Taylor. 2004. *The Welland Canal's Corridor.* St. Catherines, Ont.: Looking Back P.

Suess, E. 1875. *Die Entstehung der Alpen.* Vienna: W. Braumüller.

———. 1883–1909. *Das Antlitz der Erde.* 3 vols. In 4. Prague, Vienna, Leipzig.

Summerhayes, S.V. & C.S. Elton. 1923. Contribution to the ecology of Spitzbergen and Bear Island. *J. of Ecology* 214–268.

Swammerdam, J. 1758. *The book of nature; or, the history of insects.* Anon. Transl. L: Thomas Flloyd. 1978. NY: Arno P.

Swank, W.T., J.L. Meyer & D.A. Crossley, Jr. 2001. Long term ecological research: Coweeta history and perspectives. Pp. 143–163 in Barrett & Barrett 2001.

Sweet, J.M. 1972. Instructions to collectors: John Walker (1793) and Robert Jameson (1817); with biographical notes on James Anderson (L.L.D.) and James Anderson (M.D.). *Annals of Science* 29: 397–414.

Swift, J. 1733. *On poetry: a rhapsody.* L: J. Huggonson.

Sytin, A. 1997. *Petrus Simon Pallas, botanicus.* Moscow: KMK Scientific P.

——— & L.A. Borkin. 2007. "What a bliss it is to see nature in its very being . . . " on the Siberian journey made by Academician Peter Simon Pallas. *Science First Hand* 56–67.

Tamm, E.E. 2004. *Beyond the outer shore: the untold story of Ed Ricketts, the pioneering ecologist.* NY: Four Walls Eight Windows.

Tanner, H.H., ed. 1987. *Atlas of Great Lakes Indian history.* Norman: U. of Oklahoma P.

Tansley, A.G. 1916. Review: Clements, *Plant succession (1916). J. of Ecology* 4: 198–204.

———. 1927. Eug. Warming, in memoriam. *Botanisk Tidsskrift* 39: 54–56.

———. 1935. The use and abuse of vegetational concepts and terms. *Ecology* 16: 284–307.

———. 1940. Henry Chandler Cowles (1869–1939). *J. of Ecology* 28: 450–452.

———. 1946. Frederic Edward Clements (1874–1945). *J. of Ecology* 34: 194–196.

———. 1947. The early history of modern plant ecology. *J. of Ecology* 35: 130–137.

Taylor, F.J. 1980. Phytoplankton ecology before 1900: supplementary notes to the *Depths of the Ocean.* Pp. 509–521 in Sears & Merriman 1980.

Taylor, H. 1963. Thoreau's scientific interests as seen in his Journal. *McNeese Review* 14: 45–59.

Taylor, J.A., ed. 1984. *Biogeography: recent advances and future directions.* Totowa: Barnes & Noble.

Taylor, P.J. 1988. Technocratic optimism, H.T. Odum, and the partial transformation of ecological metaphor after World War II. *J. of the Hist. of Biol.* 21: 213–244.

Temple, S.A. 1977. Plant-animal mutualism: co-evolution with dodo leads to near extinction of plant. *Science* 197: 885–886.

Tepaske, J.J. 1996a. Jorge Juan y Santacilla (1713–73), Spanish scientist. *Latin American lives,* p. 531. NY: Macmillan Compendium.

———. 1996b. José Celestino Mutis (1732–1808). *Latin American lives,* pp. 703–704. NY: Macmillan Compendium.

———. 1996c. Antonio de Ulloa (1716–95), naval officer, scientist, royal bureaucrat. *Latin American Lives,* p. 1009.

Tétry, A. 1971. Yves Delagne (1854–1920), zoology, anatomy, physiology, embryology, general biology. *DSB* 4: 11–13.

Theophrastos. 1916. *Enquiry into plants.* A. Hort, transl. 2 vols. L: W. Heinemann/NY: G.P. Putnam's Sons.

———. 1976–90. *De causis plantarum.* B. Einarson & G.K.K. Link, transl. 3 vols. Cambridge MA: Harvard U.P.

Thienemann, A.F. 1917. Otto Zacharias: ein Nachruf. *Archiv fuer Hydrobiologie und Planktonkunde* 11: I—XXII.

———. 1925. Der See als Lebenseinheit. *Die Naturwissenschaften* 13: 589–600.

———. 1937. Einar Naumann (13 August 1891–22 September 1934), einar Forscherleben im Dienste der Limnologie. *Verhandlungen der Internationale Vereinigjung für theoretische und angewandte Limnologie* 8: 1–41.

———. 1959. *Erinnerungen und Tagebluchblätter eines Biologen: ein Leben im Dienste der Limnologie.* Stuttgart: Schweizerbart.

——— & E. Naumann. 1922. Vorschlag und Aufruf zur Grundung einer Internationalen Vereinigung für theoretische und angewandte Limnologie. *Archiv für Hydrobiologie* 13: 585–607.

Thomas, K.B. 1971. William Benjamin Carpenter (1813–85), medicine, natura l history. *DSB* 3: 87–89.

Thomas, P.D. 1976. Sir Charles Wyville Thompson (1830–82), natural history, ocenography. *DSB* 13: 360–361.

———. 1997. Henry David Thoreau (1817–62), natural philosopher. *BDAC* 774–776.

Thompson, W.R. 1939. Biological control and the theories of the interactions of populations. *Parasitology* 31: 299–388.

Thomson, A.D. 1982. A bibliography of the work of Leonard Cockayne. *New Zealand J. of Botany* 20: 205–219.

———. 1983. *The life and correspondence of Leonard Cockayne.* Christchurch: Caxton.

Thomson, C.W. 1878. *The voyage of the* Challenger: *the Atlantic, a preliminary account of H.M.S.* Challenger *during the year 1873 and the early part of the year 1876.* 2 vols. L: Macmillan. 1878: NY: Harper & Brothers.

Thomson, K. 2008. *The legacy of the Mastodon.* New Haven: Yale U.P.

———. 2012. *Jefferson's shadow: the story of his science.* New Haven: Yale U.P.

Thoreau, H.D. 1854. *Walden, or, life in the woods.* Boston: Ticknor & Fields.

———. 1881, 1884, 1888, 1892. *Early Spring, summer, winter, autumn.* 4 vols. Boston: Houghton Mifflin.

———. 1980. *The natural history essays.* R. Sattelmeyer, ed. Salt Lake City: Peregrine Smith.

———. 1993. *Faith in a seed: the dispersion of seeds and other late natural history writings.* B.P. Dean, ed. Washington DC: Island P.

Thorndike, E.L. 1898. Animal intelligence: an experimental study of the associative processes in animals. *Psychological Review Monograph Supplement* vol. 2, n. 8: 1–109. Extracts in Houck & Drickamer 1996: 45–54.

Thorpe, W.H. 1956. Ethology as a new branch of biology. Pp. 34–49 in A.A. Buzzati-Traverso, ed. *Perspectives in marine biology*. Berkeley: U. of California P. Ed. 2, 1965.

Thorsheim, P. 2006. *Inventing pollution: coal, smoke, and culture in Britain since 1800*. Columbus: Ohio State U.P.

Thurmann, J. 1849. *Essai de phytostatique appliqué à la Chaine du Jura et aus Contrées Voisines*. Berne, Switzerland: Jent & Gassmann.

Tieszen, L.L., ed. 1978. *Vegetation and production ecology of an Alaskan Arctic tundra*. NY: Springer-Verlag.

Tobey, R.C. 1976. Theoretical science and technology in American ecology. *Technology and Culture* 17: 718–728.

———. 1977. American grassland ecology, 1895–1955: the life cycle of a professional research community. Pp. 1–31 in Egerton 1977.

———. 1981. *Saving the prairies: the life cycle of the founding school of American plant ecology*. Berkeley: U. California P.

Tobien, H. 1972. Oswald Heer (1809–83), paleontology, botany. *DSB* 6: 220–222.

Tobler, F. 1925. *Biologie der Flechten*. Berlin: Borntraeger.

Tokarev, Y.N. 2004. Black Sea marine biological institutions of the Ukraine: a historical review. Pp. 223–228 in Morcos et al. 2004.

Tomczak, M., Jr. 1980. A review of Wüst's classification of the major deep-sea oceanographic research programs. Pp. 188–194 in Sears & Merriman 1980.

Topping, J.C., ed. 1989. *Second North American Conference on Preparing for Climate Change Proceedings*. Washington DC: Climate Institute.

Townsend, G.F. & E. Crane. 1973. History of apiculture. Pp. 387–406 in Smith et al. 1973.

Trager, W. 1970. *Symbiosis*. NY: Van Nostrand Reinhold.

Tramer, E.J. 1969. Bird species diversity: components of Shannon's formula. *Ecology* 50: 927–929.

Tubbs, S.A. & C.S. Jenkinson. 2003. *The Lewis and Clark companion: an encyclopedic guide to the Voyage of Discovery*. NY: Henry Holt.

Turner, R.S. 1974. Julius Robert Mayer (1814–78), physics, physiology. *DSB* 9: 235–240.

Turney, J. 2003. *Lovelock and Gaia: signs of life*. NY: Columbia U.P.

Turrill, W.B. 1963. *Joseph Dalton Hooker: botanist, explorer, and administrator*. L: Thomas Nelson & Sons.

Tuxen, S.L. 1967. The entomologist J.C. Fabricius. *Annual Review of Entomology* 12: 1–14.

———. 1973. Entomology systematizes and describes: 1700–1815. Pp. 95–118 in Smith et al. 1973.

Twing, C.E. 1983. The lumbering frontier. Pp. 121–136 in Flader 1983.

Tyler, D.B. 1968. *The Wilkes Expedition: the first United States Exploring Expedition, 1838–1842*. Philadelphia: American Philosophical Society.

Tyndall, J. 1861. The Bakerian Lecture: on the absorption and radiation of heat by gases and vapors, and on the physical connection of radiation, absorption, and conduction. *Philos. Mag*. 4: 169–194, 273–285. Excerpts in Howe 2017: 32–38.

Tyndall, L.C. 1900. John Tyndall (1820–93), natural philosopher. *DNB* 19: 1358–1363.

Tyrrell, T. 1999. The relative influences of nitrogen and phosphorus on oceanic primary production. *Nature* 400: 525–531.

———. 2013. *On Gaia: a critical investigation of the relationship between life and Earth*. Princeton: Princeton UP.

Urness, C. 1974. Thomas Pennant (1726–98), natural history. *DSB* 10: 507–510.

Van Bennekom, A.J. 2013. Political factors in the establishment of biological stations before 1900. Pp. 125–139 in Groeben 2013.

Van der Pas, P.W. 1973a. Jan Ingen-Housz (1730–99), medicine, plant physiology, physics. *DSB* 7: 11–16.

———. 1973b. Engelbert Kaempfer (1651–1716), geography, botany. *DSB* 7: 204–207.

———. 1973c. Paul Erich Otto Wilhelm Knuth (1854–1900), botany. *DSB* 7: 417–418.

Van der Valk, A.G. 2014. From formation to ecosystem: Tansley's response to Clements' climax. *J. Hist. Biol*. 47: 293–321.

Van Dobben, W.H. & R.H. Lowe-McConnell, eds. 1975. *Unifying concepts in ecology*. The Hague: W. Junk.

Vanney, J.R. 1993. *Le mystère des abysses: histoires et découvertes des profondeurs océaniques*. Paris: Fayard.

Van Oosten, J. 1923. The whitefishes (*Coregonus clupeaformis*): a study of the scales of whitefishes of known ages. *Zoologica* 2: 380–412.

———. 1929. Life history of the lake herring (*Leucichthys artedi* Le Sueur) of Lake Huron as revealed by its scales, with a critique of the scale method. *U.S. Bureau of Fisheries Bull*. 44: 265–428.

———., compiler. 1957. *Great Lakes fauna, flora, and their environment*. Ann Arbor: Great Lakes Commission.

Vaughan, T.W., ed. 1937. *International aspects of oceanography*. Washington: National Academy of Sciences.

Vaupell, C.T. 1857. *Bögens indvandring I de Danske Skove*. Copenhagen: C.A. Reitzals.

Vedder, L.A. 2006. *John James Audubon and the birds of America: a visionary achievement in ornithological achievement*. San Marino CA: Huntington Library.

Vermeij, G.I. 1978. *Biogeography and adaptation: patterns of marine life*. Cambridge MA: Harvard U.P.

———. 1997. *Privileged hands: a scientific life*. NY: W.H. Freeman/Macmillan.

Vernadsky, V.I. 1923. La composition chimique de la matière vivant et la chimie de l'ecorce Terrestre. *Revue Générale des Sciences Purers et Appliquées* 34: 51–53.

———. 1924. La matière vivant et la chimie de la mere. *Revue Générale des Sciences Pures et Appliquées* 35: 5–13, 46–54.

———. 1926a. Sur la multiplication des organisms et son role dans le mécanisme de la biosphere. *Revue Générale des Sciences Pures et Appliquées* 37: 661–668, 700–708.

———. 1926b. *Biosfera*. Leningrad.

———. 1929. *La Biosphère*. Paris: Félix Alcan.

———. 1945. The biosphere and the noosphere. *American Scientist* 33: 112.

———. 1986. *The Biosphere*. Oracle AZ: Synergetic P. 1997. D.B. Langmuir, transl. NY: Copernicus P./Springer-Verlag.

Vernberg, F. J. 2018. Its a long story. Kingston Springs TN: Westview P.

Vernberg, W.B. & F. J. Vernberg. 1972. Environmental physiology of marine animals. NY: Springer-Verlag.

Vernet, J. 1972. Francisco Hernández (1517–87), natural history. *DSB* 6: 309–310.

———. 1976. Antonio de Ulloa y la Torre Giral (1716–95), natural history. *DSB* 13: 530–531.

———. 1978. José Celestino Bruno Mutis y Bossio (1732–1808), botany, astronomy. *DSB* 15: 429–430.

Verrill, A.E. 1873. Report upon the invertebrate animals of the Vineyard Sound and the adjacent waters, with an account of the physical characters of the region. Pp. 295–778 in *Report of the U. S. Commission of Fish and Fisheries for 1871–1872*. Washington: Government Printing Office.

Verrill, G.E. 1958. *The ancestry, life, and work of Addison E. Verrill of Yale University*. Santa Barbara: Pacific Coast Publishing Company.

Vetter, J., ed. 2011. *Knowing global environments: new historical perspectives on the field sciences*. New Brunswick: Rutgers U.P.

Villiers, A. 1967. *Captain James Cook*. NY: Charles Scribner's Sons.

Vincent, W. & C. Bertola. 2012. François Alphonse Forel and the oceanography of lakes. *Archives des Sciences* 65: 51–64. Abridged: *American Society of Limnology & Oceanography Bull.* 21 (2012) 69–77.

Viola, H.J. & C. Margolis, eds. 1985. *Magnificent voyagers: the U.S. Exploring Expedition, 1838–1842*. Washington: Smithsonian Institution.

Virville, A.D. de., ed. 1954. *Histoire de la botanique en France*. Paris: Soc. d'Edition d'Enseignment Supérieur.

Vitikeinen, O. 2000. William Nylander (1822–1899), professor of botany, lichen researcher. Pp. 352–355 in U. Marjomaa, ed. *100 faces from Finland: a biographical kaleidoscope*. Helsinki: Finnish Literature Society.

Volk, T. 1998. *Gaia's body: toward a physiology of Earth*. NY: Copernicus P./Springer-Verlag. 2003. NY: Copernicus Books.

Volterra, E. 1976. Vitto Volterra (1860–1940), mathematics, natural philosophy. *DSB* 14: 85–88.

Volterra, V. 1926. Fluctuations in the abundance of a species considered mathematically. *Nature* 118: 558–560. Reprinted in Real & Brown 1991: 283–285.

Von Hagen, W. 1945. *South America called them: explorations of the great naturalists: La Condamine, Humboldt, Darwin, Spruce*. NY: Knopf.

———. 1948a. *The green world of the naturalists: a treasury of five centuries of natural history in South America*. NY: Greenberg.

———. 1948b. The immortal botanist: the strange and stirring story of José Celestino Mutis and the eighteenth-century botanical expedition to Columbia in the time of the Kingdom of New Granada. *J. of the New York Botanical Garden* 49: 177–184, 210–218.

Wagenitz, G. 1972. August Heinrich Rudolf Grisebach (1814–79), botany, taxonomy. *DSB* 5: 546–547.

Wagensberg, J. 2012. Tales of tales. Pp. 11–13 in Sagan 2012.

Wagner, M.F. 1844. *Reisen in der Regenschaft Algier in den Jahren 1836, 1837, und 1838*. Leipzig: Leopold Voss.

Wagner, P., A.M. Bauer & W. Böhme. 2012. Amphibians and reptiles collected by Moritz Wagner, with a focus on the ZFMK collection. *Bonn Zoological Bull.* 61: 216–240. Online.

Wallace, A.R. 1853a. *A narrative of Travels on the Amazon and Rio Negro, with an account of the native tribes, and observations on the climate, geology, and natural history of the Amazon Valley*. L: Reeve. Ed. 3, 1890. L: Ward & Lack.

———. 1853b. *Palm trees of the Amazon and their uses*. L: John van Voorst.

———. 1865. On the phenomena of variation and geographical distribution as illustrated by the Papilionidae of the Malayan region. *Linnean Soc. of L* 25: 1–71 + 8 plates.

———. 1869. *The Malay Archipelago, the land of the orangutan and the birds of paradise: a narrative of travel with studies of man and nature*. 2 vols. L: Macmillan/NY: Harper & Brothers.

———. 1876. *The geographical distribution of animals: with a study of the relations of living and extinct faunas elucidating the past changes of the Earth's surface*. 2 vols. L: Macmillan. 1962. NY: Hafner. Extract in Lomolino et al. 2004: 164–177.

———. 1880. *Island life, or the phenomena and causes of insula faunas and floras*. L: Macmillan. Ed. 2, 1895.

———. 1890. See Wallace 1853.

———. 1894. What are zoological regions? *Nature* 49: 610–613.

———. 1905. *My life: a record of events and opinions*. 2 vols., L: Chapman & Hall/NY: Dodd, Mead.

———. 1986. *The Malay Archipelago and the land of the orangutan, and the birds of paradise*. J. Bastin, intro. L: Oxford U.P.

———. 2002. *An anthology of his shorter writings*. C.H. Smith, ed. L & NY: Oxford U.P.

———. 2013. *Letters from the Malay Archipelago*. J. Van Wyhe & K. Rookmaakere, eds. L & NY: Oxford U.P.

———. 2015. *The annotated Malay Archipelago (1869)*. J. Van Wyhe, ed. 2 vols. in 1. Singapore City: National U. of Singapore P.

Waller, D.M. & T.R. Rooney, eds. 2008. *The vanishing present: Wisconsin's changing lands, waters, and wildlife*. Chicago: U. of Chicago P.

Waller, J. 2004. Edward Blyth (1810–73). *DNCB* 1: 225–226.

Wallroth, F.W. 1825–27. *Naturgeschichte der flechten: nachneuen Normen und in ihrem umfang bearbeitet*. 2 vols. Frankfort: Wilmans.

Walls, L.D. 1995. *Seeing new worlds: Henry David Thoreau and nineteenth-century natural science*. Madison: U. of Wisconsin P.

———, ed. 1999. *Material faith: Thoreau on science*. Boston: Houghton Mifflin.

———. 2009. *The Passage to Cosmos: Alexander von Humboldt and the shaping of America*. Chicago: U. of Chicago P.

Walters, C.J. 1971. Systems ecology: the systems approach and mathematical models in ecology. Pp. 276–292 in Odum 1971.

———, G. Steer & G. Spangler. 1980. Responses of lake trout *Salvelinus namaycush* to harvesting, stocking, and lamprey reduction. *Canadian J. of Fisheries and Aquatic Sciences* 37: 2133–2145.

Walters, S.M. 1981. *The shaping of Cambridge botany*. Cambridge: Cambridge U.P.

Ward, H.B. 1896. Biological examination of Lake Michigan in the Traverse Bay Region. *Michigan Fish Commission Bull.* 6, 99 pp.

——— & G.C. Whipple, eds. 1918. *American fresh-water biology*. NY: John Wiley & Sons.

Ward, P. & R.A. Myers. 2005. Shifts in open-ocean fish communities coinciding with the commencement of commercial fishing. *Ecology* 86: 835–847.

Ward, R. 1974. *Into the ocean world: the biology of the sea*. NY: Knopf.

Warington, R. 1851. Notice of observation on the adjustment of the relations between animal and vegetable kingdoms by which the vital functions of both are maintained. *Quart. J. of the Chem. Soc. of London* 3: 52–54.

Warming, J.E.B. 1895. *Pflantesanfund: Grundträk af den okologiske pflantegeografi*. Copenhagen: Philipsens Forlag.

———. 1896. *Lehrbuch der oekologischen pflanzengeographie: eine einführung in die kenntniss der Pflanzenverine*. Berlin: Gebrüder Borntraeger.

———. 1909. *Oecology of plants: an introduction to the study of plant communities*. I.B. Balfour & P. Groom, transl. Oxford: Clarendon P. 1977. NY: Arno P.

Warren, J.P. 2006. *John Burroughs and the place of nature*. Athens: U. of Georgia P.

Warren, L. 2004. *Constantine Samuel Rafinesque: a voice in the American wilderness*. Lexington: U.P. of Kentucky.

Waters, H.J. Spring 2020. In Grinnell's footsteps. *Audubon* 34–41, 54.

Waterson, A.P. & L. Wilkinson. 1978. *An introduction to the history of virology*. NY: Cambridge U.P.

Watson, H.C. 1843–47. Notes of a botanical tour in the Western Azores. *London J. Bot*. 2: 1–9, 125–131, 394–408; 3: 582–617; 6: 380–397.

———. 1870. Botany of the Azores. Pp. 113–282 in F.D. Godman, *Natural history of the Azores, or Western Islands*. London: John Van Voorst.

Watson, S. & G.L. Goodale, compilers. 1888. List of the writings of Dr. Asa Gray, chronically arranged, with an index. *American J. of Science* 36: 3–67.

Watterson, R.L. 1973. Frank Rattray Lillie (1870–1947), embryology, zoology. *DSB* 8: 354–360.

Weart, S.R. 2003. *The discovery of global warming: new histories of science, technology and medicine*. Cambridge MA: Harvard U.P.

Webster, C. 1975. John Ray (1627–1705), natural history. *DSB* 11: 313–318.

———. 1976. William Turner (1508–68), natural history, medicine. *DSB* 13: 501–502.

Weddell, H.A. 1875. Excursion lichenologique dans l'Ile d'Ile d'Yeu la côte de la Vendée. *Mémoire de Societé de Sciences Naturelle* 19: 251–316.

Wegener, A.L. 1929. *The origin of continents and oceans*. J.G.A. Skerl, trans. L: Methnen. 1966. NY: Dover.

Wegmann, E. 1976. Eduard Suess (1831–1914), geology. *DSB* 13: 143–149.

Weiner, D.R. 1988. *Models of nature: ecology, conservation, and cultural revolution in Soviet Russia*. Bloomington: Indiana U.P.

———. 1999. *A little corner of freedom: Russian nature protection, from Stalin to Gorbachëv*. Berkeley: U. of California P.

Weisheipl, J.A. 1980. *Albertus Magnus and the sciences*. Toronto: Pontifical Institute of Medieval Studies.

Weiss, H.B. & G.M. Ziegler. 1931. *Thomas Say: early American naturalist*. Springfield IL: C.C. Thomas. 1978. NY: Arno P.

Welker, R.H. 1955. *Birds and men: American birds in science, art, literature, and conservation*. Cambridge MA: Harvard U.P.

———. 1975. *Natural man: the life of William Beebe*. Bloomington: Indiana U.P.

Wells, B.W. 1978. *Daylight in the Swamp!* Garden City: Doubleday.

Wells, J.W. 1973. Charles-Alexandere Lesueur (1778–1846), natural history. *DSB* 8: 266–267.

Wells, L.R. 1980. Lake trout (*Salvelinus namaycush*) and sea lamprey (*Petromyzon marinus*) populations in Lake Michigan, 1971–78. *Canadian J. of Fisheries and Aquatic Sciences* 37: 2047–2051.

Wendland, F. 1992. *Peter Simon Pallas (1741–1811): Materialien einer Biographie*. Berlin: Walteer de Gruyer.

Werrett, S. 2000. Russian Academy of Sciences. Pp. 665–666 in Hessenbruch 2000a.

Westfall, R.S. 1972. Robert Hooke (1635–1708), physics. *DSB* 6: 481–488.

Westman, W.E. 1982. Eminent ecologist: Robert Harding Whittaker. *ESAB* 63: 11–12.

——— & R.K. Peet. 1982. Robertt H. Whitaker (1920–1980): the man and his work. *Vegetatio* 48: 97–122.

Wetzel, N. 2004. *Emil Otto Zacharias—ein Lebensweg als Vermittler: zur Kultur der Naturwissenschaft im 19. Jahrhundert*. Jena, Germany: M.A. thesis, Friedrich-Schiller Universität.

Wheeler, A. 1976a. John Vaughan Thompson (1799–1847), natural history. *DSB* 13: 353–356.

———. 1976b. Edward Wotton (1492–1555), medicine, natural history. *DSB* 14: 507–508.

———, ed. 1983. *Contributions to the history of North American natural history*. L: Society for Bibliography of Natural History.

Wheeler, W.M. 1907. The fungus-growing ants of North America. *American Museum of Natural History B*. 23: 669–807. 1973. NY: Dover Publications.

———. 1911. The ant colony as an organism. *J. of Morphology* 22: 307–325.

———. 1923. *Social life among the insects: being a series of lectures*. NY: Harcourt, Brace and World. 1968. NY: Johnson Reprint Corp.

White, G.E. 1981. (Nathan) Roscoe Pound (1870–1964), law professor, jurist, and botanist. *DAB* supplement 7: 624–628.

White, G.W. 1973. William Maclure (1763–1840), geology. *DSB* 8: 615–617.

White, R. 2004. Nehemiah Grew (bap. 1641–1717), botanist and physician. *ODNB* vol. 23 & online.

Whitnah, D.R. 1965. *A history of the United States Weather Bureau*. Urbana: U. of Illinois P. Ed. 1, 1961.

Whittaker, R.H. 1962. Classification of natural communities. *Botanical Review* 28: 1–239. 1977. NY: Arno P.

———. 1970. *Communities and ecosystems*. NY & L: Macmillan.

———. 1973. *Ordination and classification of communities*. The Hague: W. Junk.

———. 1994. Autobiographical letter and abbreviated bibliography. Pp. 617–621 in Burgess 1994.

Wigdor, D. 1974. *Roscoe Pound: philosopher of law*. Westport CT: Greenwood P.

Wiley, E.O. 1988. Vicariance biogeography. *Ann. Rev. of Ecology & Systematics* 19: 513–545.

Wilkinson, L. 1992. *Animals and disease: an introduction to the history of comparative medicine*. NY: Cambridge U.P.

Willdenow, K.L. 1810. *The principles of botany and of vegetable physiology*. Anon. transl. Edinburgh: William Blackwell, T. Cadell & W.L. Davies.S.

Williams, G. 2013. *Naturalists at sea: scientific travelers from Dampier to Darwin*. New Haven: Yale U.P.

Williams, R.L. 1976. *The loggers*. NY: Time-Life.

Willis, C.G. et al. 2008. Phylogenetic patterns of species loss in Thoreau's Woods are driven by change. *National Acad. of Sciences Proc*. 105 no. 44: 17029–17033.

Willis, E.O. & Y. Oniki. 1978. Birds and army ants. *Annual Review of Ecology and Systematics* 9: 243–263.

Wilson, A. (& G. Ord, vols. 8–9). 1808–14. *American ornithology: or, the natural history of the birds of the United States: illustrated with plates engraved and colored from original drawings taken from nature.* Philadelphia: Bradford & Inskeep. Many later editions.

Wilson, D.S. & E.O. Wilson. 2007. Rethinking the theoretical foundations of sociobiology. *Quarterly Review of Biology* 82: 327–348.

Wilson, E.O. 1971. *The insect societies.* Cambridge MA: Harvard U.P.

———. 1975. *Sociobiology: the new synthesis.* Cambridge MA: Harvard U.P.

———. 1991. Philip Jackson Darlington, Jr., November 14, 1904—December 16, 1983. *National Academy of Sciences Biographical Memoirs* 60: 32–94. Online.

———. 1994. *Naturalist.* Washington: Island P./Shearwater Books.

——— & G.E. Hutchinson. 1989. Robert Helmer MacArthur, April 7, 1930—November 1, 1972. *Nat. Acad. of Sci. Biog. Mem.* 57: 317–327.

——— & C.D. Michener. 1982. Alfred Edwards Emerson, December 31, 1896—October 3, 1976. *National Acad. Sci. Biographical Memoirs* 53: 158–177.

Wilson, H.F. & M.H. Doner. 1937. *The historical development of insect classification.* St. Louis: Swift, Printer for Authors.

Wilson, L.G. 1972. *Charles Lyell: the years to 1841.* New Haven: Yale U.P.

———. 1973. Charles Lyell (1797–1972), geology. *DSB* 8: 563–576.

Winslow, C.E.A. 1943. *The conquest of epidemic disease: a chapter in the history of ideas.* Princeton: Princeton U.P.

Winsor, M.P. 1972. Stephen Alfred Forbes (1844–1930). *DSB* 5: 69–71.

———. 1991. *Reading the shape of nature: comparative zoology at the Agassiz Museum.* Chicago: U. of Chicago P.

Wires, L.R. 2014. *The double-crested cormorant: plight of a feathered pariah.* New Haven: Yale U.P.

Witmer, M.C. & A.S. Cheke. 1991. The dodo and the tambulacoque tree: an obligate mutualism reconsidered. *Oikos* 61: 133–137.

Wohrle, G. 1985. *Theophrastus Methode in seinen botanischen Schriften.* Amsterdam: B.B. Gruner.

Wolf, A. 1952. *A history of science, technology, and philosophy in the 18*th *century.* NY: Macmillan.

Wood, C.A., ed. 1931. *An introduction to the literature of vertebrate zoology.* Oxford UK: Oxford U.P./L: Humphrey Milford.

Wood, R.G. 1966. *Stephen Harriman Long, 1784–1864: army engineer, explorer, inventor.* Glendale CA: Arthur H. Clark.

Woodward, B.B. 1909. Henry Nottidge Moseley (1844–1890), naturalist. *DNB* 13: 1073–1074.

Woodward, J. 1699. Some thoughts and experiments concerning vegetation. *Roy. Soc. L Philos. Trans.* 21: 193–227.

Woodwell, G.M. & E.V. Pecan, eds. 1973. *Carbon and the biosphere.* Springfield VA: Atomic Energy Commission.

Woolfenden, G.E. & J.W. Fitzpatrick. 1984. *The Florida scrub jay: demography of a cooperatively breeding bird.* Princeton: Princeton U.P.

———. 1996. Florida scrub jay *Aphelocoma coerulescens.* Pp. 1–28 in no. 228, in A. Poole & F. Gill, eds. *The birds of North America.* Philadelphia: Academy of Natural Sciences/ Washington: American Ornithologists Union.

Worland, J. & N.M. Hobbes. 18 October 2018. Oil companies see green. *Time* 23–24.

Worster, D. 1977. *Nature's economy: the roots of ecology.* San Francisco: Sierra Club Books.

———. 2008. *A passion for nature: the life of John Muir.* NY: Oxford U.P.

Wright, S.B. & N.L. Bradley. 2008. Thinking like a flower: phenology and climate change at the Leopold Shack. Pp. 41–53 in D.M. Waller & T.R. Rooney 2008.

Wulf, A. 2015. *The invention of nature: Alexander von Humboldt's new world.* NY: Knopf.

Wüst, G. 1964. The major deep-sea expeditions and research vessels, 1873–1960: a contribution to the history of oceanography. *Progress in Oceanography* 2: 1–52.

Yonge, C.M. 1956. Development of marine biological laboratories. *Science Progress* 44: 1–15.

Yount, L. 1994. *Contemporary women scientists.* NY: Facts on File.

Zanobiol, B. 1972. Girolamo Fracastoro (178–1553), medicine, philosophy. *DSB* 5: 104–107.

Zavodnik, D. 1990. The ships and marine research of Rovinj Zoological Station before 1918. Pp. 214–219 in Lenz & Deacon 1990.

Zedillo, E., ed. 2008. *Global warming: looking beyond Kyoto.* Washington DC: Center for the Study of Global Civilization/ Brookings Institution.

Zim, H.S. & L. Ingle. 1955. *Seashores: a guide to animals and plants along the beaches.* NY: Golden P.

——— & H.H. Shoemaker. 1956. *Fishes: a guide to fresh and salt-water species.* NY: Golden P.

Zimmer, C. October 2012. Ecosystems on the brink. *Scientific American* 307, no. 4: 60–65.

Zinn, D.J. 1980. Alexander Agassiz (1835–1910) and the financial support of oceanography in the United States. Pp. 83–93 in Sears & Merriman 1980.

Index

Note: Page locators in *italics* indicate a figure on the corresponding page.

C

D

E

F

G

H

I

J

N

O

P

For Product Safety Concerns and Information please contact our EU representative GPSR@taylorandfrancis.com Taylor & Francis Verlag GmbH, Kaufingerstraße 24, 80331 München, Germany

Batch number: 10731517

Printed by Printforce, the Netherlands